Methodology in Robust and Nonparametric Statistics

METHODOLOGY IN ROBUST AND NONPARAMETRIC STATISTICS

JANA JUREČKOVÁ
PRANAB KUMAR SEN
JAN PICEK

CRC Press is an imprint of the
Taylor & Francis Group, an **informa** business
A CHAPMAN & HALL BOOK

CRC Press
Taylor & Francis Group
6000 Broken Sound Parkway NW, Suite 300
Boca Raton, FL 33487-2742

Printed and bound in Great Britain by TJ International Ltd, Padstow, Cornwall
Version Date: 20120625

International Standard Book Number: 978-1-4398-4068-9 (Hardback)

Visit the Taylor & Francis Web site at
http://www.taylorandfrancis.com

and the CRC Press Web site at
http://www.crcpress.com

Contents

Preface

The present treatise on robustness and nonparametrics has its genesis in an earlier monograph by the first two authors (J. Jurečková and P. K. Sen (1996): *Robust Statistical Procedures: Asymptotics and Interrelations* Wiley, New York, JS (1996) in the sequel). The basic concepts of robustness, mostly laid down during the 1960s and early 1970s, led to novel theoretical tools, useful applicable methodology, and interaction with existing and new statistical procedures. During the 1970s and 1980s, robustness gained considerable common ground with nonparametrics, which started 25 years earlier and first seemed to cover different areas. Compared with the literature on robustness and nonparametrics prior to 1990, JS (1990) concentrated on unified study of interrelations and related asymptotic theory. The detailed historical perspectives are described in Preface to the First Edition.

The past two decades have witnessed a phenomenal growth of research literature in the composite area of robust and nonparametric statistics. Nonparametrics have accommodated semiparametrics and Bayesian methodology in more diverse and complex setups. Robustness concepts have permeated many areas of theoretical as well as applied statistics, multivariate analysis being one of the most notable ones. Noticing that we have seen a greater part of JS (1996) merits updating to suit this enhanced need. The present monograph follows this objective.

The asymptotics and interrelations treated in Part I of JS (1996) have been thoroughly updated in the present treatise (Chapters 2–7). Multivariate robust and nonparametric estimation with special emphasis on affine-equivariant procedures have been presented in Chapter 8, followed by hypotheses testing and confidence sets in Chapter 9. In this respect, some aspects in Part II of JS (1996) have been deemphasized to make room for more recent developments.

Although our treatise is largely theoretical, mathematical abstractions have been carefully kept at bay. The inclusion of omnibus Section 2.5 provides a pool of basic mathematical tools that have been used in latter chapters in course of derivations of main results. To many readers, these tools are familiar in some form or another, while for most applied statisticians, this collection may be perceived as an unnecessary probe into relevant source materials that are often presented at higher levels of abstractions. Our treatise of robustness is by no means complete. For example, there has been little emphasis on linear models with "mixed-effects" and more general nonlinear models. However, we sincerely hope that the methodology presented here with due emphasis on asymptotics and interrelations will pave way for further developments on robust statistical procedures in more complex models that are yet to be developed or could not be included in this monograph for various reasons.

Jana Jurečková, Prague, Czech Republic
Pranab K. Sen, Chapel Hill, North Carolina
Jan Picek, Liberec, Czech Republic

March 5, 2012

Preface to the First Edition

The burgeoning growth of robust statistical procedures over the past 30 years has made this area one of the fastest growing fields in contemporary statistics. Basic concepts of robustness have permeated both statistical theory and practice, creating a genuine need for a full study of their interactions.

Linearity of regression, stochastic independence, homoscedasticity, and normality of errors, typically assumed in a parametric formulation, provide access to standard statistical tools for drawing conclusions about parameters of interest, but there may not be any guarantee that such regularity assumptions are tenable in a given context. Therefore, natural interest turned toward probing into the effects of plausible departures from model-assumptions on the performance characteristics of classical statistical procedure, and this was the genesis of robust statistics. In the assessment of such departures and/or interpretations of robustness, theoretical researchers and applied statisticians have differed, often sharply, resulting in a diversity of concepts and measures of influence functions, among these, breakdown points, qualitative versus quantitative robustness, and local versus global robustness. All of these related ideas can be presented in a logical and unified manner, integrated by a sound theoretical basis, and this is the motivation for the theoretical focus of this book.

Nonparametric procedures are among the precursors of modern robust statistics. In nonparametrics, robustness is generally interpreted in a global sense, while in robust statistics, a relatively local up to an infinitesimal interpretation is more commonly adopted. The classical M-, L-, and R-procedures are the fundamental considerations in this respect; other developments include differential statistical functions, regression quantiles, and regression rank score statistics. Huber's innovation of M-procedures, followed by the contribution of Hampel, revolutionized the field; their excellent treatments of this aspect of robustness are to be found in Huber (1981) and Hampel et al. (1986). Some other books that are not entirely devoted to robustness but contain enlightening applications of specialized viewpoints are those by Shorack and Wellner (1986) and Koul (1992), which place more emphasis on weighted empiricals, and by Serfling (1980), which particularly emphasizes von Mises's statistical functions. The more recent Rieder's book (1994) is a valuable source of mathematical abstractions relating to infinitesimal concepts of robustness and nonparametric optimality based on a least-favorable local alternative approach. Some other contemporary books on robustness have excellent interpretations and sound motivations for real data analysis, and are more applications oriented. Integration of mathematical concepts with statistical interpretations often precludes the exact (finite) treatment of robustness. The incorporation of asymptotics in robustness has eliminated this impasse to a great extent, and sometimes has yielded finite sample interpretations. Yet, while the different types of robust procedures have evolved in individual patterns their ultimate goals are often the same. This fact is reflected by asymptotic relations and equivalences of various (class of) procedures, and there is a need to focus on their interrelations so as to unify the general methodology and channel the basic concepts into a coherent system. This perspective has not fully been explored in contemporary works. Thus, it is our primary objective. In this book we attempt to harmoniously blend asymptotics and its interrelations to depict robustness in a proper and broader setup.

A discussion of the organization of this book is deferred to Chapter 1. In general, Part I (Chapters 2 through 7) forms the core of theoretical foundations and deals with asymptotics and interrelations, and Part II (Chapters 8,9, and 10) is devoted to robust statistical inference, given the general theory and methodology of Part I. Admittedly, Part II has greater statistical depth and is therefore of greater interest to applied statisticians, whereas Part I provides handy resource material for the requisite theoretical background. Still both theoretical statisticians and graduate students in mathematical statistics may desire this integrated view. Although our approach is largely theoretical, mathematical abstractions have been carefully kept at bay. Section 2.5 provides a pool of basic mathematical tools that have been used in latter chapters in derivations of main results. To many readers, these tools are familiar in some form or another, while for most applied statisticians, this collection may preclude unnecessary probe into relevant source materials that are often presented at higher levels of abstractions. Our treatment of robustness is by no means complete. For example, there has been little emphasis on linear models with "mixed-effects," multivariate linear models, and more general nonlinear models. However, we sincerely hope that the methodology presented here with due emphasis on asymptotics and interrelations will pave the way for further developments on robust statistical procedures in more complex models that are yet to be developed or could not be included in this book for various reasons.

Jana Jurečková, Prague, Czech Republic
Pranab K. Sen, Chapel Hill, North Carolina

October 8, 1995

Acknowledgments

The present monograph follows up the mathematical tools and theoretical perspectives in asymptotics, nonparametrics and robustness, disseminated in an earlier monograph by the first two authors (1996). During our long cooperation, we learned much from each other, and also from a number of prominent researchers in various academic centers around the world. The impact of this omnibus research collaboration is overwhelming, and to all of our colleagues, we owe a deep sense of gratitude. A greater part of our collaborative research was supported by research grants from National Science Foundation (USA), Office of Naval Research, National Science and Engineering Research Council of Canada, Boshamer Research Foundation at the University of North Carolina, Czech Republic Grants GACR 201/09/0133 and GACR 209/10/2045, Research Project MSM 0021620839, and Hájek Center for Theoretical and Applied Statistics LC 06024. We also acknowledge the support of local facilities in our working places (Charles University in Prague, University of North Carolina at Chapel Hill, Technical University of Liberec), which enabled us to have collective work related to this project in a very convenient manner. We gratefully acknowledge these supports. Our thanks belong to the Wiley-Interscience Group in New York for not only taking active interest in the publication of the earlier 1996 monograph by the first two authors but also releasing our contract in a very courteous manner, a couple of years ago, in favor of our desire to go for an extensive updating of that monograph.

Rob Calver (Commissioning Editor, Chapman & Hall/CRC Press) has been an inspiring source for initiating the current monograph project with them. The continual support of Rachel Holt (Copy Editor), Marsha Pronin (Project Coordinator), Gillian Lindsey (Marketing Development Manager), and Michele Smith (Marketing Coordinator) have been very helpful in bringing this project to a successful completion.

We are very grateful to the reviewers for their penetrating reading and constructive comments and suggestions, which we have tried to incorporate to the extent possible in this final version. Last but not the least important acknowledgement is our deep sense of gratitude and profound thanks to our spouses (Josef, Gauri, and Helena) for their unconditional support and patience during the undertaking and completion of this project. Without their understanding, this would not have been possible.

CHAPTER 1

Introduction and Synopsis

1.1 Introduction

More than 100 years ago, *goodness of fit tests* appeared (Pearson 1900), and rank measures of association (Spearman 1904) evolved soon after. The 1930s and 1940s witnessed a phenomenal growth of literature on rank and permutation statistical procedures, some originating from the social sciences (Friedman (1937), Kendall (1938), Pitman (1937b,c, 1938), among others). Such tests used to be termed *distribution-free*, as unaffected by deviations from the assumed model. The Wald-Wolfowitz (1944) permutational central limit theorem paved the way for asymptotics. The nonparametric inference was reinforced by Wilcoxon–Mann–Whitney and Kruskal–Wallis rank tests and by Hoeffding's (1948) optimal nonparametric estimation. Pitman's (1948) local alternatives revolutionized the asymptotics for parametrics as well for nonparametrics.

Back in the 1940s and 1950s, the *Biometrika* school studied the effects of nonnormality as well as of *heteroscedasticity* on the conventional tests and estimators. This led to terminology *robust* (Box 1953). The Tukey innovation of robustness in the late 1950s was in the early 1960s followed up by Huber's formulation of robustness, more abstract and theoretical. The combined field of robust and nonparametric inference has become a part of the general statistical consciousness during the past five decades. We have witnessed a phenomenal growth of research literature in this area, also on its applications in social, engineering, physical, clinical, biomedical, public health, and environmental sciences, with a more recent annexation of bioinformatics, genomics and data mining. The robust and nonparametric methods obtained a significant role in modern statistical inference. The main objective of the present treatise is to focus on this impactful integrated area of research.

A good statistical procedure should not be sensitive to small departures from the model. The sensitivity of a statistical test pertains to the instability of its size as well of its power. The sensitivity of point estimates relates to a significant *bias*, to its inconsistency and small efficiency. For the confidence set, it may lead to an incorrect coverage probability or to an incorrect volume of the set. Robustness can be differentiated from nonparametrics so that it puts more emphasize on *local departures* from the model. A nonparametric procedure also behaves well in a local neighborhood, but not necessarily in a specific one. For some departures from the assumed model, the robust and nonparametric procedures behave almost in an isomorphic manner. In this common vein, a thorough theoretical/methodological study of robust and nonparametric procedures constitutes our main objective.

To motivate robust and nonparametric statistical inference, consider the classical problem of estimating the *location parameter* interpreted as the mean, median, or center of a symmetric distribution. Usually one assumes that the distribution associated with the population is normal with mean θ and variance σ^2, generally both unknown. Then the *maximum likelihood estimator* of the population mean θ is the sample mean $\bar{X}$, and when $\sigma^2 < \infty$, $\bar{X}$ is an optimal estimator of θ in several well-defined ways. In reality, the distribution F may differ from the normal in various ways; it can be asymmetric or to have *heavier tails.* Then $\bar{X}$ can lose its optimality and its efficiency can quickly deteriorate. However, for practical reasons, it can be quite relevant to allow departures from the assumed normality; then we should look for alternative estimators that are less sensitive to such departures. In biomedical studies, the response variable Y is typically nonnegative and with a positively skewed distribution. Therefore, often one can make a transformation $X = g(Y)$, where $g(.)$ is a monotone and continuous function [on $\boldsymbol{R}_1 = (-\infty, \infty)$]. Typical examples of $g(.)$ are

(1) $g(y) = \log y$ (*log-transformation*);
(2) $g(y) = y^{1/2}$ (*square-root transformation*);
(3) $g(y) = y^{1/3}$ (*cube-root transformation*), appropriate when Y refers to the weight/volume of some organ that regresses to the third power of the diameter x;
(4) $g(y) = y^{\lambda}$ for some $\lambda > 0$.

All of those can be modeled in the framework of the classical Box-Cox (1964) transformation. The first three transformations are commonly used in practice. In the first case, Y has a lognormal distribution when X has a normal one. In the other cases, X remains a positive random variable, and hence its normality is reasonable only when the ratio of its mean and standard deviation is large. Thus, from the theoretical standpoint, X may not always have a normal law and it is reasonable to allow for possible departures from the assumed normal law. The extent of such a departure may depend on the adopted transformation as well as on the original distribution. If we consider the sample median $\tilde{X}$ instead of the sample mean $\bar{X}$, its performance will be fairly insensitive to heavy tails and be more *robust* than the mean $\bar{X}$. On the other hand, $\tilde{X}$ is about one-third less efficient than $\bar{X}$ for normal F and may not estimate the mean for an asymmetric F. Therefore, one has to compromise between robustness and efficiency of estimators. Fortunately, there are alternative robust and nonparametric estimators that blend efficiency and robustness more coherently. A similar feature holds for tests of significance.

The word "robust" dates from a Biometrika article "Non-Normality and Tests on Variance" (Box 1953). The robustness as a subdiscipline in statistics was primarily the creation of John W. Tukey and Peter J. Huber. Tukey (1960 and before) recommended considering contaminated distributions as alternative possibilities. Huber (1964) derived an optimal robust estimate for the location parameter problem. Huber's graduate level textbook (1981) provides a good coverage of the basic theory developed during the 1960s and 1970s, further extended for new concepts and developments in the second edition (Huber and Ronchetti 2009). Two Berkeley dissertations (Hampel 1968 and Jaeckel 1969) imported the concept of influence curve and the use of the adaptive trimmed mean. Two important simulation studies, the Princeton Study (Andrews, Bickel, Hampel, Huber, Rogers, and Tukey 1972) and Stigler (1975) compared the behavior of robust and nonrobust estimates of location. Jurečková (1977) studied the asymptotic relation between M- and R-estimators in linear models. Koenker and Bassett (1978) devised the concept of *regression quantiles* in the linear model. Sen (1981a) used a unified martingale approach to the asymptotics of robust and nonparametric statistics, with an emphasis on sequential methods. The book by Hampel, Ronchetti, Rousseeuw, and Stahel (1986, 2005) further developed applications of the influence function, breakdown points, including some treatment of covariance matrices to time series, and others have been devised. More contemporary advanced graduate-level monographs discuss the properties of robust estimators and test statistics for the location as well as linear models, along with the computational aspects and applications; namely Serfling (1980, 2002), Rousseeuw and Leroy (1987, 2003), Maronna, Martin, and Yohai (2006). Other monographs, by Shorack and Wellner (1986), by Koul (1992, 2002) and by van der Vaart and Wellner (1996, 2000), concentrating on asymptotics of the *(weighted) empirical process*, have touched on robustness primarily from an asymptotic point of view. The book by Rieder (1994) is devoted solely to robust asymptotic statistics, but it emphasizes mostly the infinitesimal concept of robustness and nonparametric optimality based on the notion of least favorable local alternatives. Jurečková and Sen (1996) stressed the general asymptotics and interrelations of robust and nonparametric estimates and tests, mostly confined to univariate models. We also refer to Staudte and Sheather (1990), to Rey (1980), and to Jurečková and Picek (2006) and Jurečková and Dodge (2000) for further expositions of robustness, some at an intermediate level of presentation.

Comparing with the univariate models, the developments of robustness and nonparametrics in multivariate models have been rather piecemeal. The book by Maronna et al. (2006) captures some aspects. As Stigler (2010) pointed out, robustness concepts did not extend in a straightforward and elegant way to the multivariate models as it was originally anticipated, because some concepts do not have multivariate analogues. The same is true for the multivariate time series models. However, robustness and nonparametrics are even more important for treating the multivariate data, because in the multivariate models, the probability distributions are uncertain and normality and even elliptic symmetry are not guaranteed. The equivariance/invariance of multivariate statistical procedures is an important subject matter for discussions, but it may involve complexity of statistical modeling. Nevertheless, there has been a steady flow of research on *spatial nonparametrics* in the past two decades (Serfling 2010; Oja 2010).

A significant development in this area relates to *Monte Carlo* simulation and *resampling* plan-based studies. Although numerical evidence gathered from such studies is overwhelming, the theoretical and methodological justifications have been acquired solely from the relevant asymptotic theory. These developments took place in different models and levels of mathematical sophistication; hence, there is a need for a thoroughly unified treatment of this asymptotic theory. With this motivation, our primary objective is to focus on theoretical and methodological aspects of asymptotics of robust and nonparametric statistics in a unified and logically integrated manner. Important for the theory and methodology of robust statistics is not only mathematical ingenuity but also statistical and data-analytical insights. The primary tools are based on *M-, L-,* and *R-statistics* and estimators, the *minimum-distance* procedures, and on the corresponding *differentiable statistical functionals.* These procedures also appear in the monographs mentioned above, although with different motivations and objectives. We shall focus on a systematic and in-depth study of the *interrelationships* of diverse statistics and estimators beyond the usual first-order asymptotics, and incorporate such affine relations to explore the robustness properties to a finer extent. In our discussion, the asymptotics and empirical processes will play a unifying and vital role.

The celebrated *Bahadur* (1966) *representation* for sample quantiles provided a novel approach to the study of the asymptotic theory of order statistics, quantile functions, and a broad class of statistical functionals. The past two decades have witnessed a phenomenal growth of research literature on Bahadur representations for various statistics (which are typically nonlinear in structure). Bahadur's (1966) own results, as further extended by Kiefer (1967) and supplemented by Ghosh (1971) in a weaker and yet elegant form, led the way to various types of representations for statistics and estimators. Along this direction, we have observed that the *first-order and second-order asymptotic distributional representations*, having their roots in the Bahadur representations, have emerged as very useful tools for studying the general asymptotic theory for a broad class of robust and nonparametric estimators and statistics. Such asymptotic representations' results also provide a convenient way of assessing the interrelations of various competing estimators or statistics in a meaningful asymptotic setup. The second-order asymptotic representations which may not necessarily require stringent moment conditions may cast light on the appropriateness and motivation of various risk measures in the study of higher-order efficiency properties. The first- and second-order asymptotic representations for various robust procedures are a major objective of this book, with a thorough discussion of their role and their incorporation to present the relevant asymptotics in a unified manner.

The second-order asymptotic representations may cast light on the motivation of various risk measures in the study of higher-order efficiency properties. For instance, the Pitman (1937) is isomorphic to quadratic risk criterion in a conventional first-order optimality setup for a broad class of estimators (see Keating, Mason, and Sen 1993), while the second-order distributional representations provide a more profound comparison.

Among other noteworthy developments over the past 35 years, *regression quantiles* have steadily reshaped the domain and scope of robust statistics. It started with a humble aim of regression L-estimators by Koenker and Bassett (1978) and traversed the court of robustness onto the domain of *regression rank scores*; see Gutenbrunner and Jurečková (1992). In this way, it provides a natural link to various classes of robust estimators and tests and strengthens their interrelationships.

In our coverage of topics and presentation of (asymptotic) theory and interrelationships, we have attempted to be more comprehensive than any other source. By making this book mathematically rigorous, theoretically sound, and statistically motivating, we have primarily aimed at the advanced graduate level, for coursework on asymptotic theory of robust statistical inference with special emphasis on the interrelationships among families of (competing) statistics (or estimators). The asymptotic theory is further streamlined to match the needs of practical applications. In this respect, it would be helpful for readers to have familiarity with the basic theory of robustness, although we do provide a survey of robust statistics. The reader is, of course, expected to be familiar with the basic theory of statistical inference and decision theory including estimation theory, hypothesis testing, and classical linear models, at least, at an intermediate level, though a measure-theoretic orientation is not that essential. Advanced calculus, real analysis, linear algebra, and some matrix theory are also important mathematical prerequisites.

1.2 Synopsis

The basic theory and methodology is initiated in Section 2.2. Section 2.3 discusses qualitative, quantitative, local, and global aspects of robustness. *Minimax theory, scale-equivariance* and *studentization* from the robustness perspective are considered in Section 2.4. Results in Sections 2.2, 2.3, and 2.4 are presented without derivations. Likewise, some basic results in probability theory and large sample statistical theory are presented without derivations in Section 2.5. The final section of Chapter 2 provides a pool of basic mathematical and statistical tools, often asymptotic. These tools are useful in later chapters in the derivations of main results. These tools are familiar to many mathematically oriented readers, but applied statisticians may be relieved of the burden of probing into the relevant source materials, which are often presented at high levels of abstraction.

Chapter 3 provides a basic survey of robust and nonparametric estimators of location and regression parameters in univariate models, with a special emphasis on certain families of estimators:

1. Maximum likelihood type or M-estimators
2. Linear functions of order statistics or L-estimators, including regression quantiles
3. Rank statistics or R-estimators, with access to regression rank scores estimators
4. Minimum distance estimators
5. Pitman-type estimators, or P-estimators
6. Bayes-type or B-estimators
7. Statistical functionals

The basic motivations and formulations of these estimators and their first-order asymptotic representations are presented at a systematic sequence. These findings pave the way for the study of asymptotic normality, consistency, and other related properties. Many of the results of Chapter 3 are strengthened and presented in a comparatively more general level in subsequent chapters, dealing with individual classes. However, the implications of these first-order results are lucidly presented in this chapter. The representation of a robust statistic (which is typically nonlinear) in terms of an average or sum of independent, centered random variables, plus a remainder term converging to 0 in probability with increasing sample size, not only provides visible access to the related asymptotic normality but also to other properties, such as breakdown points, influence functions, and robust variance estimation. They all constitute an important aspect of the integrated study of robustness. The genesis of higher-order representations lies in such first-order ones. In this respect, the basic insight is due to Bahadur (1966), who formulated a very precise representation for a sample quantile, under very simple regularity conditions; he cleverly revealed the limitations of higher-order asymptotics of robust statistics, and called for the need of "smoothing" to eliminate some of these drawbacks. In our presentation, we will examine various robust statistics in the light of Bahadur representations to facilitate a comprehensive study of their intricate interrelationships.

Chapter 4 is devoted to the study of L-estimators of location and regression parameters. It demonstrates appropriate asymptotic representations for this possibly nonlinear statistic in terms of a linear statistic and a remainder term, and incorporates this representation in more refined statistical analysis as far as possible. The Bahadur representations occupy a focal point in this perspective. The representations of sample quantiles are presented in Section 4.2;

extensions to finite mixtures of sample quantiles are embraced in this mold. Some generalizations to possibly nonlinear functions of a (fixed) number of sample quantiles are also discussed in the section. L-statistics with *smooth scores* occupy a prominent place in statistical inference, and a detailed account of these statistics is given in Section 4.3, and some further refinements for general L-estimators are presented in Section 4.4. Because of the affinity of L-statistics to statistical functionals introduced in Section 3.7, we integrate the findings in Sections 4.2, 4.3, and 4.4 with those of differentiable statistical functionals, and this is presented in Section 4.5. Allied second-order representations are considered in Section 4.6. L-estimation in linear models constitutes an important topic in robust estimation. In Section 4.7, we consider an elaborate treatment of regression quantiles, with access to the trimmed least squares estimation. The last section of Chapter 4 is devoted to considerations of finite-sample breakdown points of L- and related M-estimators. The relevance of regression quantiles in R-estimation theory is relegated to Chapter 6, devoted to the ranks.

M-estimators of location, regression, and of general parameters are studied in Chapter 5. M-estimators are typically expressed as implicit statistical functionals, based on estimating equations. Their first- and second-order representations are treated separately, depending on the behavior of their score functions (continuity/discontinuity, monotonicity, and others). Unlike the L-estimators, the M-estimators are not generally scale equivariant, and hence we are led to their *studentized* versions. Due considerations are also given to the *one-step* and *adaptive scale equivariant* M-estimation procedures.

Chapter 6 develops parallel results for R-estimators of location and regression parameters. R-estimators, as inversions of rank tests, are scale equivariant and enjoy robustness on a global basis. For smaller datasets, the R-estimators in location/regression models can be computed by iterative methods, which become laborious for more observations and regressors. Their one-step versions starting with a preliminary estimator and their asymptotic properties are based on the *uniform asymptotic linearity* of pertaining empirical processes (Jurečková 1969, 1971 a,b). The last two decades have witnessed significant developments in applications of *regression quantiles* and of *regression rank scores* as their duals, and their relations to ordinary R-estimators. Robust estimation of scale parameters in the linear model is one important application.

For location as well as linear models, R-, L-, and M-estimators, following similar ideas, enjoy some interrelations. Based on findings of the earlier chapters, we explore these interrelations more thoroughly and systematically in Chapter 7. This chapter also treats the k-step versions of M- and other estimators, based on the Newton-Raphson algorithm, and their relations to the non-iterative versions.

Chapter 8 disseminates some aspects of robust and nonparametric estimation in multivariate models. This is a fast-developing and complex area, complex not only mathematically, but also conceptually, from the standpoint of criteria that would suit the purpose. Our attempt is to cover various concepts and most recent results, starting with various kinds of multivariate symmetry and then continuing from robust componentwise to affine-equivariant estimation in multivariate location and linear models. The coordinate-wise approach based on robust/nonparametric statistics focus quite favorably compared to the affine-equivariant approach, and it is supported by some numerical studies. Our treatise also includes the Stein-rule estimation, minimum risk efficiency and estimation of multivariate scatter, and some discussions of the subject.

The nonparametric and robust statistical tests and confidence sets are considered in Chapter 9. The topic is the general interpretation of robustness of testing procedures, with due emphasis on the location/scale and linear models treated in earlier chapters. Tests based on robust estimators as well as robust rank order and M-statistics criteria are considered in detail, along with the classical likelihood ratio type tests and their variants. Section 9.3 describes the theory of minimax tests of hypotheses and alternatives dominated by 2-alternating capacities.

Confidence sets are used to interpret the margin of fluctuations of the point estimators from their population counterparts. Procedures based on robust estimators and test statistics are considered, adopting suitable *resampling plans* for variance estimation problem, such as the *jackknife* and the *bootstrap*, besides the uniform asymptotic linearity results developed in earlier chapters. The proof of the uniform asymptotic linearity results is relegated to the Appendix. These derivations are interesting in their own right. A full though by no means exhaustive bibliography is provided at the end of the book with the hope that the references will facilitate additional studies in related problems.

CHAPTER 2

Preliminaries

2.1 Introduction

In this chapter we focus on robust statistical inference for the classical linear models. Section 2.2 first reminds some basic aspects of the classical theory of statistical inference. Section 2.3 outlines the concept of robustness. It discusses quantitative as well as qualitative aspects of robustness and also its local as well as global aspects. The minimax theory, scale-equivariance, and studentization are all important features related to robust statistical procedures. Their basic formulation is considered in Section 2.4. In the dissemination of the theory of robust statistical inference, we need to make use of basic results in probability theory as well as large sample theory. A systematic account of these results (mostly without proofs) is given in Section 2.5. Throughout this chapter, we emphasize the motivations (rather than the derivations). More technical derivations will be given in subsequent chapters.

2.2 Inference in Linear Models

We start with the classical normal theory of linear model in an univariate setup. Let $Y_1, \ldots, Y_n$ be n (≥ 1) independent random variables with

$$\mathbb{E}Y_i = \theta_i, \ i = 1, \ldots, n; \quad \theta_n = (\theta_1, \ldots, \theta_n)^\top. \tag{2.1}$$

Assume that

(1) each Y_i has a normal distribution,

(2) these normal distributions all have a common finite variance σ^2 (usually unknown),

(3) $\boldsymbol{\theta}_n$ belongs to a p-dimensional subspace $\boldsymbol{R}_p^\star$ of the n-dimensional Euclidean space $\boldsymbol{R}_n$, where $n > p$.

In the classical *linear model*, we have

$$\boldsymbol{\theta}_n = \mathbf{X}_n \boldsymbol{\beta}, \ \ \boldsymbol{\beta}_n = (\beta_1, \ldots, \beta_p)^\top, \tag{2.2}$$

where $\boldsymbol{\beta}$ is a vector of unknown regression parameters and $\mathbf{X}_n$ is a known design matrix (of order $n \times p$). In the particular case where all the θ_i are the same (e.g., equal to μ), we have $\mathbf{X}_n = (1, ..., 1)^\top$ and $\beta = \mu$ so that $p = 1$. Thus, the location model is a particular case of the linear model in (2.2). The linear model is often expressed as

$$\mathbf{Y}_n = (Y_1, \dots, Y_n)^\top = \mathbf{X}_n \boldsymbol{\beta} + \mathbf{e}_n; \quad \mathbf{e}_n = (e_1, \dots, e_n)^\top, \tag{2.3}$$

where the e_i are independent and identically distributed (i.i.d.) random variables with a normal distribution $\mathcal{N}(0, \sigma^2)$ and $0 < \sigma^2 < \infty$.

Consider the estimation of $\boldsymbol{\theta}_n$ in the model in (2.1), or of $\boldsymbol{\beta}$ in the model (2.2). The joint density of the Y_i is given by

$$\frac{1}{\sigma^n (2\pi)^{n/2}} \exp\Big\{ -\frac{1}{2\sigma^2} \sum_{i=1}^n (y_i - \theta_i)^2 \Big\}. \tag{2.4}$$

The *maximum likelihood estimator* (MLE) of $\boldsymbol{\theta}_n$ may be obtained by maximizing (2.4) with respect to $\boldsymbol{\theta}_n$ subject to the constraint that $\boldsymbol{\theta}_n \in \boldsymbol{R}_p^\star$. This reduces to minimizing, with respect to $\boldsymbol{\theta}_n$,

$$\|\mathbf{Y}_n - \boldsymbol{\theta}_n\|^2 = (\mathbf{Y}_n - \boldsymbol{\theta}_n)^\top (\mathbf{Y}_n - \boldsymbol{\theta}_n) \quad \text{when} \quad \boldsymbol{\theta}_n \in \boldsymbol{R}_p^\star. \tag{2.5}$$

The solution to the minimization problem in (2.5), denoted by $\widehat{\boldsymbol{\theta}}_n$, is the classical *least squares estimator* (LSE) of $\boldsymbol{\theta}_n$. For the model in (2.3), the LSE $\widehat{\boldsymbol{\beta}}_n$ of $\boldsymbol{\beta}$ is obtained by minimizing

$$\|\mathbf{Y}_n - \mathbf{X}_n \boldsymbol{\beta}\|^2 = (\mathbf{Y}_n - \mathbf{X}_n \boldsymbol{\beta})^\top (\mathbf{Y}_n - \mathbf{X}_n \boldsymbol{\beta})$$

with respect to $\boldsymbol{\beta}$, where $\mathbf{x}_i^\top$ denotes the ith row of $\mathbf{X}_n$, $i = 1, \dots, n$. It is explicitly given by

$$\widehat{\boldsymbol{\beta}}_n = (\mathbf{X}_n^\top \mathbf{X}_n)^{-1} (\mathbf{X}_n^\top \mathbf{Y}_n). \tag{2.6}$$

We assume that $\mathbf{X}_n^\top \mathbf{X}_n$ is of full rank p ($\leq n$); otherwise, a generalized inverse has to be used in (2.6). Remark that the equivalence of the LSE and the MLE holds for the entire class of *spherically symmetric distributions*, for which the joint density of $Y_1, \dots, Y_n$ is solely a function of the norm $\|\mathbf{Y}_n - \boldsymbol{\theta}_n\|^2$). This class covers the homoscedastic normal densities; however, not most of the nonnormal densities. In general, the MLE and LSE are not the same, and the algebraic form of the MLE highly depends on the form of the underlining error distribution.

The *optimality* of the LSE $\widehat{\boldsymbol{\theta}}_n$ may be interpreted with respect to various criteria. First, for an arbitrary vector $\mathbf{a}_n = (a_1, \dots, a_n)^\top$ of real constants, $\hat{\xi}_n = \sum_{i=1}^n a_i \hat{\theta}_i$ $(= \mathbf{a}_n^\top \widehat{\boldsymbol{\theta}}_n)$ uniformly minimizes the variance among the unbiased estimators of $\xi = \mathbf{a}_n^\top \boldsymbol{\theta}_n$. Second, $\hat{\xi}_n$ minimizes the *quadratic risk* $\mathbb{E}_\xi (T - \xi)^2$ among the class of all *equivariant* estimators of ξ : an estimator $T_n = T(\mathbf{X}_n)$ is equivariant, if it satisfies

$$T(\mathbf{Y}_n + \mathbf{t}_n) = T(\mathbf{Y}_n) + \mathbf{a}_n^\top \mathbf{t}_n$$

for every $\mathbf{t}_n = (t_1, \dots, t_n)^\top \in \boldsymbol{R}_p^\star$ and all $\mathbf{Y}_n$. Note that for the model in (2.3), $\widehat{\boldsymbol{\beta}}_n$ is a linear function of $\widehat{\boldsymbol{\theta}}_n$, and hence these optimality properties are also shared by $\widehat{\boldsymbol{\beta}}_n$. There are other optimality properties too.

If we drop the assumptions of normality and independence of $Y_1, \dots, Y_n$ and only assume that

$$\mathbb{E}Y_i = \theta_i, \;\; i = 1, \dots, n, \theta_n = (\theta_1, \dots, \theta_n)^\top \in \boldsymbol{R}_p^\star,$$

$$Cov(Y_i, Y_j) = \delta_{ij}\sigma^2 \quad \text{for } i, j = 1, \dots, n; \; \sigma^2 < \infty, \tag{2.7}$$

where δ_{ij} the usual Kronecker delta (i.e., $\delta_{ij} = 1$ if $i = j$, and 0, otherwise), then an optimality property of $\widehat{\boldsymbol{\theta}}_n$, can still be proved but in a more restricted sense. Among all linear unbiased estimators of $\mathbf{a}_n^\top \boldsymbol{\theta}_n$, $\mathbf{a}_n^\top \widehat{\boldsymbol{\theta}}_n$ uniformly minimizes the variance. This is provided by the classical Gauss-Markov theorem, and is referred to as the BLUE (*best linear unbiased estimator*) property. In this setup, the form of the distribution of the X_i is not that important, but the X_i need to have finite variances. If we want to remove the restriction on linear unbiased estimators and establish the optimality of the LSE in a wider class of estimators, we may need more specific conditions on these distributions. This may be illustrated by the following results of Rao (1959, 1967, 1973) and Kagan, Linnik, and Rao (1967, 1973):

> Consider the linear model in (2.3) sans the normality of the errors. Assume that the e_i are i.i.d. random variables with the distribution function F, defined on $\boldsymbol{R}$, such that
>
> $$\int_{\boldsymbol{R}} x dF(x) = 0 \text{ and } \int_{\boldsymbol{R}} x^{2s} dF(x) < \infty$$
>
> for a positive integer s. Assume that $\mathbf{X}_n$ is of rank p, and consider p linearly independent linear functions of $\boldsymbol{\beta}$, denoted by $\ell_1, \dots, \ell_p$, respectively. Let $\hat{\ell}_{n1}, \dots, \hat{\ell}_{np}$ be the LSE of $\ell_1, \dots, \ell_p$, respectively. If $\hat{\ell}_{n1}, \dots, \hat{\ell}_{np}$ are optimal estimators of their respective expectations in the class of all polynomial unbiased estimators of order less than or equal to s, then under suitable regularity conditions on $\mathbf{X}_n$, the first $(s+1)$ moments of F should coincide with the corresponding moments of a normal distribution. Specifically, F should coincide with a normal distribution if $\hat{\ell}_{n1}, \dots, \hat{\ell}_{np}$ are optimal in the class of all unbiased estimators.

Furthermore, Kagan (1970) proved that the latter characterization of the normal law is continuous in the location model in the following sense:

> Let $Y_1, \dots, Y_n$ $(n \geq 3)$ be i.i.d. random variables with distribution function $F(y - \theta)$ such that $\int_{\boldsymbol{R}} y dF(y) = 0$ and $\int_{\boldsymbol{R}} y^2 dF(y) = \sigma^2 < \infty$, and let $\bar{Y}_n = n^{-1}\sum_{i=1}^n Y_i$, the LSE of θ, be ε-*admissible* in the sense that there exists no unbiased estimator $\tilde{\theta}$ of θ satisfying
>
> $$\frac{n\mathbb{E}_\theta(\tilde{\theta} - \theta)^2}{\sigma^2} < \frac{n\mathbb{E}_\theta(\bar{Y}_n - \theta)^2}{\sigma^2} - \varepsilon, \;\; \varepsilon > 0.$$
>
> Then
>
> $$\sup_{y \in \boldsymbol{R}} \left| F(y) - \Phi(\frac{y}{\sigma}) \right| \leq K(-\log \varepsilon)^{-1/2}, \;\; \varepsilon > 0,$$

where $\Phi(.)$ stands for the standard normal distribution function and K $(0 < K < \infty)$ is some constant, independent of ε.

We refer to Kagan et al. (1973) for additional characterization results.

Let us look at the LSE $\widehat{\boldsymbol{\beta}}_n$ in (2.6). Using (2.3), we have

$$\widehat{\boldsymbol{\beta}}_n = \boldsymbol{\beta} + (\mathbf{X}_n^\top \mathbf{X}_n)^{-1}(\mathbf{X}_n^\top \mathbf{e}_n), \tag{2.8}$$

Therefore, when the elements of $\mathbf{e}_n$ are i.i.d. random variables with the normal distribution with 0 mean and a finite positive variance σ^2, we obtain that $(\widehat{\boldsymbol{\beta}}_n - \boldsymbol{\beta})$ has the p-variate normal distribution with mean vector $\mathbf{0}$ and dispersion matrix $\sigma^2(\mathbf{X}_n^\top \mathbf{X}_n)^{-1}$. Further, $(\mathbf{Y}_n - \mathbf{X}_n\widehat{\boldsymbol{\beta}}_n)^\top(\mathbf{Y}_n - \mathbf{X}_n\widehat{\boldsymbol{\beta}}_n)/\sigma^2$ has the central chi-square distribution with $n-p$ *degrees of freedom* (d.f.), independently of $\widehat{\boldsymbol{\beta}}_n$. Therefore, using the fact that $(\widehat{\boldsymbol{\beta}}_n - \boldsymbol{\beta})^\top(\mathbf{X}_n^\top \mathbf{X}_n)(\widehat{\boldsymbol{\beta}}_n - \boldsymbol{\beta})/\sigma^2$ has the central chi-square distribution with p d.f. by the Cochran theorem, we obtain from above that

$$\frac{(\widehat{\boldsymbol{\beta}}_n - \boldsymbol{\beta})^\top(\mathbf{X}_n^\top \mathbf{X}_n)(\widehat{\boldsymbol{\beta}}_n - \boldsymbol{\beta})/p}{\|\mathbf{Y}_n - \mathbf{X}_n\widehat{\boldsymbol{\beta}}_n\|^2/(n-p)} \tag{2.9}$$

has the central variance-ratio (F)-distribution with $(p, n-p)$ d.f. If we denote the upper $100\alpha\%$ quantile $(0 < \alpha < 1)$ of this F-distribution by $F_{p,n-p,\alpha}$, we obtain from (2.9) that

$$I\!P\left\{\frac{(\widehat{\boldsymbol{\beta}}_n - \boldsymbol{\beta})^\top(\mathbf{X}_n^\top \mathbf{X}_n)(\widehat{\boldsymbol{\beta}}_n - \boldsymbol{\beta})}{\|\mathbf{Y}_n - \mathbf{X}_n\widehat{\boldsymbol{\beta}}_n\|^2} \le p(n-p)^{-1}F_{p,n-p:\alpha}\right\} = 1-\alpha. \tag{2.10}$$

Then (2.10) provides a *confidence ellipsoid* for $\boldsymbol{\beta}$ with the *confidence coefficient* (or *coverage probability*) $1-\alpha$. Note that, by the Courant theorem,

$$\sup\left\{\frac{|\mathbf{a}^\top(\widehat{\boldsymbol{\beta}}_n - \boldsymbol{\beta})|}{(\mathbf{a}^\top(\mathbf{X}_n^\top \mathbf{X}_n)^{-1}\mathbf{a})^{\frac{1}{2}}} : \mathbf{a} \in \boldsymbol{R}_p\right\} = \left\{(\widehat{\boldsymbol{\beta}}_n - \boldsymbol{\beta})^\top(\mathbf{X}_n^\top \mathbf{X}_n)(\widehat{\boldsymbol{\beta}}_n - \boldsymbol{\beta})\right\}^{1/2}.$$

Thus (2.10) may equivalently be written as

$$I\!P\left\{|\mathbf{a}^\top(\widehat{\boldsymbol{\beta}}_n - \boldsymbol{\beta})| \le \left\{\frac{p}{(n-p)}F_{p,n-p:\alpha}(\|\mathbf{Y}_n - \mathbf{X}_n\widehat{\boldsymbol{\beta}}_n\|^2)(\mathbf{a}^\top(\mathbf{X}_n^\top \mathbf{X}_n)^{-1}\mathbf{a})\right\}^{1/2} \text{ for every } \mathbf{a} \in \boldsymbol{R}_p\right\} = 1-\alpha. \tag{2.11}$$

This provides a *simultaneous confidence interval* for all possible linear combinations of the $\boldsymbol{\beta}_j$. If our interest centers only on a subset of q of the p parameters $(\boldsymbol{\beta}_1, \ldots, \boldsymbol{\beta}_p)$ in (2.9), in the numerator, we may choose an appropriate quadratic form in these q estimators. It will lead to a parallel form of (2.10) or (2.11), where $F_{p,n-p:,\alpha}$ has to be replaced by $F_{q,n-p:\alpha}$. Since $F_{q,n-p:,\alpha} < F_{p,n-p:,\alpha}$ for $q < p$, this would lead to a shorter confidence region for the given subset of parameters. Like the point estimators, these confidence regions have also some optimal properties under the basic assumption that the e_i are i.i.d. random variables with a normal distribution. The exact distribution theory in (2.9) may not hold when the e_i may not have the normal distribution. As a result, the coverage probability in (2.10) or (2.11) may cease to be equal to $1 - \alpha$. The confidence intervals are more vulnerable to any departure from the assumed normality of the errors. We may have some justifications for large values of n, as will be discussed later on.

The optimal tests of linear hypotheses in the model (2.1) under the normal distribution of Y_i are typically based on the MLE or, equivalently, on the LSE. Consider the general linear hypothesis

$$H_0 : \boldsymbol{\theta}_n \in \boldsymbol{R}_s^{\star\star}, \quad \text{a linear subspace of } \boldsymbol{R}_p^{\star} \text{ with } s < p.$$

This formulation covers various hypotheses in the linear model (2.3), and it is usually referred to in the literature as the *analysis of variance* (ANOVA) model. Rather than the unrestricted LSE in (2.5), we will consider the restricted LSE $\widehat{\boldsymbol{\theta}}_n^{\star}$, which leads to a minimization of $\|\mathbf{Y}_n - \boldsymbol{\theta}_n\|^2$ under the restriction that $\boldsymbol{\theta}_n \in \boldsymbol{R}_s^{\star\star}$. If we define $F_{p-s,n-p:\alpha}$ as in (2.9), for the case where the Y_i are normally distributed with a common variance $\sigma^2 : 0 < \sigma < \infty$, we have the classical test for H_0 with the *critical region* :

$$\frac{\|\widehat{\boldsymbol{\theta}}_n^{\star} - \widehat{\boldsymbol{\theta}}_n\|^2}{\|\mathbf{Y}_n - \widehat{\boldsymbol{\theta}}_n\|^2} > \frac{p-s}{n-p} \cdot F_{p-s,n-p:\alpha}. \tag{2.12}$$

Since $\frac{n-p}{p-s} \cdot \frac{\|\widehat{\boldsymbol{\theta}}_n^{\star} - \widehat{\boldsymbol{\theta}}_n\|^2}{\|\mathbf{Y}_n - \widehat{\boldsymbol{\theta}}_n\|^2}$ has the central F-distribution with $(p-s, n-p)$ d.f. under H_0, the *level of significance* or *size* of this test is equal to $\alpha : (0 < \alpha < 1)$. Under any departure from the null hypothesis, this statistic has a noncentral F-distribution with d.f. $(p-s, n-p)$ and a nonnegative noncentrality parameter. Then the probability of the critical region in (2.12) is always greater than or equal to α, and hence, the test is *unbiased*. Moreover, the test is invariant under linear (orthogonal) transformations : $\mathbf{Y}_n \mapsto \mathbf{Y}_n^* = \mathbf{D}_n\mathbf{Y}_n$. In fact, it is the most powerful among *invariant* and *unbiased* tests of H_0 of size α. While the normality of the X_i is not that crucial for the above mentioned invariance, a negation of this normality may nevertheless distort the distribution of *maximal invariants* and take away the most powerful property of the test in (2.12). Even the size of the test may be different from α when the X_i are not necessarily homoscedastic and normal. Thus, the optimality of the classical ANOVA tests based on the LSE is confined to homoscedastic normal distributions of the X_i, and they may not be robust against any departure from this basic assumption.

To focus on this latter aspect, consider the usual linear model in (2.3) under the Gauss–Markov setup in (2.7). By (2.8), the asymptotic normality of $n^{1/2}(\widehat{\boldsymbol{\beta}}_n - \boldsymbol{\beta})$ as $n \to \infty$ may be proved under quite general conditions on $\mathbf{X}_n$, whenever the e_i are i.i.d. random variables with finite positive variance σ^2. For example, if $\mathbf{Q}$ is a positive definite matrix such that

$$\lim_{n\to\infty} n^{-1}(\mathbf{X}_n^{\top}\mathbf{X}_n) = \mathbf{Q}, \tag{2.13}$$

then the multivariate central limit theorem in Section 2.5 may be directly adapted to establish the asymptotic normality. The condition in (2.13) may be replaced by a Noether-type condition that

$$\max\{\mathbf{x}_i^\top(\mathbf{X}_n^\top\mathbf{X}_n)^{-1}\mathbf{x}_i : 1 \le i \le n\} \to 0, \text{ as } n \to \infty, \tag{2.14}$$

along with a milder growth condition on $tr(\mathbf{X}_n^\top\mathbf{X}_n)$. Thus, we have under (2.13),

$$n^{1/2}(\widehat{\boldsymbol{\beta}}_n - \boldsymbol{\beta}) \xrightarrow{\mathcal{D}} \mathcal{N}_p(\mathbf{0}, \sigma^2\mathbf{Q}^{-1}) \quad \text{as } n \to \infty. \tag{2.15}$$

Consider the classical *Cramér-Rao inequality*

$$\sigma^2 \ge [\mathcal{I}(f)]^{-1}, \tag{2.16}$$

with $\mathcal{I}(f)$ being the *Fisher information* of f,

$$\mathcal{I}(f) = \int_{\boldsymbol{R}} \left(\frac{f'(x)}{f(x)}\right)^2 dF(x) \quad \text{(and assume it to be finite)} \tag{2.17}$$

Then the strict equality sign holds only when $f'(x)/f(x) = kx$ a.e., for some nonzero finite constant k. This is the case when F is itself a normal distribution function. If we used the MLE based on the true density f instead of the LSE, then we would have (2.15) with σ^2 replaced by $\mathcal{I}(f)^{-1}$. Clearly, the true model MLE is asymptotically efficient, and the LSE is asymptotically efficient if and only if the errors e_i are i.i.d. and normally distributed. Under the same setup it can be shown that

$$\frac{\|\mathbf{Y}_n - \mathbf{X}_n\widehat{\boldsymbol{\beta}}_n\|^2}{n-p} \xrightarrow{\mathcal{P}} \sigma^2 \quad \text{as } n \to \infty, \tag{2.18}$$

so that (2.9) holds with the F-distribution being replaced by the central chi-square distribution with p d.f. Thus, (2.10) and (2.11) hold with the only change that $F_{p,n-p:\alpha}$ is replaced by $F_{p,\infty:\alpha}$ $(= p^{-1}\chi^2_{p,\alpha})$. By (2.16) and the shortness-criterion of the diameters of the confidence ellipsoids, we conclude that the LSE-based confidence region is asymptotically efficient iff (if and only if) the errors e_i are i.i.d. and normally distributed. Similarly, using (2.15) and (2.18), it can be shown that for the restricted LSE $\widehat{\boldsymbol{\beta}}_n^\star$ it holds under H_0

$$\frac{\|\widehat{\boldsymbol{\theta}}_n^\star - \widehat{\boldsymbol{\theta}}_n\|^2(n-p)}{\|\mathbf{Y}_n - \mathbf{X}_n\widehat{\boldsymbol{\beta}}_n\|^2} \xrightarrow{\mathcal{D}} \chi^2_{p-s}, \tag{2.19}$$

and that its *Pitman efficacy* (to be defined more precisely in a later section) with respect to local alternatives is proportional to σ^{-2}. For the true model MLE-based *likelihood ratio test*, we have the same asymptotic chi-square distribution under H_0 and the corresponding Pitman efficacy is proportional to $\mathcal{I}(f)$. Hence, in the light of the conventional Pitman efficiency, the test in (2.12) is asymptotically optimal iff the underlying distribution function is normal.

In the above derivations require some of the probabilistic tools discussed in Section 2.5, and hence the reader is suggested to look into that section before attempting to solve these problems.

2.3 Robustness Concepts

We have seen in the preceding section that the performance of the inference procedures based on the least squares is rather sensitive to possible variations from assumed normal distributions of the errors. It means that the LSE is *distributionally nonrobust*, i.e. *nonrobust* with respect to deviations from the assumed normal distribution. The same criticism may be made against the statistical inference based on the maximum likelihood when the assumed and the true distributions of the errors are not the same. This brings us to feel intuitively that if we are not sure of the underlying model (e.g., the form of the error distribution function), we should rather use procedures that do not show this kind of sensitivity (even if they are not quite optimal under the model) or, otherwise speaking, that are distributionally *robust*.

These ideas are not new. The historic studies by Stigler (1973, 1980) reveal that even the earlier statisticians were aware of the consequences of using an incorrect model in a statistical inference procedure. Pearson (1931) observed the sensitivity of the classical ANOVA procedures to possible departures from the assumed normal model (mostly, in the form of skewness and kurtosis), and scores of papers appeared over the next 30 years (mostly in *Biometrika*) examining the effect of such deviations on the size and power of the classical ANOVA tests. The term *robustness* was first used by Box (1953). We should also mention the work of the Princeton Statistical Research Group incorporated in the late 1940s (under the leadership of J. W. Tukey) which tried to find robust alternatives of the classical methods then in practice.

Though it is intuitively clear what robustness should be, its definition did not evolve in a universal way. Even today, there exist several (related) mathematical concepts of robustness. This should not be considered a shortcoming; rather, it illustrates the basic fact that there are diverse aspects of robustness and, hence, diverse criteria as well. Box and Anderson (1955) argued that a good statistical procedure should be insensitive to changes not involving the parameters or the hypothesis of interest, but should be sensitive to the changes of parameters to be estimated or hypotheses being tested. This idea is generally accepted. It has the natural implications of insensitivity to model departures and sensitivity to good performance characteristics under the model. At this point, it may be useful to elucidate this diverse general framework.

1. *Local and global robustness.* In a local model, we allow only small changes. For example, we may assume that the errors are normally distributed according to a given variance, but this distribution is locally *contaminated* by

a heavier tailed distribution, a case that may arise in the *outlier* or *gross error model.* In a global sense, we admit that the real distribution is not precisely known and can be a member of a large set of distributions. For example, the error distribution function may be assumed to be a member of the class of all symmetric distribution functions (where the assumption of finite variance or Fisher information may or may not be imposed). In a local model, we are naturally interested in retaining the performance optimality of a procedure (for the given model) to a maximum possible extent, allowing its insensitivity only to small departures from the assumed model. In the global case, *validity-robustness* of a procedure (over the entire class of models contemplated) dominates the picture, and there are also issues relating to typical measures of optimality or desirability of a procedure for a broader class of models.

2. *Specific property-based robustness.* We try to determine which special property of the procedure should be given priority in the robustness picture. For example, we could try to limit the sensitivity of the bias or the mean square error of the estimator (as well as its asymptotic versions) or some other characteristic of the distribution of this estimator. For testing procedure, we may similarly emphasize the size, power, or some other performance characteristic, and seek robustness specifically for it.

3. *Type of model departures.* We must specify what kinds of deviations we have in mind. One possibility is the deviations in the shape of the error distribution, discussed so far. It is equally possible to use some other measures to depict such departures. Other assumptions, such as the independence of the errors, should also be given consideration in determining the scope of robustness.

We now turn to some basic concepts of robustness in estimation for a simple model. Other illustrations can be found in Huber (1981), Huber and Ronchetti (2009), Hampel et al. (1985, 2005), and Maronna et al. (2006). Consider the simple model where $X_1, \ldots, X_n$ are i.i.d. random variables with a distribution function G, generally unknown. We have a *parametric model* $\{F_\theta : \ \theta \in \Theta\}$ formed by a dominated system of distributions and wish to estimate θ for which F_θ is as close to G as possible. Let $T_n = T_n(X_1, \ldots, X_n)$ be an estimator of θ which we express as a functional $T(G_n)$ of the *empirical* (sample) distribution function G_n of $X_1, \ldots, X_n$. We naturally let T_n be a *Fisher-consistent estimator* of θ; that is, $T(F_\theta) = \theta, \ \theta \in \Theta$. We assume that T_n has high efficiency of some kind (such as its risk, variance or bias (eventually asymptotic)) if they are reasonably small. First consider the *local* (infinitesimal) concept of robustness, introduced by Hampel (1968, 1971). Denote by $\mathcal{L}_F(T_n)$ the probability distribution (law) of T_n where the true distribution function of X_1 is F. The sequence $\{T_n\}$ of estimators or test statistics is called *qualitatively robust* at $F = F_0$ if the sequence $\{\mathcal{L}_F(T_n)\}$ is equicontinuous at F_0 with respect to the Prokhorov metric Π on a probability space, as will be considered in Section 2.5. In other words, the sequence $\{T_n\}$ is called qualitatively robust if for each $\varepsilon > 0$, there exist a $\delta > 0$ and an integer n_0 such that

$$\Pi(F, F_0) < \delta \Rightarrow \Pi(\mathcal{L}_F(T_n), \mathcal{L}_{F_0}(T_n)) < \varepsilon \tag{2.20}$$

for any distribution function F and all $n \geq n_0$. If $T_n = T(G_n)$, where G_n is the empirical distribution function of $X_1, \ldots, X_n$, then the definition in (2.20) conveys the continuity of $T(.)$ with respect to the weak topology metricized by the Prokhorov or some other metric (e.g., the Levi metric). The weak continuity of T_n and its consistency at G (the true distribution of X_1) in the sense that $T_n \to T(G)$ almost surely (a.s.) as $n \to \infty$ characterize the robustness of T_n in a neighborhood of G. Hampel (1968, 1974) introduced the concept of *influence curve* (IC) $IC(x; G, T)$ whose value at the point $x\,(\in \boldsymbol{R}_1)$ is equal to the directional derivative of $T(G)$ at G in the direction of the one-point distribution function $\delta_x(t)$:

$$IC(x; G, T) = \lim_{\varepsilon \downarrow 0}\{\varepsilon^{-1}[T((1-\varepsilon)G + \varepsilon\delta_x) - T(G)]\}, \quad x \in \boldsymbol{R}_1$$

where for every $t, x \in \boldsymbol{R}_1$,

$$\delta_x(t) = 0 \text{ or } 1 \quad \text{according as } t \text{ is} \leq x \text{ or not.}$$

Because $IC(x)$ can be understood as a measure of sensitivity of T to the single point x, we can restrict ourselves to functionals with bounded influence curves. However, not every qualitatively robust functional has a bounded IC. A simple example is the R-estimator of location [i.e., of the center of symmetry θ of the distribution function $G(x) = F(X - \theta)$], based on the classical normal-scores signed-rank statistic:

$$T_n = \frac{1}{2}(T_n^- + T_n^+), \tag{2.21}$$

where

$$T_n^- = \sup\{t : \ S_n(\mathbf{X}_n - t\mathbf{1}) > 0\}, \ T_n^+ = \inf\{t : \ S_n(\mathbf{X}_n - t\mathbf{1}) < 0\} \tag{2.22}$$

and

$$S_n(\mathbf{X}_n - t\mathbf{1}) = \sum_{i=1}^{n} \mathrm{sign}(\mathrm{X_i - t})\mathrm{a_n}(\mathrm{R_{ni}^+(t)}), \tag{2.23}$$

where $R_{ni}^+(t)$ is the rank of $|X_i - t|$ among $|X_1 - t|, \ldots, |X_n - t|$, for $i = 1, \ldots, n$, and $a_n(k)$ is the expected value of the kth smallest order statistic in a sample of size n drawn from the chi distribution with 1 d.f., for $k = 1, \ldots, n$. Denote by $\Phi(.)$ the standard normal distribution function. Then, if F possesses an absolutely continuous symmetric density function f with a finite Fisher information $I(f)$, defined by (2.17), we have

$$IC(x; F, T) = \Phi^{-1}(F(x))[\int \{-f'(y)\}\Phi^{-1}(F(y))dy]^{-1}, \quad x \in \boldsymbol{R}_1 \tag{2.24}$$

so that as $x \to \pm\infty$, $IC(x)$ goes to $\pm\infty$. We refer to Chapter 6 for a detailed study of R-estimators.

Hampel (1986, 1974) introduced the terminology *gross error sensitivity* of T_n at F by defining it as

$$\gamma^\star = \gamma^\star(T, F) = \sup\{|IC(x; F, T)| : x \in \boldsymbol{R}_1\}. \tag{2.25}$$

Hampel also looked for an estimator minimizing the asymptotic variance at the assumed model subject to the Fisher consistency and to a prescribed bound on the gross error sensitivity. The solution of such a minimization problem is an estimator asymptotically efficient among the qualitatively robust (or infinitesimally robust) estimators with influence curve bounded by a given constant.

Another natural measure of robustness of the functional $T = T(F)$ is its *maximal bias* (maxbias) over $\mathcal{F} \ni F_0$, with F_0 being the true distribution function:

$$b(\mathcal{F}) = \sup_{F \in \mathcal{F}} \{|T(F) - T(F_0)|\}. \tag{2.26}$$

The statistical functional is often monotone with respect to the stochastic ordering:

$$X \prec Y \; X \text{ stochastically smaller than } X \text{ if } F(x) \geq G(x),\ x \in \boldsymbol{R}_1,$$
$$F, G \text{ being distribution functions of } X, Y, \text{ respectively.}$$

Then it attains its maxbias either at the stochastically largest member F_∞ or at the stochastically smallest member $F_{-\infty}$ of $\mathcal{F}$.

Let us illustrate the situation on the classical location model with i.i.d. observations $X_1, \ldots, X_n$ with distribution function $G(x) = F_0(x - \theta)$, $x \in \boldsymbol{R}_1$. Since our risk functions must be invariant to the translation, we can restrict our considerations to estimators that are *translation-equivariant*; namely, they should satisfy

$$T_n(X_1 + c, \ldots, X_n + c) = T_n(X_1, \ldots, X_n) + c, \quad \forall c \in \boldsymbol{R}_1.$$

A neighborhood of the model with distribution function F_0 can be represented by one of the following cases:

(1) *Model of ε-contaminacy.* The neighborhood is defined by

$$\mathcal{P}_\varepsilon = \{F : F = (1 - \varepsilon)F_0 + \varepsilon H,\ H \in \mathcal{H}\}, \quad \varepsilon > 0, \tag{2.27}$$

where $\mathcal{H}$ is some set of distribution functions , defined on $\boldsymbol{R}_1$.

(2) The *Kolmogorov-neighborhood* of F_0 is defined by

$$\mathcal{P}_\varepsilon = \{F : \sup_{x \in \boldsymbol{R}} |F(x) - F_0(x)| < \varepsilon\}, \quad \varepsilon > 0. \tag{2.28}$$

Moreover, Huber (1968) suggested the following criterion of performance of an estimator T_n for the finite sample case: For some fixed $a > 0$, set

$$S(T;a) = \sup_{\mathcal{P}_\varepsilon}\{\max[I\!P\{T_n < \theta - a\}, I\!P\{T_n > \theta + a\}]\}. \tag{2.29}$$

The estimator T_n is called *minimax robust* if it minimizes $S(T;a)$ over the class of translation-equivariant estimators.
Another possible criterion is the *minimax asymptotic bias* or the *minimax asymptotic variance of the estimator* over the family $\mathcal{P}_\varepsilon$. This approach, first considered by Huber (1964), depends on the symmetry of the model and requires regularity conditions that ensure the asymptotic normality of the estimators under consideration. In admitting that $100\varepsilon\%$ of the observations come from a contaminating population, this approach is more realistic than that of qualitative robustness. These two performance characteristics can be reconciled by (2.29) under quite general conditions when n is large if we allow a to depend on n (e.g., $a = n^{-1/2}c$, for some fixed $c > 0$). Some further results along this line having relevance to our subsequent studies are briefly presented in the following section.

Some authors, Huber-Carol (1970); Jaeckel (1971); Beran (1977); Rieder (1978, 1979, 1980, 1981, 1994), have considered similar criteria but looked for a *minimax solution* over a *shrinking neighborhood* of the model [in the sense that in (2.27) or (2.28), ε is made to depend on n, and it converges to 0 as $n \to \infty$, e.g., $\varepsilon = \varepsilon_n = \mathrm{O}(n^{-1/2})$]. In addition to the possibility of using models in (2.27) and (2.28), they have considered the possibility of representing such shrinking neighborhoods by means of the *Hellinger* or the *total variation* metrics. Although motivations for such an approach may be obtained from *Pitman efficiency* (as will be shown later), a return to the infinitesimal approach may yield more robust results as Hampel (1968, 1974) has demonstrated. Hampel (1971), aware of the shortcomings of the infinitesimal approach, introduced the concept of the *breakdown point* as a global measure of robustness; the breakdown point of T_n with respect to the distribution function F is defined as

$$\begin{aligned} \varepsilon^\star \quad = \quad & \sup_{0<\varepsilon\leq 1}\Big\{\varepsilon : \exists \text{ a compact subset } K(\varepsilon) \text{ of the parameter space} \\ & \text{s. t. } \Pi(F,G) < \varepsilon \Rightarrow \lim_{n\to\infty} 2I\!P_G\{T_n \in K(\varepsilon)\} = 1\Big\}, \end{aligned}$$

where $\Pi(\cdot)$ is defined in (2.20). The breakdown point is then the largest possible fraction of the population F, which is replaced by arbitrary data but still allows the estimator to be informative for the parameter. Hampel emphasized that a good estimator should be qualitatively robust, have a bounded influence function, and have a high breakdown point (ideally equal to 1/2). With this in view, we present the following concepts.

2.3.1 Finite-Sample Breakdown and Tail-Performance

Donoho and Huber (1983) introduced a finite-sample version of the breakdown point that is an extension of the original idea of Hodges (1967). Jurečková (1981) introduced a finite sample measure of tail behavior of an estimator of location, which was then extended to the linear regression in He et al. (1990). These two finite sample measures of robustness, though defined differently, appear to be very close to each other, and this is a point that we will illustrate in more detail.

Consider the linear regression model

$$Y_i = \mathbf{x}_i^\top \boldsymbol{\beta} + \mathbf{e}_i, \quad \mathbf{x}_i \in \boldsymbol{R}_p, \ i = 1, \ldots, n, \tag{2.30}$$

and denote by $\mathbf{Z}$ the set of n data points,

$$\mathbf{Z} = \Big(\mathbf{z}_1 \ldots, \mathbf{z}_n\Big)^\top, \quad \mathbf{z}_i^\top = (\mathbf{x}_i^\top, y_i) = (x_{i1}, \ldots, x_{ip}, y_i),$$

$i = 1, \ldots, n$. Let $\mathbf{T}_n = \mathbf{T}_n(\mathbf{Z})$ be an estimator of $\boldsymbol{\beta}$. Consider all possible contaminated samples $\mathbf{Z}'$ that are obtained from $\mathbf{Z}$ by replacing any m of the original points $\mathbf{z}_1, \ldots, \mathbf{z}_n$ by arbitrary values. Then the *finite sample breakdown point* $m_n^\star(\mathbf{T}, \mathbf{Z})$ of estimator $\mathbf{T}_n$ is defined as

$$m_n^\star(\mathbf{T}, \mathbf{Z}) = \min\{m : \sup_{\mathbf{Z}'} \|\mathbf{T}_n(\mathbf{Z}') - \mathbf{T}_n(\mathbf{Z})\| = \infty\}.$$

Some authors instead call the ratio $\varepsilon_n^\star(\mathbf{T}, \mathbf{Z}) = m_n^\star(\mathbf{T}, \mathbf{Z})/n$ a breakdown point. For instance, if $\mathbf{T}$ is the least squares estimator, we can easily see that $m_n^\star(\mathbf{T}, \mathbf{Z}) = 1, \forall \mathbf{Z}$. On the other hand, assuming that the errors $e_1, \ldots, e_n$ are i.i.d. with a joint F such that

$$0 < F(x) < 1, \quad \forall x \in \boldsymbol{R}_1,$$

we can consider the tail-behavior measure $B(a, \mathbf{T}_n) : \boldsymbol{R}_1^+ \to \boldsymbol{R}_1^+$ of $\mathbf{T}_n$ under a fixed $\mathbf{X}$ as

$$B(a, \mathbf{T}_n) = \frac{-\log I\!P_{\boldsymbol{\beta}}(\max_i |\mathbf{x}_i^\top(\mathbf{T}_n - \boldsymbol{\beta})| > a)}{-\log\big(1 - F(a)\big)}, \quad a \in \boldsymbol{R}_1^+. \tag{2.31}$$

Naturally, for a reasonable estimator $\mathbf{T}_n$,

$$\lim_{a\to\infty} I\!P_{\boldsymbol{\beta}}(\max_{1\le i\le n} |\mathbf{x}_i^\top(\mathbf{T}_n - \boldsymbol{\beta})| > a) = 0.$$

We are interested in estimators for which this rate of convergence is as fast as possible. Since we are not able to control $B(a, \mathbf{T}_n)$ for all $a \in (0, \infty)$, we consider as a *measure of tail behavior* of $\mathbf{T}_n$ the limit $\lim_{a\to\infty} B(a, \mathbf{T}_n)$ if it exists, eventually $\limsup_{a\to\infty} B(a, \mathbf{T}_n)$ and $\liminf_{a\to\infty} B(a, \mathbf{T}_n)$. If the estimator $\mathbf{T}_n$ is such that there exist at least one positive and at least one negative residuals $y_i - \mathbf{x}_i^\top \mathbf{T}_n$, $i = 1, \ldots, n$, then we can easily verify that

$$\bar{B} = \limsup_{a\to\infty} B(a, \mathbf{T}_n) \le n.$$

In other words, the probability $\mathbb{P}_{\boldsymbol{\beta}}(\max_i |\mathbf{x}_i^\top(\mathbf{T}_n - \boldsymbol{\beta})| > a)$ will tend to zero at most n times faster than $\mathbb{P}(e_1 > a)$ as $a \to \infty$. We will refer to distributions satisfying

$$\lim_{a\to\infty} \Big[-(ca^r)^{-1} \log\{1 - F(a)\}\Big] = 1, \tag{2.32}$$

for some $c > 0$, $r > 0$, as to *exponentially tailed or to distributions of type I*, and we will refer to distributions satisfying

$$\lim_{a\to\infty} \Big[-(m \log a)^{-1} \log\{1 - F(a)\}\Big] = 1,$$

for some $m > 0$, as *algebraically tailed or type II*. It turns out that the finite-sample breakdown point is just in correspondence with the tail behavior of $\mathbf{T}_n$ for distributions of type II. Let us illustrate it on the location model. We will remark on the regression model in connection with the LSE and later in connection with some specific estimators.

In the location model

$$Y_i = \theta + e_i, \quad i = 1, \ldots, n,$$

we have $\mathbf{X} = \mathbf{1}_n$ and $\boldsymbol{\beta} = \theta$. Hence

$$B(a, T_n) = \frac{-\log \mathbb{P}_\theta(|T_n - \theta| > a)}{-\log\{1 - F(a)\}}, \quad a \in \boldsymbol{R}_1^+.$$

A close link between tail performance and breakdown point for large class of estimators is illustrated in the following theorem.

THEOREM 2.1 *Suppose that $T_n(Y_1, \ldots, Y_n)$ is a location equivariant estimator of θ such that T_n is increasing in each argument Y_i. Then T_n has a universal breakdown point $m^\star = m_n^\star(T)$, and under any symmetric, absolutely continuous F such that $0 < F(x) < 1\ \forall x \in \boldsymbol{R}_1$ and such that*

$$\lim_{a\to\infty} \frac{-\log\big(1 - F(a + c)\big)}{-\log\big(1 - F(a)\big)} = 1 \quad \forall c \in \boldsymbol{R}_1, \tag{2.33}$$

T_n satisfies the inequalities

$$m^\star \le \liminf_{a\to\infty} B(a, T_n) \le \limsup_{a\to\infty} B(a, T_n) \le n - m^\star + 1.$$

Theorem 2.1 will be proved with the aid of the following lemma:

LEMMA 2.1 *Let $Y_{(1)} \le \ldots \le Y_{(n)}$ be the order statistics corresponding to $Y_1, \ldots, Y_n$. Then, under the conditions of Theorem 2.1*
(i) $m^\star$ is universal, i.e. independent of the special choice of $Y_1, \ldots, Y_n$, and
(ii) there exists a constant A such that

$$Y_{(m^\star)} - A \le T_n \le Y_{(n-m^\star+1)} + A.$$

PROOF (i) It suffices to show that, if $m^\star(T_n, \mathbf{Y}) = m$ for an arbitrary vector

$\mathbf{Y} = (Y_1, \ldots, Y_n)$, then $m^\star(T_n, \mathbf{Y}^0)$ when $\mathbf{Y}^0 = \mathbf{0}$. Set $\mathbf{C} = \max_{1 \le i \le n} |Y_i|$. If $m^\star(T_n, \mathbf{Y}) = m$ and T_n is translation equivariant, then (a) $\forall B > 0$, there exists a contaminated sample $\mathbf{Z} = (\mathbf{V}, \mathbf{Y}^\top)$, consisting of m new elements $\mathbf{V}$ and n-m old elements from $\mathbf{Y}$ such that $|T_n(\mathbf{Z})| > B + C$ and (b) there exists $B > 0$ such that for all contaminated samples $\mathbf{Z} = (\mathbf{V}, \mathbf{Y}^\top)$ with $(m-1)$ new elements V and $n - m + 1$ old elements, $|T_n(\mathbf{Z})| < B - C$.
Let $\mathbf{Z}$ satisfy (a) for fixed B and $T_n(\mathbf{Z}) > 0$. Then, by equivariance, $T_n(\mathbf{Z} - C) > B$, and hence by monotonicity $T_n(\mathbf{V} - C, \mathbf{0}) > B$. Similarly,

$$T_n(\mathbf{Z}) < 0 \Rightarrow T_n(\mathbf{Z} + C) < -B \Rightarrow T_n(\mathbf{V} - C, \mathbf{0}) < -B.$$

A similar argument shows that if for all perturbations of $(m-1)$ elements of $\mathbf{Y}$ is $|T_n(\mathbf{Z})| < B - C$, then for all perturbations of $(m-1)$ elements of $\mathbf{Y}^0 = (0, \ldots, 0)$ is $|T_n(\mathbf{Z})| < B$.
(ii) Set $m = m^\star$ and let R_i be the rank of Y_i among $Y_1, \ldots, Y_n$; then, by equivariance,

$$\begin{aligned} T_n(Y_1, \ldots, Y_n) &= T_n(Y_1 - Y_{(m)}, \ldots, Y_n - Y_{(m)}) + Y_{(m)} \\ \ge Y_{(m)} &+ T_n((Y_1 - Y_{(m)})\mathbb{I}[R_1 \le m], \ldots, (Y_n - Y_{(m)})\mathbb{I}[R_n \le m]) \end{aligned}$$

where only $(m-1)$ arguments of T_n are nonzero in the last expression. By the definition of $m^\star$, $|T_n(0, \ldots, 0)|$ with only $(m^\star - 1)$ zeros by outliers is bounded, say by A; hence, $T_n \ge Y_{(m)} - A$. Similarly, we obtain the other inequality. □

PROOF of Theorem 2.1. By Lemma 2.1,

$$\begin{aligned} & \mathbb{P}_\theta(T_n - \theta > a) = P_0(T_n > a) \ge P_0(Y_{(m)} > a + A) \\ \ge\ & P_0(Y_1 > a + A, \ldots, Y_{n-m+1} > a + A) = \big(1 - F(a + A)\big)^{n-m+1}, \end{aligned}$$

hence
$$\begin{aligned} \limsup_{a\to\infty} B(a, T_n) &\le \limsup_{a\to\infty} \frac{-\log 2\mathbb{P}_0(T_n > a)}{-\log\big(1 - F(a)\big)} \\ &\le \lim_{a\to\infty} \frac{\log 2 + (n - m + 1)\log\big(1 - F(a + A)\big)}{\log\big(1 - F(a)\big)} = n - m + 1 \end{aligned}$$

Conversely,

$$\begin{aligned} & \mathbb{P}_0(T_n > a) \le \mathbb{P}_0(Y_{(n-m+1)} > a - A) \\ &= n\binom{n-1}{m-1}\int_{a-A}^{\infty} F^{n-m}(x)\big(1 - F(x)\big)^{m-1} dF(x) \\ &\le n\binom{n-1}{m-1}\int_{F(a-A)}^{1} (1-u)^{m-1} du = \binom{n}{m}\big(1 - F(a - A)\big)^m, \end{aligned}$$

$$\text{thus } \liminf_{a\to\infty} B(a, T_n) \ge \lim_{a\to\infty} \frac{\log\binom{n}{m^\star} + m^\star \log\big(1 - F(a - A)\big)}{\log\Big(1 - F(a)\Big)} = m^\star$$

□

Remark 2.1 *Note that Theorem 2.1 holds for both types I and II distributions. If T_n has breakdown point $m^\star = n/2$, like the sample median, then*

$$\frac{n}{2} \leq \liminf_{a\to\infty} B(a, T_n) \leq \limsup_{a\to\infty} B(a, T_n) \leq \frac{n}{2} + 1,$$

for both types of distributions. In such case, $\mathbb{P}_\theta(|T_n - \theta| > a)$ tends to zero as $a \to \infty$ at least $(n/2)$-times faster than the tails of the underlying error distributions.

The estimators of location parameter can be affected only by outliers in outcomes $\mathbf{Y}$. For this reason, there are many robust estimators of location with good breakdown and tail-behavior properties.

In the regression model (2.30), there may also be outliers among the rows of matrix $\mathbf{X}$. In fact, an influential (*leverage*) point $\mathbf{x}_i$ can completely change the direction of the regression hyperplane. There are various regression diagnostics methods to identify the leverage points, but it is generally more difficult to identify outliers in $\mathbf{X}$ than those in $\mathbf{Y}$. Mathematically the concept of leverage point is not precisely defined in the literature. While the most influential points in LS regression are the rows of $\mathbf{X}$ leading to great diagonal elements of the *hat matrix* $\mathbf{H} = \mathbf{X}(\mathbf{X}^\top\mathbf{X})^{-1}\mathbf{X}^\top$, that is, i_0 such that

$$\mathbf{x}_{i_0}^\top(\mathbf{X}^\top\mathbf{X})^{-1}\mathbf{x}_{i_0} \gg 0, \tag{2.34}$$

(Huber 1981), the most influential points in L_1-regression are apparently determined by quantities

$$\sup_{\|\boldsymbol{b}\|=1} \left\{ \frac{|\mathbf{x}_i^\top\mathbf{b}|}{\sum_{j=1}^n |\mathbf{x}_j^\top\mathbf{b}|} \right\}, \tag{2.35}$$

see Bloomfield and Steiger (1983).

Very illuminating is the effect of a leverage point on the tail performance of the least squares estimator. The LSE is optimal in the location model for the normal distribution. However, in a regression model, the leverage point causes the LSE to have very poor tail performance, even with Gaussian errors. Consider the linear regression model (2.30) and the measure of tail performance (2.31). Let $\bar{h}$ denote the maximal diagonal element of the hat-matrix $\mathbf{H} = \mathbf{X}(\mathbf{X}^\top\mathbf{X})^{-1}\mathbf{X}^\top$,

$$\bar{h} = \max_{1\leq i\leq n} h_{ii}, \quad h_{ii} = \mathbf{x}_i^\top(\mathbf{X}^\top\mathbf{X})^{-1}\mathbf{x}_i, \quad i = 1,\ldots,n. \tag{2.36}$$

Remember that $\mathbf{H}$ is a projection matrix, $\mathbf{H}^\top\mathbf{H} = \mathbf{H}$ of rank p, with the trace p and with the diagonal elements $0 \leq h_{ii} \leq 1,\ 1 \leq i \leq n$.

THEOREM 2.2 *Let $\mathbf{T}_n$ be the LSE of $\boldsymbol{\beta}$ in the model (2.30) with i.i.d. errors e_i, $i = 1,\ldots,n$, distributed according to a symmetric distribution function F with nondegenerate tails (i.e., $0 < F(x) < 1\ \forall x \in \boldsymbol{R}_1$). Denote*

$$\overline{B} = \limsup_{a\to\infty} B(a, \mathbf{T}_n), \quad \underline{B} = \liminf_{a\to\infty} B(a, \mathbf{T}_n).$$

Then

(1) *If F is of type I with $1 \leq r \leq 2$, then* $\bar{h}^{1-r} \leq \underline{B} \leq \overline{B} \leq \bar{h}^{-r} \wedge n$.

(2) *If F is of type I with $r = 1$, then* $\bar{h}^{-1/2} \leq \underline{B} \leq \bar{B} \leq \bar{h}^{-1}$.

(3) *If F is normal, then* $\underline{B} = \overline{B} = \bar{h}^{-1}$.

(4) *If F is of type II, then* $\underline{B} = \overline{B} = 1$.

PROOF Without loss of generality, assume that $\bar{h} = h_{11}$. Let $\mathbf{h}_i^\top$ be the ith row of $\mathbf{H}$ and $\hat{Y}_i = \mathbf{x}_i^\top \mathbf{T}_n = \mathbf{h}_i^\top \mathbf{Y}$, $i = 1, \ldots, n$. Then

$$\begin{aligned}
I\!\!P_{\boldsymbol{\beta}}(\max_i |\mathbf{x}_i^\top(\mathbf{T}_n - \boldsymbol{\beta})| > a) &= I\!\!P_0(\max_i |\mathbf{h}_i^\top \mathbf{Y}| > a) \\
&\geq I\!\!P_0(\mathbf{h}_1^\top \mathbf{Y} > a) \geq I\!\!P_0(\bar{h} Y_1 > a, h_{12} Y_2 \geq 0, \ldots, h_{1n} Y_n \geq 0) \\
&\geq I\!\!P_0(Y_1 > a/\bar{h}) \cdot 2^{-(n-1)} = 2^{-(n-1)} \Big(1 - F(a/\bar{h})\Big).
\end{aligned}$$

Hence,

$$\overline{B} \leq \lim_{a \to \infty} \frac{-\log\left(1 - F(a/\bar{h})\right)}{-\log\left(1 - F(a)\right)} = \lim_{a \to \infty} \frac{c(a/\bar{h})^r}{ca^r} = \bar{h}^{-r}$$

for F of type I, which gives the upper bounds in propositions (1) and (2), respectively. For F of type II,

$$\bar{B} \leq \lim_{a \to \infty} \frac{m \log(a/\bar{h})}{m \log a} = 1,$$

while $\underline{B} \geq 1$, because there are nonnegative as well as nonpositive residuals $Y_i - \hat{Y}_i$, $i = 1, \ldots, n$.

If F is normal $\mathcal{N}(0, \sigma^2)$, then $\widehat{\mathbf{Y}} - \mathbf{X}\boldsymbol{\beta}$ has an n-dimensional normal distribution $N_n(\mathbf{0}, \sigma^2 \mathbf{H})$. Hence

$$I\!\!P_0(\max_i |\hat{Y}_i| > a) \geq I\!\!P_0(\mathbf{h}_1^\top \mathbf{Y} > a) = 1 - \Phi(a\sigma^{-1}\bar{h}^{-1/2})$$

and

$$\bar{B} \leq \bar{h}^{-1}.$$

On the other hand, if F is of type I with $1 < r \leq 2$, then applying the Markov inequality, we get for any $\varepsilon \in (0, 1)$,

$$\begin{aligned}
& I\!\!P_{\boldsymbol{\beta}}(\max_i |\mathbf{x}_i^\top(\mathbf{T}_n - \boldsymbol{\beta})| > a) = I\!\!P_0(\max_i |\hat{Y}_i| > a) \\
\leq\ & I\!\!E_0[\exp\left\{(1 - \varepsilon)c\bar{h}^{1-r}(\max_i |\hat{Y}_i|)^r\right\}] / \exp\left\{(1 - \varepsilon)c\bar{h}^{1-r}a^r\right\} \qquad (2.37)
\end{aligned}$$

and if the expectation in (2.37) is finite, say $\leq C_\varepsilon < \infty$, then

$$-\log I\!\!P_0(\max_i |\hat{Y}_i| > a) \geq -\log C_\varepsilon + (1 - \varepsilon)c\bar{h}^{1-r}a^r. \qquad (2.38)$$

Thus, we obtain the lower bound in propositions (1) and (3).

We need to find the upper bound for the expectation in (2.37). Denoting $\|\mathbf{x}\|_s = \left(\sum_{i=1}^n x_i^s\right)^{1/s}$, we get by the Hölder inequality for $s = r/(r-1) \geq 2$

$$(\max_i |\hat{Y}_i|)^r = \max_i |\mathbf{h}_i^\top \mathbf{Y}|^r \leq \max_i \left(\|\mathbf{h}_i\|_s \|\mathbf{Y}\|_r\right)^r$$
$$\leq \max_i (\sum_{j=1}^n h_{ij}^2)^{r/s} \sum_{k=1}^n |Y_k|^r \leq \bar{h}^{r-1} \sum_{k=1}^n |Y_k|^r.$$

Hence,

$$\mathbb{E}_0 \exp\left\{(1-\varepsilon)c\bar{h}^{1-r}\left(\max_i |\hat{Y}_i|\right)^r\right\}$$
$$\leq \mathbb{E}_0 \exp\left\{(1-\varepsilon)c\sum_{k=1}^n |Y_k|^r\right\} = \left[\mathbb{E}_0 \exp\left\{(1-\varepsilon)c|Y_1|^r\right\}\right]^n.$$

Because of (2.32), there exists $K \geq 0$ such that

$$1 - F(x) \leq \exp\left\{-\left(1 - \frac{\varepsilon}{2}\right)bx^r\right\}, \quad \text{for } x > k. \tag{2.39}$$

Thus, dividing the integration domain of the integral

$$\mathbb{E}_0 \exp\left\{(1-\varepsilon)c|Y_1|^r\right\} = -2\int_0^\infty \exp\left\{(1-\varepsilon)bx^r\right\}d\left(1 - F(x)\right) \tag{2.40}$$

in $(0, K)$ and (K, ∞) and using (2.39) and integration by parts in the second integral, we conclude that the expectation in (2.40) is finite and that (2.38) applies. Analogously, if $r = 1$, then

$$\mathbb{P}_0\left(\max_i |\hat{Y}_i| > a\right) \leq \mathbb{E}_0 \exp\left\{(1-\varepsilon)c\bar{h}^{-\frac{1}{2}} \max_i |\hat{Y}_i|\right\} \exp\left\{-(1-\varepsilon)c\bar{h}^{-\frac{1}{2}}a\right\}$$

and

$$\max_i |\hat{Y}_i| = \max_i |\mathbf{h}_i^\top \mathbf{Y}| \leq \bar{h}^{\frac{1}{2}} \sum_{k=1}^n |Y_k|,$$

which gives the lower bound in proposition (2). □

2.4 Robust and Minimax Estimation of Location

Let $X_1, \ldots, X_n$ be i.i.d. random variables with distribution function F, where F is generally unknown and is usually assumed to be a member of a broad class of distribution functions. As a first step toward a plausible characterization of F, we try to establish some descriptive statistics (e.g., location and dispersion measure of F). Let us illustrate the location case. There are quite a few characteristics of the location or the center of the population, but none of these is universal.

Following Bickel and Lehmann (1975b), we begin by summarizing some desirable properties of a measure of location. Let $\mathcal{F}$ be a family of distribution functions (defined on $\boldsymbol{R}_1$) such that if X is distributed according to $F \in \mathcal{F}$ (denoted by $X \sim F$), then the distribution function of $aX + b$ also belongs to $\mathcal{F}$ for all $a, b \in \boldsymbol{R}_1$. A measure of location is then a functional $\theta(F)$ on $\mathcal{F}$. Writing $\theta(F)$ as $T(X)$ (when $X \sim F$), we desire

1. $T(\mathbf{X} + b) = T(\mathbf{X}) + b$, $b \in \boldsymbol{R}_1$ and $F \in \mathcal{F}$ (translation-equivariance).
2. $T(a\mathbf{X}) = aT(\mathbf{X})$, $a \in \boldsymbol{R}_1$ and $F \in \mathcal{F}$ (scale equivariance).
3. If G is stochastically larger than F [i.e., with respect to some partial ordering $\prec$, $F \prec G$, such as $F(x) \geq G(x)$ for all $x \in \boldsymbol{R}_1$], then for all such F and G belonging to $\mathcal{F}$, $\theta(F) \leq \theta(G)$.
4. T is qualitatively robust at every $F \in \mathcal{F}$, in the Hampel sense, as described in the preceding section.
5. It may also be natural to expect $T(\mathbf{X})$ to change sign under reflection with respect to the origin, $T(-\mathbf{X}) = -T(\mathbf{X})$, and this is contained in point 2 above (by letting $a = -1$). Thus, the *sign equivariance* follows from the scale equivariance in an extended sense.

There are many functionals that have the first four properties. A very notable exception is the mean $\theta(F) = \int x dF(x)$, which has properties 1 to 3 but not property 4. Some of these functionals will be characterized in Chapter 3. Here we mainly illustrate two broad classes of functionals and some criteria of choosing an appropriate one among them.

L-functionals Let $Q(t) = F^{-1}(t)$ [$= \inf\{t : F(x) \geq t\}$], $t \in (0,1)$ be the usual quantile function associated with the distribution function F, and let $K(t)$ be a distribution function on [0,1]. Then, an L-functional is typically of the form

$$\theta(F) = \int_0^1 Q(t) dK(t). \tag{2.41}$$

It is easy to verify that $\theta(F)$ in (2.41) has properties 1, 2, and 3. However, $\theta(F)$ is not qualitatively robust (in the Hampel sense) if the support of the distribution function $K(.)$ is the entire unit interval [0,1]. If, on the other hand, $[a_0, a^0]$ is the shortest support of $K(.)$ for some a_0, a^0, $0 < a_0 < a^0 < 1$, then $\theta(F)$ is qualitatively robust at F_0, provided K and F_0 have no common points of discontinuity. This provides an intuitive justification for the use of trimmed or Winsorized means in actual practice; we will make more comments on it later.

M-functionals Let $\psi : \boldsymbol{R}_1 \mapsto \boldsymbol{R}_1$ be a nondecreasing function that takes on negative as well as positive values such that $\int \psi(x) dF(x) = 0$. For simplicity of presentation, assume that ψ is monotone, and define $\theta(F)$ as

$$\theta(F) = \frac{1}{2}[\theta^-(F) + \theta^+(F)], \tag{2.42}$$

where

$$\begin{aligned} \theta^-(F) &= \sup\Big\{t : \int \psi(x-t)dF(x) > 0\Big\}, \\ \theta^+(F) &= \inf\Big\{t : \int \psi(x-t)dF(x) < 0\Big\}. \end{aligned} \tag{2.43}$$

Then $\theta(F)$ is termed an M-functional. It has properties 1 and 3, but generally it is not scale-equivariant. If ψ is a bounded function, then $\theta(F)$ has property 4. The particular case of $\psi(x) \equiv x$ leads us to the mean (which satisfies 2), but when ψ is not necessarily linear, property 2 may not hold. To make $\theta(F)$ in (2.42) scale equivariant, in (2.43) one replaces $\psi(x)$ by $\psi(x/s(F))$, $x \in \boldsymbol{R}^1$, where the functional $s(F) = s(X)$ satisfies a basic equivariance condition that $s(aX+b) = |a|s(X)$, for every $a, b \in \boldsymbol{R}_1$. However, with this amendment, $\theta(F)$ may not have property 3 any more.

Let again $X_1, \ldots, X_n$ be i.i.d. random variables with the distribution function F, defined on $\boldsymbol{R}_1$. Then the corresponding empirical (sample) distribution function F_n is a step functional, also defined on $\boldsymbol{R}_1$. If we replace F by F_n in (2.41) or (2.42), we get a natural estimator $T_n = \theta(F_n)$. Under quite general regularity conditions (to be discussed in several later chapters), $n^{-1/2}(T_n - \theta(F))$ is asymptotically normally distributed with 0 mean and a finite, positive variance $\sigma^2(T, F)$. This *asymptotic variance* (i.e., the variance of the asymptotic distribution) plays a very basic role in the study of the performance of an estimator. As before, Φ stands for the normal distribution function. Let

$$e(T, \bar{X}; F) = \sigma^2(\bar{X}, F)/\sigma^2(T, F),$$

when F is the true distribution function. A natural requirement for a suitable robust estimator T_n is

$$\begin{gathered} e(T, \bar{X}; F) \text{ is very close to 1 when } F \equiv \Phi, \\ \sup\{e(T, \bar{X}; F) : \ F \in \mathcal{F}_1\} = \infty, \quad \inf\{e(T, \bar{X}; F) : \ F \in \mathcal{F}_1\} \geq c > 0 \\ \text{for a large subfamily } \mathcal{F}_1 \subset \mathcal{F}, \text{ whereas} \\ e(T, \bar{X}; F) \geq 1 \ \ \forall F \in \mathcal{F}_1^0 \ \text{ for a nonempty subset } \mathcal{F}_1^0 \subset \mathcal{F}. \end{gathered} \tag{2.44}$$

There exists no universally optimal measure of location among the class of L-estimators and M-estimators of location, satisfying all requirements in (2.44). The R-estimators defined in (2.21) through (2.23) may have some distinct advantages in this respect. For example, if $\mathcal{F}_1$ is the class of symmetric absolutely continuous distribution functions on $\boldsymbol{R}_1$, then (2.44) holds with $c = 1$ for the classical *normal-scores* R-estimator. However, this R-estimator does not have a bounded influence function.

Confined to the class of L- and M-estimators, Bickel and Lehmann (1975b) recommended the α-trimmed mean of the form

$$\theta(F) = (1-2\alpha)^{-1}\int_{\alpha}^{1-\alpha} Q(t)dt,\ 0<\alpha<1/2,$$

with $\alpha \in [0.05, 0.10]$ and they showed that

$$\inf_{F\in\mathcal{F}_1} e(T,\bar{X};F) = (1-2\alpha)^2,\ \sup_{F\in\mathcal{F}_1} e(T,\bar{X};F) = \infty,$$

where $T_n = \theta(F_n)$ and $\mathcal{F}_1$ is the family of distribution functions F, which are absolutely continuous in an interval $(Q(\alpha)-\eta, Q(1-\alpha)+\eta)$ for some $\eta > 0$, and have positive and finite derivatives in a neighborhood of $Q(\alpha)$ and $Q(1-\alpha)$, respectively.

Intuitively, it is clear that the more restrictive is the class $\mathcal{F}_1$ of distribution functions (what is our choice), the easier it is to find a satisfactory measure of location over $\mathcal{F}_1$. To illustrate this point, consider the Huber model of ε-*contaminacy* in (2.27) with F_0 as the standard normal distribution function Φ and with a symmetric contamination. Thus, our model is of the form

$$\mathcal{E} = \{F:\ F = (1-\varepsilon)\Phi + \varepsilon H,\ H\in\mathcal{H}\},\ 0<\varepsilon<1, \tag{2.45}$$

where $\mathcal{H}$ is the set of substochastic distribution functions symmetric about zero. Let us restrict our attention to the class of functionals T that are translation equivariant (see property 1) and such that $n^{1/2}(T(F_n)-\theta(F))$ is asymptotically normally distributed. With a *squared error loss*, the risk function will be the variance $\sigma^2(T,F)$ computed from this asymptotic normal distribution (and may as well be the *asymptotic distributional risk* [ADR]. We are looking for the *minimax functional* that minimizes

$$\sup\{\sigma^2(T,F):\ F\in\mathcal{F}_\varepsilon\ \}. \tag{2.46}$$

First, we look for a solution of the above problem in the subclass of M-functionals in (2.42) and(2.43). If the distribution function F has an absolutely continuous density function f with a finite Fisher information $I(f)$ defined by (2.17), and if the score function ψ in (2.43) is square integrable (with respect to F),

$$\sigma_\psi^2 = \int \psi^2(x)dF(x) < \infty,$$

then, under quite mild regularity conditions (as we will see in Chapter 3), the asymptotic variance of $n^{1/2}(T(F_n)-\theta(F))$ is

$$\sigma^2(T,F) = \sigma_\psi^2\{\int \psi(x)\{-f'(x)/f(x)\}dF(x)\}^{-2}.$$

Let us denote

$$\mathcal{F}_\varepsilon^\star = \{F\in\mathcal{F}_\varepsilon:\ I(f)<\infty\}. \tag{2.47}$$

Huber (1964) proved that the *least favorable distribution* F_0 for which $I(f_0) \le I(f)$ for all $F\in\mathcal{F}_\varepsilon^\star$ exists and satisfies $\int dF_0 = 1$, and that the MLE of the shift θ in $F_0(x-\theta)$ minimizes the risk in (2.45). This estimator is an M-estimator (M-functional) generated by the score function

$$\psi_0(x) = \begin{cases} x & \text{for } |x| \le k, \\ k \cdot \text{sign}(x) & \text{for } |x| > k, \end{cases} \tag{2.48}$$

where k (> 0) depends on ε through

$$\int_{-k}^{+k} \phi(x)dx + k^{-1}2\phi(k); \; \phi(x) = (d/dx)\Phi(x).$$

Hence, for any M-functional T_ψ, for every $F \in \mathcal{F}_1$,

$$\sigma^2(T_{\psi_0}, F) \le \sigma^2(T_{\psi_0}, F_0) = 1/I(f_0) \le \sigma^2(T_\psi, F) \le \sup_{F\in\mathcal{F}_1} \sigma^2(T, F).$$

Noting that T_{ψ_0} is translation-equivariant and that the asymptotic risk of the Pitman estimator of θ in $F_0(x-\theta)$ is equal to the reciprocal of $I(f_0)$, we conclude that T_{ψ_0} minimizes (2.45) also in the class of translation-equivariant estimators. This estimator (or functional) has properties 3 and 4 but not 2, it is not scale-equivariant. As we will see in Chapter 4, another minimax solution to this problem is the trimmed mean in (2.44) with

$$\alpha = (1-\varepsilon)(1-\phi(k)) + \varepsilon/2 \text{ and } k \text{ satisfying (2.48).}$$

The trimmed mean is scale-equivariant, but not the estimator T_{ψ_0}. Another scale-equivariant solution lies in the class of R-estimators, and this will be considered in Chapter 6. As has been already mentioned, a scale-equivariant version of the functional T may be constructed by using the score function $\psi_0(x/s(F_0))$ instead of $\psi_0(x)$. However, this amended M-estimator loses the minimax property. This is mainly due to the fact that the asymptotic variance of this latter estimator at the model F_0 is given by

$$\sigma^2(T_{\psi_0}, F_0, s) = \frac{s^2(F_0)\int \psi_0^2(\frac{x}{s(F_0)})dF_0(x)}{\left\{\int \psi_0'(\frac{x}{s(F_0)})dF_0(x)\right\}^2}, \tag{2.49}$$

and it can be easily shown that (2.49) fails to attain the information limit $I(f_0)$ for any finite k, in view of the segmented form of ψ_0 in (2.48). We refer to Chapter 5 for a more detailed study of these properties.

2.5 Clippings from Probability and Asymptotic Theory

Consider a probability space $(\Omega, \mathcal{A}, \mathbb{P})$ where $\mathcal{A}$ is the σ-field of subsets of a nonempty set Ω and $\mathbb{P}(.)$ is a normed measure. A set of observations $\mathbf{X} = (X_1, \ldots, X_m)$ along with a distribution function $F(\mathbf{x}) = \mathbb{P}\{X < x\}$ generate a probability space $(\Omega, \mathcal{A}, \mathbb{P})$ where the range of Ω is $\boldsymbol{R}_m$ and $\mathbb{P}(.)$ is uniquely determined by $F(.)$. Thus, a random vector $\mathbf{X}$ is defined as a measurable function from $(\Omega, \mathcal{A}, \mathbb{P})$ into $\boldsymbol{R}_m$, for some $m \ge 1$; for $m = 1$, it is termed a random variable. This definition extends to more general (random) functions as well. For example, a *stochastic process* $X = \{X(t),\ t \in T\}$ (or more generally, a sequence $X_\gamma = \{X_\gamma(t),\ t \in T\},\ \gamma \in \Gamma$) is a collection of random variables defined over some probability space $(\Omega, \mathcal{A}, I\!\!P)$, where T is an *index set* such that, for each $t \in T$, $X(t)$ (or $X_\gamma(t)$) is $\mathcal{A}$-measurable. In asymptotic theory of statistical inference, such a stochastic process can often be represented as *random elements* in an appropriate *function space*. Thus, we may consider a *topological space* (T) endowed with a *(semi-)metric* $\mathbf{d} : T \times T \mapsto \boldsymbol{R}_1^+$, and we denote the class of Borel subsets of T by $\mathcal{B}(T)$. Then a function $U : \Omega \mapsto T$ is said to be a random element of T if $\{\Omega : U(\in B\} \in \mathcal{A},\ \forall B \in \mathcal{B}(T)$. Notable examples of T are the following:

1. $C = C[0,1]$, the space of continuous functions on the unit interval $[0,1]$ with $d(x,y) = \|x(t) - y(t)\| = \sup\{|x(t) - y(t)| : t \in [0,1]\}$.
2. $D = D[0,1]$, the space of functions on [0,1] having only discontinuities of the first kind (i.e., they are right continuous and have left-hand limits). Now letting $\Lambda = \{\lambda_t,\ t \in [0,1]\}$ be the class of strictly increasing, continuous mapping of [0,1] onto itself, $d(.)$ is the *Skorokhod J_1-metric (topology)* defined for $x, y \in D[0,1]$ by

$$\rho_S(x,y) = \inf\{\varepsilon > 0 : \|\lambda_t - t\| \le \varepsilon,\ \|x(t) - y(\lambda_t)\| \le \varepsilon\}. \qquad (2.50)$$

Clearly, $C[0,1]$ is a subspace of $D[0,1]$, and their extensions to $[0,K]$ for some $K \in (0,\infty)$ are straightforward. With a modified metric, the domains may also be extended to $\boldsymbol{R}_1^+$ (see, Whitt 1970; Lindvall 1973). Further, extensions to $[0,1]^p$ for some $p \ge 1$ have also been worked out; we may refer to Sen (1981a,ch.2) for a survey of these results.

Consider next a stochastic process $X_\gamma;\ \gamma \in \Gamma$, where Γ is a linearly ordered index set (and in the discrete case, Γ can be taken equivalently as $I\!\!N$,the set of nonnegative integers or any subset of it). Let $\{F_\gamma,\ \gamma \in \Gamma\}$ be a system of subsigma fields of $\mathcal{A}$, such that X_γ is F_γ-measurable, for each $\gamma \in \Gamma$. Then $\{X_\gamma,\ \gamma \in \Gamma\}$ is said to be adapted to $\mathcal{F} = \{F_\gamma,\ \gamma \in \Gamma\}$.

Further, $\{F_\gamma,\ \gamma \in \Gamma\}$ is nondecreasing or nonincreasing if for $\gamma, \gamma' \in \Gamma$ and $\gamma < \gamma'$, $F_\gamma, \subset F_{\gamma'}$ or $F_\gamma \supset F_{\gamma'}$.

(a) If F_γ is nondecreasing (or nonincreasing) in $\gamma \in \Gamma$, $I\!\!E|X_\gamma| < \infty$, $\gamma \in \Gamma$, and if for every $\gamma, \gamma' \in \Gamma$, and $\gamma' >$ (or $<$) γ,

$$I\!\!E\{X_{\gamma'}|F_\gamma\} = X_\gamma \text{ almost everywhere (a.e.)},$$

then $\{X_\gamma,\ \gamma \in \Gamma\}$ is termed a *martingale* (or *reverse martingale*). If in (2.5.2) the „$=$ " sign is replaced by „$\ge$ "(or „$\le$") sign, then $\{X_\gamma,\ \gamma \in \Gamma\}$ is a *sub-*(or *super-) martingale* or reversed ones according as F_γ is nondecreasing or nonincreasing.

(b) If $\{X_\gamma, \mathcal{F}_\gamma;\ \gamma \in \Gamma\}$ is a (forward or reverse) martingale and $g(.)$ is an integrable, *convex* function, then $\{g(X_\gamma), \mathcal{F}_\gamma;\ \gamma \in \Gamma\}$ is a (forward or reverse) submartingale; if $g(.)$ is *concave*, then the submartingale structure changes to a supermartingale structure.

(c) If the X_i are (centered) independent random variables, then S_n form a martingale ($S_n = \sum_{j=1}^{n} X_j$), and $|S_n|^p$, $p > 0$, form a submartingale or supermartingale sequence according as p is $\geq$ or < 1. If the X_i are i.i.d. random variables, then $\bar{X}_n$ $(= n^{-1}S_n)$ form a reverse martingale, and the $|\bar{X}_n|^p$ form a reverse submartingale or supermartingale according as p is $\geq$ or < 1. In this setup, the X_i does not need to be centered.

(d) *The Doob moment inequality* (Doob 1967, p. 318). Let $\{X_n, \mathcal{F}_n;\ n \geq 1\}$ be a nonnegative submartingale. Then

$$\mathbb{E}\Big\{\max_{1\leq k\leq n} X_k^\alpha\Big\} \leq \begin{array}{ll} e(e-1)^{-1}\{1+\mathbb{E}(X_n \log^+ X_n)\}, & \alpha = 1, \\ \{\alpha/(\alpha-1)\}^\alpha \mathbb{E}\{X_n^\alpha\}, & \alpha > 1, \end{array}$$

where $\log^+ x = \max\{1, \log x\}$, $x \geq 0$.

(e) *Submartinagle convergence theorem* (Doob, 1967). Let $\{X_n, \mathcal{F}_n;\ n \geq 1\}$ be a submartingale, and suppose that $\sup_n \mathbb{E}|X_n| < \infty$. Then there exists a random variable X such that

$$X_n \to X \text{ a.s.,} \quad \text{as } n \to \infty, \quad \mathbb{E}|X| \leq \sup_n \mathbb{E}|X_n|.$$

Now let $\{X_n, \mathcal{F}_n;\ n \geq 1\}$ be a submartingale or a reverse submartingale, and assume that the X_n are uniformly integrable (i.e., $\mathbb{E}\{|X_n| I(|X_n| > c)\} \to 0$ as $c \to \infty$, uniformly in n). Then there exists a random variable X such that $X_n \to X$ a.s., and in the first mean, as $n \to \infty$. In particular, if $\{X_n, \mathcal{F}_n;\ n \geq 1\}$ is a reversed martingale, then the X_n are uniformly integrable. Hence,

$$X_n \to X \text{ a.s. and in the first mean as } n \to \infty.$$

Some other inequalities will be introduced later. As will be seen in the subsequent chapters, asymptotic theory of statistical inference rests on deeper adaptations of various basic concepts and tools in probability theory. While a complete coverage of this topics is beyond the scope of this limited introduction, we present the most pertinent ones, along with their sources, so that a further indepth reading may be facilitated when necessary. Chapters 2 to 4 of Sen and Singer (1993) provide a survey of most of these results.

2.5.1 Modes of Convergence of Stochastic Elements

Unlike the elements in real analysis whose convergence is well defined, those in stochastic analysis which are in fact themselves stochastic have different convergence properties that are formulated depending on the specific model of interest. Critical in their formulation are the definitions of the probability space and of random elements on functional spaces, which were introduced earlier.

DEFINITION 2.1 *[Stochastic convergence and convergence in probability]. Let $\{X_n\}$ and X be random elements defined on a probability space $(\Omega, \mathcal{A}, \mathbb{P})$. If the X_n and X have realizations in a space T endowed with a distance or norm $d : T \times T \mapsto \boldsymbol{R}_1^+$, then we say that X_n converges in probability to X [relative to the metric $d(.)$] if for every $\varepsilon > 0$,*

$$\lim_{n \to \infty} \mathbb{P}\{d(X_n, X) > \varepsilon\} = 0. \tag{2.51}$$

In notation, we express this as

$$d(X_n, X) \xrightarrow{P} 0 \ \text{ or } \ X_n \xrightarrow{P} X, \ \text{ as } n \to \infty.$$

If the X_n and X are real valued, then $d(x, y) = |x - y|$. So $d(X_n, X) \to 0 \Rightarrow |X_n - X| \to 0$. If these elements belong to $\boldsymbol{R}_p$, for some $p \geq 1$, we can take $d(\mathbf{x}, \mathbf{y}) = \|\mathbf{x} - \mathbf{y}\|$, the Euclidean norm. In this case, it is also possible to choose some other norms, such as $d(\mathbf{x}, \mathbf{y}) = \|\mathbf{x} - \mathbf{y}\|_\infty = \max\{|x_j - y_j| : 1 \leq j \leq p\}$. Since $\|\mathbf{x} - \mathbf{y}\|_\infty \leq \|\mathbf{x} - \mathbf{y}\| \leq p\|\mathbf{x} - \mathbf{y}\|_\infty$, either definition of the metric $d(.)$ suffices. Consider next the case where X_n and X are random elements in an appropriate function space T, endowed with a (semi-) metric $d : T \times T \mapsto \boldsymbol{R}^+$. We have already introduced such function spaces and discussed their appropriate metrics $d(.)$. For example, for the $C[0, 1]$ space, we may choose the uniform metric $d(x, y) = \sup\{|x(t) - y(t)| : 0 \leq t \leq 1\}$, and the Definition 2.1 remains valid.

DEFINITION 2.2 *[Convergence with probability 1 or almost sure convergence, strong convergence, convergence almost everywhere]. In the same setup as in the preceding definition, suppose that for every $\varepsilon > 0$,*

$$\lim_{n \to \infty} \mathbb{P}\{\bigcup_{m \geq n} [d(X_m, X) > \varepsilon]\} = 0. \tag{2.52}$$

Then we say that X_n converges to X with probability 1 as $n \to \infty$. The other terminologies are synonymous. In notation, we write (2.52) as [relative to the metric $d(.)$]

$$d(X_n, X) \xrightarrow{\text{a.s.}} 0 \ \text{ or } \ X_n \to X, \text{ a.s.}, \ \text{ as } n \to \infty.$$

Note that (2.52) implies (2.51), and hence,

$$X_n \to X, \text{a.s.} \Rightarrow X_n \xrightarrow{P} X,$$

but the converse may not be true.

DEFINITION 2.3 *[Convergence in the rth mean, $r > 0$]. Suppose that the X_n and X are real- or vector-valued random variables. If*

$$\lim_{n \to \infty} \mathbb{E}\|X_n - X\|^r = 0 \ \text{ for some } r > 0, \tag{2.53}$$

we say that X_n converges in the rth mean to X; in our earlier notation

$$X_n \xrightarrow{rth} X, \ \text{ or } X_n \to X \text{ in } L_r \text{ norm as } n \to \infty.$$

The definition extends to function spaces wherein $\|x - y\|$ is to be replaced by $d(x, y)$. It is easy to verify that

$$X_n \overset{rth}{\to} X \text{ for some } r > 0 \Rightarrow X_n \overset{P}{\to} X. \tag{2.54}$$

But (2.53), unless accompanied by suitable rate of convergence or some other intrinsic properties of the X_n, may not necessarily imply (2.52). We will present some related results later.

DEFINITION 2.4 ***[Complete convergence].*** *In the setup of Definition 2.2, suppose that*

$$\lim_{n\to\infty} \sum_{N \geq n} \mathbb{P}\{d(X_N, X) > \varepsilon\} = 0 \ \forall \varepsilon > 0. \tag{2.55}$$

Then we say that X_n converges completely to X as $n \to \infty$, or that $X_n \overset{C}{\to} X$, as $n \to \infty$.

Note that

$$X_n \overset{C}{\to} X \Rightarrow X_n \to X, \quad \text{a.s., as} \quad n \to \infty.$$

The converse may not be universally true.

We have tacitly assumed that the X_n and X are defined on a common probability space. In many statistical adaptations, this coherence condition is not needed. A much weaker mode of convergence can be formulated (under weaker regularity conditions) that eliminates this coherence, and encompasses a wider domain of adaptations. Before we look at the formulation of „weak convergence," let us consider a very simple example.

Example 2.1 *(Binomial distribution)* Consider a sequence $\{Y_i;\ i \geq 1\}$ of i.i.d. random variables such that

$$\mathbb{P}(Y_i = 0) = 1 - \mathbb{P}(Y_i = 1) = 1 - \pi,\ 0 < \pi < 1.$$

Let $S_n = Y_1 + \ldots + Y_n$, $n \geq 1$. Note that $\mathbb{E}S_n = n\pi$ and $V(S_n) = n\pi(1-\pi)$. Consider the random variables:

$$X_n = \frac{S_n - n\pi}{\sqrt{n\pi(1-\pi)}}, \quad n \geq 1. \tag{2.56}$$

It is well known that S_n has the $Bin(n, \pi)$ distribution and that S_n can only assume values in $\{0, 1, \ldots, n\}$. Thus, in (2.56), X_n is a discrete random variable, although the mesh becomes dense as n increases. If we denote by $G_n(y) = \mathbb{P}\{X_n \leq y\}$, $y \in \boldsymbol{R}$, the distribution function of X_n, and if $G(Y)$ stands for the standard normal distribution function of a random variable X, then we will see that as n increases, G_n converges to G:

$$\lim_{n\to\infty} \sup_{y \in \boldsymbol{R}} |G_n(y) - G(y)| = 0. \tag{2.57}$$

We can rewrite this as

$$G_n \overset{w}{\to} G \text{ as } n \to \infty. \tag{2.58}$$

We need to keep two things in mind here: (1) X may not be defined on the same probability space as the X_n are defined, and (2) if G were not continuous everywhere, then in (2.57) we might have to exclude the points of discontinuity (or jump-points) of G (where the left- and right-hand limits may not agree). Given these technicalities, we may rewrite (2.58) as $X_n \to X$, in law/distribution, or by definition

$$X_n \xrightarrow{\mathcal{D}} X \text{ or } \mathcal{L}(X_n) \to \mathcal{L}(X), \text{ as } n \to \infty. \tag{2.59}$$

This leads to the following

DEFINITION 2.5 ***[Weak convergence or convergence in law/distribution]*** *If for a sequence $\{G_n\}$ of distribution functions, defined on $\boldsymbol{R}_k$, for some $k \geq 1$, there exists a distribution function G, such that*

(1) G is the distribution function of a random element X not necessarily defined on the same probability space and

(2) at all points of continuity of G, $G_n \to G$ as $n \to \infty$.

Then we say that X_n, which has the distribution function G_n, converges in law (or distribution) to X. We denote it by the notation of (2.58) or (2.59).

It is evident that if $X_n \xrightarrow{P} X$, then $X_n \xrightarrow{\mathcal{D}} X$ as well, but the converse is not necessarily true. The concept of weak convergence plays a most fundamental role in the asymptotic theory of statistical inference. To see the full impact of this concept, we need to extend the notion of convergence to encompass function spaces with appropriate topologies that eliminate the basic difficulties caused by discontinuities of various types and other topological metrization problems. We will deal with these issues more adequately in a later section.

2.5.2 Basic Probability Inequalities

The following elementary inequality provides the genesis of all subsequent inequalities.

1. *Chebychev's Inequality.* Let U be a nonnegative random variable such that $\mu = \mathbb{E}U$ exists. Then for every positive t,

$$\mathbb{P}\{U \geq \mu t\} \leq t^{-1}. \tag{2.60}$$

An immediate consequence of (2.60) is the Markov inequality. Let T_n be a statistic such that for some $r > 0$, for some T (possibly random), $\nu_{nr} = \mathbb{E}|T_n - T|^r$ exists. Then letting $U = U_n = |T_n - T|^r$, we have for every $t > 0$,

$$\begin{aligned} &\mathbb{P}\{|T_n - T| \geq t\} = \mathbb{P}\{U_n \geq t^r\} \\ &= \mathbb{P}\{U_n \geq \nu_{nr}(t^r/\nu_{nr})\} \leq \nu_{nr}/t^r. \end{aligned} \tag{2.61}$$

Thus, if $\nu_{nr} \to 0$ as $n \to \infty$, we obtain from (2.61) that $T_n - T \xrightarrow{P} 0$. This establishes the implification relation in (2.54). Moreover, if $\sum_{n \geq 1} \nu_{nr}$ converges, then (2.61) implies that $T_n \to T$a.s., as $n \to \infty$. In most statistical applications, this simple prescription works well. However, for (2.52) or (2.55) to hold, we do not need this series convergence criterion. Often, we may obtain some of these results by more powerful inequalities (and under more specific structures on the random variables).

Another variant of (2.60), under more stringent conditions, is the following:

2. *Bernstein inequality.* Let U be a random variable such that $M_U(t) = \mathbb{E}(e^{tU})$ exists for all $t \in [0, K]$, for some $K > 0$. Then, for every real u, we have

$$\mathbb{P}\{U \geq u\} \leq \inf_{t \in [0,K]} \left\{ e^{-tU} M_U(t) \right\}. \tag{2.62}$$

We may observe here that if U is the sum of n (centered) random variables, independent and having finite moment generating functions, then $M_U(t)$ is factorizable into n terms. So the right-hand side of (2.62) reduces to a term that exponentially converges to 0 as $n \to \infty$. This generally gives a much sharper bound than that obtainable using (2.60) or (2.61). Further, if the components of U are all bounded random variables, the moment generating function exists, and hence such sharper bounds can easily be obtained.

Example 2.2 *(Example 2.1 revisited).* Let $T_n = n^{-1}S_n$ and $U_n = S_n - n\pi$. Then $M_{U_n}(t) = [M_Y(t)]^n = \{1+\pi(e^t-1)\}^n$. Substituting this in (2.62) and following a few routine steps, we have for every $0 < \pi < 1$ and $\varepsilon > 0$,

$$\mathbb{P}\{|T_n - \pi| > \varepsilon\} \leq [\rho(\varepsilon, \pi)]^n + [\rho(\varepsilon, 1-\pi)]^n$$

where

$$\rho(\varepsilon, \pi) = \left\{ \frac{\pi}{\pi + \varepsilon} \right\}^{\pi+\varepsilon} \left\{ \frac{1-\pi}{1-\pi-\varepsilon} \right\}^{1-\pi-\varepsilon}, \tag{2.63}$$

and it is easy to verify that $0 \leq \rho(\varepsilon, \pi) < 1$, $\forall \varepsilon > 0$, $\pi \in (0, 1)$.

Hoeffding (1963) incorporated this simple probability inequality in the derivation of a probability inequality for a general class of bounded random variables. His ingenious tool is based on two other inequalities: that the geometric mean of a set of positive numbers can not be larger than their arithmetic mean, and that for $x \in [0, 1]$, $t \geq 0$, $e^{tx} \leq 1 + x(e^t - 1)$.

3. *Hoeffding inequality.* Let $\{X_k; k \geq 1\}$ be independent (but not necessarily identically distributed) random variables such that $\mathbb{P}\{0 \leq X_k \leq 1\} = 1$ $\forall k \geq 1$. Set $\mu_k = \mathbb{E}X_k$, $k \geq 1$, and $\bar{\mu}_n = n^{-1}\sum_{k=1}^n \mu_k$. Also let $\bar{X}_n = n^{-1}\sum_{k=1}^n X_k$, $n \geq 1$. Then for every $\varepsilon > 0$,

$$\mathbb{P}\{|\bar{X}_n - \bar{\mu}_n| \geq \varepsilon\} \leq [\rho(\varepsilon, \bar{\mu}_n)]^n + [\rho(\varepsilon, 1-\bar{\mu}_n)]^n, \tag{2.64}$$

where $\rho(.)$ is defined by (2.63).

The above inequality extends directly to the case where for some (a, b) : $-\infty < a < b < +\infty$, $\mathbb{P}\{a \leq X_k \leq b\} = 1$ $\forall k$. In this case, we define

$Y_k = (X_k - a)/(b-a)$, and apply (2.64), replacing ε by $\varepsilon' = \varepsilon/(b-a)$ and the μ_k by the $\mathbb{E}Y_k$, $k \geq 1$. As in the earlier cases, (2.64) holds for every $n \geq 1$ and $\varepsilon > 0$. It is also possible to allow that ε to depend on n, for example, $\varepsilon = tn^{-1/2}$, $t > 0$. Looking at (2.63), we may verify that $(\partial/\partial\varepsilon) \log \rho(\varepsilon, \pi)|_{\varepsilon=0} = 0$ and

$$(\partial^2/\partial\varepsilon^2) \log \rho(\varepsilon, \pi) = -\{(\pi+\varepsilon)(1-\pi-\varepsilon)\}^{-1} \leq -4.$$

As such, $\log \rho(\varepsilon, \pi) \leq -2\varepsilon^2$, $\forall \varepsilon > 0$, while for $\varepsilon = tn^{-1/2}$, $\log \rho(tn^{-1/2}, \pi) = -(2n)^{-1}t^2\{\pi(1-\pi)\}^{-1}\{1 + \mathrm{O}(n^{-1/2})\}$, so (2.64) yields a rate $e^{-t^2/2\pi(1-\pi)}$, that is similar to the rate obtained using the asymptotic normal distribution of $\sqrt{n}(\bar{X}_n - \bar{\mu}_n)$.

There are some other probability inequalities more intimately associated with the laws of large numbers, which will be reviewed later on.

2.5.3 Some Useful Inequalities and Lemmas

We collect some basic moment and other inequalities in probability theory. We start with those most often used.

1. *Jensen inequality.* Let X be a random variable, and let $g(x)$, $x \in \boldsymbol{R}$, be a convex function such that $\mathbb{E}g(X)$ exists. Then
$$g(\mathbb{E}X) \leq \mathbb{E}g(X), \tag{2.65}$$
where the equality sign holds only when $g(.)$ is linear a.e.
Recall that for $x \in \boldsymbol{R}$, $p \geq 1$, $|x|^p$ is convex, and hence (2.65) when applied to a sequence $\{T_n - T\}$ yields the following:
$$T_n \stackrel{rth}{\to} T, \text{ for some } r > 0 \Rightarrow T_n \stackrel{sth}{\to} T, \ \ \forall s \leq r.$$

2. *Hölder inequality.* Let X and Y be two, not necessary independent, random variables such that for $p > 0$, $q > 0$ with $p^{-1} + q^{-1} = 1$, $\mathbb{E}|X|^p$ and $\mathbb{E}|Y|^q$ both exist. Then
$$\mathbb{E}|XY| \leq (\mathbb{E}|X|^p)^{1/p}(\mathbb{E}|Y|^q)^{1/q}$$
The special case of $p = q = 2$ is known as the *Cauchy–Schwarz inequality.*

3. *Minkowski inequality.* For random variables $X_1, \ldots, X_n$, not necessary independent, and $p \geq 1$, whenever $\mathbb{E}|X_i|^p < \infty$, $\forall i$,
$$(\mathbb{E}\{|\sum_{i=1}^n X_i|^p\})^{1/p} \leq \sum_{i=1}^n [\mathbb{E}|X_i|^p]^{1/p}.$$
A variant form of this inequality is
$$[\{|\sum_{i=1}^n (X_i + Y_i)|^p\}]^{1/p} \leq (\sum_{i=1}^n \{|X_i|^p)^{1/p} + (\sum_{i=1}^n \{|Y_i|^p)^{1/p}.$$

4. *C_r-inequality.* Consider

$$|a+b|^r \leq C_r\{|a|^r + |b|^r\}, \;\; r \geq 0,$$

where

$$C_r = \begin{cases} 1, & 0 \leq r \leq 1 \\ 2^{r-1}, & r > 1 \end{cases}$$

For $m \geq 2$ items, $C_r = 1$ or m^{r-1} according as $0 \leq r \leq 1$ or $r \geq 1$.

5. *Arithmetic-geometric-harmonic mean (A.M.-G.M.-H.M.) inequality.* Let $a_1, \ldots, a_n$ be nonnegative members, $n \geq 1$. Then

$$\begin{aligned} \text{A.M.} = \bar{a}_n = \frac{1}{n}\sum_{i=1}^{n} a_i \;&\geq\; (\prod_{i=1}^{n} a_i)^{1/n} = G.M. \\ &\geq\; (\frac{1}{n}\sum_{i=1}^{n}\frac{1}{a_i})^{-1} = H.M., \end{aligned} \tag{2.66}$$

where equality holds only when $a_1 = \ldots = a_n$. Weighted and integral versions of (2.66) are also available in the literature.

6. *Entropy inequality.* Let $\{a_i\}$, $\{b_i\}$ be convergent sequence of positive numbers such that $\sum_{i=1}^{n} a_i \geq \sum_{i=1}^{n} b_i$. Then

$$\sum_{i=1}^{n} a_i \log(b_i/a_i) \leq 0,$$

where the equality sign holds only when $a_i = b_i$, $\forall i \geq 1$.

7. *Monotone convergence lemma.* If the events $\{A_n\}$ are monotone, that is, either $A_1 \subset A_2 \subset A_3 \subset \ldots$ or $A_1 \supset A_2 \supset A_3 \supset \ldots$, with limit A, then

$$\lim_{n\to\infty} \mathbb{P}(A_n) = \mathbb{P}(A).$$

(Compare the result with the limit of a monotone sequence of numbers.)

8. *Borel–Cantelli lemma.* Let $\{A_n\}$ be a sequence of events, and denote by $\mathbb{P}(A_n)$ the probability that A_n occurs, $n \geq 1$. Also let A denote the event that A_n occur infinitely often (i.o.). Then

$$\sum_{n\geq 1} \mathbb{P}(A_n) \leq +\infty \Rightarrow \mathbb{P}(A) = 0,$$

whether or not the A_n are independent. If the A_n are independent, then

$$\sum_{n\geq 1} \mathbb{P}(A_n) = +\infty \Rightarrow \mathbb{P}(A) = 1,$$

9. *Khintchine equivalence lemma, I.* Let $\{X_n\}$ and $\{Y_n\}$ be two arbitrary sequence of random variables. Then if $\sum_{n\geq 1} \mathbb{P}(X_n \neq Y_n) \leq +\infty$, then the strong law of large numbers holds for both sequences or none.

10. *Khintchine equivalence lemma, II.* If X is a random variable, then

$$\mathbb{E}|X| < \infty \iff \sum_{k\geq 1} k\mathbb{P}\{k \leq |X| < k+1\} < \infty.$$

11. *Fatou lemma.* If the X_n are nonnegative random variables, then

$$\mathbb{E}\{\liminf_{n\to\infty} X_n\} \leq \liminf_{n\to\infty} \mathbb{E}(X_n).$$

The Fatou lemma holds under a conditional setup too. That is, for $X_n|\mathcal{B}$, where $\mathcal{B}$ is a sub-sigma field of $\mathcal{A}$ and $(\Omega, \mathcal{A}, \mathbb{P})$ is the probability space.

12. *Uniform integrability.* Let $\{G_n,\ n \geq n_0\}$ be a sequence of distribution functions, defined on $\boldsymbol{R}_q$, for some $q \geq 1$, and let $h(y) : \boldsymbol{R}_q \mapsto \boldsymbol{R}_p$, $p \geq 1$, be a continuous function such that

$$\sup_{n\geq n_0} \int_{\{\|y\|>a\}} \|h(y)\| dG_n(y) \to 0.$$

Then $h(.)$ is uniformly integrable (relatively to $\{G_n\}$).

13. *Lebesgue dominated convergence theorem.* For a sequence $\{X_n\}$ of measurable functions, suppose that there exists an Y, on the same probability space $(\Omega, \mathcal{A}, \mathbb{P})$,

$$|X_n| \leq Y \text{ a.e., where } \mathbb{E}|Y| < \infty$$

and either $X_n \to X$a.e., or $X_n \xrightarrow{\mathcal{D}} X$ for a suitable random variable X. Then

$$\mathbb{E}|X_n - X| \to 0, \text{ as } n \to \infty. \tag{2.67}$$

A related version deemphasizing the dominating random variable Y is the following: Let $\{X,\ X_n,\ n \geq n_0\}$ be a sequence of random variables such that $X_n - X \xrightarrow{P} 0$ and

$$\mathbb{E}\{\sup_{n\geq n_0} |X_n|\} < \infty.$$

Then (2.67) holds, and hence $\mathbb{E}X_n \to \mathbb{E}X$ as $n \to \infty$. The Lebesgue dominated convergence theorem also holds under a conditional setup, namely, sub-sigma fields of $\mathcal{A}$ given $\mathcal{B}$.

14. *Kolmogorov three series criterion.* The series $\sum_{n\geq 1} X_n$ of independent summands converges a.s. to a random variable if and only if, for a fixed $c > 0$, the three series

$$\sum_{n\geq 1} \mathbb{P}\{|X_n| > c\},\ \sum_{n\geq 1} \text{Var}(X_n^c),\ \text{ and } \sum_{n\geq 1} \mathbb{E}(X_n^c)$$

all converge, where

$$X_n^c = X_n I(|X_n| \leq c).$$

15. *Hewit-Savage zero-one Law:* Let $\{X_i;\ i \geq 1\}$ be i.i.d. random variables. Then every exchangeable event (that remains invariant under any permutation of the indices $1, \ldots, n$ of $X_1, \ldots, X_n$) has probability either equal to 0 or 1.

16. *Fubini Theorem.*

THEOREM 2.3 *Let $(\Omega, \mathcal{A}, I\!P)$ be a product probability space with $\Omega = \Omega_1 \times \Omega_2$, $\mathcal{A} = \mathcal{A}_1 \times \mathcal{A}_2$ and $I\!P = I\!P_1 \times I\!P_2$ and let $(\bar{\mathcal{A}}_1, \bar{I\!P}_1)$, $(\bar{\mathcal{A}}_2, \bar{I\!P}_2)$, and $(\bar{\mathcal{A}}_1 \times \bar{\mathcal{A}}_2, \bar{I\!P}_1 \times \bar{I\!P}_2)$ be their completions. Let $f(w_1, w_2)$ be a measurable function with respect to $\bar{\mathcal{A}}_1 \times \bar{\mathcal{A}}_2$ and integrable with respect to $\bar{I\!P}_1 \times \bar{I\!P}_2$. Then*

(a) $f(w_1, .)$ is $\bar{\mathcal{A}}_2$-measurable and $\bar{I\!P}_2$- integrable.
(b) The integral $\int f(., w_2) d\bar{I\!P}_2(w_2)$ is $\bar{\mathcal{A}}_1$-measurable and $\bar{I\!P}_1$-integrable.
(c) For $f(., .)$ nonnegative or $I\!P_1 \times I\!P_2$ integrable,

$$\begin{aligned} &\int\int f(w_1, w_2) d\bar{I\!P}_1(w_1) d\bar{I\!P}_2(w_2) \\ &= \int [\int f(w_1, w_2) d\bar{I\!P}_2(w_2)] d\bar{I\!P}_1(w_1) \\ &= \int [\int f(w_1, w_2) d\bar{I\!P}_1(w_1)] d\bar{I\!P}_2(w_2) \end{aligned}$$

17. *Faddeev lemma.*

LEMMA 2.2 *Let the sequence $\{f_n(t, u); (t, u) \in (0, 1)^2, n \geq 1\}$ be densities in t for fixed $u \in (0, 1)$, such that for every $\varepsilon > 0$,*

$$\lim_{n \to \infty} \int_{u-\varepsilon}^{u+\varepsilon} f_n(t, u) dt = 1.$$

Further assume that there exists another sequence $\{g_n(t, u), (t, u) \in (0, 1)^2$, $n \geq 1\}$, such that $g_n(t, u)$ is $\nearrow$ or $\searrow$ in t according as t is in $(0, u)$ or $(u, 1)$, when u, n are fixed, and

$$f_n(t, u) \leq g_n(t, u), \ \forall (t, u) \in (0, 1)^2, \ \ n \geq 1;$$

$$\sup_n \int_0^1 g_n(t, u) dt < \infty.$$

Then for every integrable $\varphi(.) : [0, 1] \mapsto \boldsymbol{R}$,

$$\lim_{n \to \infty} \int_0^1 \varphi(t) f_n(t, u) dt = \varphi(u) \ \text{a.e.} \ \ (u \in [0, 1]).$$

18. *Uniform continuity and equicontinuity.* A function $f(.)$ defined on a metric space S (equipped with a metric $\rho : S \times S \mapsto \boldsymbol{R}^+$), and taking values in a metric space T (equipped with a metric $\delta(.)$), is said to be uniformly continuous on S, if for every $\varepsilon > 0$, there exists an $\eta > 0$, such that

$$\rho(x_1, x_2) < \eta \Rightarrow \delta(f(x_1), f(x_2)) < \varepsilon, \ \ \forall x_1, x_2 \in S.$$

Let us now extend this definition to the case where $f(.)$ is stochastic. We introduce $\mathcal{B}(T)$ as the class of Borel subsets of T, and consider the probability space $(T, \mathcal{B}, \mathbb{P})$ where T is equipped with the metric $\delta : T \times T \mapsto \boldsymbol{R}^+$. In this way, we introduce $\mathcal{P}$ as the space of all probability measures on $(T, \mathcal{B})$, and we also interpret uniform continuity as *equicontinuity* with respect to the space $\mathcal{P}$.

19. The *weak***-topology* in $\mathcal{P}$ is the weakest topology such that, for every bounded continuous function φ and $P \in \mathcal{P}$, the mapping

$$P \to \int \varphi dP : \mathcal{P} \mapsto \boldsymbol{R} \text{ is continuous.}$$

The space $\mathcal{P}$ of probability measures, topologized by the *weak**-topology, is itself a complete separable and metrizable space.

20. The *Prokhorov metric* (distance) between two members P_1 and $P_2 \in \mathcal{P}$ is defined by

$$\rho_P(P_1, P_2) = \inf\{\varepsilon > 0 | P_1(A) \leq P_2\{A^\varepsilon\} + \varepsilon \ \forall A \in \mathcal{B}\} \tag{2.68}$$

where A^ε is the closed ε-neighborhood of A $(= \{x \in T : \inf_{y \in A} d(x, y) < \varepsilon\})$, $\varepsilon > 0$. The *Levi distance* between two distribution functions F_1 and F_2 defined on $\boldsymbol{R}$ is given by

$$\rho_L(F_1, F_2) = \inf\{\varepsilon : F_1(x-\varepsilon)-\varepsilon \leq F_2(x) \leq F_1(x+\varepsilon)+\varepsilon, \ \forall x \in \boldsymbol{R}\}. \tag{2.69}$$

Thus, for a fixed F and ε, we have the *Levi neighborhood* $\mathbf{n}_\varepsilon^L(F)$ of distribution function F_2's that satisfy (2.69) for $F_1 = F$. Similarly, for a fixed $P \in \mathcal{P}$, (2.68) defines for every $\varepsilon : 0 < \varepsilon < 1$, a neighborhood of $\mathbf{n}_\varepsilon^P(P)$ of probability measures $(P_2 \in \mathcal{P})$ termed the *Prokhorov neighborhood* of P [see (2.28)]. [Note that on $\boldsymbol{R}$, the *Kolmogorov distance* $d_K(F, G) = \|F - G\| = \sup\{|F(x) - G(x)| : x \in \boldsymbol{R}\}$ or for general $\mathcal{P}$, the *total variation* distance $d_{TV}(P_1, P_2) = \sup\{|P_1(A) - P_2(A)| : A \in \mathcal{B}\}$ do not generate the *weak**-topology mentioned before. Similarly, a *contamination neighborhood* $\mathbf{n}_\varepsilon^c(F) = \{G : \ G = (1 - \varepsilon)F + \varepsilon H, \ H \in \mathcal{P}\}$ (see (2.27)) is not a neighborhood in a topological sense. In Section 2.3, we have referred to this as the *gross-error model*.]

2.5.4 Laws of Large Numbers and Related Inequalities

The probability inequality in (2.61), for $r = 2$, known as the Chebyshev inequality, is the precursor of the so-called laws of large numbers (LLN). They were originally developed for sums of independent random variables, but have been extended to dependent cases as well.

1. *Khintchine weak LLN.* Let $\{X_k;\ k \geq 1\}$ be a sequence of i.i.d. random variables with a finite $\theta = \mathbb{E}X$. Let $\bar{X}_n = n^{-1} \sum_{i=1}^n X_i$, $n \geq 1$. Then

$$\bar{X}_n \to \theta, \text{ in probability, as } n \to \infty. \tag{2.70}$$

In particular, if X_k can only assume values 0 and 1 with respective probabilities $1-\pi$ and π $(0<\pi<1)$, then $\theta=\pi$ and (2.70) is referred to as the *Bernoulli LLN.*

2. *Khintchine strong LLN.* If the X_i are i.i.d. random variables, then $\bar{X}_n \to c$ a.s., as $n\to\infty$, if and only if $I\!E X_i$ exists and $c = I\!E X_i$. Again, if the X_i are Bernoulli random variables, $\bar{X}_n \to \pi$ a.s., as $n\to\infty$, and this is known as the *Borel strong LLN.*

3. *Markov weak LLN.* Let $\{X_k, k\geq 1\}$ be independent random variables with $\mu_k = I\!E X_k$, and $\nu_{k,\delta} = I\!E|X_k-\mu_k|^{1+\delta} < \infty$, for some $\delta : 0<\delta\leq 1$, $k\geq 1$. If
$$n^{-1-\delta}\sum_{k=1}^{n}\nu_{k,\delta} \to 0 \text{ as } n\to\infty, \tag{2.71}$$
then
$$\bar{X}_n - I\!E\bar{X}_n \to 0, \text{ in probability, as } n\to\infty.$$
The *Markov condition* (2.71) compensates for the possible nonidentically distributed nature of the X_k.

4. *Kolmogorov strong LLN.* Let X_k, $k\geq 1$, be independent random variables such that $\mu_k = I\!E X_k$ and $\sigma_k^2 = \text{Var}(X_k)$ exist for every $k\geq 1$, and further assume that
$$\sum_{k\geq 1} k^{-2}\sigma_k^2 < \infty.$$
Then
$$\bar{X}_n - I\!E\bar{X}_n \to 0 \text{ a.s., as } n\to\infty. \tag{2.72}$$

5. *Kolmogorov maximal inequality.* Let $\{T_n, n\geq 1\}$ be a zero mean martingale, such that $I\!E T_n^2$ exists for every $n\geq 1$. Then for every $t>0$,
$$I\!P\{\max_{1\leq k\leq n}|T_k| > t\} \leq t^{-2} I\!E(T_n^2). \tag{2.73}$$
If $\{T_n, n\geq m\}$ is a (zero mean) reversed martingale, then for every $N\geq n\geq m$ and $t>0$,
$$I\!P\{\max_{n\leq k\leq N}|T_k| > t\} \leq t^{-2} I\!E(T_n^2). \tag{2.74}$$
The inequality (2.73) was originally established by Kolmogorov for T_n as the sum of (centered) independent random variables; (2.74) holds for $\bar{X}_n$ as well as for U-statistics, which are known to be reversed martingales. In this way, the Bernstein inequality (2.62) extends readily to the next case.

6. *Bernstein inequality for submartingales.* Let $\{T_n, n\geq 1\}$ be a submartingale such that $M_n(\theta) = I\!E\{e^{\theta T_n}\}$ exists for all $\theta\in(0,\theta_0)$. Then $\forall t>0$,
$$I\!P\{\max_{1\leq k\leq n}|T_k| > t\} \leq \inf_{\theta>0}\{e^{-\theta t}M_n(\theta)\}.$$
If $\{T_n, n\geq m\}$ is a reversed submartingale, and $M_n(\theta)$ exists, then
$$I\!P\{\sup_{N\leq n}|T_N| > t\} \leq \inf_{\theta>0}\{e^{-\theta t}M_n(\theta)\}, \quad \forall t>0.$$

7. *Hájek-Rényi-Chow inequality:* Let $\{T_n,\ n \geq 1\}$ be a submartingale, and let $\{c_n,\ n \geq 1\}$ be a nonincreasing sequence of positive numbers. Assume that $T_n^+ = (T_n \vee 0)$ has a finite expectation for $n \geq 1$. Then for every $t > 0$,

$$I\!P\{\max_{1\leq k\leq n} c_k T_k > t\} \leq t^{-1}\{c_1 I\!E T_1^+ + \sum_{k=2}^{n} c_k I\!E(T_k^+ - T_{k-1}^+)\}. \quad (2.75)$$

If $\{T_n,\ n \geq m\}$ is a reversed martingale, and the c_k are $\nearrow$, then

$$I\!P\{\max_{n\leq k\leq N} c_k T_k > t\} \leq t^{-1}\{c_n I\!E T_n^+ + \sum_{k=n+1}^{N} (c_k - c_{k-1}) I\!E T_k^+\}. \quad (2.76)$$

The inequality (2.73) [or (2.74)] is a particular case of (2.75) or (2.76), corresponding to the case $c_k = 1,\ k \geq 1$.

8. *Kolmogorov strong LLN for martingales.* Let $\{T_n = \sum_{k\leq n} Y_k,\ n \geq 1\}$ be a martingale such that $I\!E|Y_k|^p$ exists for some $p:\ 1 \leq p \leq 2\ \forall k \geq 1$. Thus, $\{T_n,\ n \geq 1\}$ is a zero mean L_p-martingale. Further let $\{b_n,\ n \geq 1\}$ be an increasing sequence of positive numbers, such that $b_n \nearrow \infty$ as $n \to \infty$, and

$$\sum_{n\geq 2} b_n^{-p} I\!E\{|Y_n|^p | Y_j,\ j < n\} < \infty \text{ a.s.}$$

Then

$$b_n^{-1} T_n \to 0 \text{ a.s.}, \quad \text{as } n \to \infty. \quad (2.77)$$

Note that (2.72) is a special case of (2.77) when $p = 2$ and $b_n \equiv n$.

2.5.5 Central Limit Theorems

The sample *statistics*, as the test-statistics or estimators, are the basic tools for drawing statistical conclusions from the sample observations. Thus, a thorough study relates to the distribution theory of statistics, termed the *sampling distribution*. Unfortunately, the exact sampling distribution cannot be easily derived with exception of some specialized cases, and the task becomes prohibitively laborious as the sample size becomes large. On the other hand, after a suitable normalization $T_n \mapsto (T_n - a_n)/b_n$ with suitable $\{a_n\}$ and $\{b_n\}$, the distribution of the normalized sequence can be approximated by some simple one as n becomes large, for instance Poisson, normal, chi-square, beta, gamma, etc. In other words, the notion of *weak convergence* introduced in Definition 2.5 applies. The present book will concentrate on the basic concepts of the weak convergence, at various levels of generality and in various contexts. The basic results will be presented in a logical order, with basic lemmas and theorems preceding the main theory.

Consider a sequence $\{\mathbf{T}_n,\ n \geq n_0\}$ of normalized random vectors, and let $F_n(\mathbf{t}) = I\!P\{\mathbf{T}_n \leq \mathbf{t}\}$, $\mathbf{t} \in \boldsymbol{R}_p$ for some $p \geq 1$. Let $\mathbf{T}$ be a stochastic vector with distribution function F and let $\mathbf{J} \subset \boldsymbol{R}_p$ be the set of points of continuity of F. Then, if

$$F_n(\mathbf{t}) \to F(\mathbf{t}), \ \forall \mathbf{t} \in \mathbf{J} \ \text{ as } n \to \infty, \tag{2.78}$$

we say that $F_n \xrightarrow{w} F$, or $\mathbf{T}_n \xrightarrow{\mathcal{D}} \mathbf{T}$, or $\mathcal{L}(\mathbf{T}_n \to \mathcal{L}(\mathbf{T})$. This definition will be later extended to more general stochastic elements, defined on appropriate function spaces.

1. *Helly–Bray lemma*: Let $\{F_n\}$ satisfy (2.78), and let $g(.)$ be a continuous function on a compact $\mathbf{C} \subset \boldsymbol{R}_p$ whose boundary lies in the set $\mathbf{J}$. Then

$$\int_{\mathbf{C}} g(.)dF_n(.) \to \int_{\mathbf{C}} g(.)dF(.) \ \text{ as } n \to \infty. \tag{2.79}$$

   If $g(.)$ is bounded and continuous on $\boldsymbol{R}_p$, then (2.79) also holds when $\mathbf{C}$ is replaced by $\boldsymbol{R}_p$.

2. *Sverdrup lemma*: Let $\{\mathbf{T}_n\}$ be a sequence of random vectors such that $\mathbf{T}_n \xrightarrow{\mathcal{D}} \mathbf{T}$, and let $\mathbf{g}(.) : \boldsymbol{R}_p \mapsto \boldsymbol{R}_q$, $q \geq 1$, be a continuous mapping. Then $\mathbf{g}(\mathbf{T}_n) \xrightarrow{\mathcal{D}} \mathbf{g}(\mathbf{T})$.

3. *Cramér–Wold theorem.* Let $\{\mathbf{T}_n\}$ be a sequence of random vectors. Then $\mathbf{T}_n \xrightarrow{\mathcal{D}} \mathbf{T}$ if and only if $\boldsymbol{\lambda}^\top \mathbf{T}_n \xrightarrow{\mathcal{D}} \boldsymbol{\lambda}^\top \mathbf{T}$ for every fixed $\boldsymbol{\lambda} \in \boldsymbol{R}_p$.

4. *Lévy–Cramér Theorem.* Let $\{F_n, \ n \geq n_0\}$ be a sequence of distribution functions, defined on $\boldsymbol{R}_p$, for some $p \geq 1$, with the corresponding sequence of characteristic functions $\{\phi_n(\mathbf{t}), \ n \geq n_0\}, \mathbf{t} \in \boldsymbol{R}_p$. Then a necessary and sufficient condition for $F_n \xrightarrow{w} F$ is that $\phi_n(\mathbf{t}) \to \phi(\mathbf{t}) \ \forall \mathbf{t} \in \boldsymbol{R}_p$, where $\phi(\mathbf{t})$ is continuous at $\mathbf{t} = \mathbf{0}$ and is the characteristic function of F.

   The case of $p = 1$ and T_n being a sum of independent random variables constitutes the main domain of the central limit theorem (CLT).

5. *Classical Central Limit Theorem.* Let $\{X_k, \ k \geq 1\}$ be i.i.d. random variables with $I\!\!E X = \mu$ and $\text{Var}(X) = \sigma^2 < \infty$. Let $T_n = \sum_{k=1}^n X_k$, $n \geq 1$, and let

$$Z_n = (T_n - n\mu)/(\sigma\sqrt{n}). \tag{2.80}$$

   Then $Z_n \xrightarrow{\mathcal{D}} Z$, where Z has the standard normal distribution. If the $\mathbf{X}_k$ are p-vectors with $\boldsymbol{\mu} = I\!\!E\mathbf{X}$ and $\boldsymbol{\Sigma} = \text{Var}(\mathbf{X})$, then, by incorporating the Cramér–Wold theorem, we obtain $\mathbf{Z}_n = n^{-1/2}(\mathbf{T}_n - n\boldsymbol{\mu}) \xrightarrow{\mathcal{D}} Z \sim \mathcal{N}_p(\mathbf{0}, \boldsymbol{\Sigma})$. Especially, when the X_k can only assume the values 0 and 1 with probability $1 - \pi$ and π, respectively (the Bernoulli case), we have

$$\Big(n\pi(1-\pi)\Big)^{-\frac{1}{2}} (T_n - n\pi) \xrightarrow{\mathcal{D}} \mathcal{N}(0, 1).$$

   This result is known as the *de Moivre–Laplace theorem.*

6. *Liapounov central limit theorem.* Let $X_k, \ k \geq 1$, be independent random

variables with $\mu_k = \mathbb{E}X_k$, $\sigma_k^2 = \mathrm{Var}(X_k)$, and for some $\delta \in (0,1]$,

$$\nu_{2+\delta}^{(k)} = \mathbb{E}\{|X_k - \mu_k|^{2+\delta}\} < \infty \ \forall k \geq 1. \tag{2.81}$$

Denote $T_n = \sum_{k=1}^n X_k$, $\xi_n = \mathbb{E}T_n = \sum_{k=1}^n \mu_k$, $s_n^2 = \mathrm{Var}(T_n) = \sum_{k=1}^n \sigma_k^2$, $Z_n = (T_n - \xi_n)/s_n$, and let

$$\rho_n = s_n^{-2-\delta}\Big(\sum_{k=1}^n \nu_{2+\delta}^{(k)}\Big), \quad n \geq 1. \tag{2.82}$$

Then

$$\rho_n \to 0 \Rightarrow Z_n \xrightarrow{\mathcal{D}} \mathcal{N}(0,1) \ \text{ as } n \to \infty.$$

7. *Lindeberg–Feller theorem.* Define T_n, ξ_n, s_n^2, Z_n and so on, as in Liapunov Theorem, without assuming (2.81) and (2.82). Consider the following three conditions:

A. *Uniform asymptotic negligibility condition* (UAN):

$$\max\{\sigma_k^2/s_n^2 \,:\, 1 \leq k \leq n\} \to 0, \quad \text{as } n \to \infty. \tag{2.83}$$

B. *Asymptotic normality condition* (AN):

$$\mathbb{P}\{Z_n \leq x\} \to \Phi(x) = (2\pi)^{-1/2} \int_{-\infty}^{x} e^{-\frac{1}{2}t^2} dt, \quad x \in \boldsymbol{R}.$$

C. *Lindeberg–Feller (uniform integrability) condition*: for every $\varepsilon > 0$

$$\frac{1}{s_n^2} \sum_{k=1}^n \mathbb{E}\{(X_k - \mu_k)^2 I(|X_k - \mu_k| > \varepsilon s_n)\} \to 0 \quad \text{as } n \to \infty. \tag{2.84}$$

Then **A** and **B** hold simultaneously if and only if **C** holds.
If we let $Y_{nk} = s_n^{-1}(X_k - \mu_k)$, $1 \leq k \leq n$, then (2.83) implies that for every $\varepsilon > 0$ we have $\max_{1 \leq k \leq n} \mathbb{P}\{|Y_{nk}| > \varepsilon\} \to 0$ as $n \to \infty$, that is, the Y_{nk} are *infinitesimal*. The *if* part (i.e., $\mathbf{C} \Rightarrow \mathbf{A}, \mathbf{B}$) of the theorem is due to Lindeberg, while the *only if* part is due to Feller.

8. *Bounded central limit theorem.* If all X_k are bounded random variables, then it follows readily that $Z_n \xrightarrow{\mathcal{D}} \mathcal{N}(0,1)$ if and only if $s_n^2 \to \infty$ as $n \to \infty$.

9. *Hájek–Šidák central limit theorem.* Let $\{Y_k,\ k \geq 1\}$ be a sequence of i.i.d. random variables such that $\mathbb{E}Y = \mu$ and $\mathrm{Var}(Y) = \sigma^2$ exist, and let $\mathbf{c}_n = (c_{n1}, \ldots, c_{nn})$, $n \geq 1$, be a triangular scheme of real numbers such that

$$\max\left\{\frac{c_{nk}^2}{\sum_{k=1}^n c_{nk}^2} \,:\, 1 \leq k \leq n\right\} \to 0 \ \text{ as } n \to \infty. \tag{2.85}$$

Then

$$Z_n = \frac{\sum_{k=1}^n c_{nk}(Y_k - \mu)}{\sigma(\sum_{k=1}^n c_{nk}^2)^{1/2}} \xrightarrow{\mathcal{D}} \mathcal{N}(0,1) \quad \text{as } n \to \infty.$$

By incorporating the Y_{nk} in the Lindeberg–Feller theorem or in the above theorem, we may conceive of a more general *triangular array* of row-wise independent random variables:

10. *Triangular array central limit theorem.* Consider a triangular array of row-wise independent random variables X_{nj}, $j \leq k_n$, $n \geq 1$, where $k_n \to \infty$ as $n \to \infty$. Then the X_{nk} form an infinitesimal system of random variables and

$$Z_n = \sum_{k=1}^{k_n} X_{nk} \xrightarrow{\mathcal{D}} \mathcal{N}(0,1)$$

if and only if

$$\sum_{k=1}^{k_n} I\!P\{|X_{nk}| > \varepsilon\} \to 0$$

and

$$\sum_{k=1}^{k_n} \Big\{ \int_{\{|x|<\varepsilon\}} x^2 dP_{nk}(x) - (\int_{\{|x|<\varepsilon\}} x dP_{nk}(x))^2 \Big\} \to 1,$$

for every $\varepsilon > 0$ as $n \to \infty$, where $P_{nk}(x) = I\!P\{X_{nk} < x\}$, $x \in \boldsymbol{R}$, $k \leq k_n$, $n \geq 1$.

11. *Dependent central limit theorem.* (Dvoretzky 1972). Let $\{X_{nk},\ k \leq k_n;\ n \geq 1\}$ be a triangular sequence. Let $\mathcal{B}_{nk} = \mathcal{B}(X_{nj},\ j \leq k)$, $k \geq 1$, and

$$\mu_{nk} = I\!E(X_{nk}|\mathcal{B}_{nk-1}) \text{ and } \sigma_{nk}^2 = I\!E(X_{nk}^2|\mathcal{B}_{nk-1}) - \mu_{nk}^2,$$

for $k \geq 1$. Suppose that

$$\sum_{k=1}^{k_n} \mu_{nk} \to 0, \quad \sum_{k=1}^{k_n} \sigma_{nk}^2 \to 1 \quad \text{as} \quad n \to \infty$$

and that the (conditional) Lindeberg condition holds in the form

$$\sum_{k=1}^{k_n} I\!E(X_{nk}^2 I(|X_{nk}| > \varepsilon)|\mathcal{B}_{nk-1}) \to 0,\ \forall \varepsilon > 0, \quad \text{as } n \to \infty. \tag{2.86}$$

Then $Z_n = \sum_{k \leq k_n} X_{nk} \xrightarrow{\mathcal{D}} \mathcal{N}(0,1)$.

12. *Martingale (array) central limit theorem.* If the X_{nk} are martingale differences for each n, then the μ_{nk} are 0 (a.e.), while

$$I\!E Z_n^2 = s_n^2 = I\!E(\sum_{k \leq k_n} \sigma_{nk}^2).$$

Therefore, if

$$\{\sum_{k \leq k_n} \sigma_{nk}^2\}/s_n^2 \xrightarrow{\mathcal{P}} 1, \tag{2.87}$$

and the Lindeberg condition in (2.84) holds for the X_{nk}, then by the Chebyshev, inequality (2.86) holds too. Both (2.87) and (2.84) imply that

$Z_n/s_n \xrightarrow{\mathcal{D}} \mathcal{N}(0,1)$ as $n \to \infty$. This result applies as well to a martingale sequence (a result due to Brown 1971).

13. *Reverse martingale central limit theorem.* Let $\{T_k,\ k \geq 1\}$ form a reverse martingale, and set $\mathbb{E}T_n = 0,\ \forall n \geq 1$. Put $Y_k = T_k - T_{k+1},\ k \geq 1$ and $v_k^2 = \mathbb{E}(Y_k^2|T_{k+1}, T_{k+2}, \ldots)$ and let

$$w_n^2 = \sum_{k \geq n} v_k^2 \text{ and } s_n^2 = \mathbb{E}w_n^2 = \mathbb{E}T_n^2.$$

If (a) $w_n^2/s_n^2 \xrightarrow{\mathcal{P}} 1$ as $n \to \infty$, and (b)

$$w_n^{-2} \sum_{k \geq n} \mathbb{E}(Y_k^2 I(|Y_k| > \varepsilon w_n|T_{k+1}, T_{k+2}, \ldots) \xrightarrow{\mathcal{P}} 0\ \forall \varepsilon > 0,$$

or (c)

$$w_n^{-2} \sum_{k \geq n} Y_k^2 \to 1, \text{ a.s.,} \quad \text{as } n \to \infty, \tag{2.88}$$

then $T_n/s_n \xrightarrow{\mathcal{D}} \mathcal{N}(0,1)$.

14. *Multivariate central limit theorem.* Using the Cramér–Wold theorem (3) to any version of the preceding CLT's, we conclude that the CLT extends to the multivariate case under parallel regularity conditions.

15. *Renewal (CL) theorem.* Let $\{X_k,\ k \geq 1\}$ be a sequence of identically distributed nonnegative random variables with mean $\mu > 0$ and variance σ^2 ($< \infty$). Further assume that either the X_k are independent or $\{X_k - \mu,\ k \geq 1\}$ form a martingale-difference sequence. Let

$$T_n = X_1 + \ldots + X_n,\ n \geq 1,\ T_0 = 0.$$

Define

$$N_t = \max\{k : T_k \leq t\},\ N_0 = 0 \ \text{ for every } \ t > 0.$$

Then, by definition, $T_{N_t} \leq t < T_{N_t+1},\ \forall t \geq 0$. Therefore,

$$t^{-1/2}(N_t - t/\mu) \xrightarrow{\mathcal{D}} \mathcal{N}(0, \sigma^2/\mu^3) \ \text{ as } \ t \to \infty.$$

16. *Berry–Esséen theorem.* Consider the classical CLT in (2.80) and assume that $\mathbb{E}|X - \mu|^{2+\delta} = \nu_{2+\delta} < \infty$ for some $\delta :\ 0 < \delta \leq 1$. Then there exists a constant $C > 0$ such that

$$\Delta_n = \sup_{x \in \boldsymbol{R}} |\Phi_n(x) - \Phi(x)| \leq Cn^{-\delta/2}(\nu_{2+\delta}/\sigma^{2+\delta})$$

where $\Phi_n(x) = \mathbb{P}\{Z_n \leq x\}$ and $\Phi(x)$ is the standard normal distribution function.

Various modifications of this uniform bound are available in the literature.

17. *Edgeworth expansions.* For the statistic Z_n in the preceding theorem,

assume the existence of higher-order (central) moments up to a certain order. Then

$$\begin{aligned}\Phi_n(x) &= \Phi(x) - \frac{1}{\sqrt{n}}(\mu_3/6\sigma^3)\Phi^{(3)}(x) \\ &+ \frac{1}{n}\{(\frac{\mu_4 - 3\sigma^4}{\sigma^4})\Phi^{(4)}(x) + \frac{1}{72}(\frac{\mu_3^2}{\sigma^6})\Phi^{(6)}(x)\} + O(n^{-3/2}),\end{aligned}$$

where $\Phi^{(k)}(x) = (d^k/dx^k)\Phi(x)$, $k \geq 0$, and $\mu_k = \mathbb{E}(X - \mu)^k$, $k \geq 1$, are the central moments of the random variable X, $\mu = \mathbb{E}X$. Imposing appropriate moment conditions on X, we can go up to $O(n^{-(s-2)/2})$ for some $s \geq 3$ instead of $O(n^{-3/2})$. A more general version of this expansion is given by Bhattacharya and Ghosh (1978) for a smooth function $H(.) : \boldsymbol{R}_p \mapsto \boldsymbol{R}_q$, $p, q \geq 1$.

18. *Law of the iterated logarithms.* If the X_i, $i \geq 1$ are i.i.d. random variables with finite mean μ and variance σ^2, then

$$\limsup_{n\to\infty} \frac{\sum_{i=1}^n (X_i - \mu)}{(2\sigma^2 n \log\log n)^{1/2}} = 1, \text{ with probability 1.}$$

The clause that the X_i are identically distributed can be eliminated by either a boundedness condition or a moment condition in the following manner:

18a. Let X_i, $i \geq 1$ be independent random variables with means μ_i and finite variance σ_i^2, $i \geq 1$, such that

$$B_n^2 = \sum_{i=1}^n \sigma_i^2 \to \infty \text{ as } n \to \infty.$$

Moreover, assume that either of the following two conditions hold:
Kolmogorov: There exists a suitable sequence $\{m_n\}$ of positive numbers, such that with probability 1,

$$|X_n - \mu_n| \leq m_n \text{ where } m_n = \mathrm{o}(B_n(\log\log B_n)^{-1/2}), \text{ as } n \to \infty.$$

Chung: For some $\varepsilon > 0$,

$$B_n^{-3}\{\sum_{i=1}^n \mathbb{E}|X_i - \mu_i|^3\} = \mathrm{O}((\log B_n)^{-1-\varepsilon}).$$

Then

$$\limsup_{n\to\infty} \frac{\sum_{i=1}^n (X_i - \mu_i)}{(2B_n^2 \log\log B_n)^{1/2}} = 1, \text{ with probability 1.}$$

For extensions to some dependent sequences of random variables, we may refer to Stout (1974). The derivations of these laws of iterated logarithms (LIL) are based on the CLT with a suitable rate of convergence.

2.5.6 Limit Theorems Allied to Central Limit Theorems

The present subsection shows how can we use the CLT even for nonlinear statistics, to which the CLT does not directly applies. This particularly concerns the robust and nonparametric statistics and other important cases arising in practice. The use of the CLT can be done by means of suitable approximations.

1. *Slutzky theorem.* Let $\{X_n\}$ and $\{Y_n\}$ be two sequences of random variables (not necessarily independent) such that

$$X_n \xrightarrow{\mathcal{D}} X \text{ and } Y_n \xrightarrow{\mathcal{P}} c, \text{ a constant, as } n \to \infty,$$

where X has a distribution function F (which may be degenerate at a point). Then

$$X_n \pm Y_n \xrightarrow{\mathcal{D}} X \pm c, \quad X_n Y_n \xrightarrow{\mathcal{D}} cX,$$
$$X_n / Y_n \xrightarrow{\mathcal{D}} X/c, \text{ if } c \neq 0.$$

The result extends directly to the case where the X_n or Y_n are vector valued random elements. A couple of examples can be cited that illustrate the utility of the Slutzky theorem.

Example 2.3 Let X_i, $i \geq 1$ be i.i.d. random variables with mean μ and variance σ^2. An unbiased estimator of σ^2 is

$$\begin{aligned} S_n^2 &= \frac{1}{n-1} \sum_{i=1}^{n} (X_i - \bar{X}_n)^2 \\ &= \frac{n}{n-1} \{ \frac{1}{n} \sum_{i=1}^{n} (X_i - \mu)^2 - (\bar{X}_n - \mu)^2 \}, \end{aligned} \tag{2.89}$$

where $\sqrt{n}(\bar{X}_n - \mu)^2 \xrightarrow{\mathcal{P}} 0$ by the Chebyshev inequality. On the other hand, by the classical CLT,

$$\sqrt{n}\{ \frac{1}{n} \sum_{i=1}^{n} (X_i - \mu)^2 - \sigma^2 \} \xrightarrow{\mathcal{D}} \mathcal{N}(0, \mu_4 - \sigma^4),$$

whenever $\mathbb{E}X^4 < \infty$, where $\mu_4 = \mathbb{E}(X - \mu)^4$. Finally, $n/(n-1) \to 1$ as $n \to \infty$. As a result, by the Slutzky Theorem, we claim that under $\mu_4 < \infty$,

$$\sqrt{n}(S_n^2 - \sigma^2) \xrightarrow{\mathcal{D}} \mathcal{N}(0, \mu_4 - \sigma^4),$$

although S_n^2 does not have independent summands.

Example 2.4 In the same setup as in the preceding example, let

$$t_n = \frac{\sqrt{n}(\bar{X}_n - \mu)}{S_n} = \frac{\sqrt{n}(\bar{X}_n - \mu)/\sigma}{(S_n/\sigma)},$$

and note that, $S_n^2/\sigma^2 \xrightarrow{\mathcal{P}} 1$ as $n \to \infty$, by (2.89), while the CLT applies to $\sqrt{n}(\bar{X}_n - \mu)/\sigma$. Therefore by the Slutzky theorem, we conclude that $t_n \xrightarrow{\mathcal{D}} \mathcal{N}(0,1)$ as $n \to \infty$.

2. *Transformation on statistics.* Suppose that $\{T_n\}$ is a sequence of statistics and $g(T_n)$ is a transformation on T_n such that

$$n^{1/2}(T_n - \theta)/\sigma \xrightarrow{\mathcal{D}} \mathcal{N}(0,1), \tag{2.90}$$

and suppose also that $g: \boldsymbol{R} \mapsto \boldsymbol{R}$ is a continuous function such that $g'(\theta)$ exists and is different from 0. Then

$$n^{1/2}[g(T_n) - g(\theta)]/\{\sigma g'(\theta)\} \xrightarrow{\mathcal{D}} \mathcal{N}(0,1).$$

As an example, consider S_n^2 in (2.89), and take $g(S_n^2) = S_n = \sqrt{S_n^2}$. It is easy to verify that

$$\sqrt{n}(S_n - \sigma) \xrightarrow{\mathcal{D}} \mathcal{N}(0, (\mu_4 - \sigma^4)/4\sigma^2)$$

whenever $\mu_4 < \infty$. We proceed similarly in the case where $g(.)$ and/or T_n are vector-valued functions or statistics.

3. *Variance stabilizing transformations.* Suppose that in (2.90), $\sigma^2 = h^2(\theta)$ is a nonnegative function of θ. If we choose $g(.)$ such that $g'(\theta)h(\theta) = c$, so that

$$g(\theta) = c\int_0^\theta [h(y)]^{-1}dy,$$

then we have

$$n^{1/2}[g(T_n) - g(\theta)] \xrightarrow{\mathcal{D}} \mathcal{N}(0, c^2).$$

Important examples are

(a) $g(T_n) = \sin^{-1}\sqrt{T_n}$ (binomial proportion),

(b) $g(T_n) = \sqrt{T_n}$ (Poisson random variable),

(c) $g(T_n) = \log T_n$ ($T_n \equiv S_n^2$ in (2.89)), and

(d) $g(T_n) = \tanh^{-1} T_n$ ($T_n \equiv$ sample correlation coefficient).

4. *Projection theorem* of Hoeffding–Hájek. Let $T_n = T(X_1, \dots, X_n)$ be a symmetric function of $X_1, \dots, X_n$. For each i $(= 1, \dots, n)$, let

$$T_{ni} = \mathbb{E}[T_n|X_i] - \mathbb{E}T_n; \ T_n^\star = \sum_{i=1}^n T_{ni},$$

and note that the T_{ni} are independent random variables. Moreover,

$$\mathbb{E}T_n^\star = 0 \ \text{ and } \ \text{Var}(T_n^\star) = \sum_{i=1}^n \mathbb{E}(T_{ni}^2).$$

Further, $$\mathbb{E}\{T_n^\star(T_n - \mathbb{E}T_n)\} = \sum_{i=1}^n \mathbb{E}\{T_{ni}(T_n - \mathbb{E}T_n)\}$$
$$= \sum_{i=1}^n \mathbb{E}\{[\mathbb{E}(T_n - \mathbb{E}T_n|X_i)](T_n - \mathbb{E}T_n)\}$$
$$= \sum_{i=1}^n \mathbb{E}\{[\mathbb{E}(T_n - \mathbb{E}T_n|X_i)]\mathbb{E}\{(T_n - \mathbb{E}T_n)|X_i\}\}$$
$$= \sum_{i=1}^n \mathbb{E}(T_{ni}^2) = \text{Var}(T_n^\star).$$

Therefore,

$$\mathbb{E}\Big\{[(T_n - \mathbb{E}T_n) - T_n^\star]^2\Big\} = \mathbb{E}(T_n - \mathbb{E}T_n)^2 + \text{Var}(T_n^\star) - 2\text{Var}(T_n^\star)$$
$$= \text{Var}(T_n) - \text{Var}(T_n^\star).$$

Thus, $(T_n - \mathbb{E}T_n - T_n^\star)/\sqrt{\text{Var}(T_n^\star)} \xrightarrow{\mathcal{P}} 0$ whenever

$$[\text{Var}(T_n) - \text{Var}(T_n^\star)]/\text{Var}(T_n^\star) \to 0 \quad \text{as } n \to \infty,$$

and a suitable version of the CLT can be adopted to show that

$$T_n^\star/\sqrt{\text{Var}(T_n^\star)} \xrightarrow{\mathcal{D}} \mathcal{N}(0,1).$$

Consequently, by the Slutzky theorem,

$$(T_n - \mathbb{E}T_n)/\sqrt{\text{Var}(T_n^\star)} \xrightarrow{\mathcal{D}} \mathcal{N}(0,1),$$

and it is also possible to replace $\text{Var}(T_n^\star)$ by $\text{Var}(T_n)$. Hoeffding (1948) incorporated this projection technique for U-statistics which are symmetric and unbiased estimators but may not have generally independent summands; Hájek (1968) popularized this technique for a general class of rank statistics where also the independence of the summands is vitiated. Van Zwet (1984) used this projection (termed the *Hoeffding-decomposition)* for general symmetric statistics for deriving Berry–Esséen-type bounds.

2.5.7 Central Limit Theorems for Quadratic Forms

Quadratic function(al)s play a basic role in the asymptotic statistical inference, especially in multiparameter problems. We shall first present two basic (finite sample size) results that provide an access to the desired asymptotics.

1. *Courant theorem.* Let $\mathbf{x} \in \boldsymbol{R}_p$ for some $p \geq 1$, and let $\mathbf{A}$ and $\mathbf{B}$ be two positive semidefinite matrices, $\mathbf{B}$ being nonsingular. Then

$$\lambda_p = ch_{\min}(\mathbf{A}\mathbf{B}^{-1}) = \inf\{\frac{\mathbf{x}^\top\mathbf{A}\mathbf{x}}{\mathbf{x}^\top\mathbf{B}\mathbf{x}} : \mathbf{x} \in \boldsymbol{R}_p\}$$

$$\leq \sup\{\frac{\mathbf{x}^\top \mathbf{A}\mathbf{x}}{\mathbf{x}^\top \mathbf{B}\mathbf{x}} : \mathbf{x} \in \boldsymbol{R}_p\} = ch_{\max}(\mathbf{A}\mathbf{B}^{-1}) = \lambda_1$$

2. *Cochran theorem.* Let $\mathbf{X} \sim \mathcal{N}_p(\boldsymbol{\mu}, \boldsymbol{\Sigma})$ where $rank(\boldsymbol{\Sigma}) = q \leq p$. Let $\mathbf{A}$ be a positive semidefinite matrix such that $\mathbf{A}\boldsymbol{\Sigma}\mathbf{A} = \mathbf{A}$. Then

$$\mathbf{X}^\top \mathbf{A}\mathbf{X} \sim \chi^2_{q,\Delta}; \;\; \Delta = \boldsymbol{\mu}^\top \mathbf{A}\boldsymbol{\mu},$$

where $\chi^2_{a,\delta}$ is a random variable having the noncentral chi-square distribution with a distribution function and noncentrality parameter δ. $\mathbf{X}^\top \mathbf{A}\mathbf{X}$ has the central chi-square distribution with q distribution function provided $\boldsymbol{\mu} = \mathbf{0}$.

3. *Slutzky–Cochran theorem:* Let $\{\mathbf{T}_n\}$ be a sequence of random p-vectors such that $\sqrt{n}(\mathbf{T}_n - \boldsymbol{\theta}) \xrightarrow{\mathcal{D}} \mathcal{N}_p(\mathbf{0}, \boldsymbol{\Sigma})$, where $rank(\boldsymbol{\Sigma}) = q : \; 1 \leq q \leq p$. Let $\{\mathbf{A}_n\}$ be a sequence of random matrices such that $\mathbf{A}_n \xrightarrow{\mathcal{P}} \mathbf{A}$ and $\mathbf{A}\boldsymbol{\Sigma}\mathbf{A} = \mathbf{A}$. Then

$$Q_n = n(\mathbf{T}_n - \boldsymbol{\theta})^\top \mathbf{A}_n(\mathbf{T}_n - \boldsymbol{\theta}) \xrightarrow{\mathcal{D}} \chi^2_{q,0}. \tag{2.91}$$

This result extends directly to local alternative setups, and will be considered later. For a class of statistical models, $\boldsymbol{\Sigma}$ itself is an idempotent matrix of rank q $(\leq p)$, so that

$$\boldsymbol{\Sigma} = \boldsymbol{\Sigma}^2 \;\text{ and }\; Trace(\boldsymbol{\Sigma}) = R(\boldsymbol{\Sigma}) = q \; (\leq p).$$

In that case $\mathbf{A} = \boldsymbol{\Sigma}$ satisfies the condition that $\mathbf{A}\boldsymbol{\Sigma}\mathbf{A} = \boldsymbol{\Sigma}^3 = \boldsymbol{\Sigma}^2 = \boldsymbol{\Sigma} = \mathbf{A}$, and hence (2.54) holds.

4. *Cochran partition theorem:* Suppose that Q_n is defined as in (2.91) and that there exist k (≥ 1) positive semidefinite matrices $\mathbf{A}_{n1}, \ldots, \mathbf{A}_{nk}$, such that $\mathbf{A}_{n1} + \ldots + \mathbf{A}_{nk} = \mathbf{A}_n$ and $\mathbf{A}_{ni} \to \mathbf{A}_i : \; \mathbf{A}_i\boldsymbol{\Sigma}\mathbf{A}_i = \mathbf{A}_i$, $rank(\mathbf{A}_i) = q_i$, $i = 1, \ldots, k$.

   Let Qni be defined as in (2.91) with $\mathbf{A}_n$ replaced by $\mathbf{A}_{ni}$, $i = 1, \ldots, k$. Then a necessary and sufficient condition for

   (a) $Q_{ni} \xrightarrow{\mathcal{D}} \chi^2_{q_i,0}$, $i = 1, \ldots, k$, and

   (b) $Q_{n1}, \ldots, Q_{nk}$ are asymptotically independent

   is that $\mathbf{A}_i\mathbf{A}_j = \mathbf{0}$, $\forall i \neq j = 1, \ldots, k$.

2.5.8 Contiguity of Probability Measures

An excellent account of this concept and its impact on asymptotics is available in Hájek and Šidák (1967, ch. 6) and in Hájek, Šidák, and Sen (1999).

DEFINITION 2.6 ***[Contiguity].*** *Let $\{P_n\}$ and $\{Q_n\}$ be two sequence of (absolutely continuous) probability measures on measure spaces $\{(\Omega_n, \mathcal{A}_n, \mu_n)\}$.*

Let $p_n = dP_n/d\mu_n$ *and* $q_n = dQ_n/d\mu_n$. *Then, if for any sequence of events* $\{A_n\}: A_n \in \mathcal{A}_n$,

$$[P_n(A_n) \to 0] \Rightarrow [Q_n(A_n) \to 0], \tag{2.92}$$

the sequence of measures $\{Q_n\}$ *is said to be contiguous to* $\{P_n\}$.

Note that if P_n and Q_n are L_1-norm equivalent,

$$\|P_n - Q_n\| = \sup\{|P_n(A_n) - Q_n(A_n)| : A_n \in \mathcal{A}_n\} \to 0, \tag{2.93}$$

then (2.92) holds, but the converse may not be true. If T_n is $\mathcal{A}_n$- measurable, then the contiguity ensures that

$$T_n \to 0, \text{ in } P_n\text{-probability } \Rightarrow T_n \to 0, \text{ in } Q_n\text{-probability}.$$

Define the *likelihood ratio* statistic as

$$L_n = \begin{cases} q_n/p_n, & \text{if } p_n > 0, \\ 1, & \text{if } p_n = q_n = 0, \\ \infty, & \text{if } 0 = p_n < q_n. \end{cases}$$

LeCam (1960) characterized the contiguity of $\{Q_n\}$ to $\{P_n\}$ by two basic lemmas:

First, if

$$\log L_n \xrightarrow{\mathcal{D}} \mathcal{N}(-\frac{1}{2}\sigma^2, \sigma^2) \text{ (under } \{P_n\}), \tag{2.94}$$

then $\{Q_n\}$ is contiguous to $\{P_n\}$.

Second, if

$$(T_n, \log L_n) \xrightarrow{\mathcal{D}} \mathcal{N}_2(\boldsymbol{\mu}, \boldsymbol{\Sigma}) \text{ (under } \{P_n\}), \tag{2.95}$$

where $\boldsymbol{\mu} = (\mu_1, \mu_2)^\top$ and $\boldsymbol{\Sigma} = ((\sigma_{ij}))_{i,j=1,2}$ with $\mu_2 = -\frac{1}{2}\sigma_{22}$, then

$$T_n \xrightarrow{\mathcal{D}} \mathcal{N}(\mu_1 + \sigma_{12}, \sigma_{11}) \text{ (under } \{Q_n\}). \tag{2.96}$$

An important feature of this result is that the limit distribution under $\{P_n\}$ lends itself directly to the one under $\{Q_n\}$ by a simple adjustment of the mean. There exist simpler ways of proving (2.95) in the nonparametrics, but the distribution theory under $\{Q_n\}$ can be quite involved. The beauty of the contiguity-based proof lies in the simplicity of the proof for the contiguous alternatives. For an excellent treatise of this subject matter, we refer to the classical textbook of Hájek and Šidák (1967, ch. 6). The concept of contiguity has much broader scope in statistical asymptotics, and we shall refer to some of this later. A significant development in this direction is the Hájek-Inagaki-LeCam Theorem.

2.5.9 Hájek–Inagaki–LeCam theorem and the LAN condition

Following LeCam and Yang (1990), consider a family of probability measures $\{P_{\boldsymbol{\theta},n} : \boldsymbol{\theta} \in \boldsymbol{\Theta}\}$ on some measure space $(\Omega_n, \mathcal{A}_n)$ for each n (≥ 1), where

$\boldsymbol{\Theta} \in \boldsymbol{R}_k$, for some $k \geq 1$. Let δ_n be a positive number (typically $n^{-1/2}$), and assume that

(1) the true value of $\boldsymbol{\theta}$ is interior to $\boldsymbol{\Theta}$, and

(2) for $\mathbf{t}_n \in$ bounded set B, $\{P_{\boldsymbol{\theta}+\delta_n \mathbf{t}_n, n}\}$ and $\{P_{\boldsymbol{\theta},n}\}$ are contiguous.

Denote the log-likelihood ratio by

$$\Lambda_n(\boldsymbol{\theta} + \delta_n \mathbf{t}_n, \boldsymbol{\theta}) = \log\{dP_{\boldsymbol{\theta}+\delta_n \mathbf{t}_n, n}/dP_{\boldsymbol{\theta},n}\}.$$

Then the family $\varepsilon_n = \{P_{\boldsymbol{\eta},n},\ \boldsymbol{\eta} \in \boldsymbol{\Theta}\}$ is called *locally asymptotically quadratic* (LAQ) at $\boldsymbol{\theta}$ if there exist a stochastic vector $\mathbf{S}_n$ and a stochastic matrix $\mathbf{K}_n$, such that

$$\Lambda_n(\boldsymbol{\theta} + \delta_n \mathbf{t}_n, \boldsymbol{\theta}) - \mathbf{t}_n^\top \mathbf{S}_n + \frac{1}{2}\mathbf{t}_n^\top \mathbf{K}_n \mathbf{t}_n \to 0,$$

in $P_{\boldsymbol{\theta},n}$-probability, for every $\mathbf{t}_n \in B$, where $\mathbf{K}_n$ is a.s. p.d. The LAN *(locally asymptotically normal)* conditions refer to the particular case of LAQ where $\mathbf{K}_n$ can be taken as nonstochastic, while *locally asymptotically mixed normal* (LAMN) differ from the LAQ conditions in that the limiting distribution of the matrices $\mathbf{K}_n$, if it exists, does not depend on $\mathbf{T}_n$ $(\in B)$. Note that a density $f(x, \boldsymbol{\eta})$ is DQM *(differentiable in quadratic mean)* at $\boldsymbol{\theta}$ if there exist vectors $\mathbf{V}(x)$ such that

$$\int \{|\boldsymbol{\theta}\boldsymbol{\eta}|^{-1}|\sqrt{f(x,\boldsymbol{\eta})} - \sqrt{f(x,\boldsymbol{\theta})} - (\boldsymbol{\eta} - \boldsymbol{\theta})^\top \mathbf{V}(x)|\}^2 d\mu(x) \to 0 \text{ as } \boldsymbol{\eta} \to \boldsymbol{\theta}.$$

Therefore, DQM $\Rightarrow$ LAQ.

THEOREM 2.4 ***[Convolution Theorem].*** *For the model $\mathcal{F}_{\boldsymbol{\theta},n}\{P_{\boldsymbol{\theta}+\boldsymbol{\tau}\delta_n,n};\ \boldsymbol{\tau} \in \boldsymbol{R}_k\}$, satisfying the LAMN condition, let $\mathcal{L}(\mathbf{K}_n|P_{\boldsymbol{\theta},n}) \to \mathcal{L}(\mathbf{K})$ with nonrandom $\mathbf{K}$. If $\mathbf{T}_n$ is an estimator of $\mathbf{A}\boldsymbol{\tau}$ for a given nonrandom matrix $\mathbf{A}$ such that $\mathcal{L}(\mathbf{T}_n - \mathbf{A}\boldsymbol{\tau}|\boldsymbol{\theta} + \boldsymbol{\tau}\delta_n)$ tends to a limit H independent of $\boldsymbol{\tau}$, then $H(.)$ is the distribution of $\mathbf{A}\mathbf{K}^{-1/2}\mathbf{Z} + \mathbf{U}$, where $\mathbf{Z}$ and $\mathbf{U}$ are independent, and $\mathbf{Z} \sim \mathcal{N}(\mathbf{0}, \mathbf{I})$. If $\widehat{\boldsymbol{\xi}}_n$ is a BAN estimator of $\mathbf{A}\boldsymbol{\tau}$, $\widehat{\boldsymbol{\xi}}_n \xrightarrow{\mathcal{D}} \mathbf{A}\mathbf{K}^{-1/2}\mathbf{Z}$, we may write*

$$\mathbf{T}_n \stackrel{\mathcal{D}}{=} \widehat{\boldsymbol{\xi}}_n + \mathbf{U} \quad \text{as } n \to \infty, \tag{2.97}$$

where $\widehat{\boldsymbol{\xi}}_n$ and $\mathbf{U}$ are asymptotically independent.

2.5.10 Weak Convergence of Probability Measures

During the past 40 years, weak convergence of probability measures has been studied under increasing generality for function spaces, (including the $C[0,1]$ and $D[0,1]$ spaces). An up-to-date treatment of this area is beyond the scope of the present book. We shall rather concentrate only on the parts which have a relevance to the theory outlined in subsequent chapters. More can be found in the book by Dudley (1985).

Let S be a *metric space* and let $\mathcal{S}$ be the class of Borel sets of S. Let $\{(\Omega_n, \mathcal{B}_n, \mu_n);\ n \geq n_0\}$ be a sequence of probability spaces, let T_n be a measurable mapping: $(\Omega_n, \mathcal{B}_n) \mapsto (S, \mathcal{S})$, and let P_n be the probability measure induced by T_n in $(S, \mathcal{S})$ for $n \geq n_0$. Further let $(\Omega, \mathcal{B}, \mu)$be a probability space, $T:\ (\Omega, \mathcal{B}) \mapsto (S, \mathcal{S})$, be a measurable map, and let P be the probability measure induced by T in $(S, \mathcal{S})$. The space spanned by Ω_n may not necessarily contain Ω.

DEFINITION 2.7 *The sequence $\{P_n,\ n \geq n_0\}$ of probability measures on $(S, \mathcal{S})$ is said to converge weakly to P (denoted by $P_n \Rightarrow P$), if*

$$\int g dP_n \to \int g dP, \quad \forall g \in C(S) \tag{2.98}$$

where $C(S)$ is the class of bounded, continuous, real functions on S.
The convergence (2.98) implies that

$$h(T_n) \xrightarrow{\mathcal{D}} h(T) \quad \textit{when} \quad P_n \Rightarrow P \textit{ as } n \to \infty$$

for every continuous functional h assuming values in $\boldsymbol{R}_k$ and for $k \geq 1$.

The assumed continuity of h may be replaced by assuming that the set of discontinuities of h has P-measure 0. There is a basic difference between Definitions 2.5 and 2.7. The first one entails the *convergence of finite-dimensional* laws, while the latter additionally entails the *tightness* or *relative compactness* of $\{P_n\}$.

DEFINITION 2.8 *[LeCam-Prokhorov]. A family Π of probability measures on $(S, \mathcal{S})$ is said to be tight if for every $\varepsilon > 0$, there exists a compact set K_ε such that*

$$P(K_\varepsilon) > 1 - \varepsilon, \quad \forall P \in \Pi.$$

The subsequent chapters will deal with various stochastic processes, some of which are multiparameter. One would mostly have a functional space that can be reduced to the $D[0,1]^p$ space – a multi-parameter extension of the $D[0,1]$ space, by a proper transformation. In the same way, the $C[0,1]$ space lends itself to the $C[0,1]^p$ space, where $C[0,1]^p$ is a subspace of $D[0,1]^p$. With the $C[0,1]^p$ space we associate the *uniform topology* specified by the metric

$$\rho(x, y) = \sup\{|x(\mathbf{t}) - y(\mathbf{t})| : \mathbf{t} \in [0,1]^p\}.$$

Under the metric ρ, $C[0,1]^p$ is a complete and separable metric space. However, the space $D[0,1]^p$ is the space of all real-valued functions $f:\ [0,1]^p \mapsto \boldsymbol{R}$, where f has only discontinuities of the first kind. For this we extend the metric ρ_S in (2.50) as follows: Let Λ denote the class of all strictly increasing, continuous mappings $\lambda : [0,1] \mapsto [0,1]$. Moreover, let $\boldsymbol{\lambda}(\mathbf{t}) = (\lambda_1(t_1), \ldots, \lambda_p(t_p))$, $\boldsymbol{\lambda} \in \Lambda^p$ and $\mathbf{t} \in [0,1]^p$. Then the Skorokhod distance between two elements x and y in $D[0,1]^p$ is defined by

$$\rho_S(x, y) = \inf\Big\{\varepsilon > 0 : \exists \boldsymbol{\lambda} \in \Lambda^p : \text{ s.t. } \sup_{\mathbf{t} \in [0,1]^p} \|\boldsymbol{\lambda}(\mathbf{t}) - \mathbf{t}\| < \varepsilon$$

$$\text{and} \sup_{\mathbf{t}\in[0,1]^p} |x(\mathbf{t}) - y(\boldsymbol{\lambda}(\mathbf{t}))| < \varepsilon \Big\}.$$

Consider a partition of $[0,1]^p$ formed by finitely many hyperplanes parallel to the p principal axes such that each element of this partition is a left-closed right-open rectangle of diameter at least δ (> 0). A typical rectangle is denoted by R, and the partition generated by the hyperplanes is denoted by $\mathcal{R}$. Let

$$w_f^0(R) = \sup\{|f(\mathbf{t}) - f(\mathbf{s})| : \mathbf{s}, \mathbf{t} \in R\}$$

and

$$w'_f(\delta) = \inf_{\mathcal{R}}\{\max_{R\in\mathcal{R}}[w_f^0(R)]\}.$$

Then a function $f : [0,1]^p \mapsto \boldsymbol{R}$, belongs to $D[0,1]^p$, iff,

$$\lim_{\delta\to 0} w'_f(\delta) = 0.$$

THEOREM 2.5 ***[Compactness Theorem].*** *A sequence $\{P_n\}$ of probability measures on $(D[0,1]^p, \mathcal{D}^p)$ is tight (relatively compact) iff*

(a) for every $\varepsilon > 0$, there exists a $M_\varepsilon < \infty$ such that

$$P_n(\{f \in D[0,1]^p : \|f\| > M_\varepsilon\}) \leq \varepsilon, \ \forall n \tag{2.99}$$

and

(b)

$$\lim_{\delta\downarrow 0} \limsup_{n\to\infty} P_n(\{f \in D[0,1]^p : w'_f(\delta) \geq \varepsilon\}) = 0. \tag{2.100}$$

for every $\varepsilon > 0$.

In practice, we may replace $w'_f(\delta)$ in (2.100) by $w_f(2\delta)$ where

$$w_f(\delta) = \sup\{|f(\mathbf{t}) - f(\mathbf{s})| : \mathbf{s}, \mathbf{t} \in [0,1]^p, \ \|\mathbf{s} - \mathbf{t}\| \leq \delta\},$$

because

$$w'_f(\delta) \leq w_f(2\delta), \ \forall\, 0 < \delta \leq \frac{1}{2} \text{ and } f \in D[0,1]^p.$$

Recall that (2.99) shows that f is uniformly bounded in probability on $D[0,1]^p$; this can be verified using some „maximal inequality." On the other hand, to verify (2.100) is generally more delicate; this is usually accomplished by using certain tricky inequalities known as *Billingsley-type inequalities* after Billingsley (1968) who gives an elaborate treatment for $p = 1$. Bickel and Wichura (1971) derived the following multiparameter extension.

THEOREM 2.6 ***[Billingsley Inequality].*** *Compactness theorem holds if the inequality*

$$P_n(\{m(B(\mathbf{s},\mathbf{t}], B(\mathbf{s}^\top, \mathbf{t}^\top]) \geq \lambda\}) \leq \lambda^{-\gamma}\{\mu(B(\mathbf{s},\mathbf{t}] \cup B(\mathbf{s}^\top, \mathbf{t}^\top])\}^{1+\beta} \tag{2.101}$$

is valid for every pair of neighboring blocks in $[0,1]^p$, $\lambda > 0$, and n for some $\gamma > 0$, $\beta > 0$, where $\mathbf{s}$, $\mathbf{t}\ \mathbf{s}'$, $\mathbf{t}' \in [0,1]^p$,

$$m(B(\mathbf{s},\mathbf{t}], B(\mathbf{s}',\mathbf{t}']) = \min\{|f(B(\mathbf{s},\mathbf{t}])|, |f(B(\mathbf{s}',\mathbf{t}'])|\},$$

$f(B(\mathbf{s},\mathbf{t}])$ is the increment of f around the block $B(\mathbf{s},\mathbf{t}]$, and μ is a measure on the Borel sets of $[0,1]^p$.

The following condition for (2.101) is more stringent but more easily verifiable:

$$\begin{aligned} & \mathbb{E}(|f_n(B(\mathbf{s},\mathbf{t}])|^{\alpha_1}|f_n(B(\mathbf{s}^\top,\mathbf{t}^\top])|^{\alpha_2} \\ \leq \;& M\{\mu(B(\mathbf{s},\mathbf{t}])\mu(B(\mathbf{s}^\top,\mathbf{t}^\top])\}^{(1+\beta)/2}, \end{aligned} \quad (2.102)$$

where M, α_1, α_2, β are positive constants and $\mu(.)$ is a sigma-finite measure on $[0,1]^p$.

For the convergence of *finite-dimensional distributions* of $\{P_n\}$ to P, we can usually use some simpler techniques based on the Cramér–Wold theorem and central limit theorems, but the compactness part of the weak convergence in (2.93) is generally difficult to establish. This can be simplified by using the following generalization of the Slutzky theorem:

THEOREM 2.7 *If $\{X_n\}$ and $\{Y_n\}$ both belong to a common separable space S, equipped with a topology $\rho_S(.)$, then*

$$X_n \xrightarrow{\mathcal{D}} X \;\; \textit{and} \;\; \rho(X_n, Y_n) \xrightarrow{\mathcal{P}} 0 \Rightarrow Y_n \xrightarrow{\mathcal{D}} X.$$

Verification of $\rho(X_n, Y_n) \xrightarrow{\mathcal{P}} 0$ can proceed by incorporating the following lemma:

LEMMA 2.3 ***[Convexity Lemma].*** *Let $\{Y_n(\mathbf{t}),\ \mathbf{t}\in\mathbf{T}\}$ be a sequence of random convex functions defined on a convex, open subset $\mathbf{T}\subset\boldsymbol{R}_p$, for some $p\geq 1$. Suppose that $\{\xi(\mathbf{t}),\ \mathbf{t}\in\mathbf{T}\}$ is a real-valued function on $\mathbf{T}$, such that*

$$Y_n(\mathbf{t}) \to \xi(\mathbf{t}) \;\; \textit{in probability/a.s.,} \;\; \textit{for each } \mathbf{t}\in\mathbf{T}. \quad (2.103)$$

Then for each compact set $\mathcal{C}$ of $\mathbf{T}$,

$$\sup\{|Y_n(\mathbf{t})-\xi(\mathbf{t})| : \mathbf{t}\in\mathcal{C}\} \to 0, \;\; \textit{in probability/a.s.,} \quad (2.104)$$

and $\xi(\mathbf{t})$ is necessarily convex on $\mathbf{T}$.

For nonstochastic Y_n, this is essentially Theorem 10.8 of Rockafellar (1970), and its adaptation to stochastic analysis is due to Anderson and Gill (1982), Heiler and Willers (1988), and Pollard (1991), among others. If $Y_n(\mathbf{t})$ can be written as $Y_n^{(1)}(\mathbf{t}) \pm Y_n^{(2)}(\mathbf{t})$ (or a finite mixture, not necessarily convex, of several $Y_n^{(j)}(\mathbf{t})$), where the individual $Y_n^{(j)}(\mathbf{t})$ satisfy the hypothesis of the Lemma 2.3, then (2.104) also holds for Y_n. Two important results follow as corollaries to the above lemma.

1. If $\xi(\mathbf{t})$, $\mathbf{t}\in\mathcal{C}$ attains an extrema, say, ξ_0, at an interior point $\mathbf{t}_0$ of $\mathcal{C}$, then $Y_n(\mathbf{t})$ has also an extrema that converges to ξ_0 in the same mode as (2.104).
2. If $\xi = 0$, then $Y_n(\mathbf{t}) \to 0$ uniformly in $\mathcal{C}$, in the same mode as in (2.104).

2.5.11 Some Important Gaussian Functions

DEFINITION 2.9 ***[Wiener measure].*** *Let $\boldsymbol{R}_p^+$ be the nonnegative orthant of $\boldsymbol{R}_p$, $p \geq 1$, and consider the probability space $(\Omega, \mathcal{C}^p, W)$, where $\mathcal{C}^p$ is the sigma-field generated by the open subsets of $C[\boldsymbol{R}_p^+]$ and W is the probability measure induced in $C[\boldsymbol{R}_p^+]$ by a real-valued stochastic process $X(.) = X(\mathbf{t}),\ \mathbf{t} \in \boldsymbol{R}_p^+\}$. Then W is defined to be a Wiener measure on $(C[\boldsymbol{R}_p^+], \mathcal{C}^p)$, if the following hold:*

1.
$$W\{[X(\mathbf{t}),\ \mathbf{t} \in \boldsymbol{R}_p^+] \in C[\boldsymbol{R}_p^+]\} = 1.$$

2. *Let B and C be any two rectangles in $\boldsymbol{R}_p^+$, such that $B \cap C = 0$ (i.e., they are disjoint, though they might be adjacent). Let $X(B)$ and $X(C)$ be the increments of the process $X(.)$ over the rectangles B and C, respectively. Then $X(B)$ and $X(C)$ are stochastically independent.*

3. *For every $B \subset \boldsymbol{R}_p^+$ and real x*
$$W\{X(B) \leq x\} = \frac{1}{\sqrt{2\pi\sigma^2\lambda(B)}} \int_{-\infty}^{x} \exp\Big\{ - u^2/2\sigma^2\lambda(B)\Big\} du \qquad (2.105)$$
where $0 < \sigma^2 < \infty$ and $\lambda(B)$ is the Lebesgue measure of B. In particular, if $\sigma = 1$, W is termed a standard Wiener measure on $C[\boldsymbol{R}_p^+]$.

Note that part 1 implies that the sample paths of $X(.)$ are continuous a.e., part 2 implies that $X(.)$ has independent increments, while part 3 specifies the Gaussian nature. The increment of $X(.)$ over a block $B = B(\mathbf{s}, \mathbf{t}]$ is defined as

$$X(B) = \sum_{j=1}^{p} \sum_{l_j=0,1} (-1)^{p-\Sigma_{j=1}^{p} l_j} X(l_j t_j + (1 - l_j)s_j,\ 1 \leq j \leq p). \qquad (2.106)$$

The above definition simplifies for $p = 1$ where $X(t)$ has sample paths continuous a.e. on $\boldsymbol{R}^+$. Moreover, the increments $X(0) = 0,\ X(t_1),\ X(t_2) - X(t_1),\ X(t_3) - X(t_2), \ldots$ for any partition $(0, t_1),\ [t_1, t_2),\ [t_2, t_3), \ldots$ of $\boldsymbol{R}^+$ are all independently normally distributed with 0 means and variances $\sigma^2(t_j - t_{j-1}),\ j \geq 1$, where $t_0 = 0$. We say that $X(.)$ is a *Brownian motion* or *Wiener process* (standard if $\sigma = 1$) on $\boldsymbol{R}^+$. By construction, $\{X(t),\ t \in [0, T]\}$ remains a Wiener process for every $T:\ 0 < T < \infty$. Moreover, consider a related process $X^0(.)$ defined by

$$X^0(t) = X(t) - tX(1),\ \ 0 \leq t \leq 1,$$

so that $X^0(.)$ is also Gaussian and has continuous (a.e.) sample paths. Furthermore, $\mathbb{E}X^0(t) = 0\ \forall t \in [0, 1]$, and

$$\mathbb{E}[X^0(s)X^0(t)] = \sigma^2(s \wedge t - st)$$

for every $s, t \in [0, 1]$, so that the process no longer has independent increments, but $X^0(0) = X^0(1) = 0$, with probability 1. In other words, $X^0(.)$ is tied-down at the two ends $\{0,1\}$. X^0 is termed a *Brownian bridge* or a *tied down Wiener process* (standard when $\sigma^2 = 1$). For general p (≥ 1), we define $X^0(.) = \{X^0(\mathbf{t}),\ \mathbf{t} \in [0,1]^p\}$, by letting

$$X^0(\mathbf{t}) = X(\mathbf{t}) - \|\mathbf{t}\| X(\mathbf{1}),\ \mathbf{t} \in [0,1]^p.$$

It is easy to verify that $\mathbb{E}X^0(\mathbf{t}) = 0\ \forall \mathbf{t} \in [0,1]^p$, and that

$$\mathbb{E}[X^0(B)X^0(C)] = \sigma^2[\lambda(B \cap C) - \lambda(B)\lambda(C)],$$

for every $B, C \subset [0,1]^p$, where the increments are defined as in (2.106). $X^0(.)$ is Gaussian and is tied-down (to 0, w.p. 1) at each edge of $[0,1]^p$ (standard one when $\sigma = 1$).

Consider especially the case of $p = 2$. Let $X^\star(.) = \{X^\star(t_1,t_2),\ 0 \leq t_1 \leq \infty,\ 0 \leq t_2 \leq 1\}$ be defined by

$$X^\star(t_1,t_2) = X(t_1,t_2) - t_2 X(t_1,1),\ \ \mathbf{t} \in \mathbf{T} \tag{2.107}$$

where $\mathbf{T} = \boldsymbol{R}_1^+ \times [0,1]$, and $\{X(\mathbf{t}),\ \mathbf{t} \in \boldsymbol{R}_2^+\}$ is a standard Wiener process on $\boldsymbol{R}_2^+$ (also termed a *Brownian sheet*). Then $\mathbb{E}X^\star(\mathbf{t}) = 0$, $\mathbf{t} \in \mathbf{T}$, and

$$\mathbb{E}[X^\star(s)X^\star(t)] = (s_1 \wedge t_1)(s_2 \wedge t_2 - s_2 t_2) \tag{2.108}$$

for every $\mathbf{s}, \mathbf{t} \in \mathbf{T}$. $X^\star(.)$ is termed a *Kiefer process.* This Gaussian function is also tied down at $t_1 = 0$ and $t_2 = 0, 1$. The definition extends directly to $\mathbf{T} = \boldsymbol{R}_{p_1}^+ \times [0,1]^{p_2}$, $p_1 \geq 1$, $p_2 \geq 1$, $p_1 + p_2 = p \geq 2$.

Consider next the case where $\mathbf{X}(.)$ is itself a vector (stochastic) process. We write

$$\mathbf{X}(\mathbf{t}) = [X_1(\mathbf{t}), \ldots, X_k(\mathbf{t})]',\ \ \mathbf{t} \in \mathbf{T},$$

where $k \geq 1$ and $\mathbf{T} \subset \boldsymbol{R}_p^+$ for some $p \geq 1$. Suppose that $\{X_j(\mathbf{t}),\ \mathbf{t} \in \boldsymbol{R}_p^+\}$ are independent copies of a Wiener process. Then a multivariate Gaussian process $\mathbf{M}(.) = \{\mathbf{M}(\mathbf{t}),\ \mathbf{t} \in \mathbf{T}\}$ can be defined by introducing a nonstochastic matrix-valued $\mathbf{A}(.) = \{\mathbf{A}(\mathbf{t}), \mathbf{t} \in \mathbf{T}\}$ and letting

$$\mathbf{M}(\mathbf{t}) = \mathbf{A}(\mathbf{t})\mathbf{X}(\mathbf{t}),\ \ \mathbf{t} \in \mathbf{T}.$$

Because of this representation, distributional properties of $\mathbf{X}(.)$ can be studied parallelly with those of $\mathbf{M}(.)$. We shall confine ourselves to a special case $p = 1$ and $T = \boldsymbol{R}^+$. Then we can define a *k-parameter Bessel process* $B = \{B(t), t \in \boldsymbol{R}^+\}$ by letting

$$B^2(t) = [\mathbf{X}(t)]^\top [\mathbf{X}(t)] = \|\mathbf{X}(t)\|^2,\ \ t \in \boldsymbol{R}^+. \tag{2.109}$$

We may also define $B^2(t)$ in the following way

$$B^2(t) = [\mathbf{M}(t)]^\top [\mathbf{Q}(t)]^- [\mathbf{M}(t)],\ \ t \in \boldsymbol{R}^+,$$

where

$$\mathbf{Q}(t) = [\mathbf{A}(t)][\mathbf{A}(t)]^\top,\ \ t \in \boldsymbol{R}^+,$$

where $[\ ]^-$ stands for a *generalized inverse.* Thus, (2.109) is a canonical representation for Bessel processes.

2.5.12 Weak Invariance Principles

Functional central limit theorems relate to the extensions of the basic results in Section 2.5.5 along the lines of the main themes in Section 2.5.10 and 2.5.11. Such functional CLT's are also known as *weak invariance principles.* Since the seminal work on this subject in classic texts of Parthasarathy (1967) and Billingsley (1968), there has been a steady flow of research. We only present some important results having direct relevance to subsequent chapters.

THEOREM 2.8 ***[Dvoretzky–McLeish theorem].*** *Let $\{Z_{n,k},\ 0 \le k \le k_n;\ n \ge 1\}$ be a triangular array of random variables (not necessarily independent). We set $Z_{n,0} = 0$ w.p.1 $\forall n \ge 1$. Let*

$$S_{n,k} = Z_{n,0} + \ldots, Z_{n,k},\quad k \le k_n,\ n \ge 1,$$

and let $\mathcal{F}_{n,k}$ be the sigma-field generated by $S_{n,k}$, for $k \ge 1$ and $n \ge 1$. Assume that $\mathbb{E}Z_{n,k}^2 < \infty$ for every k, n. Let

$$\mu_{n,k} = \mathbb{E}(Z_{n,k}|\mathcal{F}_{n,k-1}) \quad and \quad \sigma_{n,k}^2 = \mathrm{Var}(Z_{n,k}|\mathcal{F}_{n,k-1}),$$

for $k \ge 1,\ n \ge 1$, and assume that $k_n \to \infty$ as $n \to \infty$. Define a stochastic process $W_n = \{W_n(t),\ t \in [0,1]\}$ by letting

$$W_n(t) = S_{n,k_n(t)},\ t \in [0,1], \tag{2.110}$$

where $k_n(t)$ is an integer-valued, nondecreasing, and right-continuous function of t on [0,1], with $k_n(0) = 0,\ n \ge 1$. Suppose that the (conditional) Lindeberg condition holds: that is,

$$\sum_{k \le k_n} \mathbb{E}\{Z_{n,k}^2 I(|Z_{n,k}| > \varepsilon)|\mathcal{F}_{n,k-1}\} \xrightarrow{P} 0, \tag{2.111}$$

for every $\varepsilon > 0$, as $n \to \infty$. Further assume that

$$\sum_{k \le k_n(t)} \mu_{n,k} \xrightarrow{P} 0 \ and \ \sum_{k \le k_n(t)} \sigma_{n,k}^2 \xrightarrow{P} t$$

for each $t \in [0,1]$, as $n \to \infty$. Then

$$W_n \xrightarrow{\mathcal{D}} W \text{ in the Skorohod topology on } D[0,1], \tag{2.112}$$

where W is a standard Wiener process on [0,1].

Scott (1973) studied (2.112) when the $S_{n,k}$ form a martingale array, and formulated alternative (but equivalent) regularity conditions pertaining to this weak invariance principle. Earlier Brown (1971) considered a martingale sequence $\{S_k,\ k \ge 0\}$ and established the same result under

(i) (2.111) and under

(ii)

$$\Big(\sum_{k \le n} \sigma_k^2\Big) / \mathbb{E}\Big(\sum_{k \le n} \sigma_k^2\Big) \xrightarrow{P} 1, \text{ as } n \to \infty.$$

An important inequality developed in this context is the one listed below.

THEOREM 2.9 ***[Brown submartingale inequality].*** *Let $\{X_n, \mathcal{F}_n;\ n \geq 1\}$ be a submartingale. Then for every $\varepsilon > 0$, $n \geq 1$,*

$$\begin{aligned} \mathbb{P}\{\max_{1\leq k\leq n} |X_k| > 2\varepsilon\} &\leq \mathbb{P}\{|X_n| > \varepsilon\} + \mathbb{E}\{(\frac{1}{\varepsilon}|X_n| - 2)I(|X_n| \geq 2\varepsilon)\} \\ &\leq \varepsilon^{-1}\mathbb{E}\{|X_n|I(|X_n| \geq \varepsilon)\}. \end{aligned}$$

This inequality leads to the following theorem:

THEOREM 2.10 ***[Tightness theorem].*** *Let W_n be defined as in (2.110) for a martingale sequence $\{S_k,\ k \geq 1\}$. Then the convergence of finite dimensional distributions to those of a Gaussian process W ensures the tightness of W_n (and hence the weak convergence to W).*

For a reversed martingale sequence $\{X_n, \mathcal{F}_n;\ n \geq 1\}$, define $Z_{n,i} = X_{n+i-1} - X_{n+i}$, $i \geq 1$, and without any loss of generality we set $\mathbb{E}X_n = 0$. Let

$$V_n = \sum_{i\geq 1} \mathbb{E}(Z_{n,i}^2|\mathcal{F}_{n+1}) \text{ and } s_n^2 = \mathbb{E}V_n = \mathbb{E}(X_n^2).$$

Then we have the following

THEOREM 2.11 ***[Loynes functional central limit theorem].*** *Let $n(t) = \min\{k :\ s_k^2/s_n^2 \leq t\}$, and $W_n(t) = s_n^{-1}X_{n(t)}$, $t \in [0,1]$. Then, if $V_n/s_n^2 \xrightarrow{P} 1$ and the $Z_{n,i}$ satisfy the (conditional) Lindeberg condition, W_n weakly converges to a standard Wiener process.*

The above-mentioned results can be still strengthened. Let $q(.) = \{q(t),\ 0 < t < 1\}$ be a continuous, nonnegative function on $[0,1]$ such that for some $a > 0$, $q(t)$ is $\nearrow$ in $t \in (0,a]$, and $q(t)$ is bounded away from 0 on $(a,1]$ and $\int_0^1 q^{-2}(t)dt < \infty$. Let

$$d_q(x,y) = \sup\{q^{-1}(t)|x(t) - y(t)| :\ t \in [0,1]\}. \tag{2.113}$$

Then the weak convergence results in the preceding three theorems hold even when the J_1-topology is replaced by the $d_q(.)$-metric (Sen 1981a, Theorem 2.4.8).
Such functional CLT's also hold for another important class of empirical processes, which is considered next.

2.5.13 Empirical Distributional Processes

Let $\{\mathbf{X}_i,\ i \geq 1\}$ be a sequence of i.i.d. random variables with a distribution function F defined on $\boldsymbol{R}_p$ for some $p \geq 1$. For every $n \geq 1$, the *empirical distribution function* F_n is defined by

$$F_n(\mathbf{x}) = n^{-1} \sum_{i \leq n} I(\mathbf{X}_i \leq \mathbf{x}),\ \mathbf{x} \in \boldsymbol{R}_p.$$

Note that $\mathbb{E}F_n(\mathbf{x}) = F(\mathbf{x}),\ \mathbf{x} \in \boldsymbol{R}_p$, and that

$$Cov[F_n(\mathbf{x}), F_n(\mathbf{y})] = n^{-1}\{F(\mathbf{x} \wedge \mathbf{y}) - F(\mathbf{x})F(\mathbf{y})\}$$

where $\mathbf{x} \wedge \mathbf{y} = (x_1 \wedge y_1, \ldots, x_p \wedge y_p)^\top$. The usual *empirical process* $V_n = n^{1/2}(F_n - F)$ is defined by

$$V_n(\mathbf{x}) = n^{1/2}(F_n(\mathbf{x}) - F(\mathbf{x})),\ \mathbf{x} \in \boldsymbol{R}_p. \tag{2.114}$$

If $F_{[j]}$ is the jth marginal distribution function corresponding to the distribution function F, and we assume that $F_{[j]}$ is continuous a.e. for $1 \leq j \leq p$, then we may consider the transformation $\mathbf{X} \to \mathbf{Y} = (Y^{(1)}, \ldots, Y^{(p)})^\top$ where $Y^{(j)} = F_{[j]}(X^{(j)}),\ 1 \leq j \leq p$. The $Y^{(j)}$ will have marginally uniform (0,1) distribution function. Denote their joint distribution function by $G(\mathbf{y}) = \mathbb{P}\{\mathbf{Y} \leq \mathbf{y}\},\ \mathbf{y} \in [0,1]^p$. The reduced empirical distribution function G_n is then defined by

$$G_n(\mathbf{y}) = n^{-1} \sum_{i \leq n} I(\mathbf{Y}_i \leq \mathbf{y}),\ \mathbf{y} \in [0,1]^p. \tag{2.115}$$

Consider a p-parameter *empirical process* $U_n = \{U_n(\mathbf{t}),\ \mathbf{t} \in [0,1]^p\}$ of the form

$$U_n(\mathbf{t}) = n^{1/2}[G_n(\mathbf{t}) - G(\mathbf{t})],\ \mathbf{t} \in [0,1]^p. \tag{2.116}$$

By (2.114) and (2.116),

$$U_n(\mathbf{t}) = V\Big(((F_{[1]}(x^{(1)}), \ldots, F_{[p]}(x^{(p)}))\Big)$$

whenever $\mathbf{t} = ((F_{[1]}(x^{(1)}), \ldots, F_{[p]}(x^{(p)}))$. Clearly it suffices to study U_n alone. By (2.115), a linear combination of $U_n(\mathbf{t}_1), \ldots, U_n(\mathbf{t}_m)$ is expressible as an average of i.i.d. random variables for finitely many $\mathbf{t}$'s, say, $\mathbf{t}_1, \ldots, \mathbf{t}_m$, all belonging to $[0,1]^p$. Hence, the classical CLT applies and we conclude that

$$[U_n(\mathbf{t}_1), \ldots, U_n(\mathbf{t}_m)] \xrightarrow{\mathcal{D}} [U(\mathbf{t}_1), \ldots, U(\mathbf{t}_m)],$$

where $U(.)$ is Gaussian with null mean and

$$\mathbb{E}[U(\mathbf{s})U(\mathbf{t})] = G(\mathbf{s} \wedge \mathbf{t}) - G(\mathbf{s})G(\mathbf{t}), \tag{2.117}$$

for every $\mathbf{s}, \mathbf{t} \in [0,1]^p$. Moreover, since the summands in (2.116) are i.i.d. zero-one-valued random variables, Billingsley-type inequalities are easy to verify. This leads to the following two theorems:

THEOREM 2.12 ***[Weak convergence theorem for empirical process].*** *If F is continuous, then U_n converges in law on $D[0,1]^p$ to U.*

In particular, for $p = 1$, U reduces to a standard Brownian bridge on $[0,1]$. Also, if $F(\mathbf{x}) \equiv \prod_{j=1}^p F_{[j]}(x^{(j)})$, then $G(\mathbf{t}) = |\mathbf{t}| = t_1 \cdot \ldots \cdot t_p$ so that U reduces to a standard tied-down Brownian sheet.

Consider another $(p+1)$-parameter stochastic process $U_n^\star(s, \mathbf{t})$, $s \in \boldsymbol{R}^+$, $\mathbf{t} \in [0,1]^p$ of the form

$$U_n^\star(s, \mathbf{t}) = n^{-1/2}(n_s U_{n_s}(\mathbf{t})),\ n_s = [ns],\ \mathbf{t} \in [0,1]^p, \tag{2.118}$$

where $U_m(.)$ is defined as in (2.116) for $m \geq 1$. Consider a *Kiefer-process* $U^\star = \{U^\star(s, \mathbf{t}),\ s \in \boldsymbol{R}^+,\ \mathbf{t} \in [0,1]^p\}$ as in (2.108). Then

$$U_n^\star \xrightarrow{\mathcal{D}} U^\star, \quad \text{in the } J_1\text{-topology on } D[0,\infty) \times [0,1]^p. \tag{2.119}$$

We may also define $U_n^0 = \{U_n^0(s, \mathbf{t}) = n^{1/2} U_{n_s}(\mathbf{t}),\ s \in [0,1],\ \mathbf{t} \in [0,1]^p\}$ by letting $n_s = \min\{k : n/k \leq s\}$. Then, as in Neuhaus and Sen (1977), we have

$$U_n^0 \xrightarrow{\mathcal{D}} U^\star, \text{ in the } J_1\text{-topology on } D[0,1]^{p+1}.$$

The special case of $p = 1$ leads $U^\star$ to a Kiefer process as defined in (2.107)–(2.108). The weak convergence results for U_n and $U_n^\star$ or U_n^0 in the case of $p = 1$ can be strengthened to suitable $d_q(.)$ metrics: We have the following theorem:

THEOREM 2.13 ***[Weak convergence theorem 2].*** *Let $q_1 = \{q_1(t),\ t \in \boldsymbol{R}^+\}$ satisfy the conditions stated in (2.113). Let $q_2 = \{q_2(t),\ t \in [0,1]\}$ be a continuous, nonnegative function on [0,1], bounded away from 0 on $[\gamma, 1-\gamma]$ for some $0 < \gamma \leq 1/2$, and nondecreasing (nonincreasing) on $[0, \gamma)$ $((1-\gamma, 1])$. Consider the metric*

$$\rho_q(x, y) = \sup\{|x(s,t) - y(s,t)|/q_1(s)q_2(t) : (s,t) \in [0,1]^2\}.$$

If q_1 and q_2 satisfy the conditions

$$\int_0^1 q_1^{-2}(t)dt < \infty$$

$$\int_0^{1/2} t^{-1} \exp\{-\varepsilon t^{-1}(q_2^0(t))^2\}dt < \infty$$

$$\text{for } q_2^0(t) = q_2(t) \wedge q_2(1-t),\ 0 \leq t \leq 1$$

then $U_n \xrightarrow{\mathcal{D}} U$ in ρ_q-metric on $D[0,1]^2$.

A more easily verifiable condition (Wellner 1974) is that $\int_0^1 \int_0^1 [q(t,s)]^{-2} ds dt < \infty$, where $q(t,s) = q_1(t)q_2(s),\ (t,s) \in [0,1]^2$.
A direct consequence of the above two theorems is the following:

Corollary 2.1 *Let $p = 1$; then for every $\lambda > 0$,*

$$\begin{aligned} \mathbb{P}\{\sup_{x \in \boldsymbol{R}} |V_n(x)| \geq \lambda\} &= \mathbb{P}\{\sup_{0 \leq t \leq 1} |U_n(t)| \geq \lambda\} \\ \to \mathbb{P}\{\sup_{0 \leq t \leq 1} |U(t)| \geq \lambda\} &= 2\sum_{k \geq 1}(-1)^{k-1} \exp(-2k^2\lambda^2). \end{aligned} \tag{2.120}$$

The last expression follows from the classical boundary crossing probability results for Brownian motions and bridges due to Doob (1949). Similarly,

Corollary 2.2

$$\mathbb{P}\Big\{\sup_{x\in\boldsymbol{R}} V_n(x) \geq \lambda\} = \mathbb{P}\{U_n(t) \geq \lambda, \quad \textit{for some} \quad t \in [0,1]\Big\}$$
$$\rightarrow \mathbb{P}\{U(t) \geq \lambda, \quad \textit{for some } t \in [0,1]\} = \exp(-2\lambda^2), \qquad (2.121)$$

thus $\sup\{|V_n(x)| : x \in \boldsymbol{R}\} = \sup\{|U_n(t)| : t \in [0,1]\} = O_p(1)$.

Further, Dvoretzky, Kiefer, and Wolfowitz (1956) have shown that the left-hand sides in (2.120) or (2.121) are actually $\leq$ the right-hand side limits for every finite n. Although such a dominating bound is not precisely known for $p > 1$, Kiefer (1961) managed to show that for every $\varepsilon > 0$ ($\varepsilon < 2$), there exists a positive $c(\varepsilon) < \infty$ such that

$$\mathbb{P}\{\sup_{\mathbf{t}\in[0,1]^p} |U_n(\mathbf{t})| \geq \lambda\} \leq c(\varepsilon)\exp(-(2-\varepsilon)\lambda^2)$$

for every $n \geq 1$, $p \geq 1$, and $\lambda > 0$. The last inequality in turn yields

$$\sup\{|U_n(\mathbf{t})| : \mathbf{t} \in [0,1]^p\} = \mathrm{O}((\log n)^{1/2}) \text{ a.s.}$$

as $n \rightarrow \infty$. In case $p = 1$, more precise a.s. convergence rates are available; some of them are reported in a later subsection.

Closely related to the empirical distribution function F_n (or G_n) is the *sample quantile function* Q_n, $\{Q_n(t),\ 0 < t < 1\}$, where

$$Q_n(t) = F_n^{-1}(t) = \inf\{x : F_n(x) \geq t\}. \qquad (2.122)$$

For the uniform distribution function, induced by the mapping $X \rightarrow Y = F(X)$, we have the *uniform quantile function* $Q_n^0(t)$, $t \in (0,1)$, where

$$Q_n^0(t) = G_n^{-1}(t) = \inf\{y : G_n(y) \geq t\}. \qquad (2.123)$$

This intuitively leads to the *(reduced) sample quantile process* $U_n^Q(.)$:

$$U_n^Q(t) = n^{1/2}[Q_n^0(t) - t],\ 0 \leq t \leq 1,$$

and similarly we consider the quantile process corresponding to $Q_n(t)$ in (2.122). By definition (2.123), $Q_n^0(G_n(y)) \geq y$, $\forall y \in [0,1]$, and the equality sign holds at the n points $Y_{n:k}$, $1 \leq k \leq n$, which are the order statistics of the n i.i.d. random variables $Y_1, \ldots, Y_n$ from the uniform $R(0,1)$ distribution. By a simple argument, we obtain that

$$\sup\{|U_n(t) + U_n^Q(t)| : t \in (0,1)\} \rightarrow 0 \text{ a.s.},$$

as $n \rightarrow \infty$, so that the weak convergence principles studied for $U_n(.)$ remain applicable for $U_n^Q(.)$ as well [use (2.102 in this context].

2.5.14 Weak Invariance Principle: Random Change of Time

Using the notations of Section 2.5.10, consider the $D[0,1]$ space endowed with the Skorohod J_1-topology induced by the metric $\rho_S(.)$ in (2.50), where $\{\lambda : [0,1] \mapsto [0,1]$ onto itself. Let Λ_0 be an (extended) class of λ belonging to the $D[0,1]$ space such that λ_t is nondecreasing and $0 \le \lambda_t \le 1\ \forall t \in [0,1]$ (note that the strict monotonicity and continuity condition for Λ have been weakened for Λ_0). For $x \in D[0,1]$ and $y \in \Lambda_0$, let $x \circ y$ denote the composition: $(x \circ y)(t) = x(y(t))$, $t \in [0,1]$. Let X be a random element of $D[0,1]$ and Y a random element of Λ_0 so that (X, Y) is a random element of $D[0,1] \times \Lambda_0$. Consider a sequence $\{(X_n, Y_n);\ n \ge 1\}$ of random elements of $D[0,1] \times \Lambda_0$ such that

$$(X_n, Y_n) \xrightarrow{\mathcal{D}} (X, Y)$$

and $\mathbb{P}\{X \in C[0,1]\} = 1 = \mathbb{P}\{Y \in C[0,1]\}$. Then

$$X_n \circ Y_n \xrightarrow{\mathcal{D}} X \circ Y. \tag{2.124}$$

An important application of (2.124) relates to the partial sum processes of stochastic sizes or to the empirical distributional processes, treated in earlier subsections.

2.5.15 Embedding Theorems and Strong Invariance Principles

The basic idea of *embedding of Wiener processes* is due to Skorokhod (1956), and it was consolidated by Strassen (1964, 1967). Let $\{X_n, \mathcal{F}_n;\ n \ge 1\}$ be a martingale so that $\{Y_n = X_n - X_{n-1}, \mathcal{F}_n;\ n \ge 1\}$ form a martingale difference sequence. We take $X_0 = 0$ with probability 1 so that $\mathbb{E}X_n = 0\ \forall n \ge 1$. Also, $\mathbb{E}(Y_n|\mathcal{F}_{n-1}) = 0$ a.e. $\forall n \ge 1$. We assume that

$$v_n = \mathbb{E}(Y_n^2|\mathcal{F}_{n-1}) \ \text{ exists a.e. } \ \forall n \ge 1. \tag{2.125}$$

Denote

$$V_n = \sum_{k \le n} v_k, \quad n \ge 1, \tag{2.126}$$

and assume that $V_n \xrightarrow{\text{a.s.}} \infty$, as $n \to \infty$.

THEOREM 2.14 ***[Skorohod–Strassen theorem].*** *For a martingale $\{X_n, \mathcal{F}_n; n \ge 1\}$ satisfying (2.125)–(2.126), there exists a probability space $(\Omega^\star, \mathcal{A}^\star, P^\star)$ with a standard Wiener process W $(= \{W(t),\ t \in \boldsymbol{R}^+\})$ and a sequence $\{T_n;\ n \ge 1\}$ of nonnegative random variables such that the sequence $\{X_n\}$ has the same distribution as $\{W(\sum_{k \le n} T_k)\}$.*

Strassen (1967) prescribed a construction based on the following mapping. Let

$$S_{V_n} = X_n,\ n \ge 1,\ \ S_0 = X_0 = 0\ \ w.p.1;$$

$$S_t = S_{V_n} + Y_{n+1}(t - V_n)/v_{n+1}, \quad V_n \le t \le V_{n+1}, \quad n \ge 1.$$

Thus, $S = \{S_t, \ t \in \boldsymbol{R}^+\}$ has continuous sample paths on $\boldsymbol{R}^+$. Without loss of generality, we may set that v_n are all strictly positive, with probability 1.

THEOREM 2.15 *[Strassen theorem]. Let $S = \{S_t, \ t \in \boldsymbol{R}^+\}$ be defined as above. Let $f(.) = \{f(t), \ t \in \boldsymbol{R}^+\}$ be a nonnegative and nondecreasing function such that*

$$f(t) \text{ is } \uparrow \text{ but } t^{-1}f(t) \text{ is } \downarrow \text{ in } t (\in \boldsymbol{R}^+).$$

Suppose that V_n tends to ∞ a.s. as $n \to \infty$, and that

$$\sum_{n \ge 1} \frac{1}{f(V_n)} \mathbb{E}\left\{Y_n^2 I(|Y_n| > \sqrt{f(V_n)}) | \mathcal{F}_{n-1}\right\} < \infty \text{ a.e.} \tag{2.127}$$

Then there exists a standard Wiener process W on $\boldsymbol{R}^+$ such that

$$S_t = W(t) + o((\log t)(tf(t))^{1/4}) \text{ a.s.}, \quad \text{as} \quad t \to \infty.$$

Both theorems rest on the basic embedding result:

$$X_n = W(T_1 + \ldots + T_n) \text{ a.s.}, \quad \text{for all } n \ge 1. \tag{2.128}$$

To understand this basic representation, we need to introduce a new probability space $(\Omega^\star, \mathcal{A}^\star, P^\star)$ on which W and $\{T_k; \ k \ge 1\}$ are defined and also to consider a sequence $\{X_n^\star, \ n \ge 0\}$ of random variables on this space such that $\{X_n^\star, \ n \ge 1\} \stackrel{\mathcal{D}}{=} \{X_n, \ n \ge 1\}$; (2.128) actually holds for the $X_n^\star$.

By the law of iterated logarithms for $\{X_n^\star\}$, it suffices to choose $f(t) = t(\log\log t)^2(\log t)^{-4}$ as $t \to \infty$, while sharper choices of $f(t)$ can be made under higher-order moment conditions on the X_n. In fact, for i.i.d. random variables $\{Y_n^\star\}$ with unit variance and a finite fourth moment, (2.127) can be strengthened to

$$S(n) = W(n) + \mathrm{O}((n \log\log n)^{1/4}(\log n)^{1/2}) \text{ a.s.}, \quad \text{as } n \to \infty. \tag{2.129}$$

For independent random variables, a major contribution to the Wiener process embedding is due to the Hungarian school, and reported in Csörgö and Révész (1981). For i.i.d. random variables Y_i with 0 mean and unit variance, and with a finite moment-generating function $\mathbb{E}\{\exp(tY)\}$ $\forall t \le t_0$ for some $t_0 > 0$, these authors were able to show that

$$S(n) = W(n) + \mathrm{O}(\log n) \text{ a.s.}, \quad \text{as } n \to \infty.$$

and that (2.128) or (2.129) holds uniformly in a class of distribution functions for which the fourth moment or moment generating function are uniformly bounded. Brillinger (1969) was the first person who derived a parallel embedding theorem for the empirical process U_n in (2.116) for $p = 1$. He showed that there exits a probability space with sequence of Brownian bridges $\{B_n(t), \ 0 \le t \le 1\}$ and processes $\{\tilde{U}_n(t), \ 0 \le t \le 1\}$ such that

$$\{\tilde{U}_n(t), \ 0 \le t \le 1\} \stackrel{\mathcal{D}}{=} \{U_n(t), \ 0 \le t \le 1\},$$

and for each $n \geq 1$

$$\sup_{0\leq t\leq 1} |\tilde{U}_n(t) - B_n(t)| = \mathrm{O}(n^{-1/4}(\log n)^{1/2}(\log\log n)^{1/4}) \text{ a.s.}$$

Kiefer's (1972) extension relates to the two-parameter processes $U_n^\star$ [(2.118)] and $U^\star$ [(2.119)], with a rate $O(n^{1/3}(\log n)^{2/3})$ a.s. A variety of related results for U_n, as well as U_n^Q, are discussed in Csörgö and Révész (1981). We conclude this subsection with two a.s. results.

1. [Csörgö and Révész result]. Let $\varepsilon_n = n^{-1}(\log n)^4$. Then

$$(\log\log n)^{-1/2} \sup_{\varepsilon_n\leq t\leq 1-\varepsilon_n} \frac{|U_n(t)|}{\sqrt{t(1-t)}} \leq 2 \text{ a.s., as } n \to \infty. \tag{2.130}$$

2. [Csáki result]. For every $\varepsilon > 0$,

$$(\log\log n)^{-1/2} \sup_{0<t<1} \frac{|U_n(t)|}{\{t(1-t)\}^{1/2-\varepsilon}} = \mathrm{O}(1) \text{ a.s., as } n \to \infty. \tag{2.131}$$

2.5.16 Asymptotic Relative Efficiency: Concept and Measures

In Sections 2.2, 2.3, and 2.4, the basic concepts in statistical inference were laid down, with special emphasis on robustness and efficiency considerations. For estimators, (asymptotic) variance or mean square error can be used as a yardstick for their asymptotic efficiency. The reciprocal of its asymptotic mean square error (AMSE) is defined as the *asymptotic efficacy* of a real valued estimator T_n. In a parameter setup, the classical Fréchet–Cramér–Rao bound and the asymptotic normality of $\sqrt{n}(T_n - \theta)$ provide a good justification for this adoption. Actually, the convolution theorem 2.4 ensures that a BAN estimator $\hat{\xi}_n$ is asymptotically optimal, even in multiparameter setup. The asymptotic representation results explored in the subsequent chapters provide a similar justification for the class of robust estimators treated in this book. However, a basic question regarding the AMSE in the multiparameter case concerns a matrix of order, say, $p \times p$, and the multivariate Cramér–Rao information inequality,

$$\mathbf{V} - \mathbf{I}_{\boldsymbol{\theta}}^{-1} = \text{ positively semidefinite matrix}$$

which specifies that $\mathbf{V}$, [the AMSE of $\sqrt{n}(\mathbf{T}_n - \boldsymbol{\theta})$], dominates the reciprocal of the Information matrix $\mathbf{I}_{\boldsymbol{\theta}}$ in a matrix sense. Since $\widehat{\boldsymbol{\xi}}_n$ would have AMSE $\mathbf{I}_{\boldsymbol{\theta}}^{-1}$, comparison of $\mathbf{T}_n$ with $\widehat{\boldsymbol{\xi}}_n$ will not create any problem, and one may consider various measures of *asymptotic relative efficiency (ARE)* of $\mathbf{T}_n$ with respect to $\widehat{\boldsymbol{\xi}}_n$. For example,

1. *D-ARE:* $|\mathbf{I}_{\boldsymbol{\theta}}\mathbf{V}|^{-1/p}$ *(generalized variance)*
2. *A-ARE:* $p^{-1}Trace(\mathbf{I}_{\boldsymbol{\theta}}\mathbf{V})^{-1}$ *(trace criterion)*,

3. *E-ARE:* $Ch_{\max/\min}((\mathbf{I}_{\boldsymbol{\theta}}\mathbf{V})^{-1})$ (*extreme root*).

All of these measures will be ≤ 1 for all $\mathbf{T}_n$, but they may differ from each other. However, if we have two competing estimators, say, $\mathbf{T}_1$ and $\mathbf{T}_2$, with AMSE matrices $\mathbf{V}_1$ and $\mathbf{V}_2$, respectively, then $\mathbf{V}_1^{-1}\mathbf{V}_2$ may lead to different values for the D-, A-, and E- criteria. For univariate linear models, we have an intermediate setup where $\mathbf{V} = \sigma_V^2\mathbf{A}$, and hence all three criteria reduce to $(I_{\boldsymbol{\theta}}\sigma_V^2)^{-1}$. Another possible comparison of estimators is to consider their *large deviation probabilities*, $-n^{-1}\log P_{\boldsymbol{\theta}}\{\|\mathbf{T}_n - \boldsymbol{\theta}\| \geq \varepsilon\}$ (see Hoeffding 1965). There is opportunity for further development in this direction, although we do not place much emphasis on such large deviation probability results in this book.

Let us consider briefly the case with statistical tests. In a single parameter case, if the test statistic T_n is asymptotically normal under the null hypothesis, and the alternative hypothesis is contiguous to the null one, by (2.95)–(2.96) it has also asymptotically normal law under the alternative with only a shift in the mean. If we consider a test of significance level α, $0 < \alpha < 1$ against one-sided alternative, then the limiting power is $1-\Phi(\tau_\alpha - kc)$, where $\Phi(\tau_\alpha) = 1-\alpha$, the number k comes from the alternative hypothesis specification ($k > 0$), while the constant c (≥ 0) depends on the asymptotic mean and variance of T_n. Thus, if we equate the limiting powers of two competing tests based on $T_n^{(1)}$ and $T_n^{(2)}$, then the *asymptotic relative efficiency* (ARE) of $T_n^{(2)}$ with respect to $T_n^{(1)}$ is given by

$$e(T^{(2)}|T^{(1)}) = (\ c_2^2\ )/(\ c_1^2\ ),$$

where c_1 and c_2 are, respectively, the constants appearing in the limiting power functions for the common contiguous alternative. This ratio is termed the *Pitman ARE* in the literature. It can be justified similarly for two-sided alternatives. In the multiparameter case, if the limiting distribution of both test statistics is central chi-squared distribution with p degress of freedom, while the limiting distributions are the noncentral chi-squared under contiguous alternatives, then the Pitman ARE can be computed from their respective noncentrality parameters, both of which are suitable quadratic forms. In general, their ratio may depend on the direction of the alternatives relative to the null hypothesis. Bahadur (1960) introduced the concept of ARE based on the large deviation probabilities; it has gained popularity, despite the fact that it is difficult to derive its exact form. Its approximate form is generally insensitive to departures in the asymptotic power functions in various directions.

2.6 Problems

2.2.1 Show that (2.13) implies (2.14) and that the converse may not hold. Verify this with the example where $x_{2j-1} = -j$, $x_{2j} = +j$ for $j \geq 1$.

2.2.2 Let $\mathbf{D}_n = \mathbf{X}_n^\top \mathbf{X}_n$ and $\mathbf{\Delta}_n = diag(d_{n11}^{1/2}, \ldots, d_{npp}^{1/2})$. Assume that (2.14) holds and that the $n^{-1}d_{njj}$, $1 \le j \le p$ are all bounded away from 0. Then show that as $n \to \infty$,

$$\mathbf{D}_n^{1/2}(\widehat{\boldsymbol{\beta}}_n - \boldsymbol{\beta}) \xrightarrow{\mathcal{D}} \mathcal{N}_p(\mathbf{0}, \sigma^2 \mathbf{I}_p).$$

2.2.3 Define the e_i as in (2.3), and let $f(e)$ be the density function of e. Then note that for every real θ,

$$\theta = I\!\!E(e + \theta) = \int_{-\infty}^{\infty} (e + \theta) f(e) de = \int_{-\infty}^{\infty} e f(e - \theta) de,$$

and that by differentiating with respect to θ we get

$$1 = \int_{-\infty}^{\infty} e\{-f'(e - \theta)/f(e - \theta)\} f(e - \theta) de, \; \forall \theta.$$

Under $\theta = 0$, the Cauchy–Schwartz inequalityimplies that

$$\begin{aligned} 1 &\le \Big(\int_{-\infty}^{\infty} e^2 dF(e)\Big)\Big(\int_{-\infty}^{\infty} \{-f'(e)/f(e)\}^2 (e) \\ &= \sigma^2 \mathcal{I}(f). \end{aligned}$$

2.2.4 Using the estimating equations

$$\sum_{i \le n} \mathbf{x}_i \{f'(Y_i - \mathbf{x}_i^\top \widehat{\boldsymbol{\beta}}_n)/f(Y_i - \mathbf{x}_i^\top \widehat{\boldsymbol{\beta}}_n)\} = \mathbf{0}$$

for the MLE of $\boldsymbol{\beta}$, we can show, under some the regularity conditions, that

$$\lim_{n \to \infty} \mathbf{D}_n I\!\!E[(\widehat{\boldsymbol{\beta}}_n - \boldsymbol{\beta})(\widehat{\boldsymbol{\beta}}_n - \boldsymbol{\beta})^\top] = \mathcal{I}(f)\mathbf{I}_p.$$

2.2.5 Write $\mathbf{Y}_n - \mathbf{X}_n \widehat{\boldsymbol{\beta}}_n = \mathbf{e}_n - \mathbf{X}_n(\widehat{\boldsymbol{\beta}}_n - \boldsymbol{\beta})$ and express

$$\|\mathbf{Y}_n - \mathbf{X}_n \widehat{\boldsymbol{\beta}}_n\|^2 = \|\mathbf{e}_n\|^2 + (\widehat{\boldsymbol{\beta}}_n - \boldsymbol{\beta})^\top \mathbf{X}_n^\top \mathbf{X}_n (\widehat{\boldsymbol{\beta}}_n - \boldsymbol{\beta}) - 2\mathbf{e}_n^\top \mathbf{C}_n (\widehat{\boldsymbol{\beta}}_n - \boldsymbol{\beta}).$$

Show that $n^{-1}\|\mathbf{e}_n\|^2 \to \sigma^2$ a.s.as $n \to \infty$. Using Problem 2.2.2, we can verified that $(\widehat{\boldsymbol{\beta}}_n - \boldsymbol{\beta})^\top \mathbf{X}_n^\top \mathbf{X}_n (\widehat{\boldsymbol{\beta}}_n - \boldsymbol{\beta}) = \mathrm{O}_p(1)$ so that the third term on the right-hand-side is $\mathrm{O}_p(n^{1/2})$.

2.2.6 Show that the Pitman efficacy for the likelihood ratio test based on the density f $(= F')$ is proportional to $\mathcal{I}(f)$. For the likelihood ratio test statistic we may use the approximation by quadratic form in the MLE and Problems 2.2.2 and 2.2.4.

CHAPTER 3

Robust Estimation of Location and Regression

3.1 Introduction

In this chapter we start with a systematic account of robust estimators of location and regression parameters. We place a special emphasis on the motivations of several important classes of robust estimators and on their basic properties. We will consider two general situations.

1. *Location models.* Let $X_1, \ldots, X_n$ be $n(\geq 1)$ independent and identically distributed (i.i.d.) random variables with an unknown distribution function G, defined on the real line $\boldsymbol{R}_1$. We conceive of a parametric family of distribution functions $\{F_\theta : \theta \in \boldsymbol{\Theta}\}$ (usually a dominated system of distributions) with $\boldsymbol{\Theta} \subseteq \boldsymbol{R}_p$, for some $p \geq 1$, and we wish to estimate θ for which F_θ provides the closest approximation of G. For the location model, we assume that
$$F_\theta(x) \quad = \quad F_0(x - \theta), \tag{3.1}$$
where θ is real and F_0 belongs to a class $\mathcal{F}_0$. This model extends immediately to the multivariate case, where X_i's, x, and θ are q-vectors for some $q \geq 1$.
2. *Regression models.* Suppose that $X_1, \ldots, X_n$ are independent random variables where X_i has distribution function $F(x - \theta_i)$, for $i = 1, \ldots, n$, and the vector $\boldsymbol{\theta} = (\theta_1, \ldots, \theta_n)^\top$ of parameters satisfies the condition that
$$\boldsymbol{\theta} \in \Pi_p \;\text{ for some }\; p \; (1 \leq p \leq n), \tag{3.2}$$
where Π_p a linear p-dimensional subspace of $\boldsymbol{R}_n$. As in the location model, the form of F is unknown. We conceive of a class $\mathcal{F}$ of distributions, and we assume that $F \in \mathcal{F}$; the family $\mathcal{F}$ can either be a compact neighborhood of a fixed distribution function F_0, or it may even be a broad class of absolutely continuous distribution functions. The choice of $\mathcal{F}$ has an important bearing on the choice of robust estimators for the corresponding models. Note that the location model is a special case of the regression model for which $\theta = \theta_1$, for $\theta \in \boldsymbol{\Theta} \subset \boldsymbol{R}_1$.

Among various robust estimators, three broad classes, namely the M-, L- and R- estimators, have turned out to be the most interesting, and they have been studied extensively in the literature. A brief introduction to these estimators was given in the preceding chapter. In this as well as later chapters, we focus on various properties of these estimators. For detailed studies of some other properties of these estimators, not discussed thoroughly in this book, we refer the reader to Azencott et al. (1977), Serfling (1980), Huber (1981), Sen (1981a), Bickel (1981), Lehmann (1983), Koul (1992), Rieder (1994), Hampel et al. (2005), Huber and Ronchetti (2009), Maronna at al. (2006), among others.

In Section 3.2, we consider the basic formulation of the M-estimators, which appear to be the most flexible among the three classes mentioned above. They are well defined for a variety of models for which maximum likelihood estimators (MLE) have a sense. In fact, M-estimators cover both the MLE and least square estimators (LSE) as subclasses.

In Section 3.3, L-estimators are considered. The L-estimators were originally conceived as linear combinations of (functions of) order statistics for efficient estimation of location or scale parameters. They are generally computationally appealing and possess various desirable properties. L-estimators have also been successfully extended to linear models, what we shall describe later on.

Section 3.4 is devoted to the study of R-estimators. These estimators are generally based on the ranks or signed ranks of observations or residuals, and generally correspond to suitable rank tests for symmetry or randomness against shift or regression alternatives. We will mainly consider the R-estimators of location and regression parameters.

Besides these principal classes of robust estimators, some other notable classes will be considered in this chapter. Section 3.5 briefly treats *minimum distance estimators* and *Pitman-type estimators.* The concluding section presents a general introduction to *differentiable statistical functionals*, covering all preceding estimators, with due emphasis on their robustness properties.

3.2 M-Estimators

Consider a general single parameter model with i.i.d. observations. Suppose that $X_1, \ldots, X_n$ have common distribution function F_θ with the real but unknown parameter θ, and assume that there exists a real-valued function $\rho(x,t) : \boldsymbol{R}_2 \mapsto \boldsymbol{R}_1$ such that

$$\int \rho(x,t)dF_\theta(x) \text{ has a unique minimum at } t=\theta,\ \forall\theta\in\boldsymbol{\Theta} \tag{3.3}$$

for the model $\{F_\theta : \theta \in \Theta\}, \Theta \subseteq \boldsymbol{R}_1$. The M-estimator of θ is defined as a statistic $M_n = M_n(X_1, \ldots, X_n)$, which is a solution of the minimization of $\sum_{i=1}^n \rho(X_i, t)$ with respect to t. Therefore, we may write

$$M_n = \arg\min\{\sum_{i=1}^n \rho(X_i,t) : t \in \boldsymbol{R}_1\}. \tag{3.4}$$

The minimization problem in (3.4) often leads to the equation

$$\sum_{i=1}^{n} \psi(X_i, t) = 0 \tag{3.5}$$

where $\psi(x,t) = \frac{\partial \rho(x,t)}{\partial t}$, for all x, t. For example, for the location model pertaining to the normal family of distribution functions, the MLE minimizes $\rho(x,t) = (x-t)^2$, hence, $\psi(x,t) = x - t$ for every real x,t. The L_1-norm estimator for which $\rho(x,t) = |x-t|$ for $x, t \in \boldsymbol{R}_1$ we have $\psi(x,t) = \text{sign}(x-t)$, and the solution in (3.4) reduces to the median of $X_1, \ldots, X_n$. Other forms of $\rho(x,t)$ for location or other models may be chosen from relevant considerations. If $F_\theta(x)$ has the density $f(x,\theta)$ with respect to a sigma-finite measure μ and the density is differentiable in θ, then on choosing $\rho(x,\theta) = -\log f(x,\theta)$, we obtain that the corresponding $\psi(x,t) = \dot{f}(x,t)/f(x,t)$, and hence, the class of M-estimators in (3.4) include the MLE as a subclass. For the simple location model in (3.1), if we restrict ourselves to the class $\mathcal{F}_0$ of distribution functions symmetric around 0, then we usually consider $\rho(x,t) = h(x-t)$ where $h(-x) = h(x)$ for any x. But the picture may change if the underlying distribution function is not a member of this class $\mathcal{F}_0$. Looking from a wider perspective, the following questions immediately arise:

1. What are we really estimating in the case where F, the true distribution function of the X_i, does not belong to the class $\{F_\theta : \theta \in \Theta\}$?
2. Under what regularity conditions does there exist a consistent sequence of solutions of (3.4)?
3. What is the compensation for efficiency in the robust estimation derived from (3.4)?
4. What can we generally say about the distributional and other related properties of the M-estimators?

The answer to the first question is relatively simple. If F_n is the empirical distribution function based on $X_1, \ldots, X_n$, - $F_n(x) = n^{-1}\sum_{i=1}^{n} I(X_i \leq x), x \in \boldsymbol{R}_1$, then we have by the classical Glivenko–Cantelli lemma,

$$\|F_n - F\| = \sup_{x \in \boldsymbol{R}_1} |F_n(x) - F(x)| \to 0 \quad \text{a.s., as } n \to \infty \tag{3.6}$$

Thus, parallel to (3.3), we may consider the minimization

$$\int \rho(x,t) dF(x) = \min, \quad t \in \boldsymbol{R}_1 \tag{3.7}$$

and if it admits a solution $M(F)$, then the M-estimator in (3.4) can be described generally as an estimator of $M(F)$. However, the existence of a unique solution of (3.7) is often not guaranteed by (3.6) alone, without additional conditions on the function $\rho(.,.)$ and/or the distribution function F. Characterization of $M(F)$ plays an important role for M-estimation.

The second question reminds the classical problem of existence of a consistent solution of the likelihood equation leading to the usual MLE. Some thoughts on this question can be found in the literature: Huber (1981, ch. 6), Serfling (1980, ch. 7), and Lehmann (1983, ch. 6), among others. If $\psi(x,t)$ is monotone in t, then the existence of M-estimators can be established under very general regularity conditions, and the consistent sequence of solutions of (3.5) can easily be identified under parallel regularity conditions. However, the monotonicity of $\psi(x,t)$ in t is only a sufficient condition, while the existence of a consistent M-estimator can also be established for some nonmonotone $\psi(x,t)$, which are sufficiently smooth in t. Typically, in such case, there are multiple solutions of (3.5), and it may be difficult to identify the consistent one. This is a basic problem, and various approaches have been suggested to eliminate it. We find convenient to use a *one-step versions* of M-estimators that can be made arbitrarily close to the consistent solution of (3.4) for large n. We deter the details of this theory to Chapter 6.

The last two questions relate to some specific problems that arise in the study of asymptotic normality of M_n (when suitably normalized). For $\psi(.,t)$ monotone in t, this asymptotic normality result is generally obtained by inversion, as it will be discussed later on. The main emphasis will be put on a possible asymptotic representation of a consistent sequence of M-estimators by a sum of independent random variables. That not only yields the asymptotic normality as a by-product, but also provides a deeper insight into the asymptotic behavior of M_n and other important asymptotic results. Basically, we aim to show that if there exists a consistent sequence $\{M_n\}$, then

$$n^{1/2}(M_n - M(F)) = n^{-1/2}(\gamma(F))^{-1}\sum_{i=1}^{n}\psi(X_i, M(F)) + o_p(1), \tag{3.8}$$

under some regularity conditions on F and ψ, where $f(x) = (d/dx)F(x)$,

$$\gamma(F) = -\int \psi(x, M(F))df(x) \quad \text{or} \quad \int f(x)d\psi(x, M(F)), \tag{3.9}$$

and $\psi(x,t) = (\partial/\partial t)\rho(x,t)$. Such representations, supplemented by more precise orders for the remainder term, will be considered in Chapter 5. This representation among others implies, by means of the central limit theorem on $\psi(X_i, M(F))$, that $n^{1/2}(M_n - M(F))$ is asymptotically normal with zero mean and variance $(\gamma(F))^{-2}\text{Var}[\psi(X_1, M(F))]$. This, in turn, provides an answer to question 3. For the symmetric location model, Huber (1964) studied first the asymptotic normality of M_n, and later he gave generalizations to other problems (Huber 1967). This topic has been a subject of many studies over the past 40 years, and diverse techniques have been employed by a host of researchers toward the same goal. Our approach based on asymptotic representations of the type (3.8) has some additional advantages, which we will explore systematically in this and subsequent chapters.

Let us concentrate first on the location model, briefly introduced in Section 3.1. In this model we have $F_\theta(x) = F(x-\theta)$, where F is an unknown distribution function, symmetric around 0, and we assume that $F \in \mathcal{F}_0$, the class of absolutely continuous symmetric distribution functions. The M-estimator is then defined as a solution of the minimization

$$\sum_{i=1}^{n} \rho(X_i - t) = \min, \tag{3.10}$$

where, due to the assumed symmetry of F, it is recommended that a symmetric ρ be employed in (3.10). The *influence curve* of M_n is then given by

$$IC(x; F, M) = \psi(x)/\gamma(F), \quad \gamma(F) = \int f(x)d\psi(x).$$

Note that IC(x;ψ, M) is proportional to ψ, and the proportionality coefficient $(\gamma(F))^{-1}$ depends on both ψ and the density $f(.)$. The influence function IC(x;ψ, M) is bounded whenever ψ is bounded and $\gamma(F) \neq 0$. Moreover, whenever f admits a derivative f' for almost all x, then we obtain by partial integration that

$$\gamma(F) = \int \{-f'(x)/f(x)\}\psi(x)dF(x), \tag{3.11}$$

a form that is more suitable for manipulation in a number of situations. Recall that the Huber (1964) estimator is asymptotically minimax over a contaminated neighborhood of a fixed symmetric distribution F_0 (see Section 5.2 for some details). In particular, when F_0 is the standard normal distribution function, this procedure leads to the Huber M-estimator, generated by $\psi = \psi_0$ in (2.47).

The M-estimator M_n [the solution of (3.10)] is clearly *translation-equivariant*, since $\rho(X_i - t) = \rho((X_i + a) - (t + a))\ \forall a \in \boldsymbol{R}_1$. But it may not be *scale-equivariant*, since in general $\psi(cx) \neq c\psi(x)\ \forall x \in \boldsymbol{R}_1, c > 0$. The M-estimator can be made scale-equivariant by some modifications. Huber (1964) suggested either to use the *simultaneous M-estimation* of location and scale parameters, or to *studentize* the M-estimator of location by a scale statistic $S_n = S(X_1, \ldots, X_n)$ such that $S(aX_1, \ldots, aX_n) = |a|S(X_1, \ldots, X_n)\ \forall a \in \boldsymbol{R}_1$. The minimization of $\sum_{i=1}^{n} \rho(X_i - t)$ in (3.10) is replaced by that of $\sum_{i=1}^{n} \rho((X_i - t)/S_n)$. Unlike in the case of the location parameter (which may be interpreted as the center of symmetry of a distribution), there is perhaps no natural interpretation of a scale parameter, and hence, in a real situation the *studentization* leads to a better choice of the function $\rho(.)$. Thus, besides the aim to make M-estimators scale-equivariant, the studentization serves a better purpose in practical applications. In some situations it enables estimating the common location and the ratio of scale parameters in the two-sample model (Jurečková and Sen 1982a).

Let us return to the regression model in (3.2). As a natural extension of (3.5), we now consider a distance function

$$\sum_{i=1}^{n} \rho(X_i, t_i) : \mathbf{t} = (t_1, \ldots, t_n)^\top \in \Pi_p, \tag{3.12}$$

where Π_p is defined in (3.2). Our task is to minimize (3.12) with respect to $\mathbf{t}$. The M-estimator of $\boldsymbol{\theta} = (\theta_1, \ldots, \theta_n)^\top$ is a solution of

$$\sum_{i=1}^{n} \rho(X_i, t_i) := \min, \quad \mathbf{t} \in \Pi_p \tag{3.13}$$

For the general regression model, we choose a $n \times p$ matrix $\mathbf{X} = ((x_{ij}))$ of known regression constants and p-vector $\boldsymbol{\beta} = (\beta_1, \ldots, \beta_p)^\top$ of unknown regression parameters, and set

$$\boldsymbol{\theta} = \mathbf{X}\boldsymbol{\beta} \quad \text{and} \quad \rho(a, b) = \rho(a - b).$$

Usually, the function ρ in (3.10) is taken to be convex, and its derivative $\psi(y) = (d/dy)\rho(y)$ is often taken to be of bounded variation, to induce more robustness. If ψ is continuous, we can equivalently write (3.13) as

$$\sum_{i=1}^{n} c_{ij}\psi\Big(X_i - \sum_{k=1}^{p} c_{ik}b_k\Big) = 0, \quad j = 1, \ldots, p. \tag{3.14}$$

This equation may be regarded as a natural generalization of the location model to the regression case, and, as we will see later, the M-estimators in both the models (3.13) and (3.14) have similar properties. We will focus rather on a small difference (in terms of computational simplicity) in the two situations.

For monotone (nondecreasing) ψ, the solution in (3.10) can be expressed as

$$M_n = \frac{1}{2}(M_{n,1} + M_{n,2}); \tag{3.15}$$

$$M_{n,1} = \sup\{t : \sum_{i=1}^{n} \psi(X_i - t) > 0\}, M_{n,2} = \inf\{t : \sum_{i=1}^{n} \psi(X_i - t) < 0\}. \tag{3.16}$$

If ψ is nondecreasing, then $\sum_{i=1}^{n} \psi(X_i - t)$ is nonincreasing in $t \in \boldsymbol{R}_1$. Hence, M_n in (3.15) represents the centroid of the set of solutions of (3.14), and it eliminates the possible arbitrariness of such a solution. If $p = 1$, the situation in regression model is very similar to the location model. If $p \geq 2$ and $\mathbf{x}_i$ satisfy some concordance-discordance conditions, then the left-hand side of (3.14) may be decomposed as a finite linear combination of terms that are monotone in each of the p arguments $b_1, \ldots, b_p$. In any case, whether or not ψ is monotone, we may use an initial $\sqrt{n}$- consistent estimator and adopt an iterative procedure. Generally, the consistency of the solution of (3.14) follows when $\rho(x, t)$ is convex in t. If this is not the case but ψ is continuous in t, existence of a $\sqrt{n}$-consistent estimator of $\boldsymbol{\beta}$ can be established, although it may not coincide with the global minimum of (3.13). In this context, as well as in other asymptotic results, the asymptotic linearity of M-statistics in the location or regression parameters plays a central role, and we will explore this systematically in this book.

To establish the asymptotic normality of $n^{1/2}(M_n - \theta)$ in the location model (with monotone ψ), we must show by (3.15) and (3.16) that under $\theta = 0$ and for every fixed t,

$$n^{-1/2} \sum_{i=1}^{n} \psi(X_i - n^{-1/2} t)$$

has asymptotically a normal distribution with mean $-t\gamma(F)$ and variance σ_ψ^2, defined by (2.46). In the regression model, we generally need a different approach when $p \geq 2$, while the inversion technique is applicable when $p = 1$. The linearity approach provides an easy access for the regression model: Parallel to (3.8),

$$n^{1/2}(\mathbf{M}_n - \boldsymbol{\beta}) = n^{-1/2}(\gamma(F))^{-1} \sum_{i=1}^{n} \mathbf{x}_i \psi(Y_i - \mathbf{x}_i^\top \boldsymbol{\beta}) + o_p(1) \qquad (3.17)$$

as $n \to \infty$, where $\gamma(F)$ is defined in (3.9). Then under the Noether condition,

$$\max \left\{ \mathbf{x}_i^\top \left(\sum_{j=1}^{n} \mathbf{x}_j \mathbf{x}_j^\top \right)^{-1} \mathbf{x}_i : 1 \leq i \leq n \right\} \to 0 \quad \text{as } n \to \infty, \qquad (3.18)$$

on the $\mathbf{x}_i$, the usual multivariate central limit theorem implies

$$n^{1/2}(\mathbf{M}_n - \boldsymbol{\beta}) \xrightarrow{\mathcal{D}} \mathcal{N}_p(\mathbf{0}, \nu^2 \mathbf{X}^{\star -1}), \qquad (3.19)$$

where

$$\mathbf{X}^\star = \lim_{n\to\infty} n^{-1} \Big(\sum_{j=1}^{n} \mathbf{x}_j \mathbf{x}_j^\top \Big) \text{ (positive definite), and}$$

$$\nu^2 = \Big\{ \int_{\boldsymbol{R}_1} \psi^2(x) dF(x) \Big\} / \Big\{ \int_{\boldsymbol{R}_1} f(x) d\psi(x) \Big\}^2 = \nu^2(\psi, F). \qquad (3.20)$$

This asymptotic linear representation-based proof of asymptotic normality applies also to the simple location model, where $c_i = 1 \ \forall i, p = 1$ and hence, $C^\star = 1$. Thus, for the location model,

$$n^{1/2}(M_n - \theta) \xrightarrow{\mathcal{D}} \mathcal{N}(0, \nu^2).$$

We conclude that the choice of the score function ψ is governed by the same considerations for both the location and regression models. As such, the considerations in Sections 2.3 and 2.4 pertain to both the location and regression models. Further, the M-estimators of regression parameters are also not generally scale-equivariant, because $\psi(cx) \neq c\psi(x)$ for every $x \in \boldsymbol{R}_1$ and $c > 0$ for a general ψ. The possible scale-equivariant M-estimators of regression parameters will be considered in Chapter 5. The one-step version of M-estimators of regression parameters will be considered in Chapter 6.

The M-estimators of location or regression are *regression-equivariant* in the following sense:

$$\mathbf{M}_n(Y_1 + \mathbf{b}^\top \mathbf{x}_1, \ldots, Y_n + \mathbf{b}^\top \mathbf{x}_n) = \mathbf{M}_n(Y_1, \ldots, Y_n) + \mathbf{b}, \quad \forall \mathbf{b} \in \boldsymbol{R}_p \tag{3.21}$$

because

$$\psi(y - \mathbf{x}_i^\top \mathbf{t}) = \psi((y + \mathbf{x}_i^\top \mathbf{b}) - \mathbf{x}_i^\top (\mathbf{b} + \mathbf{t})) \tag{3.22}$$

for every $\mathbf{b} \in \boldsymbol{R}_p, \mathbf{t} \in \boldsymbol{R}_p$, and $y \in \boldsymbol{R}_1$ and $\mathbf{x}_i$. Then we can set $\boldsymbol{\beta} = \mathbf{0}$ without loss of generality. If ψ is skew-symmetric about 0, then

$$\mathbf{M}_n(Y_1, \ldots, Y_n) = -\mathbf{M}_n((-1)Y_1, \ldots, (-1)Y_n), \quad \text{w. p. } 1 \tag{3.23}$$

by virtue of (3.14). On the other hand, Y_i and $(-1)Y_i$ have the same distribution (F) under $\mathbf{b} = \mathbf{0}$, whenever the distribution function F is symmetric about 0 (as is generally assumed). Hence, when $\boldsymbol{\beta} = \mathbf{0}$, we conclude by (3.23) that $\mathbf{M}_n$ has a diagonally symmetric distribution about $\mathbf{0}$. Incorporating the regression-equivariance (3.21), we conclude that whenever the ψ-function is skew-symmetric and the distribution function F is symmetric, the M-estimator $\mathbf{M}_n$ of the vector regression parameter $\boldsymbol{\beta}$ is diagonally symmetrically distributed in the sense that

$$(\mathbf{M}_n - \boldsymbol{\beta}) \quad \text{and} \quad (\boldsymbol{\beta} - \mathbf{M}_n) \quad \text{both have the same distribution function.}$$

This implies that marginally each component of $\mathbf{M}_n$ has a distribution symmetric about the corresponding element of $\boldsymbol{\beta}$, so that

$$\mathbf{M}_n \text{ is a median-unbiased estimator of } \boldsymbol{\beta}. \tag{3.24}$$

However, the skew-symmetry of ψ and symmetry of F are crucial for the median-unbiasedness of $\mathbf{M}_n$. The median-unbiasedness may not imply unbiasedness in the classical sense, which needs extra regularity conditions on ψ and F [see Jurečková and Sen (1982b) for the special case of the location model. Note that (3.21)–(3.24) may hold even for the nonmonotone score functions, like for some Hampel-type functions, mentioned in Section 2.4. However, we generally recommend monotone score functions for their computational ease, and in this case, the M-estimators of location and regression retain their validity and robustness for a broad class of (symmetric) distributions. Within this broad class of distribution functions, optimal score functions for specific subclasses (in the sense of local robustness) can be chosen as in Sections 2.3 and 2.4.

Finally, let us consider the breakdown and tail-behavior properties of M-estimator of location parameter with bounded influence function. Assume that ρ in (3.10) is convex and symmetric and that $\psi = \rho'$ is continuous and bounded. Then M_n is translation equivariant and monotone in each Y_i, $i = 1, \ldots, n$; hence, Theorem 2.1 applies. Assume first that n is odd, and denote $r = (n+1)/2$. Let $K = \sup\{\psi(x) : x \geq 0\}$ [$= -\inf\{\psi(x) : x \leq 0\}$]. Then

$$I\!P_\theta \left(|M_n - \theta| > a\right) = I\!P_0 \left(M_n > a\right) + I\!P_0 \left(M_n < -a\right)$$

$$\geq \mathbb{P}_0\left(\sum_{i=1}^{n}\psi(X_i - a) > 0\right) + \mathbb{P}_0\left(\sum_{i=1}^{n}\psi(X_i + a) < 0\right)$$

$$\geq \mathbb{P}_0\Big(\psi(X_{n:r} - a) > \frac{n-1}{n+1}K\Big) + \mathbb{P}_0\Big(\psi(X_{n:r} + a) < -\frac{n-1}{n+1}K\Big)$$

$$\geq \mathbb{P}_0(X_{n:r} > a + c_1) + \mathbb{P}_0(X_{n:r} < -a - c_1)$$

$$\geq 2\binom{n}{r}\left(F(a+c_1)\right)^{r-1}\left(1 - F(a+c_1)\right)^{r}$$

where $c_1 = \inf\{x : \psi(x) \geq \frac{n-1}{n+1}K\} \in [0, \infty)$. Similarly,

$$\mathbb{P}_0(|M_n| > a) \leq \mathbb{P}_0(\sum_{i=1}^{n}\psi(X_i - a) \geq 0) + \mathbb{P}_0(\sum_{i=1}^{n}\psi(X_i + a) \leq 0)$$

$$\leq \mathbb{P}_0\Big(\psi(X_{n:r} - a) \geq -\frac{n-1}{n+1}K\Big) + \mathbb{P}_0\Big(\psi(X_{n:r} + a) \leq \frac{n-1}{n+1}K\Big)$$

$$\leq \mathbb{P}_0(X_{n:r} \geq a - c_2) + \mathbb{P}_0(X_{n:r} \leq -a + c_2)$$

$$\leq 2r\binom{n}{r}\left(1 - F(a - c_2)\right)^{r},$$

where $c_2 = \sup\{x : \psi(x) \leq \frac{n-1}{n+1}K\} \in [0, \infty)$. Hence,

$$\limsup_{a\to\infty} B(a, M_n) \leq \frac{n+1}{2}\lim_{a\to\infty}\frac{\log\left(1 - F(a+c_1)\right)}{\log\left(1 - F(a)\right)} = \frac{n+1}{2}, \tag{3.25}$$

provided F satisfies (2.33), and similarly

$$\liminf_{a\to\infty} B(a, M_n) \geq \frac{n+1}{2}. \tag{3.26}$$

Thus, (3.25), (3.26), and Theorem 2.1 imply that

$$\lim_{a\to\infty} B(a, T_n) = \frac{n+1}{2} = m^{\star}. \tag{3.27}$$

If n is even, we proceed analogously and obtain

$$m^{\star} = \frac{n}{2}, \ \frac{n}{2} \leq \liminf_{a\to\infty} B(a, M_n) \leq \limsup_{a\to\infty} B(a, M_n) \leq \frac{n}{2} + 1. \tag{3.28}$$

We come to the conclusion that the M-estimator of the location parameter in a symmetric model, generated by a nondecreasing bounded function, has the largest possible breakdown point and its tail performance is not affected by the tails of the distribution. Hence, the M-estimator is robust.

The above considerations can be extended in a straightforward manner to the sample median. Hence, (3.27) and (3.28) also apply to the median.

3.3 L-Estimators

L-estimators are termed after linear combinations of (or of functions of) order statistics. It was observed a long time ago that the sample order statistics may provide simple and efficient estimators in the location-scale family of distributions; recall that the vector of sample order statistics is complete. To motivate the L-estimators, let us first look at the simple location model. Let $X_1, \ldots, X_n$ be n i.i.d. random variables with a distribution function F_θ, defined on the real line $\boldsymbol{R}_1$, and let

$$X_{n:1} \leq \ldots \leq X_{n:n} \tag{3.29}$$

be the sample order statistics. Then an L-estimator of a parameter θ is defined by

$$L_n = \sum_{i=1}^{n} c_{ni} X_{n:i}, \tag{3.30}$$

where the coefficients $c_{n1}, \ldots, c_{nn}$ are known. The coefficients are chosen so that L_n is a desirable estimator of θ in the sense of unbiasedness, consistency, and other efficiency properties. If the distribution function F_θ admits a density f_θ (on its support), then the ties in (3.29) may be neglected with probability 1, so the order statistics $X_{n:1}, \ldots, X_{n:n}$ are all distinct with probability 1. As in Section 2.4, we introduce the *quantile function* $Q(t) = inf\{x : F_\theta(x) \geq t\}, 0 < t < 1$, and its sample counterpart

$$Q_n(t) = \inf\{x : F_n(x) \geq t\}, \;\; 0 < t < 1, \tag{3.31}$$

where the sample distribution function F_n is defined before in (3.6). We further conceive of a sequence $\{J_n = [J_n(u) : u \in (0,1)]\}$ of functions defined on the unit interval (0,1) such that $J_n(u)$ assumes a constant value $J_n(i/(n+1))$ on $((i-1)/n, i/n]$, for $i = 1, \ldots, n$. On letting

$$c_{ni} = \frac{1}{n} J_n\Big(\frac{i}{n+1}\Big), \;\; i = 1, \ldots, n, \tag{3.32}$$

and noting that F_n has the jumps of magnitude n^{-1} at the n-order statistics (and 0 elsewhere), we obtain from (3.30) –(3.32) that

$$\begin{aligned} L_n &= n^{-1} \sum_{i=1}^{n} Q_n(i/(n+1)) J_n(i/(n+1)) \\ &= \int_{\boldsymbol{R}_1} Q_n(F_n(x)) J_n(F_n(x)) dF_n(x). \end{aligned} \tag{3.33}$$

In this form, L_n is identifiable as the sample counterpart corresponding to the L-functional in (2.41). Due to their simple and explicit forms [(3.30) or (3.33)], the L-estimators are computationally appealing relative to the other types of estimators. It is clear that the c_{ni} or equivalently the sequence $\{J_n(.)\}$ are of major considerations in the study of properties of the L-estimators.

To motivate this study, we consider the location-scale model: X_i's are i.i.d. random variables with a density function $f_{\mu,\delta}(x), x \in \boldsymbol{R}_1$, given by

$$f_{\mu,\delta}(x) = \delta^{-1} f_0((x-\mu)/\delta), \quad x \in \boldsymbol{R}_1, \mu \in \boldsymbol{R}_1, \delta \in \boldsymbol{R}_1^+,$$

where $f_0(.)$ has a specified form, independent of μ and δ. For the time being, we assume that f_0 has a finite second moment, and without any loss of generality, we assume that the variance of f_0 is equal to 1. Then we write

$$X_{n:i} = \mu + \delta Y_{n:i}, \quad i = 1, \ldots, n, \tag{3.34}$$

where $Y_{n:i}$'s are the sample order statistics of a sample of size n from a distribution function F_0 corresponding to the density f_0. If we let

$$I\!\!E Y_{n:i} = a_{n:i} \text{ and } \mathrm{Cov}(Y_{n:i}, Y_{n:j}) = b_{nij} \quad \text{for } i, j = 1, \ldots, n, \tag{3.35}$$

then both $\mathbf{a}_n = (a_{n:1}, \ldots, a_{n:n})^\top$ and $\mathbf{B}_n = ((b_{nij}))_{i,j=1,\ldots,n}$ are independent of (μ, δ) and can be computed with the density f_0. Let $\mathbf{B}_n^{-1} = ((b_n^{ij}))$. For the sake of simplicity, let us assume that $\mathbf{B}_n^{-1}$ exists. We are naturally tempted to use the weighted least squares method (see Lloyd 1952) to obtain the best (i.e., minimum variance unbiased) estimators of (μ, δ) based on the $X_{n:i}$ in (3.34). This amounts to the minimization of

$$W(\mu, \delta) = \sum_{i=1}^{n} \sum_{j=1}^{n} b_n^{ij} (X_{n:i} - \mu - \delta a_{n:i})(X_{n:j} - \mu - \delta a_{n:j})$$

with respect to the two unknown parameters (μ, δ). The resulting estimators derived from the estimating equations:

$$(\partial/\partial s) W(s,t) = 0, (\partial/\partial t) W(s,t) = 0 \quad \text{at } (s,t) = (\hat{\mu}_n, \hat{\delta}_n), \tag{3.36}$$

are clearly linear in $X_{n:i}$ and conform to the form in (3.30) with the c_{ni} depending on $\mathbf{a}_n$ and $\mathbf{B}_n^{-1}$. In the literature, this is known as BLUE (best linear unbiased estimator) of (μ, δ). The theory sketched above can be easily extended to cover the case of censored data where some of the order statistics at the lower and/or upper extremity are omitted from consideration (on the ground of robustness or for other reasons). In this case, we need to confine ourselves to a subset of the $X_{n:i}$'s (i.e., $X_{n:j}, j \in J \subset \{1, \ldots, n\}$). Using the weighted least squares method on this subset, we again obtain an optimal BLUE (within this subset). In this case some of the c_{ni}'s in (3.30) (at the beginning or end) are equal to 0. A very similar situation arises when we have a subset of selected order statistics, $\{X_{n:[np_j]}, j = 1, \ldots, k\}$, where k ($\leq n$) is a fixed positive integer and $0 < p_1 < \ldots < p_k < 1$ are some given numbers. Again the weighted least squares theory, as applied to this specific subset, yields a BLUE that conforms to (3.30), but only k of the c_{ni}'s are different from 0. In the extreme case, we may use a single quantile $X_{n:[np]}$ or a pair $(X_{n:[np_1]}, X_{n:[np_2]})$, and these may be also regarded as L-estimators. During the 1950s and 1960s, BLUE theory for various densities and under diverse censoring patterns emerged as a very popular area of research. We may refer to Sarhan and Greenberg (1962) and David (1981) for useful accounts of the developments. Remark that the $a_{n:i}$ and b_{nij} in (3.35) depend on the underlying density f_0 in a rather involved manner, and their exact computations (for any given n) require extensive numerical integration, hence, the task may become prohibitively laborious for large n. Fortunately, the quantiles $F_0^{-1}(i/(n+1))$ often provide good approximations for the $a_{n:i}$, and similar approximations are available for the b_{nij}. Use of these approximations has greatly simplified the computational aspects leading to estimators that are termed nearly or asymptotically BLUE. A very useful account of this theory in various specific models is given in Sarhan and Greenberg (1962). From this theory, we might get a clear justification for (3.32), with the further insight that there exists a function $J = \{J(u), 0 < u < 1\}$ defined on (0,1) such that

$$J_n(u) \to J(u), \text{ as } n \to \infty, \text{ for every (fixed) } u \in (0,1). \tag{3.37}$$

The optimal J generally depends on the density f_0. We may even write

$$J(u) = J(u; f_0) \quad \text{for } u \in (0,1). \tag{3.38}$$

One important implication of (3.38) is the vulnerability of the optimal score J to possible departures from the assumed density. Even when the actual density (e.g., g_0) and the assumed one (f_0) are fairly close to each other, the b_n^{ij}'s are very sensitive to small departures, and hence, the asymptotically optimal score $J(., f_0)$ may not be that close to the true score $J(., g_0)$. This feature can be easily verified with the ,,error-contamination" model (2.27). This suggests that it may be wiser for BLUE to look into possible departures from the assumed model and to choose the score function $J(.)$ in a more robust manner. While such a robust choice of J may not lead to the asymptotic best property under the model f_0, it may have other minimaxity or similar properties for a class of distributions ,,close" to the assumed one (F_0).

With this introduction we are in a position to generalize the pure parametric BLUE theory to robust BLUE theory, and this is in line with the robust M-estimators discussed in the previous section. We start with usual location model where $F_\theta(x) = F(x - \theta), x \in \boldsymbol{R}_1$, and $\theta \in \boldsymbol{R}_1$. We assume that F is symmetric about the origin. Consider a general class of L-estimators of location (3.30) satisfying

$$\sum c_{ni} = 1 \text{ and } c_{ni} = c_{n,n-i+1}, \quad \text{for every } i. \tag{3.39}$$

The first condition in (3.39) ensures the translation-equivariance of L_n, while the second condition is associated with the assumed symmetry of the distribution function F_0. It ensures that L_n has a symmetric distribution around θ so that L_n is median-unbiased. An important class of L_n belonging to this family of L-estimators is the so-called kth order rank-weighted mean (see, Sen

1964) which can be expressed as

$$T_{nk} = \binom{n}{2k+1}^{-1} \sum_{i=k+1}^{n-k} \binom{i-1}{k} \binom{n-i}{k} X_{n:i}, \tag{3.40}$$

for $k = 0, \ldots, [(n+1)/2]$. Note that for $k = 0$, $T_{n0} = \bar{X}_n = n^{-1}\sum_{i=1}^{n} X_i$, the sample mean, while for $k = [(n+1)/2]$, (3.40) reduces to the sample median. For $k \geq 1$, T_{nk} does not need to be fully efficient for the basic model where F is a normal distribution function. Thus the class $\{T_{nk}; k \geq 0\}$ covers a scale from the sample median to the sample mean, namely, from highly robust to highly nonrobust estimators of location. Since this is a subclass of L-estimators satisfying (3.39), we conclude that the class in (3.39) has the same characterization too. Though the sample median is a robust estimator of θ, it is generally not very efficient for nearly normal distribution function. Among the L-estimators of location that are less sensitive to deviations from the normal distribution and yet are more efficient than the sample median at the normal distribution (model), we can mention the following:

1. *Trimmed mean.* For an $\alpha: \ 0 < \alpha < 1/2$, let

$$L_n(\alpha) = \frac{1}{n - 2[n\alpha]} \sum_{i=[n\alpha]+1}^{n-[n\alpha]} X_{n:i} \tag{3.41}$$

   where $[x]$ denotes the largest integer $\leq x$.

2. *Winsorized mean.* For an $\alpha: \ 0 < \alpha < 1/2$, let

$$L_n^{\star}(\alpha) = \frac{1}{n}\Big\{[n\alpha]X_{n:[n\alpha]} + \sum_{i=[n\alpha]+1}^{n-[n\alpha]} X_{n:i} + [n\alpha]X_{n:n-[n\alpha]+1}\Big\}. \tag{3.42}$$

Note that (3.39) holds for either of these estimators. Thus, both the trimmed and Winsorized means are translation-equivariant estimators of location, and they are both median-unbiased for θ for symmetric F. The idea to trim off some of the extreme observations and calculate the average of the remaining central observations is so old that we can hardly say by whom it was originally proposed. Looking at (3.40), (3.41) and (3.42), we observe that by letting k such that $k/n \approx \alpha$, all of these estimators basically aim to eliminate the effect of the extreme values, although we have a relatively constant weight in (3.41) for the central values, along with the jumps in (3.42) and a bell-shaped weight function in (3.40). For all of them

$$c_{ni} = 0 \quad \text{for } i \leq k, \ \ i \geq n - k + 1,$$

and the c_{ni}'s are all nonnegative. Each is a convex combination of the set of order statistics. Though generally the L-estimators of location are convex combinations of order statistics, there may be situations where some of the

c_{ni}'s may turn out to be negative. To stress this point, we may consider a trimmed L-estimator of the form (3.30), where

$$c_{ni} = 0 \quad \text{for } i \leq [n\alpha] \text{ and } i \geq n - [n\alpha] \text{ for some } \alpha \in (0, \frac{1}{2}). \tag{3.43}$$

We shall see from the asymptotic distribution of L_n, that will be studied in Chapter 4, that L_n under (3.43) can be asymptotically efficient only for distribution function F that has typically an exponentially decreasing tail on either side. If, however, F has tails heavier than an exponential tail, then an L-estimator to be asymptotically efficient must have negative c_{ni}'s in the two extremes. Thus we cannot impose the nonnegativity on the c_{ni} as a blanket rule for L-estimators of location.

Let us now study the consistency and asymptotic efficiency properties of L-estimators of location and, in general, of other L-estimators as well. From what has been discussed after (3.36) we gather that in a relatively more general situation, we may decompose an L-estimator in the form

$$\begin{aligned} L_n &= n^{-1}\sum_{i=1}^{n} J_{n,1}(i/(n+1))X_{n:i} + \sum_{j=1}^{k} a_{nj}X_{n:[np_j]} \\ &= L_{n1} + L_{n2}, \text{say}, \end{aligned} \tag{3.44}$$

where $J_{n,1} : (0,1) \rightarrow \boldsymbol{R}_1, 0 < p_1 < \ldots < p_k < 1$, $a_{ni}, \ldots, a_{nk}$ are given numbers, and we may assume that (3.37) holds for $J_{n,1}(.)$, while the a_{nj}'s have suitable limits. Note that Winsorized mean in (3.42) is a particular case of (3.44), where $k = 2$, $a_{n1} = a_{n2} = [n\alpha]/n$, $p_1 = 1 - p_2 = \alpha$, and $J_{n,1}(u) = 0$ outside the interval $[\alpha, 1 - \alpha]$ and it is equal to 1 otherwise. Similarly, (3.40) and (3.41) are also special cases of (3.44) where $L_{n2} = 0$. The sample median corresponds to the case of $k = 1$, $p_1 = 1/2$, and $L_{n1} = 0$, while the sample mean to the case of $L_{n2} = 0$ and $J_{n,1}(u) = 1 \forall u \in (0,1)$. Often, the *interquartile range*

$$X_{n:[3n/4]} - X_{n:[n/4]} \tag{3.45}$$

is used as a measure of the scatter of a distribution, and this also corresponds to (3.44) where $L_{n1} = 0$. We will study the consistency of L_n under the setup in (3.44). Defining the quantile function $Q(t)$ as in earlier, we assume that

$$F_\theta^{-1}(= Q(p)) \quad \text{is uniquely defined for} \quad p = p_j, j = 1, \ldots, k. \tag{3.46}$$

[Note that the strict monotonicity of F_θ^{-1} at $t = p$ ensures (3.46).] Then, for every $\varepsilon > 0$,

$$\begin{aligned} &P\Big\{X_{n:k} > F_\theta^{-1}(p) + \varepsilon\Big\} \\ &= P\Big\{(k-1 \text{ or less number of } X_1, \ldots, X_n \text{ are less than } F_\theta^{-1}(p) + \varepsilon\Big\} \\ &= P\Big\{\sum_{i=1}^{n} I(F_\theta^{-1}(p) + \varepsilon > X_i) \leq k - 1\Big\}. \end{aligned} \tag{3.47}$$

Using the Hoeffding inequality [see (2.64)], we obtain an exponential rate of convergence of (3.47) to 0 whenever $\frac{k}{n} \to p$; a similar inequality holds for the lower tail. Using this inequality, we easily obtain that under (3.46)

$$X_{n:[np_j]} \to F_\theta^{-1}(p_j) \quad \text{a.s., for each} \quad j = 1, \dots k. \tag{3.48}$$

as $n \to \infty$. Thus, if we assume that

$$\lim_{n\to\infty} a_{nj} = a_j \quad \text{exists and } |a_j| < \infty \text{ for} \quad j = 1, \dots, k, \tag{3.49}$$

we obtain from (3.48) and (3.49) that under (3.46) and (3.49),

$$L_{n2} \stackrel{w.p.1}{\to} \sum_{j=1}^{k} a_j F_\theta^{-1}(p_j), \quad \text{as } n \to \infty. \tag{3.50}$$

The treatment for L_{n1} is a bit more delicate. Further regularity conditions on $J_{n,1}$, such as in (3.37), and on the distribution function F are necessary to establish the desired result. First, we note that under some smoothness conditions on the score function J, $\{L_{n1}, n \geq n_0\}$ may form a reverse martingale sequence, and in a less stringent setup it can be approximated by another reverse martingale (in the sense that the remainder term converges a.s., or in some norm to 0 as $n \to \infty$). In such a case, the reverse martingale convergence theorem can be used to show that L_{n1} converges a.s. to its centering constant, as $n \to \infty$). Sen (1981a, ch. 7) exploited this reverse martingale property for the particular case of L_n in (3.40) to cover a general class of L_{n1} for which $L_{n1} = L_n^0 + \xi_n$, where

$$\{L_n^0\} \text{ is a reverse martingale and } \quad \xi_n \stackrel{a.s.}{\to} 0 \text{ as } n \to \infty. \tag{3.51}$$

[See Lemma 7.4.1 and Theorem 7.5.1 of Sen (1981a) for some generalizations of (3.51)]. Actually, looking at (3.33), we may also rewrite L_{n1} as

$$L_{n1} = \int_{\boldsymbol{R}_1} J_{n,1}(F_n(x)) dF_n(x), \tag{3.52}$$

where by the Glivenko–Cantelli theorem, $\|F_n - F_\theta\| \to 0$ a.s. as $n \to \infty$. If we assume that $J_{n,1}(.)$ converges to some $J_1(.)$ (as $n \to \infty$), then, under suitable conditions on $J_1(.)$ and F_θ, the consistency of L_{n1} can be studied. If $J_1(.)$ can be expressed as difference of two monotone and continuous functions [on (0,1)], then the following holds:

THEOREM 3.1 *Let r and s be two positive numbers such that $r^{-1}+s^{-1} = 1$ and $\int_{\boldsymbol{R}_1} |x|^r dF_\theta(x) < \infty$ and $\limsup_n \|J_{n,1}\|_s < \infty$. Then*

$$L_{n1} - \lambda_1(\theta) \to 0 \quad a.s., \ as \quad n \to \infty \tag{3.53}$$

where $\lambda_1(\theta) = \int_{\boldsymbol{R}_1} J_1(F_\theta(x)) dF_\theta(x)$.

[See Sen 1981a, Theorem 7.6.1]. Van Zwet (1980) relaxed the assumption of pointwise convergence of $J_{n,1}(u)$ to some $J_1(u), u \in (0,1)$ [under $\|J_{n,1}\|_s <$

∞, uniformly in n] and also obtained a parallel result for $r = \infty$ (i.e., for bounded random variables); we leave these as exercises. In the context of robust estimation of location and regression, we usually have a score function J of bounded variation, and also, as in the case of the trimmed or Winsorized mean, (3.43) may hold. In such a case, the moment condition on F may be avoided, and we have the following result (see Huber 1969, ch. 3):

THEOREM 3.2 ***[Huber 1969].*** *If $J(.)$ and $F_\theta^{-1}(.)$ have no common points of discontinuity, and for some $\alpha \in (0, 1/2)$*

(a) $J(u) = 0$ whenever u does not lie in $[\alpha, 1-\alpha]$,

(b) both J and F_θ^{-1} have bounded variation on $[\alpha, 1-\alpha]$,

then

$$L_{n1} \xrightarrow{p} \int_0^1 J(t)F_\theta^{-1}(t)dt, \text{ under } F_\theta \tag{3.54}$$

for $J_{n,1}(u) = J(u), u \in (0,1)$, as $n \to \infty$.

The proof follows directly by using (3.52) with $J_{n,1}(t) = J(t)$ and bounded variation conditions on $J(.)$ and $F_\theta^{-1}(.)$, along with the Glivenko–Cantelli Theorem. We omit it.

By virtue of (3.44), (3.50), and (3.54), writing

$$\mu(F_\theta) = \int_0^1 J(t)F_\theta^{-1}(t)dt + \sum_{j=1}^{k} a_j F_\theta^{-1}(p_j), \tag{3.55}$$

we conclude that under the regularity conditions assumed before,

$$L_n \to \mu(F_\theta), \text{ in probability (actually, a.s.), as } n \to \infty \tag{3.56}$$

So far we have assumed that X_i's have the distribution function F_θ. If X_i's have some other distribution function G on $\boldsymbol{R}_1$, we may replace F by G in (3.46) through (3.56) and parallel to (3.56), we may conclude that $L_n \to \mu(G)$, in probability, as $n \to \infty$. For the model F_θ relating to (3.29), we have that a general L_n in (3.44) is a consistent estimator of θ provided that

$$\mu(F_\theta) = \theta, \quad \text{for } \theta \in \Theta. \tag{3.57}$$

This is the case, for instance, in the location model: $F_\theta(x) = F(x-\theta), \theta \in \boldsymbol{R}_1$ with F symmetric about 0, provided

$$J(1-t) = J(t), \quad t \in (0,1), \tag{3.58}$$

$$a_j = a_{k-j+1}, \; p_j = p_{k-j+1}, \text{ for } j = 1, \ldots, k, \tag{3.59}$$

$$\text{and } \int_0^1 J(t)dt + (a_1 + \ldots + a_k) = 1. \tag{3.60}$$

The condition (3.58) is sufficient for (3.57) for all F belonging to the class of symmetric distribution functions. For a special form of F we need to verify is that (3.55) equals to 0 at $\theta = 0$. (3.58) holds for the trimmed and Winsorized means.

The trimmed L-estimators are generally preferred from the point of view of robustness. Toward this end, (3.54)–(3.56) provide a useful picture for the consistency, although various authors have tried to establish the consistency and asymptotic normality of L_n under less restrictive conditions on the score function, purely for academic interest. Some of these results are displayed in detail by Serfling (1980), Sen (1981a), and others. Keeping Equation (3.58) in mind, let us mention the following asymptotic normality result for a trimmed L-estimator.

THEOREM 3.3 *[Boos 1978]. Let $J(.)$ be bounded and continuous on [0,1] up to a set of Lebesgue measure 0 and G^{-1}-measure 0, and let $J(u)$ be equal to 0 for $u \notin [\alpha, 1-\alpha]$, for some $\alpha:\ 0 < \alpha < 1/2$. Let the distribution function G of the X_i have finite and positive derivative g at $G^{-1}(p_j)$ for $j = 1, \ldots, k$. Then*

$$n^{1/2}(L_n - \mu(G)) \xrightarrow{\mathcal{D}} \mathcal{N}(0, \sigma^2(G)), \tag{3.61}$$

provided that $0 < \sigma^2(G) < \infty$, where

$$\begin{aligned} \sigma^2(G) &= Var\Big\{ - \int [I(X_1 \le y) - G(y)] J(G(y) dy \\ &+ \sum_{j=1}^{k} \frac{a_j}{g(G^{-1}(p_j))} [p_j - I(X_1 \le G^{-1}(p_j))] \Big\}. \end{aligned} \tag{3.62}$$

If the second component in (3.44) vanishes, that is, $L_n = L_{n1}$, then $\sigma^2(G)$ in (3.62) reduces to

$$\int\int J(G(x)) J(G(y)) [G(x \wedge y) - G(x)G(y)] dx dy = \sigma_0^2(G), \text{ say.}$$

Further, if $G(x) = F_\theta(x) = F(x - \theta)$ where F is symmetric about 0 and if $J(u) = J(1-u)$, $\forall u \in (0,1)$, then [see (3.58)] $\mu(G) = \theta$ and $\sigma_0^2(G) = \sigma_0^2(F)$. Thus, if F admits first- and second-order derivatives f and f', respectively, a.e. such that $I(f) = \int [f'(x)/f(x)]^2 dF(x) < \infty$, then on letting

$$J(u) = J_F(u) = \psi'(F^{-1}(u))/I(f);\ \psi(x) = -f'(x)/f(x),\ x \in \boldsymbol{R}_1, \tag{3.63}$$

we immediately obtain that the corresponding L_n has the asymptotic variance $\sigma_0^2(F)$, where

$$\sigma_0^2(F) = \{I(f)\}^{-1}.$$

Therefore, based on the score function $J_F(.)$, L_n is asymptotically efficient. Also note that in order that $J_F(u) = 0$ for $u < \alpha$ or $u > 1 - \alpha$, for some $\alpha:\ 0 < \alpha < 1/2$, we must have $\psi(x) =$ constant for $x \le F^{-1}(\alpha)$ or $x \ge F^{-1}(1-\alpha)$. In other words, $f(x)$ should be exponentially decreasing in the two tails. The exponentially decreasing tail behavior of f characterizes the asymptotic optimality of trimmed L-estimators. (3.63) also tells us that if $\psi(x)$ is not nondecreasing (i.e., f is not log-concave), then $\psi'(.)$ may be negative somewhere, so that an asymptotically efficient L_n may not have all nonnegative c_{ni}. This may be verified by a simple example of a model where $F(x) = \pi F_1(x) + (1-\pi)F_2(x)$, both F_1 and F_2 are symmetric (but they are not the same) and $0 < \pi < 1$. The proof of (3.61) is not presented here, but in the next chapter we will develop a deeper linearity result for L-estimators that will contain the proof of this asymptotic normality as a by-product (and under more general regularity conditions). For other approaches to the asymptotic normality of L-estimators, we refer to Serfling (1980), Sen (1981a), and Shorack and Wellner (1986), among others. However, we will only elaborate the „linearity" approach.

Based on what has been discussed so far, we may summarize that for the location problem the class of L-estimators discussed before contains asymptotically efficient elements. Whenever the tails of a distribution are of exponential form (i.e., they are heavier than the normal ones), the trimmed L-estimators appear to be very desirable from robustness and asymptotic efficiency considerations. On the other hand, for even heavier type of tails (of a symmetric distribution function), such as in the Cauchy type distribution function, whenever the density possesses a finite Fisher information, one may obtain an optimal score function leading to an asymptotically optimal L-estimator. However, in such a case the density is not generally log-convex, and hence, this optimal score function may even lead to some negative scores outside an interval $[\alpha, 1-\alpha]$, where $0 < \alpha < 1/2$, although it may still be bounded and continuous (a.e.). Finally, if the true distribution function G does not belong to the model $\{F_\theta : \theta \in \Theta\}$, then the L-estimator is a consistent estimator of $\mu(G)$, defined by (3.55) with F_θ replaced G. Therefore, (3.57), namely, $\mu(G) = \theta$, provides a characterization for the consistency of an L-estimator of location when the true distribution function G may not belong to the assumed model. In particular, if G is symmetric about its location (θ), then $\mu(G) = \theta$ for all symmetric score functions (for which the integral in (3.55) exists, with F replaced by G) and hence, L-estimators may generally be considered in a more general setup. Within this broad framework, particular score functions may be chosen in such a way that they retain asymptotic efficiency for suitable subclasses of symmetric distribution functions. This also points out that for symmetric score functions a departure from the assumed symmetry of the distribution function G may have some effect, and the L-estimator may cease to be a consistent estimator of the location parameter. This aspect of the vulnerability of L-estimators to departures from the assumed symmetry of the underlying distribution function G has prompted many workers to study the robustness of L-estimators to infinitesimal asymmetry, and compare the same with other types of estimators. While there seems to be no problem in studying this in an asymptotic setup, the small sample study requires extensive numerical computations, and so will not be considered here.

In (3.45), we have considered an L-estimator of dispersion of a distribution.

Unlike in the case of location parameters, here, in general, a suitably posed measure of dispersion may not be identified with the scale parameter of a distribution function, even if the distribution function may have a location-scale form. For example, the population counterpart of (3.45), namely, the population interquartile range $G^{-1}(3/4) - G^{-1}(1/4)$, is a multiple of the scale parameter, where this multiplicity factor is itself a functional of the underlying distribution function. thus, whereas in the location model a measure of location may be identified with the point of symmetry of the distribution function for the entire class of symmetric distribution functions, in the scale model we need to proceed in a different manner. A very natural setup is to allow G to belong to a suitable family $\mathcal{F}$ of distribution function and to consider a *dispersion-functional* $T(G)$ as a suitable measure of dispersion. In this context we may naturally impose the two basic conditions on $T(.)$:

1. The functional is *translation-invariant*; that is, if the original random variables $X_1, \ldots, X_n$ are simultaneously translated by a constant a ($\in \boldsymbol{R}_1$), then for every $a \in \boldsymbol{R}_1$, the measure computed from the $X_i - a$ $(1 \le i \le n)$ is the same as that computed from $X_1, \ldots, X_n$. (The same conclusion applies to the population counterparts too, for every $a \in \boldsymbol{R}_1$.)
2. The functional is *scale-equivariant*; that is, if the X_i's are replaced by bX_i, $i = 1, \ldots, n$, for some $b \in \boldsymbol{R}_1$, then the functional is also $|b|$ times the original one (in the sample as well as population case) for every $b \in \boldsymbol{R}_1$.

Quantitatively, we also expect $T(G)$ should be greater than $T(F)$ if G is more spread out than F. When F and G both have the same functional form and differ possibly in location/scale parameters, then of course the above requirement can easily be verified for various specific measures. However, two unrelated distribution functions may not always have this inequality. One approach to this problem is to consider the set of paired differences $X_i - X_j, 1 \le i < j \le n$, and to formulate a measure of dispersion based on $M = n(n-1)/2$ observations $\{|X_i - X_j| = Z_{ij}$, say, $1 \le i < j \le n\}$. The Z_{ij}'s are clearly translation-invariant (on the X_i) and scale-equivariant, and a very natural way may be to consider an L_p-norm:

$$U_n^{(p)} = \left\{ \binom{n}{2}^{-1} \sum_{1 \le i < j \le n} Z_{ij}^p \right\}^{1/p}, \quad \text{for some } p > 0. \tag{3.64}$$

For the particular case of $p = 2$, (3.64) is a scalar multiple of the usual standard deviation (and is known to be a nonrobust estimator), while for $p = 1$, it is related to the so-called Gini's mean difference. In any case $(U_n^{(p)})^p$ is a simple U-statistic, and hence, the basic theory developed by Hoeffding (1948), and others, may easily be incorporated to study the optimality and other properties of this estimator. Another possibility is to order the $M = n(n-1)/2$ values of the Z_{ij}'s, denote them by $Z_{M:1}, \ldots, Z_{M:M}$, and consider a linear combination of these order statistics as a suitable measure of the dispersion of the original X_i's. In the above formulation, it is also possible to use the order statistics $g(Z_{M:k})$, $k = 1, \ldots, M$, where $g(z)$ is monotone on $\boldsymbol{R}^+$ [e.g., $g(z) = z^2$]. Bickel and Lehmann (1976, 1979) have advocated the use of the trimmed mean of $g(Z_{M:k})$ as a suitable measure of dispersion. The basic difference between the two sets of order statistics $\{X_{n:i}\}$ and $\{Z_{M:k}\}$ is that whereas $X_{n:i}$'s are generated by n independent random variables, the $Z_{M:k}$'s correspond to the Z_{ij}'s that are not all independent. This difference has been taken into account in the formulation of the *generalized L-estimators* (see Serfling 1980), and we will briefly comment on them later. Concerning measures of dispersion for symmetric G, Bickel and Lehmann (1979) suggested a broad class of measures of the form

$$\triangle(G) = \left[\int_{1/2}^{1} \left\{ G^{-1}(t) - G^{-1}(1-t) \right\}^p d\Lambda(t) \right]^{1/p}, \quad p > 0, \tag{3.65}$$

where $\Lambda(.)$ is any finite measure on $[1/2, 1]$. Replacing $G^{-1}(t)$ by the sample quantile function $Q_n(t)$ [see (3.3.3)], we arrive at an estimator of $\triangle(G)$. For $p = 1$, this is a linear function of the order statistic $\{X_{n:k}\}$, although for $p \neq 1$, it is no longer the case. Nevertheless, such estimators can well be approximated by a linear combination of functions of order statistics, and hence, they will also be termed L-estimators.

Rivest (1982) combined M- and L-estimators of location into a form (allowing for studentization) that can be called L-M-estimators. Such an estimator (T_n) is defined as a solution of the equation

$$\sum_{i=1}^{n} J_n(i/(n+1)) \psi((X_{n:i} - t)/s_n) = 0 \quad \text{(with respect to } t\text{)}, \tag{3.66}$$

where s_n is a suitable estimator of the scale parameter, and $J_n(.)$ and $\psi(.)$ are defined as in Section 3.2 and in this section. For $J(.) \equiv 1$, T_n reduces to an M-estimator (considered in Section 3.2), while for $\psi(x) \equiv x$, T_n is an L-estimator. Like the generalized L-estimators, these L-M-estimators may also be characterized in terms of some empirical distribution functions, and the theory to be developed in Section 6.3 will be applicable to these estimators as well.

Before we pass on to the regression model, we may remark that there is an extensive literature on L-estimators and allied order statistics. For some early developments on order statistics, we may refer to Daniell (1920), Lloyd (1952), as well as the monograph edited by Sarhan and Greenberg (1962) (where other references are cited). The monograph of David (1981) contains a very thorough treatment of important statistical uses of order statistics. In the literature L-estimators arise very frequently, and the early need for the treatment of asymptotic theory for L-estimators was stressed by Bennett (1952), Jung (1955), Chernoff, Gastwirth, and Johns (1965), Birnbaum and Laska (1967), among others. More general (and diverse) asymptotic theory appeared in later works by Stigler (1969, 1973, 1974), Sen (1977, 1980, 1981a)

Helmers (1977, 1980, 1981), Boos (1979), Boos and Serfling (1979), and others. Some detailed (and unified) studies of the asymptotic theory of L-estimators are made by Shorack and Wellner (1986), while various Monte Carlo studies were reported by the Princeton group, Andrews et al. (1972). We will discuss some aspects of this asymptotic theory in subsequent text.

In a regression model, the observations are not all identically distributed, but the errors are so. Bickel's (1973) idea sparked the activity incorporating the residuals based on some preliminary estimators in the formulation of suitable quantile processes. Bickel's estimators have good efficiency properties, but they are generally computationally complex and are not invariant under a reparametrization of the vector space spanned by the columns of $\mathbf{X}$, defined after (3.13). However, Bickel's ingenuity lies in the demonstration of regression counterparts of location L-estimators. Ruppert and Carroll (1980) proposed another L-estimator of regression parameters along the same vein, but its asymptotic properties depend on the initial estimator.

A more promising direction is due to Koenker and Bassett (1978). They extended the concept of quantiles to linear models (termed *regression quantiles*). To illustrate their methodology, let us consider the usual linear model:

$$X_i = c_{i1}\beta_1 + \cdots + c_{ip}\beta_p + e_i, \quad i = 1, \ldots, n \ (> p),$$

where $\beta_1, \ldots, \beta_p$ are the unknown regression parameters, the c_{ij}'s are known regression constant, and the error components e_i's are i.i.d. random variables with a distribution function G. For a fixed $\alpha: \ 0 < \alpha < 1$, we define

$$\psi_\alpha(x) = \alpha - I(x < 0) \quad \text{and} \quad \rho_\alpha(x) = x\psi_\alpha(x), \quad \text{for } x \in \boldsymbol{R}_1. \tag{3.67}$$

Then the α-regression quantile is defined as a solution $\widehat{\boldsymbol{\beta}}(\alpha)$ of the minimization

$$\sum_{i=1}^{n} \rho_\alpha\Big(X_i - \sum_{j=1}^{n} c_{ij}t_j\Big) := \min \tag{3.68}$$

with respect to $\mathbf{t} = (t_1, \ldots, t_p)^\top$. Notice that $\widehat{\boldsymbol{\beta}}(\alpha)$ is in fact an M-estimator: Koenker and Bassett (1978) proved that $\widehat{\boldsymbol{\beta}}(\alpha)$ is a consistent estimator of $(\beta_1 + G^{-1}(\alpha), \beta_2, \ldots, \beta_p)^\top$ whenever $c_{i1} = 1$, for every $i = 1, \ldots, n$, and they also showed that the asymptotic distribution of $\widehat{\boldsymbol{\beta}}(\alpha)$ is quite analogous to that of the sample α-quantile in the location case. This regression quantile seems to provide a reasonable basis for L-estimation in linear models. Koenker and Bassett (1978) also suggested the use of trimmed least squares estimators (TLSE) in linear models. This idea may be posed as follows: First, consider two numbers α_1, α_2, such that $0 < \alpha_1 < \alpha_2 < 1$ (typically $\alpha_1 = 1 - \alpha_2$), and calculate the regression quantiles $\widehat{\boldsymbol{\beta}}(\alpha_1)$ and $\widehat{\boldsymbol{\beta}}(\alpha_2)$ along the lines of (3.67) and (3.68)]. Second, trim off all observations such that

$$X_i \leq \sum_{j=1}^{P} c_{ij}\hat{\beta}_j(\alpha_1) \text{ or } X_i \geq \sum_{j=1}^{P} c_{ij}\hat{\beta}_j(\alpha_2), \quad i = 1, \ldots, n.$$

Finally, compute the ordinary LSE from the remaining (untrimmed) observations. Koenker and Bassett (1978) made a conjecture that the trimmed LSE may be regarded as a natural extension of the classical trimmed mean (for the location model) to the regression model. Later on, this conjecture was supported by Ruppert and Carroll (1980) who derived the asymptotic distribution of the trimmed LSE and illustrated the close resemblance between the two trimmed estimators. Based on a Monte Carlo study, Antoch, Collomb, and Hassani (1984) demonstrated some other good properties of the trimmed LSE. Besides the trimmed LSE, other estimators and tests may also be effectively based on the regression quantiles in (3.68). We will have occasion to discuss some of these estimators and tests in greater depths in Chapters 4 and 6. For completeness, we may mention the work of Antoch (1984), Bassett and Koenker (1978, 1982, 1984), Jurečková (1983 a,b,c, 1984c), Jurečková and Sen (1984), Koenker and Bassett (1984), Portnoy (1983), among others. Incorporating some of these developments in a broader setup, we will study the first- and second-order asymptotic relations of L-estimators to other estimators (in general linear models) and utilize these results in the study of asymptotic properties of L-estimators and L-tests in linear models.

We see easily that the finite sample breakdown point of the L-estimator (3.30) of location is $k+1$ when k extremes are trimmed off on either side, that is,

$$c_i = 0, \quad i = 1, \ldots, k, n-k+1, \ldots, n, \quad k \leq \frac{n}{2},$$

On the other hand, L_n is translation equivariant and monotone in each argument provided that $c_i \geq 0$, $i = 1, \ldots, n$, and hence, Theorem 2.1 applies. We can show that for a distribution of type II,

$$\lim_{a \to \infty} B(a, L_n) = k+1. \tag{3.69}$$

hence, we say that $\mathbb{P}_\theta(|L_n - \theta| > a)$ converges to 0 as $a \to \infty$, $(k+1)$ times faster than $\mathbb{P}(|X_1| > a)$. The sample median is certainly the most robust in this context. To prove (3.69), let us assume that exactly k points are trimmed off on each side, while $c_{n,k+1} = c_{n,n-k} > 0$. Then we can write

$$\begin{aligned}
&\mathbb{P}_0(L_n > a) \\
&\quad \geq \mathbb{P}_0(X_{n:i} > -a, \ i = k+1, \ldots, n-k-1) \\
&\quad \geq \mathbb{P}_0\Big(X_{n:i} > -a, \ i = k+1, \ldots, n-k-1, \ X_{n:n-k} > a(\frac{2}{c_{n-k}} - 1)\Big) \\
&\quad \geq \mathbb{P}_0\Big(X_1 > -a, \ldots, X_{n-k-1} > -a, X_{n-k} > a(\frac{2}{c_{n-k}} - 1), \ldots \\
&\qquad\qquad \ldots, X_n > a(\frac{2}{c_{n-k}} - 1)\Big) \\
&\quad = (F(a))^{n-k-1}\Big[1 - F(a(\frac{2}{c_{n-k}} - 1))\Big]^{k+1}.
\end{aligned}$$

We obtain the same inequality for $\mathbb{P}_0(L_n < -a)$. thus,

$$\begin{aligned}
&\limsup_{a\to\infty} B(a, L_n) \\
&\leq \limsup_{a\to\infty} \frac{-(k+1)\log\left[1 - F\{a(2/c_{n-k} - 1)\}\right]}{-\log\left(1 - F(a)\right)} \\
&\leq \lim_{a\to\infty} (k+1)\frac{m \log a(2/c_{n-k} - 1)}{m \log a} = k+1,
\end{aligned}$$

while $\liminf B(a, L_n) \geq k+1$ by Theorem 2.1, and we arrive at (3.69). For breakdown properties of regression L-estimators, we refer to Chapter 4.

3.4 R-Estimators

Note that the ranks $(R_1, \ldots, R_n)$ of the observations $(X_1, \ldots, X_n)$ are invariant under a large class of monotone transformations. This invariance property yields robustness of rank-based tests against outliers and other distributional departures, and hence, estimators of location based on rank tests are expected to enjoy parallel robustness properties. This basic observation lays the foundation of rank (R-) estimators of location, and the same idea extends to the linear models as well. Recall that the sample median is essentially tied down to the classical sign test, and a very similar phenomenon holds for general rank statistics and their derived estimators. In the literature this is referred to as the *alignment principle.* That is, the observations may be so aligned that a suitable rank statistic based on these aligned observations is least significant, and from such aligned observations the R-estimators can easily be derived. Dealing with rank statistics, we have generally much less restrictive assumptions on the underlying distributions, and hence, from robustness perspective, rank tests are *globally robust* (compared to the M- and L-estimators, which are generally *locally robust*). The picture remains the same in the location as well as regression models, although the computational aspects for the R-estimators in linear models may be far more complicated than in the location model. In general, though an algebraic expression for an R-estimator in terms of the observations may not exist, there are already fast computational algorithms for ranking. The robustness and efficiency considerations sometimes make the R-estimators preferable to other competing ones.

We will consider the various properties of these R-estimators in a more unified manner. For example, we will begin with R-estimators of location based on the sign statistics and then move on to more general case. Let $X_1, \ldots, X_n$ be n i.i.d. random variables with a continuous distribution function F_θ, where θ is the median of the distribution and is assumed to be unique (so that F is strictly monotone at θ). Then we write

$$\theta = F_\theta^{-1}(0.5). \tag{3.70}$$

For testing the null hypothesis

$$H_0 : \theta = \theta_0 \text{ against } H : \theta \neq \theta_0, \tag{3.71}$$

the simple sign test statistic is given by

$$S_n^0 = \sum_{i=1}^{n} \text{sign}(X_i - \theta_0). \tag{3.72}$$

Note that under H_0, S_n^0 has a completely specified distribution with mean 0 and variance n. For notational simplicity, we write

$$S_n(t) = S(X_1 - t, \dots, X_n - t) = \sum_{i=1}^{n} \text{sign}(X_i - t), \quad t \in \boldsymbol{R}_1. \tag{3.73}$$

$S_n(t)$ is nonincreasing in $t \in \boldsymbol{R}_1$, while $S_n(0)$ has a distribution centered at 0. hence, we define an estimator $\hat{\theta}_n$ of θ by equating $S_n(t)$ to 0. We may note that $S_n(t) = 0$ may not have an exact root or the root may not be unique. To illustrate this point, let $X_{n:1}, \dots, X_{n:n}$ be the order statistics corresponding to $X_1, \dots, X_n$. Consider then a typical t lying between $X_{n:k}$ and $X_{n:k+1}$, for $k = 0, \dots, n$, where $X_{n:0} = -\infty$, and $X_{n:n+1} = +\infty$. Then it is easy to show that

$$S_n(t) = (n - 2k) \quad \text{for } X_{n:k} < t < X_{n:k+1}, \ k = 0, \dots, n, \tag{3.74}$$

and, if we let $\text{sign}(0) = 0$, then $S_n(X_{n:k}) = (n - 2k + 1)$ for $k = 1, \dots . n$. Thus, if n is odd ($= 2m+1$, say), we obtain that $\hat{\theta}_n = X_{n:m+1}$, while if n is even ($=2m$, e.g.), we have by (3.74) that $S_n(t) = 0$ for every $t \in (X_{n:m}, X_{n:m+1})$. In the latter case, we may take the centroid of the set of admissible solutions as the estimator $\hat{\theta}_n = (X_{n:m} + X_{n:m+1})/2$, and this agrees with the conventional definition of the sample median. Note that we do not need to assume that F_θ is symmetric about θ in this setup. If, however, we are able to make this additional assumption on the distribution function, then we can use a general class of signed-rank statistics for the testing problem in (3.71), and these in turn provide us a general class of R-estimators of location. thus, we assume that

$$F_\theta(x) = F(x - \theta), \quad \text{where } F(y) + F(-y) = 1 \ \forall y \in \boldsymbol{R}_1, \tag{3.75}$$

so that (3.70) holds and θ is median as well as the center of symmetry of the distribution function F_θ. For the hypothesis testing problem in (3.71), we may consider a general signed rank statistic

$$S_n(\mathbf{X}_n - \theta_0 \mathbf{1}_n) = \sum_{i=1}^{n} \text{sign}(X_i - \theta_0) a_n(R_{ni}^+(\theta_0)), \tag{3.76}$$

where $R_{ni}^+(\theta_0)$ is the rank of $|X_i - \theta_0|$ among $|X_1 - \theta_0|, \dots, |X_n - \theta_0|$ for $i = 1, \dots, n$, and $a_n(1) \leq a_n(2) \leq \dots a_n(n)$ are given scores. The particular case of $a_n(k) = k/(n+1)$, $k = 1, \dots, n$, leads to the *Wilcoxon signed-rank statistic*. Further, for $a_n(k)$ = expected value of the k-th smallest observation in a sample of size n from the chi distribution with 1 degree of freedom, for $k = 1, \ldots, n$, we have the *one-sample normal score statistic.* Note that under H_0, $X_i - \theta_0$ has a distribution function F symmetric about 0, and hence, $\text{sign}(X_i - \theta_0)$ and $R^+_{ni}(\theta_0)$ are stochastically independent. The vector of signs assumes all possible 2^n realizations with the common probability $1/2^n$, and the vector of ranks assumes all possible $n!$ realizations [the permutations of $(1, \ldots, n)$] with the common probability $1/(n!)$. hence, under H_0, the statistic in (3.76) has a distribution with mean 0, independently of the underlying continuous F. As in (3.73), we may write $S_n(t)$ for $S_n(\mathbf{X}_n - t\mathbf{1}_n), t \in \boldsymbol{R}_1$. Then it is easy to see that whenever the $a_n(k)$'s are monotone (in k), $S_n(t)$ is nonincreasing in $t (\in \boldsymbol{R}_1)$. Thus, as in the case of the sign statistic, we may equate $S_n(t)$ to 0 to obtain an estimator of θ. In general, $S_n(t) = 0$ does not have a unique root. We can set

$$R_n^- = \sup\{t : S_n(t) > 0\}, \; R_n^+ = \inf\{t : S_n(t) < 0\}, \tag{3.77}$$

$$R_n = \frac{1}{2}\{R_n^- + R_n^+\}. \tag{3.78}$$

Then R_n is an R-estimator of θ. Note that for $a_n(k) = 1, \; k = 1, \ldots, n$, (3.76) reduces to the sign statistic in (3.72), and hence, R_n reduces to the sample median. thus, the sample median may be interpreted as an L-estimator as well as an R-estimator. It may also be interpreted as an M-estimator when in (3.2.15)–(3.2.16) we take $\psi(t) = \text{sign}(t), \; t \in \boldsymbol{R}_1$. Similarly, for the case of the Wilcoxon signed rank statistic [in (3.76)], we have by arguments similar to the case of the sign statistic

$$R_n(W) = \text{median}\Big\{\frac{1}{2}(X_i + X_j) : 1 \leq i \leq j \leq n\Big\}. \tag{3.79}$$

However, for general $a_n(k)$ (which are nonlinear in $k/n, \; k = 1, \ldots, n$), such an algebraic expression for the R-estimator may not be available. But, using the sample median or the median of the mid-averages [i.e., (3.79)] as an initial estimator, an iterative solution for (3.77)–(3.78) can be obtained in a finite number of steps. For large samples, a one-step version may also be considered.

Note that for every real c and positive d, $\text{sign}(d(X_i + c - t - c)) = \text{sign}(X_i - t)$, for $t \in \boldsymbol{R}_1$, while the rank of $d|X_i + c - t - c|$ among $d|X_1 + c - t - c|, \ldots, d|X_n + c - t - c|$ is the same as the rank of $|X_i + c - t - c|$ among $|X_1 + c - t - c|, \ldots, |X_n + c - t - c|$ for $i = 1, \ldots, n, \; t \in \boldsymbol{R}_1$. hence, it is easy to verify that

$$R_n(\mathbf{X}_n + c\mathbf{1}_n) = R_n(\mathbf{X}_n) + c \quad \text{for every real } c, \tag{3.80}$$

that is, R_n is a translation-equivariant estimator of θ;

$$R_n(d\mathbf{X}_n) = dR_n(\mathbf{X}_n) \quad \text{for every } d, \in \boldsymbol{R}_1^+ \tag{3.81}$$

that is, R_n is scale-equivariant estimator of θ. Further, by the monotonicity of $S_n(t)$ in t and by (3.77)–(3.78), we obtain that for $t, \theta \in \boldsymbol{R}_1$,

$$P_\theta\{S_n(\mathbf{X}_n - t\mathbf{1}_n) < 0\} \;\; \leq \;\; P_\theta\{R_n \leq t\}$$

$$\leq \ P_\theta\{S_n(\mathbf{X}_n - t\mathbf{1}_n) \leq 0\}. \qquad (3.82)$$

Note that the statistic $S_n(\mathbf{X}_n - t\mathbf{1}_n)$ does not have a continuous distribution even if the distribution function F is continuous. However, when θ holds, $S_n(\mathbf{X}_n - \theta\mathbf{1}_n)$ has a distribution function independent of F such that

$$P_\theta\{S_n(\mathbf{X}_n - \theta\mathbf{1}_n) = 0\} = P_0\{S_n(\mathbf{X}_n) = 0\} = \eta_n,$$

where independently of F, $0 \leq \eta_n < 1, \forall n$ and $\lim_{n\to\infty} \eta_n = 0$. It follows that

$$\frac{1}{2}(1 - \eta_n) \leq P_\theta\{R_n < \theta\} \leq P_\theta\{R_n \leq \theta\} \leq \frac{1}{2}(1 + \eta_n), \ \ \forall\theta \in \boldsymbol{R}_1. \qquad (3.83)$$

Actually, for symmetric F, when θ holds, $X_i - \theta$ and $\theta - X_i$ both have the same distribution function F. On the other hand, by (3.76) and the fact that $\text{sign}(X_i - t) = -\text{sign}(t - X_i)$, we obtain that $R_n((-1)\mathbf{X}_n) = -R_n(\mathbf{X}_n)$ so that $R_n - \theta$ has a distribution function symmetric about 0 (although this distribution function may not be continuous everywhere). We may conclude that for a symmetric F,

$$R_n \text{ is a median-unbiased estimator of } \theta. \qquad (3.84)$$

Since, generally, R_n is not a linear estimator, evaluation of its moments requires more refined analysis. However, it will be shown later that R_n possesses finite mean and second moment under quite general conditions on the scores and the distribution function F to have finite variance for R_n to have so. For some details, we may refer to Sen (1980).

For location and shift parameters, the idea of using rank tests to derive R-estimators was elaborated in Hodges and Lehmann (1963) (see also Sen 1963). However, we should mention that in Walker and Lev (1953, ch. 18), Lincoln Moses coined the same principle for the derivation of distribution-free confidence intervals for location and shift parameters; Moses (1965) also contains an useful graphical device to obtain such estimators. The asymptotic normality and other related properties of R-estimators of location can be studied using (3.82) and the asymptotic normality of signed rank statistic for local (contiguous) alternatives (on letting $t = \theta + n^{-1/2}u, \ u \in \boldsymbol{R}_1$). However, this direct technique may not work out that well for the general linear model. As in the earlier sections, it will be convenient to develop an asymptotic representation of the form

$$n^{1/2}(R_n - \theta) = n^{-1/2}(\gamma(F))^{-1}\sum_{i=1}^{n}\phi(F(X_i - \theta)) + o_p(1), \qquad (3.85)$$

where the scores $a_n(k)$ are generated by the score function $\phi = \{\phi(t), \ 0 < t < 1\}$ and $\gamma(F)$ is a functional of the distribution function F and ϕ. This asymptotic representation extends directly to the regression model, and will be studied in Chapter 6. Generally, ϕ is a skew-symmetric function on (0,1), and hence, using the central limit theorem on the principal term on the right-hand side of (3.85) we get the asymptotic normality of R_n as an immediate corollary. This representation also justifies the adaptation of the Pitman efficiency of rank tests for the study of the asymptotic efficiency of R-estimators. We will discuss this in Chapter 6.

For two distributions differing only in location parameters, the usual two-sample rank tests can similarly be used to derive R-estimators of the shift parameter. This has been done by Hodges and Lehmann (1963), Sen (1963), and others. However, we may characterize the two sample problem as a particular case of the simple regression model where the regression constants c_i can only assume two values 0 and 1 with respective frequencies n_1 and n_2, and $n = n_1+n_2$. As such, the estimator of regression slope based on Kendall (1938) tau statistic (see Sen 1968a) contains the two-sample R-estimator based on Wilcoxon rank sum statistic as a particular case. Similarly, R-estimators of the regression parameter based on general linear rank statistics, initiated by Adichie (1967), pertain to the two-sample model as well. We will not enter into the detailed discussion on the two-sample models. Rather, we shall first present a general regression model and then append some simplifications in a simple regression model and in a two-sample model. Notable works in this direction are due to Jurečková (1969,1971a,b), Sen (1969), Koul (1971), and Jaeckel (1972), and some good accounts are given in Humak (1983; ch. 2) and Puri and Sen (1985, ch. 6).

Consider the usual linear model:

$$\mathbf{Y}_n = (Y_1, \dots, Y_n)^\top = \mathbf{X}\boldsymbol{\beta} + \mathbf{e}_n : \ \mathbf{e}_n = (e_1, \dots, e_n)^\top \tag{3.86}$$

where $\mathbf{X} = (\mathbf{x}_1^\top, \dots, \mathbf{x}_n^\top)^\top$ is a known design matrix of order $n \times p$, $\boldsymbol{\beta} = (\beta_1, \dots, \beta_p)^\top$ is the vector of unknown (regression) parameters, and the e_i's are i.i.d. random variables with the distribution function F. For the time being, we are assuming that none of the n rows of $\mathbf{X}$ has equal elements. Later we will take the first column of $\mathbf{X}$ as $(1, \dots, 1)^\top$ and introduce the necessary modifications. In the linear model in (3.86), for testing the null hypothesis $H_0 : \boldsymbol{\beta} = \boldsymbol{\beta}_0$ (specified) against $H : \boldsymbol{\beta} \neq \boldsymbol{\beta}_0$, one considers a vector of linear rank statistics

$$\mathbf{S}_n(\boldsymbol{\beta}_0) = (S_{n1}(\boldsymbol{\beta}_0), \dots, S_{np}(\boldsymbol{\beta}_0))^\top, \tag{3.87}$$

where

$$S_{nj}(\boldsymbol{\beta}_0) = \sum_{i=1}^{n} (c_{ij} - \bar{c}_{nj}) a_n(R_{ni}(\boldsymbol{\beta}_0)), \ \ j = 1, \dots, p; \tag{3.88}$$

$R_{ni}(\mathbf{t})$ is the rank of the residual

$$\delta_i(\mathbf{t}) = Y_i - \mathbf{x}_i^\top \mathbf{t} \tag{3.89}$$

among $\delta_1(\mathbf{t}), \dots, \delta_n(\mathbf{t})$, $\mathbf{t} \in \boldsymbol{R}_p$, $\bar{c}_{nj} = n^{-1} \sum_{i=1}^{n} c_{ij}$, $j = 1, \dots, p$, and $a_n(1) \leq \dots \leq a_n(n)$ are given scores. Note that $R_{ni}(\boldsymbol{\beta}_0)$ are exchangeable random variables, $i = 1, \dots, n$ under H_0, and the vector of these ranks takes on each of the $n!$ permutations of $(1, \ldots, n)$ with the common probability $(n!)^{-1}$. With respect to the uniform permutation distribution, we may enumerate the actual null hypothesis distribution (and moments) of the statistics in (3.88), and verify that

$$\mathbb{E}[\mathbf{S}_n(\boldsymbol{\beta}_0)|H_0] = \mathbf{0} \quad \text{and} \quad \operatorname{Var}[\mathbf{S}_n(\boldsymbol{\beta}_0)|H_0] = \mathbf{X}_n^{\star} \cdot A_n^2, \tag{3.90}$$

where

$$\mathbf{X}_n^{\star} = \sum_{i=1}^{n} (\mathbf{x}_i - \bar{\mathbf{x}}_n)(\mathbf{x}_i - \bar{\mathbf{x}}_n)^{\top}, \ A_n^2 = \frac{1}{n-1} \sum_{i=1}^{n} [a_n(i) - \bar{a}_n]^2, \tag{3.91}$$

with $\bar{a}_n = n^{-1} \sum_{k=1}^{n} a_n(k)$ and $\bar{\mathbf{x}}_n = n^{-1} \sum_{i=1}^{n} \mathbf{x}_i$. As in the location model, we may define here $\mathbf{S}_n(\mathbf{t})$, $\mathbf{t} \in \boldsymbol{R}_p$ by replacing $\boldsymbol{\beta}_0$ by $\mathbf{t}$, and are naturally tempted in equating $\mathbf{S}_n(\mathbf{t})$ to $\mathbf{0}$ to solve for an R-estimator of the regression parameter (vector) $\boldsymbol{\beta}$. For the case of the simple regression model (i.e., p=1), we can show that $S_{n1}(t_1)$ is nonincreasing in t_1, and hence, the justification for equating $S_{n1}(t)$ to 0 to obtain an estimator of β_1 follows the same alignment principle as in the location model. However, for $p \geq 2$, though $S_{nj}(\mathbf{t})$ is nonincreasing in t_j for each $j = 1, \ldots, p$, its behavior with the variation in t_s, $s \neq j$, may depend a lot on the design matrix $\mathbf{X}$. For example, if the p columns of $\mathbf{X}$ are pairwise concordant, then $S_{nj}(\mathbf{t})$ is nonincreasing in each of the elements of $\mathbf{t}$, and if the sth column of $\mathbf{X}$ is discordant to the jth column, then $S_{nj}(\mathbf{t})$ is nonincreasing in t_s. In the usual (fixed-effects) linear models, it is possible to parametrize in such a manner that each c_{ij} can be decomposed into $c_{ij}^{(1)} + c_{ij}^{(2)}$, such that the $\mathbf{x}_j^{(1)}$, $j = 1, \ldots, p$ are concordant, $\mathbf{x}_j^{(2)}$, $j = 1, \ldots, p$ are concordant, while $\mathbf{x}_j^{(1)}$ and $\mathbf{x}_j^{(2)}$ are discordant, for $j \neq s = 1, \ldots, p$. Under this setup, it is possible to express $S_{nj}(\mathbf{t})$ as a finite mixture of terms, each one monotone in the arguments ($\mathbf{t}$), and with this the alignment principle can be invoked to solve for $S_n(\mathbf{t}) = \mathbf{0}$ to derive the R-estimators in a meaningful way. This was essentially the idea of Jurečková (1971) for generating R-estimators in the multiple regression model. Again, since a unique solution may not generally be available, she defined an R-estimator $\mathbf{R}_n$ of $\boldsymbol{\beta}$ as a solution of the minimization problem

$$\sum_{j=1}^{p} |S_{nj}(\mathbf{t})| = (\text{ minimum with respect to } \mathbf{t} \in \boldsymbol{R}_p). \tag{3.92}$$

For the simple regression model, in view of the monotonicity of $S_{n1}(t_1)$ in t_1, the situation is parallel to that in (3.77)–(3.78). In this context, the asymptotic linearity of $S_{ni}(t)$ (uniformly in t in a small neighborhood of 0), in probability, established by Jurečková (1969), provides an easy access to the study of the asymptotic properties of R-estimators of regression. Parallel results for the general regression model (under additional regularity conditions on the design matrix $\mathbf{X}$) were also obtained by Jurečková (1971a, b). Later on, the regularity conditions on $\mathbf{X}$ were slightly relaxed by Heiler and Willers (1988). Jaeckel (1972) defined an R-estimator of regression as a solution of the minimization:

$$\sum_{i=1}^{n}\{a_n(R_{ni}(\mathbf{t}) - \bar{a}_n\}\delta_i(\mathbf{t}) = \text{ minimum (with respect to } \mathbf{t} \in \boldsymbol{R}_p). \quad (3.93)$$

Jaeckel interpreted (3.93) as a measure of dispersion of the residuals $\delta_i(t)$, $i = 1, \ldots, n$, and advocated the use of this measure instead of the usual residual variance which is generally used in the method of the least squares. He pointed out that both the solutions in (3.92) and (3.93) are asymptotically equivalent in probability, and hence, they share the same asymptotic properties. Another version of R_n was proposed by Koul (1971). Instead of the L_1-norm in (3.92), he suggested that the use of an L_2-norm involving a quadratic form in $\mathbf{S}_n(\mathbf{t})$ with $(\mathbf{X}_n^\star)^{-1}$ (or some other positive definite matrix) as the discriminant. All these three versions are asymptotically equivalent, and we will discuss the first version more thoroughly than others. It is difficult to say which of the three minimizations is computationally the simplest one. However, in each case an iterative procedure is generally needed. Using some initial estimator of $\boldsymbol{\beta}$ in defining the residuals as well as the aligned rank statistics, a Newton–Raphson method may then be used in the desired minimization problem. For large samples, this idea leads us to the adaptation of the so-called one-step R-estimators of regression, which will be considered in detail in Chapter 7.

It is assumed that the scores $a_n(k)$ are generated by a score function $\phi : (0,1) \to \boldsymbol{R}_1$ [i.e., $a_n(k) = \phi(k/(n+1))$, $k = 1, \ldots, n$], where usually ϕ is taken to be nondecreasing and square integrable. We set

$$\sigma^2(\phi, F) = \Big\{\int_0^1 \phi^2(u)du - \Big(\int_0^1 \phi(u)du\Big)^2\Big\}\Big\{\int_{\boldsymbol{R}_1} \phi(F(x))f'(x)dx\Big\}^{-2}, \quad (3.94)$$

where we assume that the distribution function F admits an absolute continuous density function f (with derivative f') almost everywhere, such that the Fisher information $I(f)$ is finite. Then, parallel to (3.85), we would have an asymptotic representation of R-estimators of regression in the form:

$$\mathbf{X}_n^\star(\mathbf{R}_n - \boldsymbol{\beta}) = \mathbf{X}_n^\star(\gamma(F,\phi))^{-1}\sum_{i=1}^{n}(\mathbf{x}_i - \bar{x}_n)\phi(F(Y_i - \mathbf{x}_i^\top\boldsymbol{\beta})) + o_p(1), \quad (3.95)$$

where $\mathbf{X}_n^\star$ is defined by (3.91) and $\gamma(F,\phi) = \int_{R_1}\phi(F(x))f'(x)dx$; a more detailed discussion of this representation will be considered in Chapter 6. From (3.95) and the multivariate central limit theorem, we immediately obtain that

$$(\mathbf{X}_n^\star)^{1/2}(\mathbf{R}_n - \boldsymbol{\beta}) \xrightarrow{\mathcal{D}} \mathcal{N}(\mathbf{0}, \sigma^2(\phi, F)\mathbf{I}), \quad \text{when } \boldsymbol{\beta} \text{ holds.} \quad (3.96)$$

Note that for the location model, (3.96) holds with $\mathbf{X}_n^\star = n$, $\beta = \theta$, and $I = 1$ (all scalar). As a result $\sigma^2(\phi, F)$ emerges as the key factor in the study of the asymptotic efficiency and other related properties of R-estimator. In this setup, we put

$$\psi(u) = -f'(F^{-1}(u))/f(F^{-1}(u)), \;\; u \in (0,1). \quad (3.97)$$

Then, noting that $\int \psi(u)du = 0$ and setting (without any loss of generality) that $\int_0^1 \phi(u)du = 0$, we obtain from (3.94) and (3.97) that

$$\sigma^2(\phi, F)I(f) = \frac{\langle\psi,\psi\rangle\,\langle\phi,\phi\rangle}{\langle\psi,\phi\rangle^2} \geq 1 \;\; \forall\phi \tag{3.98}$$

where $\langle\;,\;\rangle$ stands for the inner product (in the Hilbert space) and the equality sign in (3.98) holds only when $\psi(u)$ and $\phi(u)$ agrees up to any multiplicative constant. Thus, for $\phi \equiv \psi$, we have $\sigma^2(\phi, F) = (I(f))^{-1}$ (the Cramér–Rao information limit), and hence, the class of R-estimators contains an asymptotically efficient member. Drawing a parallel to the results discussed in Section 2.4, we also see that the class of R-estimators contains an asymptotically minimax member over a contaminated family of distributions (an analogue of the Huber minimax M-estimator). This also brings us to the need to study the interrelationships of M-, L-, and R-estimators for the common (location or regression) model, and this will be taken up in Chapter 7. We may remark here that the very fact that R-estimators are based on the ranks rather than on the observations themselves guarantees that they are less sensitive to the gross errors as well as to most heavy-tailed distributions.

Let us now consider a natural extension of the model in (3.86), namely,

$$\mathbf{Y}_n = \theta\mathbf{1}_n + \mathbf{X}\boldsymbol{\beta} + \mathbf{E}_n, \;\; \theta \in \boldsymbol{R}_1, \tag{3.99}$$

where all the other symbols have the same interpretation as in (3.86). Our task is to provide R-estimators of θ and $\boldsymbol{\beta}$. Note that the ranks $R_{ni}(\mathbf{t})$ remain invariant under any translation of the X_i (by a real $c \in \boldsymbol{R}_1$), and hence, $\mathbf{S}_n(\mathbf{t})$ in (3.87) fails to provide an estimator of θ. On the other hand, the R-estimators considered earlier for the estimation of $\boldsymbol{\beta}$ does not require F to be symmetric. Under the additional assumption that F is symmetric, we may use R-estimators of $\boldsymbol{\beta}$ to construct residuals, on which we may use the signed rank statistics to obtain R-estimators of θ. For the simple regression model, this was proposed by Adichie (1967), while for the general linear model, a more thorough treatment is due to Jurečková (1971b). For estimating the regression parameter vector $\boldsymbol{\beta}$ in the model (3.99), we use the linear rank statistic in (3.87)–(3.88) and denote the derived R-estimator by $\widehat{\boldsymbol{\beta}}_{n(R)}$. Consider then the residuals

$$\hat{Y}_i(t) = Y_i - t - \mathbf{x}_i^\top \widehat{\boldsymbol{\beta}}_{n(R)}, \;\; i = 1, \ldots, n, \;\; t \in \boldsymbol{R}_1. \tag{3.100}$$

On these residuals, we construct the signed rank statistic $S_n(t)$, as defined in (3.76) [with θ_0 replaced by t and $\mathbf{X}_n$ by $\mathbf{Y}_n - \mathbf{X}\widehat{\boldsymbol{\beta}}_{n(R)}$]. Then the R-estimator of θ is given by (3.77)–(3.78), as adapted to these residuals. Note that the residuals in (3.100) are not generally independent of each other, nor are they marginally identically distributed. hence, some of the exact properties of the R-estimators of location discussed earlier may not be generally true. (3.80) and (3.81) hold for this aligned estimator of θ, while in (3.82) the inequalities continue to hold if we replace $\mathbf{X}_n$ by $\mathbf{Y}_n - \mathbf{X}\widehat{\boldsymbol{\beta}}_{n(R)}$. However, under true $(\theta, \boldsymbol{\beta})$, the distributions of $S_n(\mathbf{Y}_n - \mathbf{X}\widehat{\boldsymbol{\beta}}_{n(R)} - \theta\mathbf{1}_n)$ may no longer be independent of the underlying F, so (3.83) and the median unbiasedness in (3.84) may no longer hold. This may also call for a more refined proof of the asymptotic normality of R-estimators of θ (taking into account the possible dependence of the residuals and their nonhomogeneity). Again, in this context, an asymptotic representation of R-estimators of the form of a mixture of (3.85) and (3.95) is available in the literature (viz., Jurečková 1971a, b), and this may be used with advantage to study the asymptotic properties of such aligned R-estimators. We will discuss such representations in more detail in Chapter 6. We may add here that for the simultaneous estimation of $(\theta, \boldsymbol{\beta})$, we need to assume that F is symmetric. Under this assumption it is possible to use more general signed rank statistics

$$S_{nj}(t_0, \mathbf{t}) = \sum_{i=1}^{n} \operatorname{sign}(Y_i - t_o - \mathbf{x}_i^\top \mathbf{t}) a_n(R_{ni}^+(t_0, \mathbf{t})), \; t_0 \in \boldsymbol{R}_1, \mathbf{t} \in \boldsymbol{R}_p,$$

for $j = 0, 1, \ldots, p$, where $c_{i0} = 1$, for $i = 1, \ldots, n$, and $R_{ni}^+(t_0, \mathbf{t})$ is the rank of $|Y_i - t_0 - \mathbf{x}_i^\top \mathbf{t}|$ among $|Y_1 - t_0 - \mathbf{x}_i^\top \mathbf{t}|, \ldots, |Y_n - t_0 - \mathbf{x}_i^\top \mathbf{t}|$, for $i = 1, \ldots, n$; the rest of the symbols have the same meaning as in before. At the true $(\theta, \boldsymbol{\beta})$, the vector $(S_{n0}(t_0, \mathbf{t}), \ldots, S_{np}(t_0, \mathbf{t}))^\top$ has a joint distribution independent of F and has the mean vector $\mathbf{0}$. Thus, by an adaptation of the same alignment principle as in the case of linear rank statistics, we have an R-estimator of $(\theta, \boldsymbol{\beta})$ as a solution of the minimization problem:

$$\sum_{j=0}^{p} |S_{nj}(t_0, \mathbf{t})| = \text{ minimum (with respect to } t_0 \in \boldsymbol{R}_1, \; \mathbf{t} \in \boldsymbol{R}_p).$$

Computational algorithm of (3.62) is naturally more involved than (3.92), although some simple approximations (based on iterations) are available for large samples. For the case of Wilcoxon signed rank statistics, this method was suggested by Koul (1969), although a more general situation is covered in Jurečková (1971a, b). In this context, the idea of using one-step R-estimators of location and regression parameters is appealing. For such R-estimators based on general signed rank statistics, results parallel to (3.95) and (3.96) hold, wherein we need to replace $\mathbf{X}_n^\star$ by

$$\mathbf{X}_n^{0\star} = \sum_{i=1}^{n} (1, \; \mathbf{x}_i^\top)^\top (1, \; \mathbf{x}_i) \quad [\text{of order } (p+1) \times (p+1)].$$

The discussions on asymptotically efficient scores and asymptotically minimax R-estimators made after (3.4.50) also pertain to these estimators.

In the case of the simple regression model,

$$X_i = \theta + \beta c_i + e_i, \; \; i = 1, \ldots, n,$$

the linear rank statistic $S_{n1}(t_1)$ is invariant under translation, and monotone nonincreasing in t_1, and hence, the R-estimator of slope β may be obtained in a convenient form (i.e., as the centroid of a closed interval. In this case, the R-estimator of intercept parameter θ may also be obtained in a convenient form by using (3.77)–(3.78) based on the residuals $X_i - \widehat{\beta}_n c_i, i = 1, \ldots, n$. On the other hand, a simultaneous solution for the intercept and slope based on (3.62) will be computationally heavier. This feature is generally true for the multiple regression model as well. For the R-estimator of regression based on linear rank statistics, the symmetry of F is not that necessary, while for the other R-estimators based on the signed rank statistics, symmetry of F constitutes an essential part of the basic assumptions. Any departure from this assumed symmetry of F may therefore have some impact on the R-estimators based on signed statistics. Hence, for the estimation of the slope alone, the linear rank statistics based R-estimators seem to have more robustness (against plausible asymmetry of the error distribution function) than the ones based on signed rank statistics (and the M-estimators considered in Section 3.2), where F is also assumed to be symmetric. A similar feature will be observed on L-estimators in the linear model (see Chapter 4), where the asymmetry leads to the bias only in estimation of the intercept and not of the slopes. Unlike the case of least square estimators, the M- and R-estimators of location or regression parameters are generally nonlinear, and hence, a re-parametrization of the model in (3.99), such as (for a nonsingular $\mathbf{B}$)

$$\mathbf{Y}_n = \theta \mathbf{1}_n + \mathbf{D}\boldsymbol{\gamma} + \mathbf{E}_n, \quad \text{where } \mathbf{D} = \mathbf{X}\mathbf{B} \quad \text{and } \boldsymbol{\gamma} = \mathbf{B}^{-1}\boldsymbol{\beta},$$

may lead to R-estimators of $\boldsymbol{\gamma}$ that do not agree (exactly) with the corresponding ones by premultiplying by $\mathbf{B}^{-1}$ the R-estimator of $\boldsymbol{\beta}$ when $p > 1$. For $p = 1$, the scale-equivariance property holds. This lack of invariance property is also shared by the M-estimators. This drawback of R- and M-estimators is particularly serious when the basic model is not be written in a unique way, and one may therefore want to preserve this invariance (under re-parametrization). However, as we will see later, this problem ceases to be of any significance in large samples. The asymptotic representation of the form (3.95), among other things, also ensures this re-parametrization invariance up to the principal order. The equivariance property holds in the simple regression model where the re-parametrization reduces to the translation and scale changes (see (3.80)–(3.81)). For a general linear model (not necessarily of full rank) there may not be a unique way of defining a reduced model of full rank, and in such a case, the particular basis should be chosen with care (unless n is large). The choice of canonical form for (3.99) eliminates this problem. Both R_n and M_n are affine-equivariant: If $\mathbf{d}_i^\top = \mathbf{x}_i^\top \mathbf{B}, \; i = 1, \ldots, n$, then $\sum_i \mathbf{d}_i \psi(X_i - \mathbf{d}_i^\top \mathbf{t}) = \mathbf{0} \Rightarrow \mathbf{B} \sum_i \mathbf{d}_i \psi(X_i - \mathbf{d}_i^\top \mathbf{t}) = \mathbf{0}$, and hence $\mathbf{B}\mathbf{M}_n^\star = \mathbf{M}_n$, similar for $\mathbf{R}_n$.

For the two-sample case, a linear rank statistic may equivalently be written in terms of a two-sample rank test statistics. Often this permits an easier way of solving for the minimization problem involved in the computation of the R-estimators. Based on the Wilcoxon two-sample rank-sum statistic, the

R-estimator of the difference of locations is given by

$$R_n = \text{ median } \Big\{Y_j - X_i : 1 \leq j \leq n_2,\ 1 \leq i \leq n_1\Big\},$$

where $Y_j = X_{n_1+j}, j = 1, \ldots, n_2$. However, for general rank statistic, the derived R-estimator may not have such a closed form even in the two-sample problem.

The multisample model of the shift in location is also a special case of the general regression model in (3.86). Here, for some fixed k (≥ 2), we have positive integers $n_1, \ldots, n_k$ such that $n = n_1 + \ldots + n_k$, and the vectors $\mathbf{x}_i$ can have only k possible realizations $(1, 0, \ldots, 0)^\top$, $(0, 1, \ldots, 0)^\top, \ldots, (0, \ldots, 0, 1)^\top$ (with respective frequencies $n_1, \ldots, n_k$). Our main interest lies in the estimation of a *contrast* $\lambda = \mathbf{l}^\top \boldsymbol{\beta}$ where $\mathbf{l}^\top \mathbf{1} = 0$. Such a contrast may also be equivalently expressed as a linear combination of all possible paired differences $\beta_j - \beta_{j'}$, $1 \leq j < j' \leq k$. For each $\beta_j - \beta_{j'}$, an R-estimator can be obtained from the relevant pair of samples, and hence, the corresponding linear combinations of these R-estimators may be taken as an estimator. However, for different combinations of paired differences, we may have different R-estimators (although these will be asymptotically equivalent, in probability). This nonuniqueness is mainly due to the nonlinear structure of the R-estimators. Lehmann (1963) suggested some *compatible* R-estimators in one-way analysis of variance models. Such compatible estimators are also relevant to the M-estimation theory in the usual analysis of variance models. An alternative solution to this compatibility criterion is to work with the individual sample (R- or M-)estimators of locations and to form a natural contrast in these estimators. In the context of biological assays, parallel line or slope ratio assays and other practical application, it is often possible to partition the set of observations $\mathbf{X}_n$ into subsets where within each subset we have a simple regression model. In such a case one may develop the theory of compatible R- (or M-)estimators by using the usual R- (or M-) estimation theory for each subset and then pooling these subset estimators by the usual least squares or other methods.

Summarizing, we see that the R-estimators in linear models may not be compatible under re-parametrization but can be made so with some additional manipulations. Generally, the R-estimators have good global robustness and efficiency properties, and they are translation and scale equivariant. On the other hand, they are usually calculated by some iterative procedure, though for large samples they may be approximated by one-step versions that require only one ranking. Further properties of such good R-estimators of location and regression parameters will be studied in Chapter 6.

Gutenbrunner (1986) and Gutenbrunner and Jurečková (1992) developed the concept of regression rank scores (which are extensions of the ordinary rank scores to linear models) and incorporated them in estimation in linear models, invariant to additional nuisance regression. Such estimators are based on appropriate linear programming algorithms, and usually computationally simpler than the usual R-estimators. Further developments due to Jurečková (1991, 1992) and Jurečková and Sen (1993) will be incorporated in a unified treatment of regression rank scores estimators in Chapter 6.

Example 3.1 Let us illustrate the estimators on the real datasets–daily precipitation amounts measured at two stations in the Czech Republic. The period covered is 1961–2007 and the samples of maximum annual 3-day precipitation amounts were drawn from the each station data. We focus on station in Valaške Meziříčí and Velké Meziříčí. The observations in Valašské Meziříčí are illustrated in Fig. 3.1, where we can see the outliers represented by extraordinarily high 1997 precipitation totals. No outliers were observed in Velké Meziříčí.

Table 3.1 Comparison of estimates based on dataset of stations Valašské Meziříčí and Velké Meziříčí

Estimator	Valašské	Velké
Mean $\bar{X}_n$	72.48	50.68
Median $\tilde{X}_n$	61.60	48.00
5%-trimmed mean $\bar{X}_{.05}$	66.28	49.81
10%-trimmed mean $\bar{X}_{.10}$	64.37	48.92
5%-Winsorized mean $\bar{W}_{.05}$	68.51	50.73
10%-Winsorized mean $\bar{W}_{.05}$	66.59	49.87
Huber M-estimator M_H	64.40	49.41
Hodges–Lehmann estimator HL	63.40	50.31
Sen weighted mean $k = [0.05n]$ $\mathcal{S}_k$	61.98	48.27
Sen weighted mean $k = [0.1n]$ $\mathcal{S}_k$	60.82	47.94
Sample standard deviation S_n	44.21	17.15
Interquartile range R_I	26.40	26.60
Median absolute deviation MAD	16.31	18.98
Gini mean difference G_n	34.51	19.20

When we wish to see the behavior of an estimator under various models, we usually simulate the model and look in the resulting values of the estimator of interest. As an example, 50 observations were simulated from the following probability distributions: Normal distribution $\mathcal{N}(0, 1)$ and $\mathcal{N}(10, 2)$ (symmetric and exponentially tailed distribution), Exponential $Exp(5)$ distribution(skewed and exponentially tailed distribution), Cauchy with the density

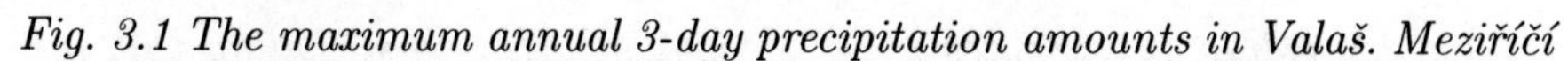

Fig. 3.1 The maximum annual 3-day precipitation amounts in Valaš. Meziříčí

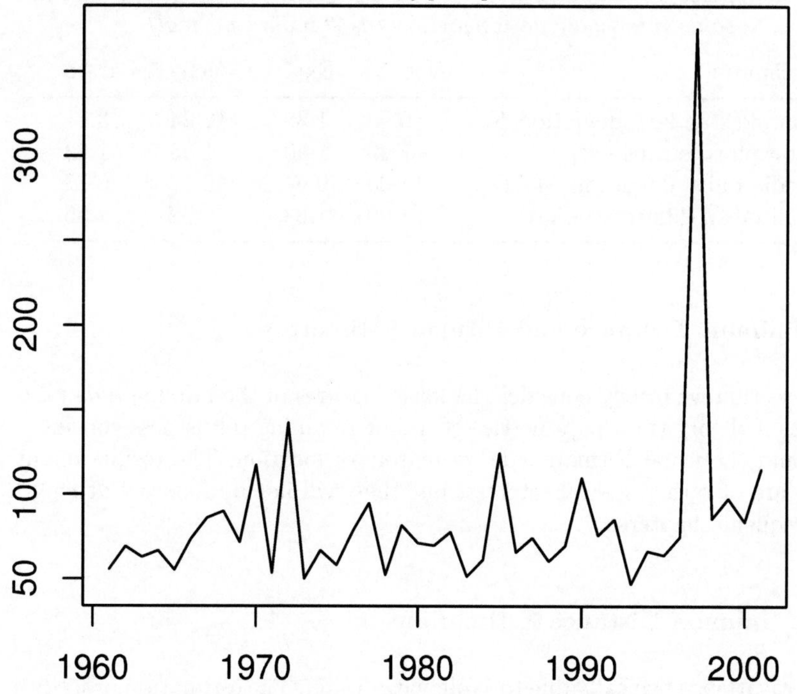

(symmetric and heavy-tailed distribution), Pareto with the density (skewed and heavy tailed distribution).

The values of various estimates under the above distributions are given in Table 3.2.

Table 3.2 Values of estimates under various models

Estimator	$\mathcal{N}(0,1)$	$\mathcal{N}(10,2)$	$Exp(5)$	Cauchy	Pareto
Mean $\bar{X}_n$	0.06	9.92	4.49	1.77	12.19
Median $\tilde{X}_n$	-0.01	9.73	2.92	-0.25	2.10
5%-trimmed m. $\bar{X}_{.05}$	0.05	9.89	4.01	-0.23	3.42
10%-trimmed m. $\bar{X}_{.10}$	0.04	9.88	3.75	-0.29	2.84
5%-Winsor. m. $\bar{W}_{.05}$	0.07	9.92	4.30	-0.10	4.17
10%-Winsor. m. $\bar{W}_{.05}$	0.05	9.91	4.05	-0.30	3.32
Huber M-est. M_H	0.05	9.89	3.89	-0.29	2.87
Hodges–Lehmann est.	0.04	9.87	3.73	-0.24	2.73
Sen weigh. m. $\mathcal{S}_{[0.05n]}$	0.02	9.75	3.16	-0.22	2.24
Sen weigh. m. $\mathcal{S}_{[0.1n]}$	0.02	9.73	3.10	-0.21	2.17
Midrange $\mathcal{R}_m$	0.05	10.37	11.85	146.94	525.34

Table 3.3 Values of estimates under various models of distribution with the same inter-quartile range 1.34 (as standard normal)

Estimator	$\mathcal{N}(0,1)$	Exp.	Cauchy	Pareto
Sample standard deviation S_n	0.98	1.38	12.24	3.11
Interquartile range R_I	1.29	1.40	1.55	1.53
Median abs. deviation MAD	1.00	0.94	0.77	0.77
Gini mean difference G_n	1.09	1.33	6.72	2.56

3.5 Minimum Distance and Pitman Estimators

In this section we briefly consider the basic features of the minimum distance estimation theory (in a parametric setup but retaining robustness considerations) and the usual Pitman-type estimators of location. The results of this section are of rather special interest, but they will not be followed at length in subsequent chapters.

3.5.1 Minimum Distance Estimation

The motivation is basically due to Wolfowitz (1957). The estimation procedure is set in a parametrized framework, although it can easily be modified to suit a broader class of models. For any given data set and a model to be fitted to the data set, a natural way to estimate the parameters may be to minimize the distance between the data and the fitted model. With a proper choice of this distance norm, the resulting estimators can be rendered robust, consistent, as well as efficient. A full bibliography of the minimum distance method was prepared by Parr (1981), and an excellent review appears in Beran (1984). Koul (1992) gives an account of asymptotic theory of MDE in the linear regression model based on weighted empirical processes.

The MDE method can briefly be described as follows. Suppose that $X_1, \ldots, X_n$ are i.i.d. random variables with a distribution function G defined on $\boldsymbol{R}_1$, and let

$$G_n(x) = n^{-1} \sum_{i=1}^{n} I(X_i \leq x), \; x \in \boldsymbol{R}_1,$$

be the corresponding empirical (sample) distribution function. We conceive of a parametric family $\mathcal{F} = \{F_\theta : \; \theta \in \Theta\}$ of distribution functions (called the *projection family*) as an appropriate model, and for every pair of distribution functions G, F, we conceive of an appropriate measure of distance

$$\delta(G, F). \tag{3.101}$$

Then the minimum distance estimator (MDE) $\tilde{\theta}_n$ of θ is defined as the value of t for which $\delta(G_n, F_t)$ is a minimum over $t \in \Theta$:

$$\tilde{\theta}_n : \delta(G_n, F_{\tilde{\theta}_n}) = \inf\left\{\delta(G_n, F_t) : t \in \Theta\right\}. \tag{3.102}$$

For example, for a location family $F_\theta(x) = F(x - \theta)$, $x, \theta \in \boldsymbol{R}_1$, we may take $\delta(G, F_t) = \int [x - \int y dF_t(y)]^2 dG(x)$, and the resulting estimator in (3.102) is the sample mean $\bar{X}_n = n^{-1}\sum_{i=1}^n X_i$ (the least square estimator of θ). A very popular measure of the distance is the so-called weighted Cramér–von Mises distance, which may be defined as

$$\delta(G_n, F) = \int [G_n(x) - F(x)]^2 w(F(x)) dF(x), \tag{3.103}$$

where $w(t)$, $t \in (0, 1)$ is a suitable (nonnegative) weight function. Again, for the location family, (3.103) relates to the minimization of the distance $\int [G_n(x) - F(x)]^2 w(F(x)) dF(x)$ (with respect to $t \in \boldsymbol{R}$). Other possibilities for this measure of distance are the well-known chi-squared distance, *Kolmogorov–Smirnov distance* (i.e., $\sup\{|G_n(x) - F_\theta(x)|,\ x \in \boldsymbol{R}_1\}$), and the *Hellinger distance* (i.e.,$\int [g_n^{1/2}(x) - f_\theta^{1/2}(x)]^2 dx$ where g_n is a suitable (nonparametric) estimator of the density f_θ), among others. Even the maximum likelihood estimator can be characterized as a MDE with a suitable distance function (essentially related to *Kullback–Leibler information*). With specific distance functions relating to the parametric estimators, the MDE are generally nonrobust. The MDE procedure is not restricted to the location model and/or to the univariate distribution functions. In a parametric setup it can as well be considered for more general forms. With the Cramér-von Mises or Kolmogorov–Smirnov distance, the resulting MDE's are generally more robust, while the Hellinger distance makes it even more natural and leads to a class of more general minimum divergence estimators, studied by Vajda (1984a-e). In this context, the work of Parr and Schucany (1980) on minimum distance and robust estimation is illuminating. Millar (1981) has studied the robustness of MDE (in the minimax sense over a shrinking neighborhood of a distribution). Robustness of Cramér–von Mises and Hellinger MDE was also demonstrated by Donoho and Liu (1988 a,b). Related asymptotics are discussed by Koul (1992).

Computationally, the MDE may not be particularly convenient. No closed expression for the MDE is generally available (apart from some well-known parametric cases), and iterative solutions are usually required. A unique solution does not exit in all cases, and additional regularity conditions on the model are therefore required to ensure this. In a general framework (see Beran 1984), the MDE may not be asymptotically normal. An example of asymptotically normal MDE is that of Blackman (1955) whose asymptotic normality can be proved by incorporating the weak convergence of the empirical distributional process. This technique may be applicable in a variety of other models. In a general setup the minimization of the distance may result in a highly nonlinear form (even in an asymptotic representation) for the MDE, and hence, asymptotic normality results may not be tenable. From the robustness considerations, the class of MDE's is quite flexible: With a proper choice of the weight function $w(.)$ in the Cramér-von Mises distance function, one can practically get estimators with arbitrary influence functions. A similar feature is also observed in some other measures of distance. The most appealing feature of the MDE is its clear statistical motivation and easy interpretation. Even if the true distribution function G may not belong to the class $\mathcal{F}$, we still know what we are estimating (i.e., the value of θ corresponding to $F_\theta \in \mathcal{F}$ nearest to G). Parallel to the MDE, minimum distance tests based upon the shortest distance between the data set and the fitted model (which are governed by appropriate null hypotheses) were studied. These tests share some of the robustness aspects of the MDE.

3.5.2 Pitman Estimators

For the location parameter model $F_\theta(x) = F(x - \theta)$, $x, \theta \in \boldsymbol{R}_1$ with F having density f and finite first-order moment, the Pitman (1939) estimator (PE) of location (θ) is defined by

$$T_n = \frac{\left\{ \int_{\boldsymbol{R}_1} t \prod_{i=1}^n f(X_i - t) dt \right\}}{\left\{ \int_{\boldsymbol{R}_1} \prod_{i=1}^n f(X_i - t) dt \right\}}. \tag{3.104}$$

If the true density function $f(.)$ is known, the Pitman estimator is translation-equivariant (i.e., $T_n(\mathbf{X}_n + c\mathbf{1}_n) = T_n(\mathbf{X}_n) + c \; \forall c \in \boldsymbol{R}_1$, and T_n has the minimal risk with respect to quadratic loss among equivariant estimators). The PE are unbiased for θ, although they need not be UMV (as UMV estimators may not generally exist for all f). The PE are consistent under essentially the same regularity conditions as pertaining to the maximum likelihood estimators of θ, and in fact, they are asymptotically equivalent, in probability. By virtue of this stochastic equivalence, from robustness considerations, the PE shares the nonrobustness properties of the usual MLE, and hence, they may not be that desirable. For various large sample properties of PE, we may refer to Strasser (1981a-c).

3.5.3 Pitman-Type Estimators of Location

A new class of robust PE, termed Pitman-type estimators (PTE), was suggested by Johns (1979). Suppose that $X_1, \ldots, X_n$ are i.i.d. random variables with a distribution function $G_\theta(x) = G(x - \theta)$, $x, \theta \in \boldsymbol{R}_1$, where the form of G is not necessarily known. For an appropriate nonnegative $h(.)$, [i.e.,

$h : \boldsymbol{R}_1 \to \boldsymbol{R}_1^+]$, the PTE of θ is defined as

$$T_n^\star = \frac{\left\{ \int_{\boldsymbol{R}_1} t \prod_{i=1}^n h(X_i - t)dt \right\}}{\left\{ \int_{\boldsymbol{R}_1} \prod_{i=1}^n h(X_i - t)dt \right\}}. \tag{3.105}$$

Whereas the PE are „close to" the MLE, the PTE are „close to" M-estimators generated by $\rho(x) = -\log h(x)$. Asymptotic equivalence of $T_n^\star$ with an appropriate M-estimator (with the pertaining order) was proved by Hanousek (1988). The analogy between (3.104) and (3.105) is quite clear. Whereas (3.104) uses the assumed form of the density f, the function $h(.)$ is of more arbitrary form, and in this way a class of robust competitors can be generated. The influence curve of $T_n^\star$ naturally depends on this choice of $h(.)$. It is given by

$$IC(x; G, T^\star) = \frac{\psi(x)}{\int_{\boldsymbol{R}_1} \psi'(x)dG(x)}, \tag{3.106}$$

where

$$\psi(x) = (d/dx) \log h(x). \tag{3.107}$$

The idea behind this type of estimators (i.e., the PTE) is that for the PTE in (3.104), we have $h(x) = (d/dx)G(x)$. This PE minimizes the quadratic risk among the translation-equivariant estimators of θ (when $F = G$). thus, the PTE class contains an efficient member and is robust by appropriate choice of the function $h(.)$ in (3.105). Huber (1984) studied the finite sample breakdown points of $T_n^\star$ (which may be in some cases). We refer to Section 2.3 for the computation of the influence curve in (3.75) and the breakdown points. Note that $\prod_{i=1}^n f(X_i - t)$ is the likelihood function at the point $\theta = t$. Under the local asymptotic normality (LAN) condition, the log-likelihood function is quadratic in $t - \widehat{\theta}_n$ in an $n^{1/2}$-neighborhood of the MLE $\hat{\theta}_n$, and this leads to the IC (3.106) of T_n with $h(.) = f(.)$. A similar expansion may be used for the product $\prod_{i=1}^n h(X_i - t)$ in (3.105) to verify (3.106)–(3.107).

We may remark that $T_n^\star$ in (3.105) can be expressed as a functional of the empirical distribution function F_n, and hence, the von Mises functional (to be studied in Section 3.6) can be used for the study of the relevant properties of these PTE. In fact, the general class of M-, L-, and R-estimators, as well as the estimators considered in this section, can be expressed in a general form of *differentiable statistical functionals.* thus, with a view to unify some of the diverse results presented in this and earlier sections, we will present an introduction to such functionals and some of their basic properties.

3.5.4 Bayes-Type Estimators of General Parameter

Following the idea of Pitman-type estimators, a new class of robust Bayesian-type (or B-) estimators was suggested by Hanousek (1990). Suppose that $X_1, \ldots, X_n$ are i.i.d. random variables with a distribution function $G_{\theta_0}(x)$, x,

$\theta_0 \in \Theta \subseteq \boldsymbol{R}_1$. The B-estimator of θ_0 corresponding to the prior density $\pi(\theta)$ on Θ and generated by the function $\rho(x,\theta) : \boldsymbol{R}_1 \times \Theta \to \boldsymbol{R}_1$ is defined as

$$T_n^\star = \frac{\int_{\boldsymbol{R}_1} \theta \exp\{-\sum_{i=1}^n \rho(X_i,\theta)\}\pi(\theta)d\theta}{\int_{\boldsymbol{R}_1} \exp\{-\sum_{i=1}^n \rho(X_i,\theta)\}\pi(\theta)d\theta}. \tag{3.108}$$

As in the case of PTE, Hanousek (1990) showed that $T_n^\star$ is asymptotically equivalent to the M-estimator generated by ρ for a large class of priors and under some conditions on ρ and G_θ. This is in correspondence with the asymptotic equivalence of Bayes and ML estimators. For a study of this relation, we refer the reader to LeCam (1953), Stone (1974), Fu and Gleser (1975), Ibragimov and Hasminskii (1981), among others.

3.6 Differentiable Statistical Functions

A parameter or a parametric function of a probability distribution $P \in \mathcal{P}$ can be often considered as a functional $T(P)$, $P \in \mathcal{P}$, defined on a space $\mathcal{P}$ of probability measures on the space $(\mathcal{X}, \mathcal{B})$. Assume that $\mathcal{X}$ is a complete separable metric space with metric d and $\mathcal{B}$ is the system of Borel subsets of $\mathcal{X}$. Equivalently, we can consider $T(P)$ as a functional $T(F)$ of the corresponding distribution function F. A natural estimator of $T(P)$ based on observations $X_1,\ldots,X_n$ is $T(P_n)$, where P_n is the empirical probability distribution of the vector $X_1,\ldots,X_n$, what is the (discrete) uniform distribution on the set $\{X_1,\ldots,X_n\}$:

$$P_n(A) = \frac{1}{n}\sum_{i=1}^n I[X \in A], \quad A \in \mathcal{B}.$$

Then $T(P_n)$ can be also described as a functional $T(F_n)$ of the empirical distribution function F_n. Naturally, we want $T(P_n)$ to be close to $T(P)$ for P_n close to P; this closeness is characterized by the continuity of $T(\cdot)$ with respect to a suitable distance on $\mathcal{P}$; this can be the Prokhorov, Hellinger, or Kolmogorov distances, or the total variation. The statistical functionals were introduced by von Mises (1936, 1947), who recognized their importance and considered not only the convergence of $T(P_n)$ to $T(P)$ in a suitable topology, but also a possible expansion of $T(\cdot)$ around P with the first-order term being linear functional and the remainder term negligible in an appropriate sense. The first linear term provides an access to the first-order asymptotic theory. The possibility of such expansion rests on some *differentiability* properties of statistical functionals. The differentiability of a statistical functional provides a unified access to its characteristics, as the influence function, asymptotic normality, and other related properties.

Before we enter into that discussion, we will briefly consider the connection of M-, L-, and R-estimators with such differentiable statistical functionals. We start with the location model, for which the M-estimator of θ, defined by (3.4) with $\psi(x,t) = \psi(x-t)$, $x, t \in \boldsymbol{R}_1$, can be written in the form

$$\int_{\boldsymbol{R}_1} \psi(x - T(F_n))dF_n(x) = 0.$$

The corresponding functional is a root $T(F)$ of the equation

$$\int_{\boldsymbol{R}_1} \psi(x - T(F))dF(x) = 0. \tag{3.109}$$

The functional counterpart of the L-estimator (3.44) can be written as

$$\begin{aligned} T(F) &= \int_0^1 F^{-1}(u)J(u)du + \sum_{j=1}^{k} a_j F^{-1}(p_j) && (3.110) \\ &= \int_{\mathbb{R}_1} xJ(F(x))dF(x) + \sum_{j=1}^{k} a_j F^{-1}(p_j). \end{aligned}$$

Unlike the M-estimators, the L-estimators are defined explicitly.

The R-estimators of location, considered in Section 3.4, are defined implicitly, like the M-estimators. In the rank statistic (3.76), let $a_n(k) = \varphi^+(k/(n+1))$ for simplicity, $k = 1, \ldots, n$, where $\varphi^+ : [0,1] \mapsto \boldsymbol{R}^+$ and $\varphi^\star(u) = \varphi^+(u)\,\mathrm{sign}(u)$, $u \in (-1, 1)$. We may write (3.76) equivalently as

$$n \int_{\boldsymbol{R}_1} \varphi^\star\Big(F_n(x) - F_n(2T(F_n) - x)\Big)dF_n(x). \tag{3.111}$$

thus, the functional corresponding to (3.111) is defined implicitly as a root $T(F)$ of the equation

$$\int_{\boldsymbol{R}_1} \varphi^\star\Big(F(x) - F(2T(F) - x)\Big)dF(x) = 0. \tag{3.112}$$

Let us fix a distance δ in the system $\mathcal{P}$ of probability measures on $(\mathcal{X}, \mathcal{B})$, where $\mathcal{X}$ is a metric space with distance d. For two distribution P, $Q \in \mathcal{P}$, consider the *contamination of P by Q in ratio $t \in [0,1]$* as follows:

$$P_t(Q) = (1-t)P + tQ. \tag{3.113}$$

For an expansion of $T(\cdot)$ around P, we need to introduce the concept of the derivative of the statistical functional.

DEFINITION 3.1 *(i) The Gâteaux derivative of the functional $T(\cdot)$ in P in direction Q is defined as the limit*

$$T'_Q(P) = \lim_{t \downarrow 0} \frac{T(P_t(Q)) - T(P)}{t} \tag{3.114}$$

provided the limit exists.

(ii) The functional $T(\cdot)$ is called Hadamard (or compact) differentiable in P,

if there exists a linear functional $L_P(Q-P)$ *such that*

$$L_P(Q-P) = \lim_{t \downarrow 0} \frac{T(P+t(Q-P)) - T(P)}{t} \tag{3.115}$$

and the convergence is uniform over $Q \in \mathcal{C}$ *for every compact subset* $\mathcal{C} \subset \mathcal{P}$. *The linear functional* $L_P(Q-P)$ *is called the Hadamard derivative of* $T(\cdot)$ *in* P *in direction* Q.

(iii) The functional $T(\cdot)$ *is called Fréchet differentiable in* P, *if there exists a linear functional* $L_P(Q-P)$ *such that*

$$L_P(Q-P) = \lim_{t \downarrow 0} \frac{T(P+t(Q-P)) - T(P)}{t} \tag{3.116}$$

and the convergence is uniform over $\{Q: \ \delta(P,Q) \le K\}$ *for any fixed* $K > 0$. *The linear functional* $L_P(Q-P)$ *is called the Fréchet derivative of* $T(\cdot)$ *in* P *in direction* Q.

Because $L_P(\cdot)$ in (ii) or (iii) is a linear functional, there exists a function $\phi : \mathcal{X} \mapsto \mathbb{R}$ such that

$$L_P(Q-P) = \int_{\mathcal{X}} \phi(x) d(Q-P).$$

If T is differentiable in the Fréchet sense, then it is differentiable in the Hadamard sense, and then, in turn, it is differentiable in the Gâteaux sense, and $T'_Q(P) = L_P(Q-P)$. The *influence function* $IF(x;\ T,P)$, $x \in \mathcal{X}$ of functional $T(\cdot)$ is the Gâteaux derivative of T in the direction of δ_x, where δ_x is the Dirac probability measure assigning probability 1 to the one-point set $\{x\}$.

It follows from the above definition that the mode of differentiability is dependent on the topology of the space involved. The Fréchet mode is more stringent than the Hadamard mode which in turn implies the Gâteaux mode. While it often suffices to use the Gateaux differentiability method, the Hadamard differentiability appears to serve more purposes in the asymptotic statistical inference; on the other hand, the Fréchet mode is often too stringent. For this reason, for our statistical analysis, we will confine ourselves to the Hadamard mode of differentiability.

Fernholz (1983) gives an excellent account of the differentiability properties of various statistical functionals. For a general nonlinear functional analysis, we recommend the monograph by Fabian et al. (2001).

THEOREM 3.4 *Let* P_n *be the empirical distribution corresponding to vector* $(X_1, \ldots, X_n)$. *Then*

$$L_P(P_n - P) = \frac{1}{n} \sum_{i=1}^{n} T'_{\delta_{X_i}}(P) = \frac{1}{n} \sum_{i=1}^{n} IF(X_i;\ T, P).$$

PROOF Indeed, because L_P is a linear functional, then $L_P(P_n - P) = \frac{1}{n}\sum_{i=1}^n L_P(\delta_{X_i} - P) = \frac{1}{n}\sum_{i=1}^n T'_{\delta_{X_i}}(P)$. □

Following the von Mises (1947) ideas, we try to write

$$T(F_n) = T(F) + L_F(F_n - F) + R_n(F_n, F), \tag{3.117}$$

where L_F is the Hadamard derivative of $T(\cdot)$ at F and $R_n(F_n, F)$ is the remainder term. Looking at Definition 3.1, we observe that the remainder in (3.117) satisfies

$$R_n(F_n, F) = o\Big(\|F_n - F\|\Big) \text{ on the set } \|F_n - G\| \to 0. \tag{3.118}$$

Here $\|.\|$ stands for the Kolmogorov supnorm (i.e., $\sup\{|F_n(x) - F(x)| : x \in \boldsymbol{R}_1\}$), although other measures introduced in Section 2.4 could also be used. By the classical results on the Kolmogorov–Smirnov statistics, we have

$$n^{1/2}\|F_n - F\| = O_p(1) \tag{3.119}$$

so that (3.118) is $o_p(n^{-1/2})$ as $n \to \infty$. On the other hand, using Theorem 3.4, we may equivalently write the linear functional $L_F(F_n - F)$ as

$$\int T^{(1)}(F; x)d[F_n(x) - F(x)] = n^{-1}\sum_{i=1}^{n} T^{(1)}(F; X_i) \tag{3.120}$$

where we set $\int T^{(1)}(F; x)dF(x) = 0$ without loss of generality. Hence, by means of the classical central limit theorem for the i.i.d. random variables, this yields a direct proof of the asymptotic normality of

$$\frac{\sqrt{n}(T_n - T(F))}{\sigma_{(F)}}, \text{ where } \sigma^2_{(F)} = \text{Var}[T^{(1)}(F; X)]. \tag{3.121}$$

Moreover, $T^{(1)}(F; x) = IF(x; T, F)$ is the influence function of $T(\cdot)$. If we only want to prove the consistency of $T(F_n)$ as an estimator of $T(F)]$, it suffices to assume the *Hadamard continuity*, $T(F) \to T(G)$ as $\|F - G\| \to 0$.

Similarly, we can introduce a functional of degree $m \geq 1$ by incorporating a *kernel* $\phi(.)$ of degree m and letting

$$T(F) = \int \dots \int \phi(x_1, \dots, x_m)dF(x_1)\dots dF(x_m). \tag{3.122}$$

von Mises (1936, 1947) considered the natural estimator of $T(F)$

$$T(F_n) = \int \dots \int \phi(x_i, \dots, x_m)dF_n(x_1)\dots dF_n(x_m). \tag{3.123}$$

For $m = 1$ (i.e., for linear functionals), $T(F_n)$ is an unbiased estimator of $T(F)$, but for $m \geq 2$, it may not be so. Halmos (1946) and Hoeffding (1948) considered unbiased and optimal (symmetric) estimators of $T(F)$; these are known as U-statistics in the literature. Typically, a U-statistic can be expressed as a linear combination of some $T(F_n)$ and vice versa. Detailed treatment of

U-statistics and von Mises functionals are given in Serfling (1980) and Sen (1981a), among other places.

By making use of the simple identity that $dF_n(X_i) = dF(X_i) + d[F_n(X_i) - F(X_i)]$, $i = 1, \ldots, m$, we obtain the classical Hoeffding (1961) decomposition:

$$T(F_n) = T(F) + \sum_{k=1}^{m} \binom{m}{k} T^{(k)}(F_n), \tag{3.124}$$

where

$$T^{(k)}(F_n) = \int \ldots \int \phi_k(x_1, \ldots, x_k) d[F_n(x_1) - F(x_1)] \ldots d[F_n(x_k) - F(x_k)],$$

and ϕ_k a kernel of degree k, is given by

$$\phi_k((x_1, \ldots, x_k) = \int \ldots \int \phi(x_1, \ldots, x_k, y_{k+1}, \ldots, y_m) dF(y_{k+1}) \ldots dF(y_m),$$

for $k = 1, \ldots, m$. The asymptotic normality of $T(F_n)$ can be deduced under the second moment conditions on the ϕ_k, $k = 1, \ldots, m$. For the asymptotic theory of the von Mises functionals, we refer to von Mises (1947), Hoeffding (1948), Miller and Sen (1972), and Sen (1974a,b,c), among others.

Hoeffding (1948, 1961) introduced an idea of L_2-projection of U-statistics into sum of independent random variables. This technique can be used for a general statistical functional in the following way: Denote

$$T_{ni} = \mathbb{E}[T(F_n)|X_i], i = 1, \ldots, n; \; \hat{T}_n = \sum_{i=1}^{n} T_{ni}.$$

Then, under quite general regularity conditions, it can be shown that $\hat{T}_n$ and $T_n - \hat{T}_n$ are uncorrelated and $V(T_n)/V(\hat{T}_n) \to 1$ as $n \to \infty$. On the other hand, $\hat{T}_n$ is a sum of independent summands on which the central limit theorem and other standard asymptotic theory can be easily adapted. thus, the asymptotic properties of $T(F_n)$ can be studied with the help of the quadratic mean approximation by $\hat{T}(F_n)$. This approach was applied by Hájek to the rank statistics in his pioneering work Hájek (1968). van Zwet (1984) elegantly pursued this approach for a general symmetric statistic, in the context of Berry–Esseen type bounds. If the statistical functional $T(F_n)$ is Hadamard differentiable, the L_2-projection method leads to $\hat{T}_n$, the same principal term $n^{-1} \sum_{i=1}^{n} T^{(1)}(F; X_i)]$.

3.7 Problems

3.2.1 Assume that (3.17) holds (a formal proof is given in Chapter 5). Because the $\psi(Y_i - \mathbf{x}_i^\top \boldsymbol{\beta})$ are independent and identically distributed random variables with 0 mean and variance σ_ψ^2, we can apply the classical central limit theorem for the right-hand side of (3.17), when the Noether condition in (3.18) is satisfied. Consider the particular case of $\psi(x) \equiv \text{sign}(x)$ so that $\mathbf{M}_n$ reduces to the L_1-norm estimator of $\boldsymbol{\beta}$. Work out the expression for γ^2 in (3.19) in this case. For $\psi(x) \equiv x$, (3.17) is an identity for the least square estimator of $\boldsymbol{\beta}$.

3.2.2 If $\mathbf{X}$ and $(-1)\mathbf{X}$ both have the same distribution function F, defined on $\boldsymbol{R}_p$, then F is said to be diagonally symmetric about $\mathbf{0}$. Show that if F is diagonally symmetric about $\mathbf{0}$, then for every $\ell \in \boldsymbol{R}_p$, $\ell^\top \mathbf{X}$ has a distribution function symmetric about 0. This property is related to the multivariate median unbiasedness: If for every $\ell \in \boldsymbol{R}_p$, $\ell^\top(\mathbf{T}_n - \theta)$ has median 0, then $\mathbf{T}_n$ is multivariate median unbiased for θ. Verify (3.17). [Sen 1990]

3.2.3 Show that the M-estimator in the location model corresponding to the score function $\psi(x) = \text{sign}(x)$, $x \in \boldsymbol{R}$, is the sample median. Suppose now that for some $\alpha > 0, I\!\!E_F|X|^\alpha < \infty$, where α need not be greater than or equal to 1. Then verify that $I\!\!E M_n = \theta$ whenever $(n+1)\alpha \geq 2$. thus, for large n, the unbiasedness of M_n may not require the existence of the first moment of F. [Sen 1959].

3.2.4 Consider the Huber (1964) estimator, for which $\psi(x) = xI(|x| \leq k) + k \cdot \text{sign(x)I(|x|} > \text{k)}$ for some finite $k > 0$. Suppose that $I\!\!E_F|X|^\alpha < \infty$ for some $\alpha > 0$ (not necessarily ≥ 1). What is the minimum sample size n, such that the Huber estimator has a finite rth moment, for some $r > 0$? Compare the situation with that of the sample mean.

3.2.5 Consider the Cauchy density $f(x) = \frac{1}{\pi(1+x^2)}$, $x \in \boldsymbol{R}$. Find the MLE of θ for a sample size 3 from the Cauchy density $f(x-\theta)$, and discuss whether it has a finite first moment or not. Is there any unbiased estimator of θ for $n = 3$?

3.2.6 Consider a mixture model:

$$f(x) = \frac{1}{2}[\phi(x) + \ell(x)], \; x \in \boldsymbol{R},$$

where ϕ is the standard normal density and $\ell(.)$ is the Laplace density $(\frac{1}{2}e^{-|x|})$. For a sample of n observations from $f(x-\theta)$, which of the estimators: mean or median, would you prefer on the ground of robustness and/or asymptotic variance?

3.2.7 For the uniform $[-\theta, \theta]$ distribution function, $\theta > 0$, which ψ-function would you recommend for the M-estimation of θ ? What about a mixture of two uniform distributions ?

3.3.1 Show that $t(\mathbf{X}_n) = \{X_{n:1}, \ldots, X_{n:n}\}$ is a symmetric function of $\mathbf{X}_n = (X_1, \ldots, X_n)$ and, conversely, that any symmetric function of $\mathbf{X}_n$ is a function of $t(\mathbf{X}_n)$. Also show that $t(\mathbf{X}_n)$ is complete for the class of distribution functions on $\boldsymbol{R}_n$ when F is (a) any discrete distribution or (b) any absolutely continuous distribution function. Is $t(\mathbf{X}_n)$ a sufficient statistics? Is it minimal sufficient in general? [Lehmann 2005]

3.3.2 Define the $a_{n:i}$ and b_{nij} as in (3.35). Show that

$$a_{n:i} = i\binom{n}{i}\int_{\boldsymbol{R}_1} x[F_0(x)]^{i-1}[1-F_0(x)]^{n-i}dF_0(x),$$

for every $i = 1,\ldots,n, n \geq 1$. Show that the $a_{n:i}$'s all exist whenever $\mathbb{E}_{F_0}(X)$ exists. What happens when $\mathbb{E}_{F_0}|X|^\alpha < \infty$ for some $\alpha: \ 0 < \alpha < 1$? Similarly, show that the b_{nij}'s all exist whenever $\mathbb{E}_{F_0}(X^2) < \infty$. What happens when $\mathbb{E}_{F_0}|X|^\alpha < \infty$ for some $\alpha: \ 0 < \alpha < 2$?

3.3.3 Write down the estimating equations in (3.36) in terms of the $a_{n:i}$, b_{nij}, and $X_{n:i}$. Under which conditions on the $a_{n:i}$ and b_{nij} are the estimators $\hat{\mu}_n$ and $\hat{\delta}_n$ uncorrelated?

3.3.4 Whenever $\mathbb{E}_{F_0}|X|^\alpha < \infty$ for some $\alpha > 0$, there exists a sample size n_0 such that for all $n \geq n_0$, one can obtain the BLUE of (μ, δ) as in (3.36), with the set J replacing $\{1,\ldots,n\}$. Treat the set J as $\{k,\ldots,n-k+1\}$ with $k: \ k\alpha \geq 2$.

3.3.5 Rewrite $a_{n:i}$ as $i\binom{n}{i}\int_0^1(F_0^{-1}(t))t^{i-1}(1-t)^{n-i}dt$. Use the convergence of the beta distribution to a normal one to claim that $a_{n:i} \to F_0^{-1}(p)$ as $i/n \to p: \ 0 < p < 1$, with $n \to \infty$. Show that the continuity and the strict monotonicity of $F_0^{-1}(t)$ at $t = p$ suffice in this respect.

3.3.6 Show that if $f(F^{-1}(p_1))$ and $f(F^{-1}(p_2))$ are both positive and $i/n \to p_1, \ j/n \to p_2: \ 0 < p_1 < p_2 < 1$, as $n \to \infty$, then

$$nb_{nij} \to \frac{p_1(1-p_2)}{f(F^{-1}(p_1))f(F^{-1}(p_2))}.$$

What happens when $p_1 = 0$ or $p_2 = 1$?

3.3.7 Under (3.39), take $\theta = 0$ without loss of generality. Then we can show that

$$L_n(X_1,\ldots,X_n) = -L_n(-X_1,\ldots,-X_n) \stackrel{\mathcal{D}}{=} -L_n(X_1,\ldots,X_n),$$

since the distribution function F_0 is symmetric about 0. hence, L_n has a distribution function symmetric about 0. Using the result in Problem 3.2.3, we can claim that L_n is median unbiased estimator of θ.

3.3.8 (3.48) and (3.50) can be verified with the aid of the Hoeffding inequality and (3.49).

3.3.9 Let $X_1,\ldots,X_n$ be a random sample from the distribution with distribution function $F_{(n)}(x-\theta)$, where

$$F_{(n)} \equiv (1-cn^{-1/2})G + cn^{-1/2}H,$$

G and H are fixed distribution functions, $G(x)+G(-x) = 1, \ \forall x \in \boldsymbol{R}_1$. Let M_n be an M-estimator defined as a solution of $\sum_{i=1}^n \psi(X_i - M) = 0$, where ψ is a nondecreasing odd function such that ψ' and ψ'' are continuous and bounded up to a finite number of points. Then

$$n^{\frac{1}{2}}(M_n-\theta) \xrightarrow{\mathcal{D}} \mathcal{N}(b, \nu^2(\psi, G)),$$

where $b = cA/\mathbb{E}_G\psi'$, $A = \int \psi(x)dH(x)$, and $\nu^2(\psi, G)$ is defined in (3.20). hence, M_n has asymptotic bias b and the asymptotic mean square error $\nu^2(\psi, G) + b^2$. [Jaeckel 1971]

3.3.10 Consider the model of Problem 3.3.9, and assume that H puts all its mass to the right of $G^{-1}(1-\alpha)$ for a fixed $0 < \alpha < 1/2$. Then the α-trimmed mean has the asymptotic bias $b(\alpha) = cG^{-1}(1-\alpha)/(1-2\alpha)$, and the asymptotic mean square error

$$\sigma^2(\alpha) + b^2(\alpha) = \frac{1}{(1-2\alpha)^2}\Big\{2\int_0^{G^{-1}1-\alpha} x^2 dG(x) + (2\alpha + c^2)(G^{-1}(\alpha))^2\Big\}.$$

[Jaeckel 1971]

3.3.11 Let $G(x) = G_0\big(\frac{x-\theta}{\delta}\big)$, $\theta \in \boldsymbol{R}_1$, $\delta \in \boldsymbol{R}_1^+$ and G_0 does not depend on (θ, δ). Then show that

$$G^{-1}(3/4) - G^{-1}(1/4) = \delta\{G_0^{-1}(3/4) - G_0^{-1}(1/4)\} = \delta T(G_0),$$

where the functional $T(G_0)$ does not depend on (θ, δ). Study the nature of $T(G_0)$ for G_0 normal, Laplace, and Cauchy, and comment on the errors in the definition of the interquartile range if the true F_0 and assumed G_0 are not the same.

3.3.12 $\frac{1}{2}(U_n^{(2)})^2$ defined in (3.64) is equal to the sample variance (unbiased form) and is an optimal nonparametric estimator of the functional $T_2(F) = \int_{\boldsymbol{R}_1}\int_{\boldsymbol{R}_1}(x_1 - x_2)^2/2dF(x_1)dF(x_2) = \sigma_F^2$. Verify that $\frac{1}{2}(U_n^{(2)})^2$ is a U-statistic with a kernel of degree 2.

3.3.13 $U_n^{(1)}$ in (3.64) is a U-statistic with a kernel of degree 2 and is an optimal nonparametric estimator of the functional

$$T_1(F) = \int_{\boldsymbol{R}_2} |x_1 - x_2| dF(x_1)dF(x_2).$$

3.4.1 Let $T_n = \frac{1}{2}(n + S_n^0)$ with S_n^0 defined in (3.72). Then $T_n \sim \text{Bin}(n, 1/2)$ under H_0, and hence, S_n^0 is distribution-free under H_0 and $n^{-1/2}S_n^0 \xrightarrow{\mathcal{D}} \mathcal{N}(0,1)$ as n increases.

3.4.2 Suppose that $X_1, \ldots, X_n$ are i.i.d. random variables with a continuous distribution function F, symmetric about 0. Let $R_{ni}^+ =$ be the rank of $|X_i|$ among $|X_1|, \ldots, |X_n|$, and $S_i = \text{sign}(X_i)$, for $i = 1, \ldots, n$. Show that the two vectors $\mathbf{R}_n^+ = (R_{n1}^+, \ldots, R_{nn}^+)^\top$ and $\mathbf{S}_n = (S_1, \ldots, S_n)^\top$ are stochastically independent. $\mathbf{R}_n^+$ can assume all possible $n!$ permutations of $1, \ldots, n)$, with the common probability $\frac{1}{n!}$ and $\mathbf{S}_n$ takes on all possible 2^n sign-inversions (i.e., $(\pm 1, \ldots, \pm 1)^\top$) with the common probability 2^{-n}. What happens if F is not symmetric about 0?

3.4.3 Define $S_n(t) = S_n(\mathbf{X}_n - t\mathbf{1}_n)$ as in after (3.76). Show that

(a) $\text{sign}(X_i - t)$ is nonincreasing in $t \in \boldsymbol{R}_1$,

(b) $R_{ni}^+(t)$ is nonincreasing in t when $X_i > t$ and is nondecreasing in t when $X_i < t$.

hence, $S_n(t)$ is nonincreasing in $t \in \boldsymbol{R}_1$. For the particular case of $a_n(k) = 1 \, \forall 1 \le k \le n$, (3.74) leads to the same result for the sign statistic.

3.4.4 The Wilcoxon signed-rank statistic W_n is given by $\sum_{i=1}^n \text{sign}(X_i) R_{ni}^+$. Verify that

$$\begin{aligned} W_n^+ &= \sum_{i=1}^n \Big\{ \frac{1}{2}\{1 + \text{sign}(X_i)\} \Big\} R_{ni}^+ = \frac{n(n+1)}{4} + \frac{1}{2} W_n \\ &= \sum_{1 \le i \le j \le n} I(X_i + X_j > 0). \end{aligned}$$

As such, $W_n(t) = 2 \sum_{1 \le i \le j \le n} I(X_i + X_j - 2t > 0) - \binom{n+1}{r}$, $t \in \boldsymbol{R}_1$.

3.4.5 Show that for S_n defined by (3.76), $P_0(S_n = 0) = \eta_n > 0$ under $\theta_0 = 0$ for every finite $n \ (\ge 1)$. However, $\eta_n \to 0$ as $n \to \infty$.

3.4.6 For the simple regression model

$$Y_i = \theta + \beta x_i + e_i, \quad i = 1, \ldots, n,$$

consider the Kendall tau statistic

$$K_n = \sum_{1 \le i < j \le n} \text{sign}(x_i - x_j) \text{sign}(Y_i - Y_j),$$

and let $K_n(t)$ be the aligned statistic when the Y_i's are replaced by $Y_i - tx_i$, $1 \le i \le n$. Show that $K_n(\beta)$ is distribution-free with location 0, and that $K_n(t)$ is nonincreasing in t. The derived R-estimator of β is

$$R_n = \text{median}\Big\{ \frac{Y_i - Y_j}{x_i - x_j} : 1 \le i < j \le n, \ x_i \ne x_j \Big\}.$$

If the x_i's are binary, then R_n reduces to the two-sample Wilcoxon scores estimator of the difference of location parameters.

3.4.7 Show that

$$P_{H_0}\{R_{ni_1}(\beta_0) = j_1, \ldots, R_{ni_k}(\beta_0) = j_k\} = \frac{1}{n^{[k]}}$$

for every $\{j_1 \ne \ldots \ne j_k\} \subset \{1, \ldots, n\}$, where $n^{[k]} = n \ldots (n - k + 1)$.

3.5.1 The Pitman estimator in (3.104) is translation-equivariant for the location model. Moreover, if there exists a sufficient statistic for θ, then T_n is a function of that statistic.

3.5.2 Examine under which relation between $h(.)$ in (3.105) and $\rho(.)$ in (3.108) both estimators have a common influence curve.

3.6.1 The U-statistic $U_n = \binom{n}{m}^{-1} \sum_{1 \le < i_1 < \ldots < i_m \le n} \phi(X_{i_1}, \ldots, X_{i_m})$ with a symmetric kernel $\phi(.)$ of degree $m \ (\ge 1)$ is a symmetric, unbiased, and optimal nonparametric estimator of $T(G)$. Defining $T(F_n)$ as in (3.123), then (a) $T(F_n) = U_n$ for $m = 1$, but (b) $T(F_n) \ne U_n$ for $m \ge 2$. However, $|T(F_n) - U_n| = O(n^{-1})$ a.s. as $n \to \infty$.

CHAPTER 4

Asymptotic Representations for L-Estimators

4.1 Introduction

In the preceding two chapters we observed that the robustification brings certain structural changes. The resulting robust estimators (or test statistics) are typically nonlinear functions of observations. The standard probabilistic and asymptotic tools presented in Section 2.5 do not provide a unified approach toward the study of (asymptotic) distribution theory and related properties of general robust estimators and test statistics. Inspired by the effectiveness of the Hoeffding-Hájek projection theorems for nonlinear statistics (Section 2.5), we can try to extend the projection technique to general robust statistics. When we are not able to get its explicit form and estimate the remainder term, we can accept another, perhaps less optimal, and reasonably simpler linear approximation of a nonlinear statistic whose remainder term is negligible and can be handled. The past 35 years have witnessed significant developments along this line, and our emphasis will be primarily on such asymptotic representations.

Let us recall the motivation for statistical functionals treated in Section 3.6. The representation in (3.117) is a first step in this direction, while the simplification in (3.120) provides the access to the desired asymptotic normality of the estimator. If, however, we want to proceed beyond the asymptotic normality (e.g., the law of iterated logarithms, weak and strong invariance principles, etc.), we need to study the order of the remainder term in a relatively more refined manner, which will demand extra regularity conditions on the functional. For a special class of L-estimators, namely the sample quantiles, such linear representations were elegantly formulated by Bahadur (1966) under very mild regularity conditions, and are known as *Bahadur representations* for sample quantiles. The Bahadur representations were a precursor to the general asymptotic representations which will be considered in this and subsequent chapters.

As such, we motivate our presentation as follows: Let $X_1, \ldots, X_n$ be i.i.d. random variables with a distribution function F, defined on $\boldsymbol{R}_1$, and let $F^{-1}(p) = \xi$ be uniquely defined for a given $p:\ 0 < p < 1$. When $\frac{k_n}{n}$ is close to p, we may consider the sample quantile $X_{n:k_n}$ as an estimator of ξ, where $X_{n:1} \leq \ldots \leq X_{n:n}$ are the sample order statistics. Bahadur (1966) showed that if F is twice differentiable at ξ and the derivative $f(\xi)$ is positive, then for any $k_n : \left|\frac{k_n}{n} - p\right| = o(n^{-1/2})$,

$$X_{n:k_n} - \xi = \{nf(\xi)\}^{-1} \sum_{i=1}^{n} \{p - I(X_i \leq \xi)\} + r_n, \tag{4.1}$$

where we have

$$r_n = O\left(n^{-3/4}(\log n)^{1/2}(\log\log n)^{1/4}\right) \text{ a.s.,} \quad \text{as } n \to \infty. \tag{4.2}$$

The representation in (4.1) expresses $X_{n:k_n} - \xi$ as an average of n i.i.d. random variables, up to an asymptotically negligible remainder term r_n; (4.1) and (4.2) imply the asymptotic normality of $n^{1/2}\{X_{n:k_n} - \xi\}$.

As we will illustrate in the next two chapters, the utility of this linear representation goes far beyond the asymptotic normality. Kiefer (1967) obtained an exact order for $r_n = r_n(p)$ in (4.1) and derived the asymptotic (nondegenerate) distribution of $n^{3/4}r_n(p)$:

$$\lim_{n\to\infty} \mathbb{P}\left\{n^{3/4}f(\xi)r_n(p) \leq x\right\}$$
$$= 2\left\{p(1-p)\right\}^{-\frac{1}{2}} \int_0^\infty \Phi(t^{-1/2}x)\phi(t\{p(1-p)\}^{-1/2})dt, \tag{4.3}$$

$\forall x \in \boldsymbol{R}$, and every $p \in (0,1)$, where Φ and ϕ stand for the random variables and density function of a standard normal variable. Duttweiler (1973) studied the moment convergence results and showed that for any $\varepsilon > 0$ as $n \to \infty$

$$\mathbb{E}\left\{n^{3/4}f(\xi)r_n(p)\right\}^2 = \left\{2p(1-p)/\pi\right\}^{1/2} + o(n^{-1/4+\varepsilon}). \tag{4.4}$$

Representations similar to (4.1) hold for a general class of estimators and related statistics, and will be systematically presented in this and the next two chapters. For a statistic $T_n = T(X_1, \ldots, X_n)$ estimating θ, based on n i.i.d. random variables $X_1, \ldots, X_n$ with a distribution function $F_\theta(x)$, $\theta \in \Theta \subset \boldsymbol{R}_1$, we would like to show that there exists a function $\xi : \boldsymbol{R}_1 \times \Theta \mapsto \boldsymbol{R}_1$, such that

$$T_n - \theta = n^{-1} \sum_{i=1}^{n} \xi(X_i, \theta) + r_n(\theta), \text{ where } r_n(\theta) = o_p(n^{-1/2}) \tag{4.5}$$

under suitable regularity conditions. The representation of the type (4.5) we will term as the *first-order* or *weak Bahadur-type representation.* Equation (4.5) can often be strengthened to

$$r_n(\theta) = o_p(n^{-1/2}), \text{ uniformly in } \theta \in K, \text{ for every compact } K \subseteq \Theta. \tag{4.6}$$

The estimator T_n satisfying (4.5) and (4.6), such that $n^{1/2}(T_n - \theta)$ is asymptotically normal, is termed the *CLUAN (consistent, linear, uniformly asymptotically normal)* estimator of θ (Bickel (1981)). Thus, a CLUAN estimator T_n satisfies (4.5) and

$$\mathbb{P}_\theta\Big\{n^{1/2}|r_n(\theta)| > \varepsilon\Big\} \to 0 \quad \text{as } n \to \infty, \tag{4.7}$$

uniformly on every compact $K \subseteq \Theta$, where

$$\mathbb{E}_\theta\{\xi(X_1,\theta)\} = 0 \quad \text{and} \quad \mathbb{E}_\theta\{\xi(X_1,\theta)\}^2 < \infty,\ \theta \in \Theta \tag{4.8}$$

and

$$n^{-1/2}\sum_{i=1}^{n} \xi(X_i,\theta) \xrightarrow{\mathcal{D}} \mathcal{N}\Big(0, \mathbb{E}_\theta\{\xi(X_1,\theta)\}^2\Big),\ \ \theta \in \Theta. \tag{4.9}$$

Clearly, if (4.5) holds uniformly in θ over compact subspaces of Θ, then the CLUAN estimators can be most conveniently studied with the help of the scores $\xi(X_i,\theta),\ i \geq 1$. Such representations have also other useful applications. For example, if two sequences $\{T_n\}$ and $\{T_n^\star\}$ of estimators of θ follow representations of the type (4.5), then not only we can easily derive the joint asymptotic distribution of $(T_n, T_n^\star)$, but the remainder terms can gather useful information on the *rate of stochastic equivalence* of $\{T_n\}$ and $\{T_n^\star\}$ as n becomes large. This representation also enables study of the quality of a *one-step version* ($T_n^\star$) of T_n, which is roughly of the form

$$T_n^\star = \hat{\theta}_n + n^{-1}\sum_{i=1}^{n} \hat{\xi}(X_i, \hat{\theta}_n), \tag{4.10}$$

where $\hat{\theta}_n$ is a consistent initial estimator of θ and $\hat{\xi}$ is either an estimator of ξ involving an estimation of nuisance parameters, or equal to ξ when ξ does not involve any nuisance parameter.

In the present chapter, we will systematically study the Bahadur-type representations of L-estimators and R-estimators; the M-estimators will be taken up in the next chapter. The higher is the order of r_n in (4.5), the more stringent are the regularity conditions on the weight functions. The highest possible rate in (4.5) is $r_n = \mathrm{O}_p(n^{-1})$, although for nonsmooth scores we typically have a rate of $n^{-3/4}$. We shall study the two cases of nonsmooth and smooth weights separately, and append a general result on a combined case.

4.2 The Bahadur Representations for Sample Quantiles

Ghosh (1971) obtained the weaker version (4.5) for the sample quantile $X_{n:k_n}$ (4.1), $k_n = np + \mathrm{o}(n^{1/2})$, under a weaker condition that F is once differentiable at ξ_p with $f(\xi_p) > 0$:

THEOREM 4.1 *If F is differentiable at ξ_p with $f(\xi_p) > 0$ and $k_n = np + o(n^{1/2})$, then the representation in (4.1) holds with $r_n = o_p(n^{-1/2})$.*

PROOF The proof is adapted to the sample distribution function F_n. Note that for a real t,

$$\begin{aligned} I\!\!P\Big\{X_{n:k_n} \leq \xi_p - n^{-1/2}t\Big\} &= I\!\!P\Big\{F_n(\xi_p - n^{-1/2}t) \geq n^{-1}k_n\Big\} \\ = I\!\!P\Big\{n^{1/2}\Big[F_n(\xi_p - n^{-1/2}t) &- F(\xi_p - n^{-1/2}t)\Big] \\ &\geq n^{1/2}\Big[n^{-1}k_n - F(\xi_p - n^{-1/2}t)\Big]\Big\}. \end{aligned} \tag{4.11}$$

Since $n^{1/2}[n^{-1}k_n - F(\xi_p - n^{-1/2}t)] = tf(\xi_p) + \text{o}(1)$ while $n^{1/2}[F_n(y) - F(y)]$ is asymptotically normal with zero mean and variance $F(y)[1 - F(y)]$ ($\leq 1/4$), we can choose t so large (but finite) that the right-hand side of (4.11) can be made arbitrarily small for all n sufficiently large. A similar result holds for the right hand side tail. Hence, we conclude that whenever $f(\xi_p)$ is positive,

$$n^{1/2}|X_{n:k_n} - \xi_p| = \text{O}(1), \quad \text{in probability.} \tag{4.12}$$

If F were continuous a.e., the process $\{n^{1/2}[F_n(F^{-1}(t)) - t],\ t \in [0,1]\}$ would converge weakly to a standard Brownian bridge on $D[0,1]$. This weak convergence in turn ensures the tightness of the same sequence of empirical processes. By virtue of the assumed differentiability of F at ξ_p (and its continuity), we may consider a small neighborhood of ξ_p and the weak convergence of the above process only on this small neighborhood; then we obtain

$$\sup\Big\{n^{1/2}|F_n(x) - F(x) - F_n(\xi_p) + F(\xi_p)| : \ |x - \xi_p| \leq \eta\Big\} = \text{o}_p(1), \tag{4.13}$$

as $n \to \infty$ for every (small) $\eta > 0$. Since $F_n(X_{n:k_n}) = n^{-1}k_n = p + \text{o}(n^{-1/2})$, the desired result follows from (4.12) and (4.13) by the first-order Taylor expansion of $F(X_{n:k_n})$ around $F(\xi_p) = p$. □

The first-order asymptotic representation in Theorem 4.1 has been strengthened to a second-order representation by Bahadur (1966) under an additional differentiability condition on F. We have the following

THEOREM 4.2 *If F is twice differentiable at ξ_p with $f(\xi_p) > 0$ then (4.1) and (4.2) hold.*

PROOF (an outline). First, we replace t in (4.11) by $t(\log n)^{1/2}$, and for $[F_n(y) - F(y)]$ (bounded random variable) we use the Hoeffding inequality (see Section 2.5). Then, parallel to (4.12), we have

$$|X_{n:k_n} - \xi_p| = \text{O}\Big((n^{-1}\log n)^{1/2}\Big) \quad \text{a.s.,} \quad \text{as } n \to \infty. \tag{4.14}$$

Consider the interval $[\xi_p - Cn^{-1/2}(\log n)^{1/2},\ \xi_p + Cn^{-1/2}(\log n)^{1/2}]$ for some arbitrary $C,\ 0 < C < \infty$, and divide this interval into $2b_n$ subintervals of width $Cn^{-1/2}(\log n)^{1/2}b_n^{-1}$ each. Choose $b_n \sim n^{1/4}$. Note that for any $a \leq x \leq b$,

$$F_n(a) - F(b) \leq F_n(x) - F(x) \leq F_n(b) - F(a).$$

Hence, $\sup\Big\{|F_n(x) - F(x) - F_n(\xi_p) + F(\xi_p)| : \ |x - \xi_p| \le C(n^{-1}\log n)^{1/2}\Big\}$ can be bounded by

$$\max\Big\{|F_n(x_j) - F(x_j) - F_n(\xi_p) + F(\xi_p)| : \ j = 1, \ldots, 2b_n\Big\} + \mathrm{O}(n^{-3/4}(\log n)),$$

where the x_j stand for the grid points

$$\xi_p - C(n^{-1}\log n)^{1/2} + jC(n^{-1}\log n)^{1/2} b_n^{-1}, \ j = 1, \ldots, 2b_n. \tag{4.15}$$

We use the Hoeffding inequality to $\mathbb{P}\{|F_n(X_j) - F(X_j) - F_n(\xi_p) + F(\xi_p)| > Cn^{-3/4}\log n(\log\log n)^{1/4}\}$ for $j = 1, 2, \ldots, 2b_n$, and complete the proof by showing that the sum of these $2b_n$ probabilities is $\mathrm{o}(n^{-m})$, for some $m > 1$. Hence, by the Borel–Cantelli lemma, we have

$$\begin{aligned}&\sup\{n^{1/2}|F_n(x) - F(x) - F_n(\xi_p) + F(\xi_p)| : |x - \xi_p| \le (n^{-1}\log n)^{1/2} C\}\\ &= \mathrm{O}\Big(n^{-3/4}(\log n)^{1/2}(\log\log n)^{1/4}\Big) \ a.s., \text{ as } n \to \infty.\end{aligned} \tag{4.16}$$

The desired result follows from (4.14) and (4.16). □

We are also interested in an intermediate result where in (4.2) we like to have an $\mathrm{O}_p(n^{-3/4})$ instead of an $a.s.$ order of $n^{-3/4}(\log n)^{1/2}(\log\log n)^{1/4}$. Keeping in mind (3.45), consider a relatively more general form

$$L_{n2} = \sum_{j=1}^{k} a_j X_{n:[np_j]+1}, \tag{4.17}$$

where $0 < p_1 < \ldots < p_k < 1$, k is a fixed positive integer, and $a_1, \ldots, a_k$ are given real numbers. We wish to establish a representation of the form

$$L_{n2} - \mu = n^{-1}\sum_{i=1}^{n} \psi_2(X_i) + R_n, \tag{4.18}$$

supplemented by the stochastic order of R_n, with some μ depending on F and $a_1, \ldots, a_k$, and where $\psi_2 : \boldsymbol{R}_1 \mapsto \boldsymbol{R}_1$ is such that $\int_{-\infty}^{\infty} \psi_2(x)dF(x) = 0$.

THEOREM 4.3 *Suppose that F is twice differentiable at $F^{-1}(p_j)$ and $F'(F^{-1}(p_j)) > 0$ for each $j = 1, \ldots, k$. Then*

$$L_{n2} - \sum_{j=1}^{k} a_j F^{-1}(p_j) = n^{-1}\sum_{i=1}^{n} \psi_2(X_i) + r_n \tag{4.19}$$

where r_n satisfy (4.2) and

$$\psi_2(x) = \sum_{j=1}^{k} a_j \Big[f(F^{-1}(p_j))\Big]^{-1}\Big\{p_j - I(x \le F^{-1}(p_j))\Big\}, \ x \in \boldsymbol{R}_1. \tag{4.20}$$

PROOF (outline). For each j $(= 1, \ldots, k)$ we make use of (4.12). In (4.15) and (4.16) we concentrate on the interval $[F^{-1}(p_j) - C(n^{-1}\log n)^{1/2}, F^{-1}(p_j) +$

$C(n^{-1}\log n)^{1/2}]$, for some arbitrary C, and use b_n ($\sim n^{1/4}$) subintervals, as in the proof of Theorem 4.2. The rest of the proof again follows by using the Hoeffding inequality. □

For the case of a single sample quantile, (4.3) depicts the asymptotic distribution of the remainder term in (4.19). In the next chapter we will consider a generalization of this result for M-estimators with discontinuous score functions. However, in the case of more than one quantiles, an asymptotic distribution may not come in a very handy form. In fact it may depend heavily on k as well as the a_j. For the quantile process related to the uniform distribution, Kiefer (1970) considered the distance between the empirical and quantile processes [i.e., $\sup\{|n^{3/4}(\log n)^{-1/2}f(F^{-1}(p))r_n(p)| : 0 < p < 1\}$] and demonstrated its close alliance with that of the Kolmogorov–Smirnov statistic. He actually showed that the limiting distribution of this distance measure is

$$1 + 2\sum_{k=1}^{\infty}(-1)^k e^{-2k^2t^4},\ t \geq 0. \tag{4.21}$$

For related results we refer to the monograph of Csörgő and Révész (1981) (see also Csörgő and Horváth (1993)). In the case of M-estimators with discontinuous score functions (see Chapter 5), we are able to find the asymptotic distribution of the remainder term. However, the remainder term in (4.19) with arbitrary weights $a_1, \ldots, a_k$ may not have simple the second-order asymptotic distribution.

Remarks. It can be shown that the reminder term in Theorem 4.2 is $O_p(n^{-3/4})$. The same argument may be used for each $X_{n:[np_j]+1}$, $1 \leq j \leq k$, to show that the remainder term is $O_p(n^{-3/4})$ under the hypothesis of Theorem 4.3. When we do not require the second derivative, we can at least claim the representation with $r_n = o_p(n^{-1/2})$. As we will see later, for smooth functions we can obtain the order $r_n = O_p(n^{-1})$, faster than for a combination of single quantiles. This is the idea behind "smoothing" sample quantiles; see (4.43)–(4.44) for more details.

In applications we often have a statistic $T_n = h(X_{n:k_n})$ with a smooth function $h(.)$:

$$\frac{h(X_{n:k_n}) - h(\xi_p)}{X_{n:k_n} - \xi_p} \to h'(\xi_p),\ \text{in probability},$$

whenever $h'(.)$ exists at ξ_p, and $X_{n:k_n} \to \xi_p$, in probability. Under this additional differentiability condition on $h(.)$ we obtain the first-order asymptotic representation for T_n . Let moreover, $h'(x)$ exist in a neighborhood of ξ_p and satisfy a Lipschitz condition of order $\alpha > 0$,

$$|h'(x) - h'(\xi_p)| \leq K|x - \xi_p|^{\alpha}\ \forall x : |x - \xi_p| \leq \delta, \tag{4.22}$$

where $0 < K < \infty$ and $\delta > 0$. If $\alpha > 1/2$, then Theorem 4.2 holds for T_n as well. If $\alpha : 0 < \alpha \leq 1/2$, then Theorem 4.2 still holds for T_n, but with the remainder term

$$r_n = \mathrm{O}(n^{-(1+\alpha)/2}(\log n)^{1/2}(\log\log n)^{1/4+\alpha/2}) \text{ a.s.,} \tag{4.23}$$

as $n \to \infty$.

More generally, consider the function t$T_n = h(X_{n:[np_1]+1}, \dots, X_{n:[np_k]+1})$ of k quantiles for some smooth $h : \boldsymbol{R}_k^> \to \boldsymbol{R}$, where $\boldsymbol{R}_k^> = \{\mathbf{x} \in \boldsymbol{R}_k : x_1 \leq \dots \leq x_k\}$. Assume that $\boldsymbol{\xi} = (\xi_{p_1}, \dots, \xi_{p_k})^\top$ lies in the interior of $\boldsymbol{R}_k^>$ and that $(\partial/\partial\mathbf{x})h(\mathbf{x}) = \mathbf{h}'$ exists in a neighborhood of $\boldsymbol{\xi}$. Then Theorem 4.1 extends directly to T_n and (4.18) holds with $\mathbf{a} = (a_1, \dots, a_k)^\top = \mathbf{h}'_{\boldsymbol{\xi}}$. Finally, if we consider a natural extension of (4.22) and assume that

$$\|\mathbf{h}'_{\mathbf{x}} - \mathbf{h}'_{\boldsymbol{\xi}}\| \leq K\|\mathbf{x} - \boldsymbol{\xi}\|^\alpha, \tag{4.24}$$

for some $\alpha : 0 < \alpha \leq 1$ and $0 < K < \infty$, whenever $\|\mathbf{x} - \boldsymbol{\xi}\| \leq \delta$ for some $\delta > 0$, then Theorem 4.2 holds for T_n with R_n in (4.18) being (4.2) or (4.23) according as α is $> 1/2$ or $0 < \alpha \leq 1/2$. We see that the degree of smoothness of $h(.)$, as reflected by the order of convergence of the *gradient* of $h(.)$, dictates the rate of convergence of the remainder term, as well as its mode of convergence. This remainder term is not generally $\mathrm{O}_p(n^{-1})$. More second-order asymptotic representations will be derived in Section 4.3, under alternative smoothness conditions.

4.3 Representations for Smooth Scores L-Statistics

Consider a linear combination of order statistics of the form

$$L_{n1} = \sum_{i=1}^{n} c_{ni} X_{n:i}, \tag{4.25}$$

where the scores c_{ni} are generated by a smooth function $J : (0,1) \mapsto \boldsymbol{R}_1$ in either of the following two ways:

$$c_{ni} = n^{-1} J(i/(n+1)), \quad i = 1, \dots, n; \tag{4.26}$$

$$c_{ni} = \int_{(i-1)/n}^{i/n} J(u)du, \quad i = 1, \dots, n. \tag{4.27}$$

Our goal is to consider a representation for L_{n1} in the form of (4.18) with specific order for R_n. We will see that $R_n = \mathrm{O}_p(n^{-r})$ for some $r \in (\frac{1}{2}, 1]$ and that the higher the value of r that we achieve, the greater is the stringency of the smoothness conditions on J.

The representation for L_{ni} will be considered under the following smoothness conditions on J and F:

A. J is continuous on (0,1) up to a finite number of points $s_1, \dots, s_m$ where m is a nonnegative integer and $0 < s_1 < \dots < s_m < 1$.

B. F is continuous *a.e.*, and $F^{-1}(s) = \inf\{x : F(x) \geq s\}$ satisfies the Lipschitz condition of order 1 in a neighborhood of $s_1, \ldots, s_m$.

Moreover, either of the following conditions is assumed:

C1. *Trimmed J.* $J(u) = 0$ for $u \in [0, \alpha) \cup (1-\alpha, 1]$ for some positive α, such that $\alpha < s_1$, $s_m < 1 - \alpha$, and J satisfies the Lipschitz condition of order ν (≤ 1) in $(\alpha, s_1), (s_1, s_2), \ldots, (s_m, 1-\alpha)$. Moreover, for some $\tau > 0$, $F^{-1}(1-\alpha+\tau) - F^{-1}(\alpha - \tau) < \infty$.

C2. *Untrimmed J.* For some $\beta > 0$, $\sup\{|x|^\beta F(x)[1-F(x)] : x \in \boldsymbol{R}_1\} < \infty$, and J satisfies the Lipschitz condition of order $\nu \geq (2/\beta) + \triangle - 1$ (for some $\triangle \in (0,1)$) in each of the intervals $(0, s_1), (s_1, s_2), \ldots, (s_m, 1)$.

A variation of **C2** will also be considered later. For the moment we have the following:

THEOREM 4.4 *Define L_{n1} with the scores c_{ni} in (4.27). Then, under* **A**, **B**, *and* **C1** *or* **C2**,

$$L_{n1} - \mu = n^{-1} \sum_{i=1}^{n} \psi_1(X_i) + R_n \text{ with } R_n = O_p(n^{-r}), \tag{4.28}$$

where

$$r = \frac{\nu+1}{2} \wedge 1, \tag{4.29}$$

$$\mu = \mu(J, F) = \int_0^1 J(u) F^{-1}(u) du, \tag{4.30}$$

$$\psi_1(x) = -\int_{\boldsymbol{R}_1} \{I(y \geq x) - F(y)\} J(F(y)) dy, \; x \in \boldsymbol{R}_1. \tag{4.31}$$

PROOF Let $U_i = F(X_i)$ and $U_{n:i} = F(X_{n:i})$, $i = 1, \ldots, n$, and let

$$F_n(x) = n^{-1} \sum_{i=1}^{n} I(X_i \leq x), \; x \in \boldsymbol{R}_1,$$

$$U_n(t) = n^{-1} \sum_{i=1}^{n} I(U_i \leq t), \; t \in [0,1].$$

Then, the function

$$\phi(s) = -\int_0^1 \{I(u \geq s) - u\} J(u) du, \; s \in [0,1], \tag{4.32}$$

is bounded, absolutely continuous with $\phi'(s) = J(s)$, $s \in (0,1)$. By virtue of (4.27), (4.31), and (4.32), we have

$$L_{n1} - \mu = \int_{\boldsymbol{R}_1} x d[\phi(F_n(x)) - \phi(F(x))] \tag{4.33}$$

$$= -\int_{\boldsymbol{R}_1} [\phi(F_n(x)) - \phi(F(x))]dx = -\int_0^1 [\phi(U_n(s)) - \phi(s)]dF^{-1}(s).$$

On the other hand, by (4.31) and (4.32)

$$n^{-1}\sum_{i=1}^{n}\psi_1(X_i) = \int_{\boldsymbol{R}_1} \psi_1(x)d[(F_n(x) - F(x)]$$

$$= -\int_{\boldsymbol{R}_1} [F_n(x) - F(x)]\phi'(F(x))dx = -\int_0^1 \{U_n(s) - s\}\phi'(s)dF^{-1}(s).$$

Hence,

$$R_n = L_{n1} - \mu - n^{-1}\sum_{i=1}^{n}\psi_1(X_i) = \int_0^1 V_n(s)dF^{-1}(s), \tag{4.34}$$

where

$$V_n(s) = \phi'(s)[U_n(s) - s] - [\phi(U_n(s)) - \phi(s)],\ s \in (0,1).$$

Let us fix an η (> 0) such that

$$0 < \eta < \frac{\Delta}{2(\gamma+1)} \text{ under } \mathbf{C2}, \text{ and } 0 < \eta < \frac{1}{2}, \text{ under } \mathbf{C1}. \tag{4.35}$$

Recall that $\left\{\frac{U_n(s)-s}{1-s};\ s \in (0,1)\right\}$ is a martingale for each n (see Section 2.5), so that using the Hájek–Rényi–Chow inequality for submartingales, we readily obtain

$$D_n = \sup\left\{n^{\frac{1}{2}}\frac{|U_n(s) - s|}{(s(1-s))^{\frac{1}{2}-\eta}} : 0 < s < 1\right\} = \mathrm{O}_p(1). \tag{4.36}$$

[We may refer to (3.48) of Csáki (1984) for an alternative derivation of (4.36)]. Thus, given $\varepsilon > 0$, there exist finite C (> 0) and n_0 such that

$$\mathbb{P}\{D_n > C\} \le \varepsilon/2, \text{ for every } n \ge n_0. \tag{4.37}$$

Without any loss of generality we may simplify the presentation by assuming that there is a single point s_0 ($0 < s_0 < 1$) of discontinuity of J. Then, using (4.34), we may decompose

$$R_n = \int_0^{s_0-\delta_n} V_n(s)dF^{-1}(s) + \int_{s_0-\delta_n}^{s_0+\delta_n} V_n(s)dF^{-1}(s)$$

$$+ \int_{s_0+\delta_n}^{1} V_n(s)dF^{-1}(s) = R_{n1} + R_{n2} + R_{n3},$$

where we take

$$\delta_n = 2Cn^{-\frac{1}{2}} \text{ with } C \text{ defined in (4.37).}$$

We will study the order of each component R_{nj}, $j = 1, 2, 3$ separately. First, by virtue of (4.37),

$$\mathbb{P}\Big\{|R_{n1}| \ge Kn^{-r}\Big\}$$

$$= \mathbb{P}\Big\{|R_{n1}| \ge Kn^{-r},\ D_n \le C\Big\} + \mathbb{P}\Big\{|R_{n1}| \ge Kn^{-r},\ D_n > C\Big\}$$

$$\leq \mathbb{P}\Big\{|R_{n1}| \geq Kn^{-r},\ D_n \leq C\Big\} + \frac{\varepsilon}{2}\ \forall n \geq n_0. \tag{4.38}$$

Since $J(u) = 0$ for $u \in (0, \alpha)$ under **C1** and $s_0 > \alpha$, the first term on the right-hand side of (4.38) is bounded from above by

$$\mathbb{P}\Big\{\int_{s_0-\delta_n}^{s_0+\delta_n} (I[U_n(s) > s] \int_0^{U_n(s)-s} |J(u+s) - J(s)| du$$

$$+ \quad I[U_n(s) \leq s] \int_{U_n(s)-s}^{0} |J(u+s) - J(s)| du) dF^{-1}(s) \geq \frac{K}{n^r}, D_n \leq C\Big\}$$

$$\leq \quad \mathbb{P}\Big\{MC^2 n^{-r}(F^{-1}(s_0 - \delta_n) - F^{-1}(\alpha - \tau)) \geq Kn^{-r}\Big\} \leq \frac{\varepsilon}{2}$$

for sufficiently large $K > 0$ and all $n \geq n_1$ ($\geq n_0$); here M is a positive constant. Hence, under **A**, **B**, and **C1**, $R_{n1} = O_p(n^{-r})$.

Let us next consider the case where **C1** is replaced by **C2** (i.e. the trimmed J by untrimmed one). Notice that under **C2**,

$$|F^{-1}(s)| \leq M_1[s(1-s)]^{-1/\beta},\ 0 < s < 1,\ 0 < M_1 < \infty,$$

and hence, using (4.35), we can bound the first term on the right hand side of (4.38) by

$$\mathbb{P}\Big\{M \int_0^1 |U_n(s) - s|^{1+\nu} dF^{-1}(s) \geq Kn^{-r},\ D_n \leq C\Big\}$$

$$\leq \mathbb{P}\Big\{M^{1+\nu} n^{-(1+\nu)/2} \int_0^1 \{s(1-s)\}^{-1+\Delta/2} ds \geq Kn^{-r}\Big\} \leq \frac{\varepsilon}{2},$$

for sufficiently large K (> 0) and for $n \geq n_2$ ($\geq n_0$). This implies that under **A**, **B**, and **C2**, $R_{n1} = O_p(n^{-r})$. The same treatment holds for R_{n3}. Thus, it remains to prove that $R_{n2} = O_p(n^{-r})$. Toward this end, note that for $s \in (s_0 - \delta_n, s_0 + \delta_n)$, both $J(s)$ and $F^{-1}(s)$ are bounded, and hence,

$$\mathbb{P}\Big\{|R_{n2}| \geq Kn^{-r}\Big\}$$

$$\leq \quad \mathbb{P}\Big\{|R_{n2}| \geq Kn^{-r},\ D_n \leq C\Big\} + \mathbb{P}\Big\{D_n > C\Big\}$$

$$\leq \quad \mathbb{P}\Big\{\sup_{s_0-\delta_n \leq s \leq s_0+\delta_n} |V_n(s)| \cdot [F^{-1}(s_0+\delta_n) - F^{-1}(s_0 - \delta_n)] \geq Kn^{-r}\Big\}$$

$$+\frac{\varepsilon}{2}$$

$$\leq \quad \mathbb{P}\Big\{MCn^{-1/2}[F^{-1}(s_0+\delta_n) - F^{-1}(s_0-\delta_n)] \geq Kn^{-r}\Big\} + \frac{\varepsilon}{2}$$

$$\leq \quad \varepsilon$$

for sufficiently large K (> 0) and $n \geq n_3$ (≥ 0). □

Let us now consider the case where the scores are defined by (4.26). When J is trimmed (see **C1**), the results in Theorem 4.4 hold for the scores in (4.26) as well. If J is not trimmed, then a representation of the type (4.28) with $r = 1$ for the scores (4.26) may demand more stringent conditions on J as well as F. We may have to assume that

$$\sup\left\{|x|^{1+a}F(x)[1-F(x)] : x \in \boldsymbol{R}_1\right\} < \infty \quad \text{for some positive } a. \tag{4.39}$$

Note that (e.g., Sen 1959) (4.39) implies that $\mathbb{E}|X|^b < \infty$, for some $b \geq 1$, so that (4.39) entails the existence of the first order moment of X. In fact, (4.39) is equivalent to the existence of $\mathbb{E}|X|^c$ for some $c > 1$. As a variant of **C2**, we then formulate the following:

C3. J satisfies the Lipschitz condition of order 1 in $(0, s_1), \ldots, (s_k, 1)$ and (4.39) holds for some $a > 0$.

THEOREM 4.5 *Assume either* **A**, **B**, *and* **C1** *with* $\nu = 1$, *or* **A**, **B**, **C3**. *Then, for the scores defined by (4.26), the representation in (4.28) holds with* $r = 1$ *for* R_n.

PROOF Denote L_{n1} by $L^\star_{n1}$ when the scores are given (4.26), and let

$$\begin{aligned} J_n(s) &= J(i/(n+1)) \quad \text{for } (i-1)/n < s \leq i/n,\ i = 1, \ldots, n, \\ \mu_n &= \textstyle\int_0^1 J_n(s)dF^{-1}(s), \\ \phi_n(s) &= -\textstyle\int_0^1 \{I(u \geq s) - u\}J_n(u)du, \quad s \in (0,1). \end{aligned} \tag{4.40}$$

Then, proceeding as in (4.33), we obtain that for L_{n1} defined by the scores in (4.27),

$$\begin{aligned} &|L^\star_{n1} - \mu_n - L_{n1} + \mu| \\ &= \left|\int_0^1 [\phi_n(U_n(s)) - \phi_n(s) - \phi(U_n(s)) + \phi(s)]dF^{-1}(s)\right| \\ &\leq \int_0^1 \Big\{I(U_n(s) \geq s)\int_0^{U_n(s)-s} |J_n(s+t) - J(s+t)|dt \\ &+ I(U_n(s) < s)\int_{U_n(s)-s}^0 |J_n(s+t) - J(s+t)|dt\Big\}dF^{-1}(s). \end{aligned} \tag{4.41}$$

Note that (4.40) and the assumed first-order Lipschitz condition on J imply

$$|J_n(s) - J(s)| \leq Mn^{-1}, \text{ for every } s \in (0,1),\ 0 < M < \infty. \tag{4.42}$$

Further, by (4.36) and (4.37), we have $|U_n(s) - s| = \mathrm{O}_p(n^{-1/2})$, uniformly in $s \in (0,1)$. As such, in the trimmed case, e.g., **C1**, the right-hand side of (4.41) is $\mathrm{O}_p(n^{-1})$. Under **C3**, the right-hand side of (4.41) is bounded by

$$2Mn^{-1}\int_0^1 |U_n(s) - s|dF^{-1}(s) \leq 2Mn^{-1}\int_{\boldsymbol{R}_1} |x|d[F_n(x) - F(x)] = \mathrm{O}_p(n^{-1}),$$

where the last step follows by using the Khintchine law of large numbers (under **C3**). Thus, under the hypothesis of the theorem, (4.41) is $\mathrm{O}_p(n^{-1})$.

On the other hand, for J being Lipschitz of order one (under **C1** or in **C2** allowing $\nu = 1$), we see that in (4.30) r reduces to 1, so (4.28) holds for L_{n1} with $R_n = \mathrm{O}_p(n^{-1})$. Hence we obtain

$$\begin{aligned} L_{n1}^{\star} &= L_{n1} - \mu + \mu_n + \mathrm{O}_p(n^{-1}) \\ &= (\mu_n - \mu) + n^{-1} \sum_{i=1}^{n} \psi_1(X_i) + \mathrm{O}_p(n^{-1}). \end{aligned}$$

Finally, by (4.42) we obtain under **C3**

$$|\mu_n - \mu| \leq Mn^{-1} \int_0^1 |F^{-1}(u)| du = Mn^{-1} \mathbb{E}|X| = \mathrm{O}(n^{-1}).$$

Under **C1**,

$$|\mu_n - \mu| \leq Mn^{-1} \int_{\alpha}^{1-\alpha} |F^{-1}(u)| du = \mathrm{O}(n^{-1}).$$

□

Remark 4.1 *If X_i's are i.i.d. random variables with the distribution function $F_\theta(x) = F(x - \theta)$, $x \in \boldsymbol{R}_1$, such that*

1. $F(x) + F(-x) = 1 \forall x \in \boldsymbol{R}_1$ *(i.e., F symmetric about 0),*
2. $J(u) = J(1 - u)\ \forall u \in (0, 1)$, *and* $\int_0^1 J(u) du = 1$,

then under the hypothesis of Theorem 4.4,

$$L_{n1} - \theta = n^{-1} \sum_{i=1}^{n} \psi_1(X_i - \theta) + O_p(n^{-r}),$$

where r and ψ_1 are defined by (4.30) and (4.31). Under the hypothesis of Theorem 4.5, we have

$$L_{n1}^{\star}(or\ L_{n1}) = \theta + n^{-1} \sum_{i=1}^{n} \psi_1(X_i - \theta) + O_p(n^{-1}). \tag{4.43}$$

This provides a general representation for L-estimators of location parameters when the score function is smooth. As notable examples, we may consider the following:

1. *Sample mean* $\bar{X}_n = n^{-1} \sum_{i=1}^{n} X_i$. This corresponds to (4.26) or (4.27) with $J(u) = 1$, for every $u \in (0, 1)$.
2. *Trimmed mean* $T_n = (n - 2k)^{-1} \sum_{j=k+1}^{n-k} X_{n:j}$ where k is a positive integer with $k/n \to \alpha$, for some $\alpha \in (0, \frac{1}{2})$. Here (C1) holds with $J(u) = (1 - 2\alpha)^{-1}$, $\alpha \leq u \leq 1 - \alpha$, and $J(u) = 0$ for $u \in (0, \alpha) \cup (1 - \alpha, 1)$.

3. *Rank-weighted mean* (Sen 1964)

$$T_{n,k} = \binom{n}{2k+1}^{-1} \sum_{i=k+1}^{n-k} \binom{i}{k}\binom{n-i-1}{k} X_{n:i+1}, \quad \text{for } k \in (0, n/2).$$

Note that for $T_{n,0} = \bar{X}_n$, while for $k = [(n+1)/2]$, it reduces to the sample median. For any fixed k (≥ 1), although $\binom{n}{2k+1}^{-1}\binom{i}{k}\binom{n-i-1}{k}$ $(i = 1, \ldots, n,)$ may not correspond to (4.26) or (4.27), they may be approximated by a smooth J (up to the order n^{-1}) where $J(u) = \{u(1-u)\}^k$, $u \in (0, 1)$.

4. Harrell and Davis (1982) *quantile estimator* defined by

$$T_n = \sum_{i=1}^{n} c_{ni} X_{n:i} \tag{4.44}$$

$$c_{ni} = \frac{\Gamma(n+1)}{\Gamma(k)\Gamma(n-k+1)} \int_{(i-1)/n}^{i/n} u^{k-1}(1-u)^{n-k} du, \; i = 1, \ldots, n,$$

where $k \sim np$, and p $(0 < p < 1)$ relates to the order of the quantile to be estimated. An alternative estimator, due to Kaigh and Lachenbruch (1982), also belongs to this class (with more affinity to the rank weighted mean). The scores in (4.27) with a bounded and continuous J are relevant for (4.44).

5. *BLUE estimator* of location (Jung 1955, 1962; Blom 1956). For a distribution function $F_\theta(x) = F(x - \theta)$, $x \in \boldsymbol{R}_1$, when F admits an absolutely continuous density function f (with derivative f'), an asymptotically best unbiased linear estimator of the location parameter θ is given by L_{n1} with the scores defined by (4.26) and where

$$J(F(x)) = \gamma'(x) \quad \text{and} \quad \gamma(x) = -\frac{f'(x)}{f(x)}, \; x \in \boldsymbol{R}_1. \tag{4.45}$$

For a large class of location-scale family of distributions, (4.45) leads to bounded J which satisfies **C1**, **C2** or **C3**, and **A**, **B**, and hence, the theorems in this section pertain to these BLUE estimators of locations. A similar case holds for the BLUE of the scale parameters.

4.4 Asymptotic representations for general L-Estimators

Let us consider the general L-estimator of the form

$$L_n = L_{n1} + L_{n2} \;\text{ or }\; L^\star_{n1} + L_{n2} \tag{4.46}$$

where L_{n2} is defined by (4.17) and L_{n1} or $L^\star_{n1}$ by (4.25)–(4.27), respectively. A notable example of this type is the *Winsorized mean* defined by $W_n =$

$\frac{1}{n}\left\{\sum_{i=k+1}^{n-k} X_{n:i} + k(X_{n:k} + X_{n:n-k+1})\right\}$ for some positive integer $k \leq n/2$. As in the case of the trimmed mean, we may let $\frac{k}{n} \to \alpha$ for $0 < \alpha < 1/2$. For $L_{n1} = \frac{1}{n}\sum_{i=k+1}^{n-k} X_{n:i}$, we have the scores defined by (4.26) or (4.27) with $J(u) = 1$ for $u \in (\alpha, 1-\alpha)$ and 0 otherwise. L_{n2} corresponds to the midrange $\frac{k}{n}(X_{n:k} + X_{n:n-k+1})$. thus, Theorem 4.3 applies to L_{n2} and Theorem 4.5 to L_{n1}. However, the remainder term for $L_n = L_{n1} + L_{n2}$ can at best be of the order $n^{-3/4}$. Define

$$\psi(x) = \psi_1(x) + \psi_2(x), \quad x \in \boldsymbol{R}_1,$$

where ψ_1 and ψ_2 are defined by (4.31) and (4.20), respectively. Then we have the following theorem [where all the notations are borrowed from (4.19) and (4.31)]

THEOREM 4.6 *Under the hypotheses of Theorems 4.3 and 4.4,*

$$L_n - \mu^0 = n^{-1}\sum_{i=1}^{n} \psi(X_i) + O_p(n^{-r}); \quad r = \frac{\gamma+1}{2} \wedge \frac{3}{4}. \tag{4.47}$$

where

$$\mu^0 = \sum_{j=1}^{k} a_j F^{-1}(p_j) + \int_0^1 J(u)du$$

Under the hypothesis of Theorem 4.3, (4.47) holds with an order of the remainder term as $O_p(n^{-3/4})$.

PROOF The proof is direct adaptation of the relevant theorems under the hypotheses and hence is omitted. □

The use of discrete quantiles (i.e., of L_{n2}) leads to a slower order of the reminder, even when the remainder term in the smooth part (i.e., L_{n1}) may be $O_p(n^{-1})$ for . The choice of the normalizing factor will depend on the component L_{n2}. The desired normalizing factor can be attained only for smooth scores with some further regularity conditions. The study of the smooth L-estimator is facilitated by the use of statistical functionals that are differentiable in a certain sense. Generally, the L-estimators of location with smooth and bounded scores are an example of such differentiable functionals. The next section is devoted to this important topic.

4.5 Representations for Statistical Functionals

Let $X_1, \ldots, X_n$ be i.i.d. random variables with the distribution function F and let F_n be the sample distribution function. Consider a parameter $\theta = T(F)$, expressible as a functional of the distribution function F. Then its natural estimator is $T_n = T(F_n)$. If $T(.)$ is first-order Hadamard differentiable at F

(cf. Definition 3.1), we have by virtue of (3.117) and (3.118),

$$\begin{aligned} T_n &= T(F_n) + \int T^{(1)}(F;x)d[F_n(x) - F(x)] + R_n(F_n, F) \\ &= T(F) + n^{-1}\sum_{i=1}^{n} T^{(1)}(F;X_i) + o(\|F_n - F\|), \end{aligned} \tag{4.48}$$

where $T^{(1)}(F;x)$ is the compact derivative of $T(.)$ at F (known as the *influence function*). Note that

$$\mathbb{E}T^{(1)}(F;X) = \int T^{(1)}(F;x)dF(x) = 0. \tag{4.49}$$

The second moment σ_F^2 of $T^{(1)}(F;X_i)$ is defined by (3.121), and we assume that it is finite and positive. Recall that for every real t (≥ 0),

$$\mathbb{P}\Big\{\|F_n - F\| > tn^{-\frac{1}{2}}\Big\} \leq 2\exp\{-2t^2\} \quad \text{for every } n \geq 1,$$

so that

$$\|F_n - F\| = O_p(n^{-\frac{1}{2}}). \tag{4.50}$$

If $T(.)$ is first-order Hadamard differentiable, then (4.48) and (4.50) imply the first-order asymptotic representation (4.5) with $\xi(X_i, \theta) = T^{(1)}(F;X_i)$. In the same spirit as in Sections 4.2 and 4.3, we will study more refined orders for the remainder term in (4.48), under additional regularity conditions on the functionals. For this reason we introduce the following notation and terminology.

For two topological vector spaces V and W, let $L_1(V,W)$ be the set of continuous linear transformations from V to W, and $\mathcal{C}(V,W)$ be the set of continuous functionals from V to W, and

$$L_2(V,W) = \{f : f \in \mathcal{C}(V,W),\ f(tH) = t^2 f(H)\ \forall H \in V,\ t \in \mathbf{R}_1\}.$$

Let $\mathcal{A}$ be an open set of V. A functional $T : \mathcal{A} \to W$ is second-order Hadamard (or compact) differentiable at $F \in \mathcal{A}$, if there exist $T_F{}' \in L_1(V,W)$ and $T_F{}'' \in L_2(V,W)$ such that for any compact set Γ of V,

$$\lim_{t\downarrow 0} t^{-2}\{T(F+tH) - T(F) - T_F'(tH) - \frac{1}{2}T_F''(tH)\} = 0, \tag{4.51}$$

uniformly for $H \in \Gamma$. Then T_F' and T_F'' are called the *first-* and *second-order Hadamard* (or *compact) derivative* of T at F, respectively.

Let $Rem_2(tH)$ be the remainder term of the second-order expansion in (4.51), so that $Rem_2(tH) = T(F+tH) - T(F) - T_F'(tH) - \frac{1}{2}T_F''(tH)$. Then (4.51) may also be represented as

$$\lim_{t\downarrow 0} t^{-2} Rem_2(F+tH) = 0$$

for any sequence $\{H_n\}$ with $H_n \to H \in V$.

Remark 4.2 *This definition of the second-order Hadamard-differentiability is adapted from Ren and Sen (1995b) and is consistent with the one given by Sen (1988a).*

The existence of the second-order Hadamard derivative implies the existence of the first-order one. We may always normalize T_F' and T_F'' in such a way that (4.49) holds, with

$$T^{(2)}(F;x,y) = T^{(2)}(F;y,x) \quad \text{a.e.}[F] \tag{4.52}$$

$$\int T^{(2)}(F;x,y)dF(y) = 0 = \int T^{(2)}(F;y,x)dF(x) \quad \text{a.e.}[F]$$

Using (4.51)–(4.52), we obtain that for a functional $T(.)$, second-order Hadamard differentiable at F,

$$T_n = T(F) + n^{-1}\sum_{i=1}^{n} T^{(1)}(F;X_i) + \frac{1}{2n^2}\sum_{i=1}^{n}\sum_{j=1}^{n} T^{(2)}(F;X_i,X_j)$$

$$+ o_p(n^{-1}) \quad \text{as } n \to \infty. \tag{4.53}$$

This yields a second order asymptotic expansion of $T(F_n)$. The third term on the right-hand side of (4.53) is in fact the leading term in the remainder for the first-order expansion for T_n. If we are able to find the asymptotic distribution of the normalized version of this term, we obtain the second order asymptotic distributional expansion for T_n.

Let us denote

$$R_n^{\star\star} = n^{-2}\sum_{i=1}^{n}\sum_{j=1}^{n} T^{(2)}(F;X_i,X_j)$$

$$= \int\int T^{(2)}(F;x,y)dF_n(x)dF_n(y). \tag{4.54}$$

Then, $R_n^{\star\star}$ is a von Mises functional of degree 2. Moreover, denote

$$\bar{T}_n^{(2)} = n^{-1}\sum_{i=1}^{n} T^{(2)}(F;X_i,X_i),$$

$$U_n^{(2)} = \binom{n}{2}^{-1} \sum_{\{1\le i<j\le n\}} T^{(2)}(F;X_i,X_j), \tag{4.55}$$

$$\tau^{(2)}(F) = \int T^{(2)}(F;x,x)dF(x)$$

and assume that $\tau^{(2)}(F)$ is well defined at F (and is finite). We can assume that

$$I\!\!E_F\Big[T^{(2)}(F;X_1,X_2)\}\Big]^2 < \infty. \tag{4.56}$$

Then,

$$nR_n^{\star\star} = \bar{T}_n^{(2)} + nU_n^{(2)}. \tag{4.57}$$

Using the Khintchine strong law of large numbers,we obtain from (4.55) that

$$\bar{T}_n^{(2)} \to \tau^{(2)}(F) \quad \text{a.s., as} \quad n \to \infty, \tag{4.58}$$

and from (4.57) and (4.58) we conclude that

$$nR_n^{\star\star} - \tau^{(2)}(F) \text{ has the same asymptotic distribution as } nU_n^{(2)}, \tag{4.59}$$

if they have any at all. Recall that

$$\mathbb{E}_F\Big[T^{(2)}(F; X_1, X_2)T^{(2)}(F; X_1, X_3)\Big] = 0, \tag{4.60}$$

by (4.52) and (4.56), so the U-statistic $U_n^{(2)}$ corresponds to a parameter that is stationary of order 1 (in the sense of Hoeffding 1948). In this case the asymptotic distribution of $nU_n^{(2)}$ is nondegenerate, and hence it follows from (4.59) that $nR_n^{\star\star} - \tau^{(2)}(F)$ has a nondegenerate asymptotic distribution. thus, it is more convenient to study first the asymptotic distribution theory of $nU_n^{(2)}$ and then to move on to the case of $nR_n^{\star\star}$.

Under the assumption (4.56), $T^{(2)}(.) \in L_2(F)$, and hence there exists a set of eigenvalues $\{\lambda_k\}$ of $T^{(2)}(.)$ (finite or infinite) corresponding to orthonormal functions $\{\tau_k(.);\ k \geq 0\}$ such that

$$\int T^{(2)}(F; x, y)\tau_k(x)dF(x) = \lambda_k\tau_k(y) \quad \text{a.e. } (F),\ \forall k \geq 0$$

$$\int \tau_k(x)\tau_q(x)dF(x) = \delta_{kq}, \tag{4.61}$$

where δ_{kq}, the Kronecker delta, is 1 or 0 according as $k = q$ or not, $k, q \geq 0$.

THEOREM 4.7 *Under (4.56) and under the convention in (4.52),*

$$nU_n^{(2)} \xrightarrow{\mathcal{D}} \sum_{\{k\geq 0\}} \lambda_k\{Z_k^2 - 1\} \tag{4.62}$$

as $n \to \infty$, where the Z_k's are i.i.d. random variables with the standard normal distribution, and the λ_k's are defined by (4.61).

PROOF Let us denote

$$Q_{(M)}(x, y) = \sum_{k\leq M} \lambda_k\tau_k(x)\tau_k(y),\ x, y \in \boldsymbol{R}_1,$$

for every positive integer M and let

$$Q(x, y) = Q_{(\infty)}(x, y) = \sum_{k\geq 0} \lambda_k\tau_k(x)\tau_k(y),\ x, y \in \boldsymbol{R}_1$$

and

$$U_{n(M)}^{\star} = n^{-1} \sum_{\{1\leq i\neq j\leq n\}} Q_{(M)}(X_i, X_j)$$

$$= \sum_{k \leq M} \lambda_k \Big[n^{-\frac{1}{2}} (\sum_{i=1}^{n} \tau_k(X_i))^2 - n^{-1} \sum_{i=1}^{n} \tau_k^2(X_i) \Big]. \tag{4.63}$$

The $\tau_k(.)$ form an orthonormal system, hence by the central limit theorem,

$$n^{-\frac{1}{2}} \sum_{i=1}^{n} \tau_k(X_i) \xrightarrow{\mathcal{D}} \mathcal{N}(0,1) \quad \text{as } n \to \infty, \text{ for every fixed } k. \tag{4.64}$$

Moreover, by the Khintchine strong law of large numbers

$$n^{-1} \sum_{i=1}^{n} \tau_k^2(X_i) \to 1 \text{ a.s., for every fixed } k \tag{4.65}$$

as $n \to \infty$. Therefore, we conclude from (4.63), (4.64), and (4.65) that for every fixed M,

$$U^{\star}_{n(M)} \xrightarrow{\mathcal{D}} \sum_{k \leq M} \lambda_k \{Z_k^2 - 1\}, \quad \text{as} \quad n \to \infty.$$

On the other hand,

$$\mathbb{E}\Big[U^{\star}_{n(M)} - U^{\star}_{n(\infty)}\Big]^2 \to 0 \quad \text{as } M \to \infty,$$

hence for every $\varepsilon > 0$, there exists an M $(= M_{\varepsilon})$ such that

$$\mathbb{P}\Big\{|U^{\star}_{n(M_{\varepsilon})} - U^{\star}_{n(\infty)}| > \varepsilon\Big\} \leq \varepsilon. \tag{4.66}$$

Therefore, identifying $nU_n^{(2)}$ as equivalent to $U^{\star}_{n(\infty)}$, the desired result follows by using (4.49), (4.66), and choosing M appropriately large. □

By virtue of (4.59) and (4.62), we immediately arrive at the following

THEOREM 4.8 *For a second-order Hadamard differentiable functional $T(.)$, under the hypothesis of Theorem 4.7,*

$$nR_n^{\star\star} - \tau^{(2)}(F) \xrightarrow{\mathcal{D}} \sum_{k \geq 0} \lambda_k \{Z_k^2 - 1\} \quad \textit{as } n \to \infty, \tag{4.67}$$

where the λ_k and Z_k are defined as in Theorem 4.7.

The term $\tau^{(2)}(F)$, which stands for the location of the asymptotic distribution in (4.67), is generally referred to as the *first-order bias* term in the literature. The asymptotic distribution in (4.67) is generally referred to as the *second-order asymptotic distribution.* thus, the representation in (4.53) with (4.67) may be termed a *second-order asymptotic distributional representation* for T_n.

Regarding (4.53), and (4.54), we obtain the following corollary of Theorem 4.8:

Corollary 4.1 *If the functional $T(F)$ is stationary of order 1 at F [i.e., $T^{(1)}(F;x) = 0$ a.e. $[F]$], then*

$$n\Big[T_n - T(F) - n^{-1}\tau^{(2)}(F)\Big] \xrightarrow{\mathcal{D}} \sum_{k\geq 0} \lambda_k\{Z_k^2 - 1\} \quad \text{as } n \to \infty,$$

where the λ_k and Z_k are defined as in Theorem 4.7.

It is clear from the above results that if $T(.)$ is a second order Hadamard differentiable functional, then all the necessary information on the form of the remainder term in (4.48) and its asymptotic distribution is contained in the second order compact derivative $T^{(2)}(.)$ and in its Fourier series representation relative to an orthonormal system. This procedure also enables consideration of the second-order asymptotic distributional representations results in a unified manner. However, not all functionals are necessarily second-order Hadamard differentiable, and hence the theorems in this section may not apply to a very wide class of functionals. This point will be made more clear in the next chapter dealing with M-estimators. Even in the context of L-estimators, an estimator of the type L_{n2} may not be, in general, Hadamard differentiable of the second-order, and hence the theorems in this section may not be applicable for the L_{n2} in Section 4.2. Actually, the normalizing factor for the remainder term r_n for L-statistics of the type L_{n2} is $n^{3/4}$ [and not n^1 as in (4.67)], and hence the end product will be a degenerate random variable even if we apply a similar expansion. It seems appropriate to incorporate these theorems only to the case where the rate r in Section 4.3 is equal to 1 [comparable to (4.67)], as will be explained in the next section. However, not all smooth L-functionals are covered in this formulation. A very glaring example is the population mean $T(F) = \int x dF(x)$, where for simplicity we assume that F is defined on the real line $\boldsymbol{R}_1$. Recall that

$$|T(F) - T(F_n)|/\|F_n - F\| = |\int_{\boldsymbol{R}_1} x d[F(x) - F_n(x)]|/\|F_n - F\|.$$

When the distribution function F has infinite support [i.e., $0 < F(x) < 1$, for all real and finite x], (4.49) cannot be bounded a.e. $[F]$, even for very large n. thus, $T(F)$ is not generally Hadamard differentiable, although as an L-functional it corresponds to the smooth case with $J(u) = 1$ for every $u \in (0,1)$.

4.6 Second-Order Asymptotic Distributional Representations for Some Smooth L-Functionals

Consider a general L-statistic of the form L_n in (4.46) where the scores for L_{n1} are defined by (4.27). Then we can write $L_n = T(F_n)$, where

$$T(F) = \int_0^1 F^{-1}(u)J(u)du + \sum_{j=1}^{k} a_j F^{-1}(p_j).$$

Introduce a function $K = K_1 + K_2 = \{K_1(t) + K_2(t),\ t \in [0,1]\}$ by letting

$$K_1(t) = \int_0^t J(s)ds,\ t \in (0,1)$$

$K_2(t)$ is a pure step function with jumps a_j at $t = p_j,\ 1 \le j \le k$.

We can write, for example,

$$T(F) = \int_0^1 F^{-1}(t)dK(t) = \int_0^1 F^{-1}(t)dK_1(t) + \int_0^1 F^{-1}(t)dK_2(t)$$
$$= T_1(K) + T_2(K). \tag{4.68}$$

The influence function ψ_1 for $T_1(.)$ is given by (4.31) and the influence function for $T_2(.)$ is given by ψ_2 in (4.20). We rewrite these in the following form:

$$\psi_1(F^{-1}(t)) = \int_0^1 \Big[\{s - I(t \le s)\}/f(F^{-1}(s))\Big]dK_1(s),$$
$$\psi_2(F^{-1}(t)) = \int_0^1 \Big[\{s - I(t \le s)\}/f(F^{-1}(s))\Big]dK_2(s). \tag{4.69}$$

Since $K_2(.)$ is a step function, the second term in (4.69) can be written as a finite sum of k terms. thus, the influence function for $T(F)$ in (4.68) can be written as

$$\psi(F^{-1}(t)) = \int_0^1 \Big[\{s - I(t \le s)\}/f(F^{-1}(s))\Big]dK(s).$$

The second-order Hadamard differentiability of $T(.)$ at F naturally depends on the positivity of $f(F^{-1}(t))$, on its differentiability, and the differentiability of $K(.)$. If the component $K_2(.)$ is nondegenerate, then there is at least one jump point for $J(.)$. Hence, it can be shown that the second-order Hadamard derivative of $T_2(.)$ at F does not exist. The results in the preceding section may be used only for the class of L-functionals of type $T_1(F)$ and even under further regularity conditions. Toward this end, assuming that $K_1(t)$ is twice differentiable, we set

$$K_1(t+a) = K_1(t) + aK_1'(t) + \frac{a^2}{2}K_1''(t+ha)$$
$$= K_1(t) + aJ(t) + \frac{a^2}{2}J'(t+ha), \quad 0 < h < 1;\ \ t, t+a \in (0,1),$$

where $J'(.)$ stands for the derivative of the score function $J(.)$. If we assume that the condition (C1) in Section 4.3 holds with $\nu = 1$ and that $J'(u)$ is continuous in $u \in (\alpha, 1-\alpha)$, then we can set

$$T_1(F_n) - T_1(F) - \int \psi_1(F;x)d[F_n(x) - F(x)]$$
$$= \int xdK_1(F_n(x)) - \int xdK_1(F(x)) + \int [F_n(x) - F(x)][f(x)]^{-1}dK_1(F(x))$$

$$= \int [K_1(F(x)) - K_1(F_n(x))]dx + \int [F_n(x) - F(x)]J(F(x))dx$$

$$= \int [K_1(F(x)) - K_1(F_n(x)) + J(F(x))[F_n(x) - F(x)]]dx \qquad (4.70)$$

$$= \int \Big\{ \frac{[K_1(F(x)) - K_1(F_n(x))]}{[F(x) - F_n(x)]} - J(F(x)) \Big\} [F(x)) - F_n(x)]dx$$

$$= \frac{1}{2} \int [F_n(x) - F(x)]^2 J(F_n^\star(x))dx,$$

where $F_n^\star \in (F_n, F)$. If we define

$$T_{2n}^{0^\star} = \frac{1}{2} \int [F_n(x) - F(x)]^2 J'(F(x))dx \qquad (4.71)$$

then we have by (4.70) and (4.71)

$$T_1(F_n) = T_1(F) + \int \psi_1(F;x)d[F_n(x) - F(x)] + T_{2n}^{0^\star}$$
$$+\frac{1}{2} \int [F_n(x) - F(x)]^2 \Big\{ J'(F_n^\star(x)) - J'(F(x)) \Big\} dx. \qquad (4.72)$$

Since $J(u)$ [and $J'(u)$] vanish outside $[\alpha, 1-\alpha]$ $(0 < \alpha < \frac{1}{2})$, and $F^{-1}(1-\alpha) - F^{-1}(\alpha)$ is bounded, $J'(F_n^\star) - J'(F)$ is uniformly continuous in the norm $\|F_n - F\|$, and hence, the last term on the right-hand side of (4.72) is $o(\|F_n - F\|^2)$. This shows that the second-order Hadamard differentiability of $T_1(.)$ at F holds under **C1** with $\nu = 1$ and under an additional continuity condition on $J'(.)$ (which exists under the first order Lipschitz condition on J). We ignored a possibility that $J(.)$ may have jump at $u = \alpha$ and $u = 1-\alpha$; however, the above proof is capable of handling any finite number of jump points for $J'(.)$ on $[\alpha, 1-\alpha]$. For simplicity of presentation, we will sacrifice this minor refinement.

Let us illustrate the above methods on some specific smooth L-functionals, described in Section 4.3:

- The trimmed mean
- The Harrell-Davis quantile estimator
- The rank-weighted mean

I. The trimmed mean described in Section 4.3 [following (4.43)] has the weight function

$$J(u) = (1-2\alpha)^{-1} I(\alpha \le u \le 1-\alpha), \;\; u \in (0,1), \;\; 0 < \alpha < \tfrac{1}{2}.$$

Some direct manipulations lead to the following explicit formulation of the associated influence function:

$$T^{(1)}(P;x) = \begin{cases} (1-2\alpha)^{-1}[F^{-1}(\alpha) - F^{-1}(\frac{1}{2})], & x < F^{-1}(\alpha), \\ (1-2\alpha)^{-1}[F^{-1}(1-\alpha) - F^{-1}(\frac{1}{2})], & x > F^{-1}(1-\alpha), \\ (1-2\alpha)^{-1}[x - F^{-1}(\frac{1}{2})], & \text{otherwise.} \end{cases}$$

Put for $\lambda > 0$

$$\Big(F + \lambda(G - F)\Big)^{-1}(u) = t, \ \text{ then } u = F(t) + \lambda[G(t) - F(t)],$$

hence,

$$F^{-1}(u) = F^{-1}(F(t) + \lambda[G(t) - F(t)]) = t + \lambda[G(t) - F(t)]/f(F^{-1}(t))$$
$$-\frac{1}{2}\lambda^2[G(t) - F(t)]^2 f'(F^{-1}(t))/f^3(F^{-1}(t)) + o(\lambda^2), \text{ as } \lambda \downarrow 0.$$

Hence,

$$\begin{aligned}
&\lim_{\lambda \downarrow 0} \frac{1}{\lambda^2}[T(F) - 2T(F + (G - F)) + T(F + 2(G - F))] \\
&= -\frac{1}{(1-2\alpha)}\Big\{\int_{\boldsymbol{R}_1} [G(x) - F(x)]^2 f^{-3}(x) f'(x) dK_1(F(x))\Big\} \\
&= -\frac{1}{(1-2\alpha)}\Big\{\int_{\boldsymbol{R}_1} [G(x) - F(x)]^2 f^{-2}(x) f'(x) J(F(x))\Big\}.
\end{aligned}$$

If $G = F_n$, then

$$T^{(2)}(F; X_i, X_j) = \tag{4.73}$$

$$-\frac{1}{1-2\alpha}\int_{\boldsymbol{R}_1} \frac{[I(X_i \le x) - F(x)][I(X_j \le x) - F(x)]}{f(x)f(x)} \frac{f'(x)}{f(x)} J(F(x)) dF(x).$$

With this formulation we can use Theorem 4.7, which would directly lead to the second-order asymptotic distributional representation of the trimmed mean. However, we should be still be sure that $f(x)$ is strictly positive over $F^{-1}(\alpha) \le x \le F^{-1}(1-\alpha)$, that $f'(x)/f(x)$ behaves well in the same domain, and that the second moment of the right-hand side of (4.73) is finite.

II. Consider the Harrell–Davis quantile estimator defined in (4.44). For simplicity of presentation, we consider the specific value of $p = 1/2$, thus, T_n corresponds to a version of the median. Here, the c_{ni} do not vanish at the two tails. However, the c_{ni} converge to 0 at an exponential rate when i does not lie in the range $[np_1, np_2]$ where $0 < p_1 < \frac{1}{2} < p_2 < 1$. Thus, if the underlying distribution function F admits a finite νth order absolute moment for some $\nu > 0$ (not necessarily greater than or equal to 1), then we may concentrate on the range $np_1 \le i \le np_2$ in (4.44), and the remainder term can be made $o_p(n^{-1})$ for n adequately large. The conditions **A**, **B**, **C1** in Section 4.3 hold for this truncated version, however though $J_n(u)$ is smooth, it depends on n:

$$J_n(u) = n^{\frac{1}{2}}(2\pi)^{-\frac{1}{2}}[4u(1-u)]^{\frac{n}{2}} u^{-1}\{1 + O(n^{-1})\}, \ u \in (0,1). \tag{4.74}$$

For $u = \frac{1}{2} + n^{-1/2}t$, t real, we get

$$[4u(1-u)]^{\frac{n}{2}} = \left[1 - \tfrac{4}{n}t^2\right]^{\frac{n}{2}} = \exp(-2t^2)[1 + O(n^{-1})]. \tag{4.75}$$

This strong concentration of $J_n(u)$ around $u = 1/2$ [in a neighborhood of the order $n^{-1/2}$] leads us to the following approximation:

$$\left| T_n - X_{n:[\frac{n}{2}+1]} \right| = \mathrm{O}(n^{-3/4}(\log n)^{3/4}) \quad \text{a.s., as} \quad n \to \infty. \tag{4.76}$$

We refer to Yoshizawa et al. (1985) for the details of this proof. Since the second-order asymptotic distributional representation in (4.67) does not hold for the sample median $X_{n:[\frac{n}{2}+1]}$ [the rate of convergence being $\mathrm{O}(n^{-3/4})$], the approximation (4.76) may suggest that the same result does not hold for the T_n, too. However, we can show, using (4.74) and (4.75) and a bit more intricate analysis, that T_n has enough smoothness to possess a second order asymptotic distributional representation.

Indeed, let us define

$$K_n(t) = \int_0^t J_n(u)du, \quad t \in [0,1],$$

where $J_n(u)$ is based on the scores in (4.44) and is equivalently expressed in (4.74) for large n. Also let

$$\theta_n = \int_{\boldsymbol{R}_1} x dK_n(F(x)).$$

Note that if F is symmetric about the median θ, then $\theta_n = \theta$ for every $n \geq 1$. Even otherwise, an asymptotic expansion of θ_n around θ can be made using the density function f and its first derivative f'. Thus, whenever we assume that f' exists at θ and is continuous in a neighborhood of θ, we may write $\theta_n = \theta + n^{-1}\delta + \mathrm{o}(n^{-1})$, where δ depends on the density f and f'. Then we have

$$\begin{aligned} T_n - \theta_n &= \int_{\boldsymbol{R}_1} x d[K_n(F_n(x)) - K_n(F(x))] \\ &= -\int_{\boldsymbol{R}_1} [K_n(F_n(x)) - K_n(F(x))] dx \\ &= -\int [F_n(x) - F(x)] J_n(F(x)) dx - \tfrac{1}{2}\int [F_n(x) - F(x)]^2 J_n'(F_n^{\star}) dx, \end{aligned} \tag{4.77}$$

where $F_n^{\star}[= hF + (1-h)F_n$ for some $h \in (0,1)]$ lies between F_n and F. Define $W_n = \{W_n(x),\ x \in \boldsymbol{R}_1\}$ as

$$W_n(x) = n^{\frac{1}{2}}[F_n(x) - F(x)], \quad x \in \boldsymbol{R}_1.$$

The first term on the right hand side of (4.77) we can write as

$$\begin{aligned} &\Big[(2\pi)^{-\frac{1}{2}} \int_{\boldsymbol{R}_1} W_n(\theta + n^{-\frac{1}{2}}y)[F(\theta + n^{-\frac{1}{2}}y)]^{-1} \\ &\exp\{-2n[F(\theta + n^{-\frac{1}{2}}y) - F(\theta)]^2\} n^{-\frac{1}{2}} dy\Big][1 + \mathrm{O}(n^{-1})]. \end{aligned} \tag{4.78}$$

Moreover,

$$c_n = c_n(F) \tag{4.79}$$

$$= (2\pi)^{-\frac{1}{2}} \int_{\boldsymbol{R}_1} [F(\theta + n^{-\frac{1}{2}}y)]^{-1} \exp\Big\{ -2n[F(\theta + n^{-\frac{1}{2}}y) - F(\theta)]^2 \Big\} dy$$

$$= \frac{1}{f(\theta)} + n^{-\frac{1}{2}} c_1^\star(F) + \mathrm{o}(n^{-\frac{1}{2}}), \text{ as } n \to \infty,$$

where $c_1^\star(F)$ depends on the density f and its first derivative f' at θ. As such, we have from (4.77)–(4.79) that

$$T_n - \theta_n + c_n[n^{-\frac{1}{2}} W_n(\theta)] = \tag{4.80}$$

$$(2\pi n)^{-\frac{1}{2}} \int_{\boldsymbol{R}_1} \frac{W_n(\theta) - W_n(\theta n^{-\frac{1}{2}})}{F(\theta + n^{-\frac{1}{2}}y)} \exp\Big\{ -2n[F(\theta + n^{-\frac{1}{2}}y) - F(\theta)]^2 \Big\} dy$$

$$+ \mathrm{O}_p(n^{-\frac{3}{4}}) - (2n)^{-1} \int_{\boldsymbol{R}_1} W_n^2(x) J_n'(F_n^\star(x)) dx.$$

In view of the symmetry of $J(u)$ around $u = \frac{1}{2}$ and the uniform boundedness of $W_n(.)$ in probability, the last term on the right hand side of (4.80) is $\mathrm{o}_p(n^{-1})$. On the other hand, $\{n^{1/4}[W_n(\theta) - W_n(\theta + n^{-1/2}y)],\ y \in \boldsymbol{R}_1\}$ is asymptotically Gaussian, so the first term on the right-hand side of (4.80), when adjusted by the normalizing factor $n^{3/4}$, has asymptotically normal distribution with zero mean and a finite variance. Hence,

$$n^{\frac{3}{4}} \Big\{ T_n - \theta_n + c_n[n^{-\frac{1}{2}} W_n(\theta)] \Big\} \xrightarrow{\mathcal{D}} Z \sim \mathcal{N}(0, \gamma^2), \tag{4.81}$$

as $n \to \infty$, where γ^2 is a finite positive number. This provides a simple second order asymptotic distributional representation result for the estimator T_n. However, (1) the rate is still $n^{3/4}$ (and not n), and (2) unlike in (4.3), here the limiting law is normal (not a mixed normal). thus, the smoothness of the score function effectively simplifies the second order asymptotic distributional representation result, although it does not lead to the rate of convergence usual for smooth L-functionals (i.e., n). Finally, looking at (4.79) and (4.81), we may conclude that c_n in (4.81) may be replaced by $\{f(\theta)\}^{-1}$; however, the replacement of θ_n by θ may demand the symmetry of f around θ (at least in a neighborhood).

III. Consider the rank-weighted mean $\{T_{n,k}\}$, defined after (4.43). Let $\phi_k(X_1, \dots, X_{2k+1})$ be the median of $X_1, \dots, X_{2k+1}$ for any fixed $k \geq 0$. Then we may equivalently write

$$T_{n,k} = \binom{n}{2k+1}^{-1} \sum_{\{1 \leq i_1 < \dots < i_{2k+1} \leq n\}} \phi_k(X_{i_1}, \dots, X_{i_{2k+1}}). \tag{4.82}$$

(Sen 1964). For each h, $0 \leq h \leq 2k+1$, let

$$\phi_{k,h}(x_1, \dots, x_h) = \mathbb{E}\Big\{ \phi_k(x_1, \dots, x_h, X_{h+1}, \dots, X_{2k+1}) \Big\}, \tag{4.83}$$

and

$$\phi_{k,h}^\star(x_1, \dots, x_h) = \phi_{k,h}(x_1, \dots, x_h) \tag{4.84}$$

$$-\sum_{j=1}^{h} \phi_{k,h-1}(x_1, \ldots, x_{j-1}, x_{j+1}, \ldots, x_h) + \ldots + (-1)^h \phi_{k,0}(\cdot).$$

Denote $U^{\star}_{n,k}(h)$ the U-statistic based on $X_1, \ldots, X_n$, corresponding to the kernel (4.84), $h = 0, 1, \ldots, 2k+1$. Then we obtain by (4.82)–(4.84) that

$$T_{n,k} = \sum_{h=0}^{2k+1} \binom{2k+1}{h} U^{\star}_{n,k}(h).$$

Note that $U^{\star}_{n,k}(0) = \theta_{(k)}$ (say) is the expected value of the median of $(X_1, \ldots, X_{2k+1})$, and hence, is nonstochastic. On the other hand, $U^{\star}_{n,k}(1)$ is an average of n independent, identically distributed random variables with mean 0 and $U^{\star}_{n,k}(2)$ is a U-statistic of degree 2 with expectation 0, and it corresponds to the parameter which is stationary of order 1 [in the sense of (4.60)]. For every $h \geq 3$, the statistic $U^{\star}_{n,k}(h)$ has a variance of the order n^{-h}. thus, the above decomposition implies

$$n\Big[T_{n,k} - \theta_{(k)} - (2k+1)U^{\star}_{n,k}(1)\Big] = \binom{2k+1}{2} nU^{\star}_{n.k}(2) + \mathrm{O}_p(n^{-\frac{1}{2}}).$$

In this representation we assume that the kernel $\phi_k(.)$ has a finite second moment (i.e., the median of a sample of size $2k+1$ has a finite variance). This is satisfied if the underlying distribution function F of X has a finite absolute moment of order $r > 0:\ r(k+1) \geq 2$ (see Sen, 1959). Then we can directly use Theorem 4.7 for $nU^{\star}_{n.k}(2)$, and receive the desired second-order asymptotic distributional representation for a fixed k. The situation is different when $k\ (= k_n)$ is made to depend on n in such a way that $\lim_{n\to\infty} k_n = \infty$, when it very much depends on the order of k_n. For example, if we let $k_n \cong \frac{n+1}{2}$, then T_{n,k_n} reduces to the sample median, and hence, we arrive at the second-order asymptotic result given by (4.3) (with an order $n^{3/4}$ instead of n). Similarly, if we let $k_n = \frac{n+1}{2} \pm q$ for some fixed positive integer q, then the weights $\binom{n}{2k+1}^{-1}\binom{i-1}{k}\binom{n-i}{k}$ are concentrated only in the range $\frac{n+1}{2} - q \leq i \leq \frac{n+1}{2} + q$, so the second-order asymptotic result will be close to (4.3) with the order $n^{3/4}$. For this reason, consider the case of k_n increasing with n in such a way that $n^{-1}k_n \leq t < 1/2$. Then we can derive the following second-order asymptotic distributional representation. For simplicity of presentation, we take $n = 2m+1$. Then for $i = m+1 \pm r$, we have

$$\begin{aligned}
&\binom{2m+1}{2k+1}^{-1}\binom{i-1}{k}\binom{2m+1-i}{k} \\
&= \binom{2m+1}{2k+1}^{-1}\binom{m+r}{k}\binom{m-r}{k} \\
&= \left[\binom{2m+1}{2k+1}^{-1}\binom{m}{k}^{2}\right]\cdot\left[\binom{m+r}{k}\binom{m-r}{k}\binom{m}{k}^{-2}\right]
\end{aligned}$$

$$= \binom{2m+1}{2k+1}^{-1} \binom{m}{k}^2 \frac{(m^2-r^2)\dots\{(m-k+1)^2-r^2\}}{m^2\dots(m-k+1)^2}.$$

Moreover,

$$\frac{(m^2-r^2)\dots\{(m-k+1)^2-r^2\}}{m^2\dots(m-k+1)^2} = \exp\Big\{\sum_{i=1}^{k}\log(1-\frac{r^2}{(m-i+1)^2})\Big\}$$

$$= \Big\{\exp(-\frac{r^2k}{m(m-k)})\Big\}\Big\{1+\mathrm{O}((m-k+1)^{-1})\Big\}.$$

If we put

$$r = r_n = \Big(\frac{m(m-k_n)}{k_n}\Big)^{\frac{1}{2}} y \text{ and } \psi(n) = \Big(\frac{n(n-2k_n)}{2k_n}\Big)^{\frac{1}{2}}, \quad y \in \boldsymbol{R}_1,$$

then we have a representation very similar to that in (4.80) with $W_n(\theta + n^{-1/2}y)$ and $F(\theta + n^{-1/2}y)$ replaced by $W_n\big(\theta + \frac{y}{\psi(n)}\big)$ and $F\big(\theta + \frac{y}{\psi(n)}\big)$, respectively. Therefore, proceeding as in (4.80) through (4.81), we obtain that as $n \to \infty$,

$$(n\psi(n))^{\frac{1}{2}}\Big\{T_{n,k_n} - \theta_{(k_n)} + \{f(\theta_{(k_n)})\}^{-1}[n^{-\frac{1}{2}}W_n(\theta_{(k_n)})]\Big\}$$
$$\xrightarrow{\mathcal{D}} Z^\star \sim \mathcal{N}(0, \gamma^{\star 2}) \quad \text{for some } \gamma^\star : 0 < \gamma^\star < \infty. \tag{4.85}$$

If $k_n = \mathrm{O}(n)$, then $n\psi(n) = \mathrm{O}(n^{3/2})$, so the rate of convergence is comparable to (4.81). On the other hand, if $k_n = \mathrm{o}(n)$, then $n\psi(n) = \mathrm{O}(n^2/k_n^{1/2})$, so in (4.85) we have $(n\psi(n))^{1/2} = \mathrm{O}(nk_n^{-1/4})$, and a better rate of convergence is achieved. In practice, if we are able to assume that the density $f(.)$ is symmetric about the median θ, then we are able to replace $\theta_{(k_n)}$ by θ for all n, and hence, we are able to achieve a second order asymptotic distributional representation almost of the order n with a slowly increasing $\{k_n\}$. This explains the relative merits of the rank-weighted average over the sample median or the other quantile estimators.

4.7 L-Estimation in Linear Model

Consider the linear regression model

$$\mathbf{Y} = \mathbf{X}\boldsymbol{\beta} + \mathbf{Z} \tag{4.86}$$

where $\mathbf{Y} = (Y_1, \dots, Y_n)^\top$ is the vector of observations, $\mathbf{X} = \mathbf{X}_n$ is an $n \times p$ matrix of known constants with the rows $\mathbf{x}_i^\top$ such that $x_{1i} = 1,\ i = 1, \dots, n$ and $\mathbf{Z} = (Z_1, \dots, Z_n)^\top$ is the vector of errors which are i.i.d. random variables with distribution function F.

While M- and R-estimators in the linear model have been extensively studied, the extension of L-estimators from the location to the linear regression model suffered from the lack of a successful extension of the sample quantile to the regression case. Bickel's (1973) analogues of L-estimators, which were apparently the first ones, had good efficiency properties but were computationally complex and were not equivariant to reparametrization.

A suitable definition of regression quantile is due to Koenker and Bassett (1978). They defined the α-regression quantile $\widehat{\boldsymbol{\beta}}_n(\alpha)$, $0 < \alpha < 1$, as the solution of the minimization

$$\widehat{\boldsymbol{\beta}} = \arg\min\Big\{\sum_{i=1}^{n}\rho_\alpha(Y_i - \mathbf{x}_i^\top\mathbf{t}):\ \mathbf{t} \in \boldsymbol{R}_p\Big\} \tag{4.87}$$

where

$$\rho_\alpha(x) = |x|\Big\{(1-\alpha)I[x<0] + \alpha I[x>0]\Big\}. \tag{4.88}$$

Koenker and Bassett (1978) pointed out that $\widehat{\boldsymbol{\beta}}(\alpha)$ could be found as an optimal solution $\widehat{\boldsymbol{\beta}}$ of the linear program

$$\begin{aligned} &\alpha\sum_{i=1}^{n} r_i^+ + (1-\alpha)\sum_{i=1}^{n} r_i^- = \min \quad \text{subject to} \\ &\sum_{j=1}^{p} x_{ij}\beta_j + r_i^+ - r_i^- = Y_i, \ \ i = 1,\ldots,n \qquad (4.89) \\ &\beta_j \in \boldsymbol{R}_1,\ j = 1,\ldots,p;\ r_i^+ \geq 0,\ r_i^- \geq 0,\ \ i = 1,\ldots,n \end{aligned}$$

where r_i^+ and r_i^- are positive and negative parts of the residuals $Y_i - \mathbf{x}_i^\top\boldsymbol{\beta}$, $i = 1,\ldots,n$, respectively. The fact that the regression quantiles could be found as a solution of (4.89) is not only important for its computational aspect, but it reveals the structure of the concept. It implies that the set $B_n(\alpha)$ of solutions of (4.89) [and hence, also that of (4.87)] is nonempty, convex, compact, and polyhedral. If there are no additional restrictions, we may fix a version $\widehat{\boldsymbol{\beta}}(\alpha)$ of α-regression quantile as the lexicographically maximal element of $B_n(\alpha)$. Moreover, the dual program to (4.89) is of interest; it takes on the form

$$\begin{aligned} &\sum_{i=1}^{n} Y_i\Delta_i = \max \quad \text{subject to} \qquad (4.90) \\ &\sum_{i=1}^{n} x_{ij}\Delta_i = 0, \ \ j = 1,\ldots p; \ \ \alpha - 1 \leq \Delta_i \leq \alpha, \ \ i = 1,\ldots n. \end{aligned}$$

Transforming $\Delta_i + (1-\alpha) = \hat{a}_i(\alpha)$, $i = 1,\ldots,n$, we obtain an equivalent version of the dual program (4.90):

$$\begin{aligned} &\sum_{i=1}^{n} Y_i\hat{a}_i(\alpha) = \max \ \ \text{subject to} \qquad (4.91) \\ &\sum_{i=1}^{n} x_{ij}\hat{a}_i(\alpha) = (1-\alpha)\sum_{i=1}^{n} x_{ij}, \ \ j = 1,\ldots p. \end{aligned}$$

The optimal solution $\hat{\mathbf{a}}_n(\alpha) = (\hat{a}_{n1}(\alpha), \ldots, \hat{a}_{nn}(\alpha))^\top$ of (4.91), dual regression quantiles, has been termed the *regression rank scores* by Gutenbrunner (1986). If $\mathbf{X}_n$ is of rank p and $0 < \alpha < 1$, then the vector $((1-\alpha)\sum_{i=1}^n x_{i1}, \ldots, (1-\alpha)\sum_{i=1}^n x_{ip})^\top$ is an interior point of the set $\{\mathbf{X}_n^\top \mathbf{z} : \mathbf{z} \in [0,1]^n\}$. Hence, the optimal solution of (4.91) satisfies the following inequalities for $i = 1, \ldots, n$:

$$\hat{a}_{ni}(\alpha) = \begin{cases} 1, & \text{if } Y_i > \mathbf{x}_i^\top \widehat{\boldsymbol{\beta}}_n(\alpha), \\ 0, & \text{if } Y_i < \mathbf{x}_i^\top \widehat{\boldsymbol{\beta}}_n(\alpha). \end{cases} \tag{4.92}$$

As we will show, the asymptotic behavior of $\widehat{\boldsymbol{\beta}}_n(\alpha)$ is analogous to that of the sample quantile in the location model. Among the asymptotic properties of $\widehat{\boldsymbol{\beta}}_n(\alpha)$, the asymptotic distribution and the Bahadur-type representation are of primary interest. Besides that, it could be shown that the regression quantile process

$$\mathbf{L}_n(\alpha) = n^{\frac{1}{2}}\left(\widehat{\boldsymbol{\beta}}_n(\alpha) - \widetilde{\boldsymbol{\beta}}(\alpha)\right),\ 0 < \alpha < 1$$

with $\widetilde{\boldsymbol{\beta}}(\alpha) = (\beta_1 + F^{-1}(\alpha), \beta_2, \ldots, \beta_p)^\top$, considered as a random element of $D^p[0,1]$, has weak convergence properties analogous to the one-dimensional quantile process. This enables to consider a broad class of estimators of $\boldsymbol{\beta}$ in the form of linear combinations of regression quantiles:

$$\mathbf{T}_n^\nu = \int_0^1 \widehat{\boldsymbol{\beta}}_n(\alpha) d\nu(\alpha) \tag{4.93}$$

where ν is a finite signed measure with a compact support $\subset (0,1)$. Specific choices of ν lead to extensions of various location L-estimators. thus, an atomic ν leads to a combination of single regression quantiles. An extension of (α_1, α_2)-trimmed mean is obtained if ν is absolutely continuous with respect to the Lebesgue measure with the density

$$J(u) = (\alpha_2 - \alpha_1)^{-1} I[\alpha_1 \le u \le \alpha_2],\ 0 < \alpha_1 < \alpha_2 < 1. \tag{4.94}$$

From the computational point of view, a more suitable definition of the trimmed least square estimator $\mathbf{T}_n(\alpha_1, \alpha_2)$ is that of Koenker and Bassett (1978): Let

$$a_i = a_{ni} = I\left[\mathbf{x}_i^\top \widehat{\boldsymbol{\beta}}_n(\alpha_1) < Y_i < \mathbf{x}_i^\top \hat{\boldsymbol{\beta}}_n(\alpha_2)\right] \tag{4.95}$$

and calculate the weighted least squares estimator with the weights a_i, $i = 1, \ldots, n$. Denoting $\mathbf{A}_n = diag(a_i)$, we get the following explicit expression for the trimmed LSE:

$$\mathbf{T}_n(\alpha_1, \alpha_2) = (\mathbf{X}_n^\top \mathbf{A}_n \mathbf{X}_n)^{-1} \mathbf{X}_n^\top \mathbf{A}_n \mathbf{Y}_n. \tag{4.96}$$

If the inverse in (4.96) does not exist, it should be replaced by a generalized inverse. However, under mild regularity conditions, $n^{-1}(\mathbf{X}_n^\top \mathbf{A}_n \mathbf{X}_n)^{-1}$ converges in probability to a positive definite matrix as $n \to \infty$. The estimator (4.96) differs from the one defined in (4.93) and (4.94) in the weights of observations for which $Y_i = \mathbf{x}_i^\top \widehat{\boldsymbol{\beta}}(\alpha_1)$, $Y_i = \mathbf{x}_i^\top \widehat{\boldsymbol{\beta}}(\alpha_2)$, or $\mathbf{x}_i^\top \widehat{\boldsymbol{\beta}}(\alpha_1) \geq \mathbf{x}_i^\top \widehat{\boldsymbol{\beta}}(\alpha_2)$, $i = 1, \ldots, n$. The probability of the last inequality tends to 0 as $n \to \infty$.

In the present section, we will derive the asymptotic representations of the sample quantiles and of the trimmed LSE of (4.96). The technique used for regression quantiles will be later applied to M-estimators generated by a discontinuous ψ-function.

For that, we will impose the following regularity conditions on the matrix $\mathbf{X}_n$:

A1 $\lim_{n\to\infty} \mathbf{Q}_n = \mathbf{Q}$, where $\mathbf{Q}_n = n^{-1}\mathbf{X}_n^\top \mathbf{X}_n$ and $\mathbf{Q}$ is a positive definite matrix.

A2 $n^{-1}\sum_{i=1}^n x_{ij}^4 = O(1)$, as $n \to \infty$, for $j = 1, \ldots, p$.

A3 $x_{i1} = 1$, $i = 1, \ldots n$.

THEOREM 4.9 *Suppose that the distribution function F of Z_i in model (4.86) ($i = 1, \ldots, n$) is continuous and twice differentiable in a neighborhood of $F^{-1}(\alpha)$ and that $F'(F^{-1}(\alpha)) = f(F^{-1}(\alpha)) > 0$, $0 < \alpha < 1$. Then, under the conditions* **A1** - **A3**,

$$\widehat{\boldsymbol{\beta}}_n(\alpha) - \widetilde{\boldsymbol{\beta}}(\alpha) = \frac{1}{nf(F^{-1}(\alpha))}\mathbf{Q}^{-1}\sum_{i=1}^n \mathbf{x}_i \psi_\alpha(Z_i - F^{-1}(\alpha)) + \mathbf{R}_n(\alpha), \tag{4.97}$$

where $\|\mathbf{R}_n(\alpha)\| = O_p(n^{-3/4})$ as $n \to \infty$ and

$$\psi_\alpha(x) = \alpha - I[x < 0], \quad x \in \boldsymbol{R}_1.$$

PROOF Denote $\mathbf{S}(\mathbf{t}) = (S_1(\mathbf{t}), \ldots, S_p(\mathbf{t}))^\top$ by

$$S_j(\mathbf{t}) = n^{-\frac{1}{2}}\sum_{i=1}^n x_{ij}\Big[\psi_\alpha(Z_i - F^{-1}(\alpha) - n^{-\frac{1}{2}}\mathbf{x}_i^\top \mathbf{t}) - \psi_\alpha(Z_i - F^{-1}(\alpha))\Big],$$

$\mathbf{t} \in \boldsymbol{R}_p$, $j = 1, \ldots, p$. By (A.2.1)–(A.2.8) in the Appendix,

$$\sup_{\|\mathbf{t}\|\leq C} \|\mathbf{S}(\mathbf{t}) + f(F^{-1}(\alpha))\mathbf{Q}\mathbf{t}\| = O_p(n^{-\frac{1}{4}}) \tag{4.98}$$

for every $C > 0$ as $n \to \infty$.

Let $\widehat{\boldsymbol{\beta}}(\alpha)$ be a solution of the minimization (4.87). Then we claim that

$$\|n^{-\frac{1}{2}}\sum_{i=1}^n \mathbf{x}_i\psi_\alpha(Y_i - \mathbf{x}_i^\top \widehat{\boldsymbol{\beta}}(\alpha))\| = O(n^{-\frac{1}{4}}) \quad \text{a.s.} \tag{4.99}$$

Indeed, let $G_j^+(\varepsilon)$ be the right derivative of the function

$$n^{-\frac{1}{2}}\sum_{i=1}^n \rho_\alpha\Big(Y_i - \mathbf{x}_i^\top(\widehat{\boldsymbol{\beta}}(\alpha) + \varepsilon \mathbf{e}_j)\Big) \tag{4.100}$$

for a fixed j, $1 \le j \le p$, where $\mathbf{e}_j \in \boldsymbol{R}_p$, $e_{jk} = \delta_{jk}$, $j,k = 1,\ldots,p$. Then $G_j^+(\varepsilon) = -n^{-1/2}\sum_{i=1}^n x_{ij}\psi_\alpha(Y_i - \mathbf{x}_i^\top\widehat{\boldsymbol{\beta}}(\alpha) + \varepsilon\mathbf{e}_j))$ is nondecreasing in ε. Hence, for $\varepsilon > 0$, $G_j^+(-\varepsilon) \le G_j^+(0) \le G_j^+(\varepsilon)$ and $G_j^+(-\varepsilon) \le 0$ and $G_j^+(\varepsilon) \ge 0$, because (4.100) attains the minimum at $\varepsilon = 0$. thus, $|G_j^+(0)| \le G_j^+(\varepsilon) - G_j^+(-\varepsilon)$, and letting $\varepsilon \downarrow 0$, we obtain

$$|G_j^+(0)| \le n^{-\frac{1}{2}}\sum_{i=1}^n |x_{ij}| I\Big[Y_i - \mathbf{x}_i^\top\hat{\boldsymbol{\beta}}(\alpha) = 0\Big]. \tag{4.101}$$

Because the matrix $\mathbf{X}_n$ is of the rank p for every $n \ge n_0$ (see condition **A1**) and the distribution function F is continuous, at most p indicators on the right-hand side of (4.101) are nonzero a.s.; this together with condition **A2** leads to (4.99).

The representation (4.97) will follow from (4.98), if we let $\mathbf{t} \to n^{1/2}(\widehat{\boldsymbol{\beta}}(\alpha) - \widetilde{\boldsymbol{\beta}}(\alpha))$, provided that we show that

$$\|n^{\frac{1}{2}}(\widehat{\boldsymbol{\beta}}(\alpha) - \widetilde{\boldsymbol{\beta}}(\alpha))\| = \mathrm{O}_p(1), \text{ as } n \to \infty. \tag{4.102}$$

Regarding (4.99), it is sufficient to prove that, given $\varepsilon > 0$, there exist $C > 0$, $\eta > 0$ and a positive integer n_0 so that for $n > n_0$,

$$I\!P\Big\{\inf_{\|\mathbf{t}\|\ge C} \|n^{-\frac{1}{2}}\sum_{i=1}^n \mathbf{x}_i\psi_\alpha(Z_i - F^{-1}(\alpha) - n^{-\frac{1}{2}}\mathbf{x}_i^\top\mathbf{t})\| < \eta\Big\} < \varepsilon. \tag{4.103}$$

To prove (4.103), let us first note that there exist $K > 0$ and n_1 such that

$$I\!P\Big\{\|n^{-\frac{1}{2}}\sum_{i=1}^n \mathbf{x}_i\psi_\alpha(Z_i - F^{-1}(\alpha))\| > K\Big\} < \varepsilon/2. \tag{4.104}$$

for $n > n_1$. Take C and η such that

$$C > \frac{2K}{\lambda_0 f(F^{-1}(\alpha)} \quad \text{and} \quad \eta < \frac{K}{2},$$

where λ_0 is the minimum eigenvalue of $\mathbf{Q}$. Then there exists n_2 so that for $n \ge n_2$,

$$I\!P\Big\{\inf_{\|\mathbf{t}\|=C}\Big[-n^{-\frac{1}{2}}\sum_{i=1}^n \mathbf{t}^\top\mathbf{x}_i\psi_\alpha\big(Z_i - F^{-1}(\alpha) - n^{-\frac{1}{2}}\mathbf{x}_i^\top\mathbf{t}\big)\Big] < C\eta\Big\} < \varepsilon. \tag{4.105}$$

Indeed, the left-hand side of (4.105) is less than or equal to

$$I\!P\Big\{\inf_{\|\mathbf{t}\|=C}\Big[-n^{-\frac{1}{2}}\sum_{i=1}^n \mathbf{t}^\top\mathbf{x}_i\psi_\alpha\big(Z_i - F^{-1}(\alpha) - n^{-\frac{1}{2}}\mathbf{x}_i^\top\mathbf{t}\big)\Big] < C\eta,$$

$$\inf_{\|\mathbf{t}\|\ge C}\Big[-n^{-\frac{1}{2}}\sum_{i=1}^n \mathbf{t}^\top\mathbf{x}_i\psi_\alpha(Z_i - F^{-1}(\alpha)) + f(F^{-1}(\alpha))\mathbf{t}^\top\mathbf{Q}\mathbf{t}\Big] \ge 2C\eta\Big\}$$

$$+\mathbb{P}\Big\{\inf_{\|\mathbf{t}\|=C}\Big[-n^{-\frac{1}{2}}\sum_{i=1}^{n}\mathbf{t}^{\top}\mathbf{x}_i\psi_\alpha(Z_i-F^{-1}(\alpha)) \qquad (4.106)$$
$$+f(F^{-1}(\alpha))\mathbf{t}^{\top}\mathbf{Q}\mathbf{t}\Big]<2C\eta\Big\}.$$

The first term of (4.106) is less than or equal to

$$\mathbb{P}\Big\{\sup_{\|\mathbf{t}\|=C}\Big[\mathbf{t}^{\top}\mathbf{S}(\mathbf{t})+f(F^{-1}(\alpha))\mathbf{t}^{\top}\mathbf{Q}\mathbf{t}\Big]\geq 2C\eta\Big\}$$
$$\leq \mathbb{P}\Big\{\sup_{\|\mathbf{t}\|=C}\|\mathbf{S}(\mathbf{t})+f(F^{-1}(\alpha))\mathbf{Q}\mathbf{t}\|\geq\eta\Big\}<\tfrac{\varepsilon}{2}$$

for $n\geq n_2$. The second term of (4.106) is less than or equal to

$$\mathbb{P}\Big\{\inf_{\|\mathbf{t}\|=C}\Big[-n^{-\frac{1}{2}}\sum_{i=1}^{n}\mathbf{t}^{\top}\mathbf{x}_i\psi_\alpha(Z_i-F^{-1}(\alpha))+f(F^{-1}(\alpha))C^2\lambda_0\Big]<2C\eta\Big\}$$

$$\leq \mathbb{P}\Big\{-C\|n^{-\frac{1}{2}}\sum_{i=1}^{n}\mathbf{x}_i\psi_\alpha(Z_i-F^{-1}(\alpha))\|\leq -KC\Big\}<\tfrac{\varepsilon}{2} \qquad (4.107)$$

for $n>n_1$, by (4.104). Combining (4.106)–(4.107) with $n_0=\max(n_1,n_2)$, we arrive at (4.105).

Now, take $\mathbf{s}\in\boldsymbol{R}_p$ such that $\|\mathbf{s}\|=C$, and put $x_i^{\star}=-n^{-\frac{1}{2}}\mathbf{x}_i^{\top}\mathbf{s}$, $i=1,\ldots,n$. Then $M(\tau)=\sum_{i=1}^{n}x_i^{\star}\psi_\alpha(Z_i-F^{-1}(\alpha)+\tau x_i^{\star})$ is nondecreasing in $\tau\in\boldsymbol{R}_1$, so that for $\tau\geq 1$,

$$M(\tau)=-n^{-\frac{1}{2}}\sum_{i=1}^{n}\mathbf{s}^{\top}\mathbf{x}_i\psi_\alpha\big(Z_i-F^{-1}(\alpha)-n^{-\frac{1}{2}}\tau\mathbf{x}_i^{\top}\mathbf{s}\big)$$
$$\geq M(1)=-n^{-\frac{1}{2}}\sum_{i=1}^{n}\mathbf{s}^{\top}\mathbf{x}_i\psi_\alpha\big(Z_i-F^{-1}(\alpha)-n^{-\frac{1}{2}}\mathbf{x}_i^{\top}\mathbf{s}\big).$$

If $\|\mathbf{t}\|\geq C$, then $\mathbf{t}=\tau\mathbf{s}$ with $\mathbf{s}=C\frac{\mathbf{t}}{\|\mathbf{t}\|}$ and $\tau=\frac{\|\mathbf{t}\|}{C}\geq 1$. Hence,

$$\mathbb{P}\Big\{\inf_{\|\mathbf{t}\|\geq C}\|-n^{-\frac{1}{2}}\sum_{i=1}^{n}\mathbf{x}_i\psi_\alpha\big(Z_i-F^{-1}(\alpha)-n^{-\frac{1}{2}}\mathbf{x}_i^{\top}\mathbf{t}\big)\|<\eta\Big\}$$
$$\leq \mathbb{P}\Big\{\inf_{\|\mathbf{t}\|\geq C}\Big[-n^{-\frac{1}{2}}\sum_{i=1}^{n}\mathbf{t}^{\top}\mathbf{x}_i\psi_\alpha\big(Z_i-F^{-1}(\alpha)-n^{-\frac{1}{2}}\mathbf{x}_i^{\top}\mathbf{t}\Big]\frac{C}{\|\mathbf{t}\|}<C\eta\Big\}$$
$$\leq \mathbb{P}\Big\{\inf_{\|\mathbf{s}\|=C}\Big[-n^{-\frac{1}{2}}\sum_{i=1}^{n}\mathbf{s}^{\top}\mathbf{x}_i\psi_\alpha\big(Z_i-F^{-1}(\alpha)-n^{-\frac{1}{2}}\mathbf{x}_i^{\top}\mathbf{s}\big)\Big]<C\eta\Big\}<\varepsilon,$$

for $n>n_0$, and this completes the proof of (4.102). The proposition of the theorem follows from (4.98) where we insert $\mathbf{t}\to n^{1/2}(\widehat{\boldsymbol{\beta}}(\alpha)-\widetilde{\boldsymbol{\beta}}(\alpha))$ and take (4.99) into account. $\square$

Remark 4.3 *The technique used in the proof of Theorem 4.9 leads to an asymptotic representation of a broad class of M-estimators. We will often refer to it in the subsequent text.*

Corollary 4.2 *The sequence $\{n^{1/2}(\widehat{\boldsymbol{\beta}}(\alpha) - \widetilde{\boldsymbol{\beta}}(\alpha))\}$ has asymptotic p-dimensional normal distribution*

$$\mathcal{N}_p\Big(\mathbf{0}, \frac{\alpha(1-\alpha)}{f^2(F^{-1}(\alpha))}\mathbf{Q}^{-1}\Big).$$

We now turn to the trimmed least squares estimator $\mathbf{T}_n$ in the form defined in (4.96). Fix $\alpha_1, \ \alpha_2, \ 0 < \alpha_1, < \alpha_2 < 1$. We will derive an asymptotic representation of $\mathbf{T}_n$ with the remainder term of order $O_p(n^{-1})$, which is in correspondence with the rate of the trimmed mean in the location case. We should impose slightly stronger condition on $\mathbf{X}_n$ and on F. Namely, the standardization of elements of $\mathbf{X}_n$ given in condition **A2** is strengthened to

A2′ $\max\big\{|x_{ij}| : 1 \le i \le n; \ 1 \le j \le p\big\} = O(1)$ as $n \to \infty$.

Moreover, the following conditions will be imposed on the distribution function F:

B1 F is absolutely continuous with density f and $0 < f(x) < \infty$ for $F^{-1}(\alpha_1) - \varepsilon < x < F^{-1}(\alpha_2) + \varepsilon, \ \varepsilon > 0$.

B2 The density f has a continuous and bounded derivative f' in neighborhoods of $F^{-1}(\alpha_1)$ and $F^{-1}(\alpha_2)$.

Let $\mathbf{T}_n = \mathbf{T}_n(\alpha_1, \alpha_2)$ be the trimmed LSE:

$$\mathbf{T}_n = (\mathbf{X}_n^\top \mathbf{A}_n \mathbf{X}_n)^{-1} \mathbf{X}_n^\top \mathbf{A}_n \mathbf{Y}_n,$$

where $\mathbf{A}_n$ is the $n \times n$ diagonal matrix with the diagonal $(a_1, \ldots, a_n)$ of (4.95). The asymptotic representation for $\mathbf{T}_n$ is given in following theorem:

THEOREM 4.10 *Suppose that the sequence $\{\mathbf{X}_n\}$ of matrices satisfies conditions* **A1**, **A2′** *and* **A3**, *and that the distribution function F of errors Z_i, $i = 1, \ldots, n$, satisfies conditions* **B1** *and* **B2**. *Then*

$$\mathbf{T}_n(\alpha_1, \alpha_2) - \boldsymbol{\beta} - \frac{\eta}{\alpha_2 - \alpha_1}\mathbf{e}_1 = \frac{1}{n(\alpha_2 - \alpha_1)}\mathbf{Q}_n^{-1}\sum_{i=1}^n \mathbf{x}_i \psi(Z_i) + \mathbf{R}_n$$

where $\quad \|\mathbf{R}_n\| = O_p(n^{-1}) \quad$ *as* $n \to \infty$,

$$\eta = (1-\alpha_2)F^{-1}(\alpha_2) + \alpha_1 F^{-1}(\alpha_1), \ \ \mathbf{e}_1 = (1, 0, \ldots, 0)^\top \in \boldsymbol{R}_p,$$

and
$$\psi(z) = \begin{cases} F^{-1}(\alpha_1) & \text{if } z < F^{-1}(\alpha_1), \\ z & \text{if } F^{-1}(\alpha_1) \le z \le F^{-1}(\alpha_2), \\ F^{-1}(\alpha_2) & \text{if } z > F^{-1}(\alpha_2). \end{cases}$$

Equivalently,

$$\mathbf{T}_n(\alpha_1,\alpha_2) - \boldsymbol{\beta} - \mathbf{e}_1\delta = \frac{1}{n(\alpha_2-\alpha_1)}\mathbf{Q}_n^{-1}\sum_{i=1}^n \mathbf{x}_i(\psi(Z_i) - I\!E\psi(Z_i)) + \mathbf{R}_n,$$

where

$$\delta = \frac{1}{\alpha_2-\alpha_1}\int_{\alpha_1}^{\alpha_2} F^{-1}(u)du. \tag{4.108}$$

Notice that the first component of $\mathbf{T}_n(\alpha_1,\alpha_2)$ is generally asymptotically biased while the other components are asymptotically unbiased. $\mathbf{T}_n$ is asymptotically unbiased in the symmetric case. Let us formulate this special case as a corollary.

Corollary 4.3 *If F is symmetric around 0, and $\alpha_1 = \alpha,\ \alpha_2 = 1-\alpha,\ 0 < \alpha < \frac{1}{2}$, then as $n \to \infty$,*

$$\mathbf{T}_n(\alpha) - \boldsymbol{\beta} = \frac{1}{n(1-2\alpha)}\mathbf{Q}_n^{-1}\sum_{i=1}^n \mathbf{x}_i\psi(Z_i) + O_p(n^{-1}).$$

The asymptotic representation in (4.108) immediately implies the asymptotic normality of $n^{1/2}(\mathbf{T}_n(\alpha_1,\alpha_2) - \boldsymbol{\beta} - \mathbf{e}_1\delta)$:

Corollary 4.4 *Under the condition of Theorem 4.10, $n^{1/2}(\mathbf{T}_n(\alpha_1,\alpha_2) - \boldsymbol{\beta} - \mathbf{e}_1\delta)$ has asymptotic p-dimensional normal distribution with expectation $\mathbf{0}$ and with the covariance matrix $\mathbf{Q}^{-1}\sigma^2(\alpha_1,\alpha_2)$ where*

$$\begin{aligned}\sigma^2(\alpha_1,\alpha_2,F) &= \frac{1}{(\alpha_2-\alpha_1)^2}\Big\{\int_{\alpha_1}^{\alpha_2} \alpha_2(F^{-1}(u)-\delta)^2du \\ &+ \alpha_1(F^{-1}(\alpha_1)-\delta)^2 + (1-\alpha_2)(F^{-1}(\alpha_2)-\delta)^2 \\ &- \Big[\alpha_1(F^{-1}(\alpha_1-\delta) + (1-\alpha_2)(F^{-1}(\alpha_2)-\delta)\Big]^2\Big\}.\end{aligned}$$

If, moreover, $F(x)+F(-x) = 1,\ x \in \boldsymbol{R}_1$, and $\alpha_1 = \alpha,\ \alpha_2 = 1-\alpha,\ 0<\alpha<\frac{1}{2}$, then $n^{1/2}(\mathbf{T}_n(\alpha) - \boldsymbol{\beta})$ has p-dimensional asymptotic normal distribution with expectation $\mathbf{0}$ and with the covariance matrix $\sigma^2(\alpha,F)\mathbf{Q}^{-1}$ where

$$\sigma^2(\alpha,F) = \frac{1}{(1-2\alpha)^2}\Big\{\int_\alpha^{1-\alpha}(F^{-1}(u))^2du + 2\alpha(F^{-1}(\alpha))^2\Big\}.$$

Remark 4.4 *Notice that $\sigma^2(\alpha_1,\alpha_2,F)$ coincides with the asymptotic variance of the (α_1,α_2)-trimmed mean.*

PROOF of Theorem 4.10. We shall first prove the following useful approximation: For $\alpha \in (0,1)$,

$$\mathbf{x}_i^\top(\widehat{\boldsymbol{\beta}}(\alpha) - \boldsymbol{\beta}) = F^{-1}(\alpha) + \mathrm{O}_p(n^{-\frac{1}{2}}) \tag{4.109}$$

as $n \to \infty$, uniformly in $i = 1, \ldots, n$. To prove it, start from Theorem 4.9 and write

$$\mathbf{x}_i^\top(\widehat{\boldsymbol{\beta}}(\alpha) - \boldsymbol{\beta}) = F^{-1}(\alpha) + A_{in} + B_{in}, \quad \text{where}$$

$$A_{in} = n^{-1}\sum_{k=1}^{n} h_{ik}\psi_\alpha(Z_k - F^{-1}(\alpha)), \quad h_{ik} = \mathbf{x}_i^\top \mathbf{Q}^{-1}\mathbf{x}_k,$$

$$B_{in} = \mathbf{x}_i^\top \mathbf{R}_n(\alpha), \; i = 1, \ldots, n; \; k = 1, \ldots, n.$$

Using the Chebyshev inequality, we conclude from **A1**, **A2′**, regarding that $I\!\!E A_{in} = 0$ and that

$$\mathrm{Var}\, \mathrm{A_{in}} = \alpha(1-\alpha)\mathrm{n}^{-2}\sum_{\mathrm{k}=1}^{\mathrm{n}} \mathrm{h}_{\mathrm{ik}}^2 = \alpha(1-\alpha)\mathrm{n}^{-2}\mathbf{x}_{\mathrm{i}}^\top \mathbf{Q}^{-1}(\mathbf{X}_{\mathrm{n}}^\top \mathbf{X}_{\mathrm{n}})\mathbf{Q}^{-1}\mathbf{x}_{\mathrm{i}} = \mathrm{O}(\mathrm{n}^{-1})$$

uniformly in $i = 1, \ldots, n$, that $A_{in} = O_p(n^{-\frac{1}{2}})$, uniformly in $i = 1, \ldots, n$. Moreover, (4.97) and **A2′** imply that $B_{in} = O_p(n^{-3/4})$ uniformly in $i = 1, \ldots, n$; hence, we obtain (4.109).

Now fix $0 < \alpha_1 < \alpha_2 < 1$, and denote

$$a_i^- = I\Big[Y_i > \mathbf{x}_i^\top \widehat{\boldsymbol{\beta}}(\alpha_1)\Big], \quad a_i^+ = I\Big[Y_i \geq \mathbf{x}_i^\top \widehat{\boldsymbol{\beta}}(\alpha_2)\Big]$$

and

$$b_i^- = I\Big[Z_i > F^{-1}(\alpha_1)\Big], \; b_i^+ = I\Big[Z_i \geq F^{-1}(\alpha_2)\Big], \; b_i = b_i^- - b_i^+, \tag{4.110}$$

$i = 1, \ldots, n$. By (4.95), $a_{ni} = a_i^- - a_i^+$, $i = 1, \ldots, n$. Hence, recalling (4.91) and (4.92), we have

$$\begin{aligned}
\Big|\sum_{i=1}^{n} a_i^+ x_{ij} - (1-\alpha_2)\sum_{i=1}^{n} x_{ij}\Big| &= \Big|\sum_{i=1}^{n}(a_i^+ - \hat{a}_i(\alpha_2)x_{ij}\Big| \\
&= \Big|\sum_{i=1}^{n}(1 - \hat{a}_i(\alpha_2))I[Y_i = \mathbf{x}_i^\top \widehat{\boldsymbol{\beta}}(\alpha_2)]x_{ij}\Big| \\
&\leq p \cdot \max\Big\{|x_{ij}| : 1 \leq i \leq n; \; 1 \leq j \leq p\Big\} \\
&= p \cdot O(1) \quad \text{as } n \to \infty.
\end{aligned}$$

An analogous inequality we obtain for $|\sum_{i=1}^{n}(a_i^- - (1-\alpha_1))x_{ij}|$, $j = 1, \ldots, p$.

In the next step we prove the following approximation:

$$\frac{1}{n}\sum_{i=1}^{n} x_{ij}[a_i Z_i - \psi(Z_i)] = \frac{\eta}{n}\sum_{i=1}^{n} x_{ij} + O_p(n^{-1}), \; j = 1, \ldots, p. \tag{4.111}$$

Indeed,

$$n^{-1}\sum_{i=1}^{n} x_{ij}[a_i Z_i - \psi(Z_i)] = n^{-1}\sum_{i=1}^{n} x_{ij}[(a_i^- - b_i^-) - (a_i^+ - b_i^+)]Z_i$$

$$-n^{-1}\sum_{i=1}^{n} x_{ij}[b_i^+ F^{-1}(\alpha_2) + (1-b_i^-)F^{-1}(\alpha_1)].$$

It follows from (4.109) that $Z_i = F^{-1}(\alpha_1) + \mathrm{O}_p(n^{-1/2})$ if $|a_i^- - b_i^-| = 1$, and that $Z_i = F^{-1}(\alpha_2) + \mathrm{O}_p(n^{-1/2})$ if $|a_i^+ - b_i^+| = 1$, respectively, uniformly in $i = 1, \ldots, n$. Hence, by (4.95), **A1**, (4.110) and by the Chebyshev inequality,

$$\begin{aligned}
&\frac{1}{n}\sum_{i=1}^{n} x_{ij}[a_i Z_i - \psi(Z_i)] \\
&= \frac{1}{n}\Big(F^{-1}(\alpha_1) + \mathrm{O}_p(n^{-\frac{1}{2}})\Big)\Big[\sum_{i=1}^{n}(1-\alpha_1-b_i^-)x_{ij} + \mathrm{O}_p(1)\Big] \\
&-\frac{1}{n}\Big(F^{-1}(\alpha_2) + \mathrm{O}_p(n^{-\frac{1}{2}})\Big)\Big[\sum_{i=1}^{n}(1-\alpha_2-b_i^+)x_{ij} + \mathrm{O}_p(1)\Big] \\
&-\frac{1}{n}F^{-1}(\alpha_1)\sum_{i=1}^{n}(1-b_i^-)x_{ij} - \frac{1}{n}F^{-1}(\alpha_2)\sum_{i=1}^{n} b_i^+ x_{ij} \\
&= -\frac{\eta}{n}\sum_{i=1}^{n} x_{ij} + \Big\{\frac{1}{n}\sum_{i=1}^{n}[(1-\alpha_1-b_i^-) - (1-\alpha_2-b_i^+)]x_{ij}\Big\}\cdot \mathrm{O}_p(n^{-\frac{1}{2}})) \\
&= -\frac{\eta}{n}\sum_{i=1}^{n} x_{ij} + \mathrm{O}_p(n^{-1});
\end{aligned}$$

this gives (4.111).

Returning to definitions of a_i^+, a_i^-, b_i^+, and b_i^- and making use of **A1** and **A2$'$**, we may analogously prove that

$$\frac{1}{n}\sum_{i=1}^{n} x_{ij}x_{ik}(a_{in} - (\alpha_2-\alpha_1)) = \mathrm{O}_p(n^{-\frac{1}{2}}),\ j,k = 1,\ldots,p, \tag{4.112}$$

or, equivalently,

$$\frac{1}{n}\mathbf{X}_n^\top \mathbf{A}_n \mathbf{X}_n = \frac{1}{n}(\alpha_2-\alpha_1)\mathbf{X}_n^\top \mathbf{X}_n + \mathrm{O}_p(n^{-\frac{1}{2}}).$$

Then it follows from (4.111), (4.112), and from **A1**, **A2$'$** and **A3** that

$$\mathbf{T}_n - \boldsymbol{\beta} - \mathbf{e}_1\delta = \frac{1}{\alpha_2-\alpha_1}(\mathbf{X}_n^\top \mathbf{X}_n)^{-1}\sum_{i=1}^{n}\mathbf{x}_i[\psi(Z_i) - I\!E\psi(Z_i)] + \mathrm{O}_p(n^{-1}),$$

and this completes the proof of the theorem. □

4.8 Breakdown Point of Some L- and M-Estimators

Let us return to Section 2.3 with more detailed comments. The breakdown point is one measure of robustness of an estimator. The breakdown point introduced by Hampel (1971) and Donoho and Huber (1983) is the minimum proportion of outliers $(\mathbf{x}_i^\top, Y_i)$ pairs that can make the estimator arbitrarily large. The maximum possible asymptotic breakdown point of equivariant regression estimates is $1/2$. As estimators with high-breakdown for linear regression models we can mention the least median of squares (Rousseeuw and Yohai (1984)), least trimmed squares (Rousseeuw (1984)), S-estimators (Rousseeuw and Yohai (1984)), MM-estimators (Yohai (1987)) and τ -estimators (Yohai and Zamar (1988)). If we define the breakdown point of an estimator $\mathbf{T}_n$ of $\boldsymbol{\beta}$ in the model (4.86) by perturbing $(\mathbf{x}_i^\top, Y_i)$ pairs, then the L_1-estimator as well as any other M-estimator has breakdown $\frac{1}{n}$ (Maronna et al. (1979), Bloomfield and Steiger (1983)).

This behavior considerably changes under a definition of breakdown point that considers regressors error-free, non-stochastic and we consider the breakdown with respect to outliers in $\mathbf{Y}$ only. Müller (1995, 1997) proved that such breakdown point m^* for any regression equivariant estimator $\mathbf{T}_n$ cannot exceed the upper bound $\frac{1}{n}\left\lfloor\frac{n-\mathcal{N}(\mathbf{X})+1}{2}\right\rfloor$ where $\mathcal{N}(\mathbf{X}) = \max_{\mathbf{b}\neq\mathbf{0}} \#\{i : \mathbf{x}_i^\top\mathbf{b} = 0\}$. The possible attainability of this upper bound by an estimator was studied by Mizera and Müller (1999).

Let $m_\star$ denote the largest integer m such that for any subset M of $N = \{1, 2, \ldots, n\}$ of size m,

$$\inf_{\|\mathbf{b}\|=1}\left\{\frac{\sum_{i\in N\setminus M}|\mathbf{x}_i^\top\mathbf{b}|}{\sum_{i\in N}|\mathbf{x}_i^\top\mathbf{b}|}\right\} > \frac{1}{2}.$$

A surprising close relation between $m^\star$ and $m_\star$ in the case of L_1-estimator and some M-estimators, including the Huber estimator, is described in the following theorem:

THEOREM 4.11 *(i) Let $\mathbf{T}_n$ be the L_1-estimator of $\boldsymbol{\beta}$ defined through the minimization*

$$\sum_{i=1}^{n}|Y_i - \mathbf{x}_i^\top\mathbf{t}| := \min, \quad \mathbf{t}\in\boldsymbol{R}_p.$$

Then

$$m_\star + 1 \leq m^\star \leq m_\star + 2. \tag{4.113}$$

(ii) The relations (4.113) are also true for the L_1-type estimators of $\boldsymbol{\beta}$ defined

through the minimization

$$\sum_{i=1}^{n} \rho(Y_i - \mathbf{x}_i^\top \mathbf{t}) := \min, \tag{4.114}$$

with the continuous function ρ *satisfying*

$$\big|\rho(u) - |u|\big| \leq K, \ \ u \in \boldsymbol{R}_1, \ 0 < K < \infty. \tag{4.115}$$

Remark 4.5 *We do not know the upper bound for* $m_\star/n$*. In the special case of regression line coming through the origin and equidistant values* x_i*'s running over [0,1],* $m_\star/n \to 1 - 2^{-1/2} \approx 0.29289$*. Generally the breakdown point of M-estimators in the linear model may be quite low, even when calculated with respect to outliers in* $\mathbf{Y}$ *only.*

PROOF Without loss of generality, put $\boldsymbol{\beta} = \mathbf{0}$.
(i) If all but $m \leq m^\star$ of the Y's are bounded (e.g., by 1), then $\|\mathbf{T}_n\|$ will be uniformly bounded. Actually, by the triangle inequality,

$$\begin{aligned}\sum_{i\in N} |Y_i - \mathbf{x}_i^\top \mathbf{b}| \ &\geq \ \sum_{i\in N\setminus M} |\mathbf{x}_i^\top \mathbf{b}| - \sum_{i\in N\setminus M} |Y_I| - \Big(\sum_{i\in M} |\mathbf{x}_i^\top \mathbf{b}| - \sum_{i\in M} |Y_i|\Big) \\ &= \ \sum_{i\in N\setminus M} |\mathbf{x}_i^\top \mathbf{b}| - \sum_{i\in M} |\mathbf{x}_i^\top \mathbf{b}| + \sum_{i\in N} |Y_i| - 2 \sum_{i\in N\setminus M} |Y_i|.\end{aligned}$$

By definition of $m_\star$, there exists $c > 1/2$ such that

$$\sum_{i\in N\setminus M} |\mathbf{x}_i^\top \mathbf{b}| \geq c \sum_{i\in M} |\mathbf{x}_i^\top \mathbf{b}|, \ \text{ and } \sum_{i\in M} |\mathbf{x}_i^\top \mathbf{b}| \leq (1-c) \sum_{i\in N} |\mathbf{x}_i^\top \mathbf{b}|$$

for all subsets M of size $m \leq m_\star$. Therefore,

$$\sum_{i\in N} |Y_i - \mathbf{x}_i^\top \mathbf{b}| - \sum_{i\in N} |Y_i| \geq (2c-1) \sum_{i\in N} |\mathbf{x}_i^\top \mathbf{b}| - 2 \sum_{i\in N\setminus M} |Y_i|,$$

and if $|Y_i| \leq 1$ for $i \in N \setminus M$, there exists a constant C such that if $\|\mathbf{b}\| > C$, then

$$\sum_{i\in N} |Y_i - \mathbf{x}_i^\top \mathbf{b}| - \sum_{i\in N} |Y_i| \geq 0.$$

thus, $m^\star \geq m_\star + 1$.

By the definition of $m_\star$, there exists a subset M of size $m_\star + 1$ and a vector $\mathbf{b}_0$, $\|\mathbf{b}_0\| = 1$ such that

$$\sum_{i\in N\setminus M} |\mathbf{x}_i^\top \mathbf{b}_0| \leq \sum_{i\in M} |\mathbf{x}_i^\top \mathbf{b}_0|.$$

Thus, for $m = n_\star + 2$, there exists a subset $M^\star \subset N$ such that

$$\sum_{i\in N\setminus M^\star} |\mathbf{x}_i^\top \mathbf{b}_0| < \sum_{i\in M^\star} |\mathbf{x}_i^\top \mathbf{b}_0| \ \text{ with strict inequality.}$$

Then, $\eta(\mathbf{b}_0) > 0$, where

$$\eta(\mathbf{b}) = \sum_{i \in M^\star} |\mathbf{x}_i^\top \mathbf{b}| - \sum_{i \in N \setminus M^\star} |\mathbf{x}_i^\top \mathbf{b}|.$$

Suppose that $Y_i = 0$ for $i \in N \setminus M^\star$, and $Y_i = c\mathbf{x}_i^\top \mathbf{b}_0$ for $i \in M^\star$, $c > 0$. Then

$$\sum_{i \in N} |Y_i - c\mathbf{x}_i^\top \mathbf{b}_0| = \sum_{i \in N \setminus M^\star} |c\mathbf{x}_i^\top \mathbf{b}_0| = \sum_{i \in M^\star} |c\mathbf{x}_i^\top \mathbf{b}_0| - c\eta(\mathbf{b}_0) = \sum_{i \in N} |Y_i| - c\eta(\mathbf{b}_0).$$

On the other hand, for a bounded $\boldsymbol{\beta}$ and sufficiently large c,

$$\begin{aligned} \sum_{i \in N} |Y_i - \mathbf{x}_i^\top \boldsymbol{\beta}| &\geq \sum_{i \in N} |Y_i| - \sum_{i \in N} |\mathbf{x}_i' \boldsymbol{\beta}| \\ &\geq \sum_{i \in N} |Y_i| - n\|\boldsymbol{\beta}\| \|(\max_i \|\mathbf{x}_i\|) > \sum_{i \in N} |Y_i - c\mathbf{x}_i' \mathbf{b}_0|. \end{aligned}$$

This means that $m_\star + 2$ outliers in $\mathbf{Y}$ may lead to a breakdown of $\mathbf{T}_n$.

(ii) Let $\mathbf{T}_1$ be an M-estimator minimizing (4.114) with ρ satisfying (4.115), and let $\mathbf{T}$ be the L_1-estimator. Then there exists a constant $c > 0$ such that

$$\|\mathbf{T}_1 - \mathbf{T}\| \leq (2nK)/c. \tag{4.116}$$

Similarly, as in (4.100) and (4.101), we have for $\|\mathbf{b}\| = 1$ (the directional derivative of the objective function a $\mathbf{T}_n$ in direction $\mathbf{b}$)

$$-\sum_{i \notin h} \text{sign}(Y_i - \mathbf{x}_i^\top \mathbf{T}) \mathbf{x}_i^\top \mathbf{b} + \sum_{i \in h} |\mathbf{x}_i^\top \mathbf{b}| > 0,$$

where $h = \{i : \, Y_i = \mathbf{x}_i^\top \mathbf{t}_n\}$. Then

$$c = \inf_{\|\mathbf{b}\|=1} \left\{ \min \left[\varepsilon_i = \pm 1, \, i \notin h : -\sum_{i \notin h} \varepsilon_i \mathbf{x}_i^\top \mathbf{b} + \sum_{i \in h} |\mathbf{x}_i^\top \mathbf{b}| > 0 \right] \right\} > 0,$$

and for $\|\boldsymbol{\beta} - \mathbf{T}\| \geq 2nK/c$,

$$\begin{aligned} \sum_{i=1}^n \rho(Y_i - \mathbf{x}_i^\top \boldsymbol{\beta}) &\geq \sum_{i=1}^n |Y_i - \mathbf{x}_i^\top \boldsymbol{\beta}| - nK \\ &\geq \sum_{i=1}^n |Y_i - \mathbf{x}_i^\top \mathbf{T}| + \frac{2nK}{c} \cdot c - nK \\ &\geq \sum_{i=1}^n \rho(Y_i - \mathbf{x}_i^\top \mathbf{T}) \geq \sum_{i=1}^n \rho(Y_i - \mathbf{x}_i^\top \mathbf{T}_1). \end{aligned}$$

Hence, (4.116) follows. Together with proposition (i), this result implies proposition (ii). □

Remark 4.6 *For a broad class of robust estimators in linear models, there is a close connection of the breakdown point and of the tail performance measure, defined in Chapter 3 (see He et al. (1990)).*

4.9 Further Developments

1. Welsh (1987) proposed the one-step estimator

$$\breve{\boldsymbol{\beta}}_n^{(1)} = \breve{\boldsymbol{\beta}}_n + \Big[T(\hat{G}_n) - \mathbf{Q}_w^{-1}\sum_{i=1}^n \mathbf{x}_i\{\int I(\hat{Y}_i \le y)h(\hat{G}_n(y))dy + \sum_{j=1}^m w_j\phi_n(q_j)I(\hat{Y}_i \le \hat{G}_n^{-1}(q_j))\}\Big].$$

with the matrix

$$\mathbf{Q}_w = \sum_{i=1}^n \mathbf{x}_i\mathbf{x}_i^\top\{h(\hat{G}_n(Y_i)) + \sum_{j=1}^m w_j I(Y_i \le q_j)\} \tag{4.117}$$

where $\breve{\boldsymbol{\beta}}_n$ is a preliminary estimator of $\boldsymbol{\beta}$, $\hat{Y}_i = Y_i - \mathbf{x}_i^\top\breve{\boldsymbol{\beta}}_n,\ 1 \le i \le n$ are the residuals,

$$\hat{G}_n(y) = n^{-1}\sum_{i=1}^n I(\hat{Y}_i \le y),\ \ y \in \boldsymbol{R}_1,$$

and $\phi_n(t)$ is a pointwise consistent estimator of $\{f'(F^{-1}(t))\}^{-1},\ t \in (0,1)$. For a possible $\phi_n(.)$ we refer, e.g., to Welsh (1987), and to Dodge and Jurečková (1995). Further, $h(.) = \{h(t),\ 0 \le t \le 1\}$ is a smooth weight function, $w_1, \ldots w_m$ are nonstochastic weights, $0 < q_1 < \ldots < q_m < 1,\ m < \infty$, and

$$T(\hat{G}_n) = \int_0^1 \hat{G}_n^{-1}(t)h(t)dt + \int_0^1 \hat{G}_n^{-1}(t)dm(t),$$

where $m(t) = \sum_{i=1}^m w_i I(q_i \le t),\ t \in (0,1)$. Asymptotic properties of $\breve{\boldsymbol{\beta}}_n^{(1)}$ were studied by Jurečková and Welsh (1990) under **A1**, **A2**, and **A3** in Theorem 4.9.

Ren (1994) followed a Hadamard differentiability approach and considered the estimator

$$\breve{\boldsymbol{\beta}}_n^{(2)} = \breve{\boldsymbol{\beta}}_n + \Big[T(\hat{G}_n) - \mathbf{Q}_n^{\star-1}n^{-1}\sum_{i=1}^n \mathbf{x}_i^\star\{\int I(\hat{Y}_i \le y_i)h(\hat{G}_n(y))dy + \sum_{j=1}^m w_j\phi_n(q_j)I(\hat{Y}_i \le \hat{G}_n^{-1}(q_j))\}\Big]$$

where $\breve{\boldsymbol{\beta}}_n$ and other quantities are defined as before. Note that $\mathbf{Q}_w$ in (4.117) depends on the preliminary estimator $\breve{\boldsymbol{\beta}}_n$, while $\mathbf{Q}_n^\star$ does not. For this estimator, Ren proved that

$$n^{1/2}(\breve{\boldsymbol{\beta}}_n^{(2)} - \boldsymbol{\beta}) \xrightarrow{\mathcal{D}} \mathcal{N}_p(\mathbf{0}, \sigma_\psi^2\mathbf{Q}^{-1}),\quad \sigma_\psi^2 = \int_{\boldsymbol{R}} \psi^2(x)dF(x) < \infty.$$

2. The *α-autoregression quantile* for the linear autoregression model was introduced by Koul and Saleh (1995), and later it was studied by Hallin and Jurečková (1999).

3. Jurečková and Picek (2005) constructed the *two-step α-regression quantile* in the model $\mathbf{Y} = \beta_0 \mathbf{1}_n + \mathbf{X}\boldsymbol{\beta} + \mathbf{e}$, $\boldsymbol{\beta} \in \mathbb{R}^p$, starting with an appropriate R-estimate (rank-estimate) of the slope parameters $\boldsymbol{\beta}$. Then, ordering the residuals of the Y_i with respect to the initial R-estimate of the slopes leads to an estimator of $F^{-1}(\alpha) + \beta_0$. More precisely, let $\widehat{\boldsymbol{\beta}}_{nR}$ denote the initial R-estimator of the slope parameters $\beta_1, \ldots, \beta_p$ generated by a specific score function φ and $\hat{\beta}_{n0}$ denote the $[n\alpha]$-th order statistic of the residuals $Y_i - \mathbf{x}_i^\top \widehat{\boldsymbol{\beta}}_{nR}$, $i = 1, \ldots, n$. Then $\hat{\beta}_{n0}$ is an analog of $e_{n:[n\alpha]} + \beta_0$ where $e_{n:1} \leq \ldots \leq e_{n:n}$ are the order statistics of unobservable errors $e_1, \ldots, e_n$. Moreover, $\widehat{\boldsymbol{\beta}}_{nR}$ and $\hat{\beta}_{n0}$ are asymptotically independent, and the whole $(p+1)$-dimensional vector $\left(\hat{\beta}_{n0}, \widehat{\boldsymbol{\beta}}_{nR}^\top\right)^\top$ is asymptotically equivalent to the α-regression quantile of Koenker and Bassett. A similar approach applies also to the AR(p) model. The simulation study illustrates that the two-step form of the regression quantile is numerically very close to that of Koenker and Bassett. Both versions exactly coincide in the extremal case ($\alpha = 0, 1$) [Jurečková (2007)].

4. Rousseeuw (1984, 1985) introduced the *least median of squares* (LMS) and the *least trimmed squares* (LTS) estimators which are robust not only to outliers in $\mathbf{Y}$, but also to outliers in $\mathbf{X}$. While the rate of convergence of the former is only $n^{-\frac{1}{3}}$, the latter has the rate $n^{-\frac{1}{2}}$ and is asymptotically normal [Butler (1982) and Rousseeuw (1985), independently]. The LMS is defined as a solution of the minimization

$$\mathbf{T}_{LMS} = \arg\min_{\boldsymbol{\beta}} \left\{ \mathrm{med}_{1 \leq i \leq n}\, r_i^2(\boldsymbol{\beta}) \right\}, \quad r_i(\boldsymbol{\beta}) = Y_i - \mathbf{x}_i^\top \boldsymbol{\beta}, \quad i = 1, \ldots, n.$$

The LTS first orders the squared residuals, say $r_{[1]}^2(\boldsymbol{\beta}) \leq \ldots \leq r_{[n]}^2(\boldsymbol{\beta})$ for every $\boldsymbol{\beta} \in \mathbb{R}^p$ and then minimizes the sum of squares of the first h_n of them for a suitable h_n, i.e.

$$\mathbf{T}_{LTS} = \arg\min_{\boldsymbol{\beta}} \sum_{i=1}^{h_n} r_{[i]}^2(\boldsymbol{\beta}).$$

Rousseeuw and Leroy (1987) noticed that the LTS has the same asymptotic behavior at symmetric distributions as the Huber-type skipped mean, what further motivates its breakdown point $\frac{1}{2}$. To combine the high breakdown point with a high efficiency, some authors constructed one-step robust M-estimators or GM-estimators which have good efficiency properties while retaining the high breakdown point [Jurečková and Portnoy (1987) and Simpson, Ruppert, and Carroll (1992)]. Other estimators combining a high breakdown point and a bounded influence function with good asymptotic efficiency are the S-estimator of Rousseeuw and Yohai (1984), the τ-estimator of Yohai and Zamar (1988) and of Lopuhaä (1992).

Kang-Mo Jung (2005) extended the LTS to the multivariate regression model and defined the least-trimmed Mahalanobis squares distance (LTMS) estimator; the breakdown properties of LTMS follow those of the LTS. For another multivariate LTS we refer to Agulló et al. (2008).

The papers following-up these estimators either develop weighted and adaptive modifications or deal with the computation aspects of the LTS. Let us refer to Čížek (2010, 2011), the recent papers of the former group of papers and to Giloni and Padberg (2002) and to Hofmann and Kontoghiorghes (2010), the recent ones in the latter group, and to the references cited therein.

5. The estimator, which trimms-off the extreme observations or residuals, is inadmissible with respect to any convex and continuous loss function, under any finite sample size, whatever is the distribution of errors. This covers also the median, the trimmed mean, the trimmed LSE and the least trimmed squares. Such estimators even cannot be Bayesian [Jurečková and Klebanov (1997, 1998)].

4.10 Problems

4.2.1 Verify that Theorem 4.1 holds for $h(X_{n:[np_1]+1}, \ldots, X_{n:[np_k]+1})$ where its gradient $\mathbf{h}'_{\xi}$ exists.

4.3.1 For the rank weighted mean $T_{n,k}$, consider the asymptotic situation where $k \sim n\alpha$ for some $0 < \alpha < 1/2$. Use the Stirling approximation to the factorials and obtain the expression for the limiting score function $J(u)$, $0 < u < 1$. Compare this score function with that arising when k is held fixed. Which, if any, conforms to (4.27)? Comment on the case where $k = n/2 - \mathrm{O}(1)$. Finally for $k \sim np$, compare the rank-weighted mean and (4.44).

4.3.2 Consider the logistic distribution function $F(x) = \{1+e^{-(x-\theta)}\}^{-1}$, $x \in \boldsymbol{R}$. Show that $-f'(x)/f(x) = 2F(x) - 1$, $x \in \boldsymbol{R}$. Find the asymptotically best linear unbiased estimator of θ.

4.3.3 Consider the Laplace distribution function: $f(x) = (1/2)e^{-|x-\theta|}$, $x \in \boldsymbol{R}$. What is the BLUE of θ ? Verify whether or not **C1**, **C2**, or **C3** holds.

4.3.4 Consider the density $f(x)$ given by

$$I[x < \theta]\frac{1}{\sigma_1}\phi\Big(\frac{x-\theta}{\sigma_1}\Big) + I[x > \theta]\frac{1}{\sigma_2}\phi\Big(\frac{x-\theta}{\sigma_2}\Big), \ x \in \boldsymbol{R},$$

where $\theta \in \Theta \subset \boldsymbol{R}$, $\sigma_1 > \sigma_2 > 0$. Note that f has a jump discontinuity at θ. Find out the asymptotically best linear unbiased estimator of θ when $\sigma_1/\sigma_2 = c$ is known.

4.4.1 Verify whether or not the representation in (4.47) holds for the density $f(.)$ in Problem 4.3.4. Show that (4.47) does not hold for the sample median.

4.4.2 In a life-testing model, let $f(x) = \frac{1}{\theta}e^{-x/\theta}$, $x \geq 0$, $\theta > 0$, be the density, and suppose that corresponding to a given $r : r \leq n$ and $r/n \sim p : 0 < p < 1$, the observed data set relates to the order statistics $X_{n:1}, \ldots, X_{n:r}$. Obtain the MLE $(\hat{\theta}_{nr})$ of θ, and show that it is of the form (4.46). Verify that (4.47) holds.

4.4.3 For Problem 4.4.2, consider a type I censoring at a point $T : 0 < T < \infty$, and let r_T be the number of failures during the time $(0, T)$. Obtain the MLE of θ. What can you say about the representation in (4.47) for this MLE of θ?

4.5.1 Consider the sample median $\tilde{X}_n = T(F_n) = \inf\{x : F_n(x) \geq 1/2\} = F_n^{-1}(0.5)$. thus, $T(F) = F^{-1}(0.5)$ is the population counterpart of $\tilde{X}_n$. Is $T(F_n)$ first-order Hadamard differentiable?

4.5.2 Consider the following estimator of the scale parameter:

$$U_n = \binom{n}{2}^{-1} \sum_{\{1 \leq i < j \leq n\}} |X_i - X_j|.$$

Show that U_n may be also written as

$$\binom{n}{2}^{-1} \sum_{i=1}^{n} (n - 2i + 1) X_{n:n-i+1}$$

$$= \binom{n}{2}^{-1} \sum_{i \leq (n+1)/2} (n - 2i + 1)[X_{n:n-i+1} - X_{n:i}].$$

Verify whether or not U_n is first-/second-order Hadamard differentiable. What happens when the distribution function F of X has a compact support?

4.5.3 Consider a general von Mises' functional

$$T(F_n) = \int \cdots \int \phi(x_1, \ldots, x_m) dF_n(x_1) \ldots dF_n(x_m),$$

where $m \geq 1$ and $\phi(.)$ is a kernel of degree m. Show that when $\phi(.)$ is bounded a.e., Hadamard differentiability of $T(F)$ (at F) holds.

4.5.4 Let $f(.)$ be a normal density with mean θ and variance σ^2. Consider the asymptotically best linear unbiased estimator of σ based on n observations $X_1, \ldots, X_n$ and verify whether this estimator satisfies the Hadamard-differentiability condition in (4.51).

4.6.1 If $\mathbb{E}_F|X|^r < \infty$ for some $r > 0$, show that for every $k \geq 1$ such that $(k + 1)r \geq p$, $\mathbb{E}|\tilde{X}_{2k+1}|^p < \infty$. In particular, for $p = 2$, comment on the Cauchy distribution function, and show that $k \geq 2$ is sufficient for the sample median $\tilde{X}_{2k+1}$ to have a finite variance.

4.7.1 Define $\mathbf{A}_n = Diag(a_i)$ as in (4.95), and verify that

$$n^{-1}\mathbf{X}_n^\top \mathbf{A}_n \mathbf{X}_n \xrightarrow{P} \mathbf{Q}^\star$$

where $\mathbf{Q}^\star$ is a positively definite matrix. What is the explicit expression for $\mathbf{Q}^\star$?

4.7.2 Show that

$$\mathbb{P}\{\mathbf{x}_i^\top \widehat{\boldsymbol{\beta}}(\alpha_1) \geq \mathbf{x}_i^\top \widehat{\boldsymbol{\beta}}(\alpha_2)\} \to 0$$

as $n \to \infty$, for $\alpha_1 < 1/2 < \alpha_2$.

CHAPTER 5

Asymptotic Representations for M-Estimators

5.1 Introduction

In Chapter 4 we studied the interrelationships of L-estimators, L-functionals, and differentiable statistical functionals established their asymptotic representations from some basic results. M-estimators are also expressible as statistical functionals but in an implicit manner. This in turn requires a more elaborate treatment. The approach based on the Hadamard (or compact) differentiable functionals requires bounded score functions as well as other regularity conditions. The bounded condition is often justified for robust functionals, but excludes the maximum likelihood estimators, the precursors of M-estimators. thus, we shall concentrate on alternative methods, based mostly on the uniform asymptotic linearity of M-statistics in the associated parameter(s); this method constitutes the main theme of the current chapter. A variety of one-step and studentized versions of M-estimators of location/regression parameters will be considered and their related first-order and second-order asymptotic representations will be systematically presented.

Section 5.2 is devoted to the study of M-estimators of general parameters. Section 5.3 probes into deeper results for the location model, under somewhat simpler regularity conditions, when the scale parameter is treated as fixed (i.e., not as nuisance). Studentized M-estimators of location are presented in Section 5.4. As in the preceding chapter, the linear (regression) model is treated fully, and discussion on M-estimators for the fixed scale and their studentized versions is deferred to Section 5.5. Section 5.6 examines the pertinent studentizing scale statistics in this development. For completeness, a broad review of the Hadamard differentiability of M-functionals is provided in Section 5.7. This section also considers some comparisons of the alternative methods.

5.2 M-Estimation of General Parameters

Let $\{X_i;\ i \geq 1\}$ be a sequence of i.i.d. random variables with a distribution function $F(x, \boldsymbol{\theta})$ where $\boldsymbol{\theta} \in \boldsymbol{\Theta}$, an open set in $\boldsymbol{R}_p$. The true value of $\boldsymbol{\theta}$ is denoted by $\boldsymbol{\theta}_0$. Let $\rho(x, \mathbf{t}) : \boldsymbol{R} \times \boldsymbol{\Theta} \mapsto \boldsymbol{R}$ be a function, absolutely continuous in the elements of $\mathbf{t}$ (i.e., $t_1, \ldots, t_p$), and such that the function

$$h(\mathbf{t}) = \mathbb{E}_{\boldsymbol{\theta}_0} \rho(X_1, \mathbf{t}) \tag{5.1}$$

exists for all $\mathbf{t} \in \boldsymbol{\Theta}$ and has a unique minimum over $\boldsymbol{\Theta}$ at $\mathbf{t} = \boldsymbol{\theta}_0$. Then the M-estimator $\mathbf{M}_n$ of $\boldsymbol{\theta}_0$ is defined as the point of global minimum of $\sum_{i=1}^n \rho(X_i, \mathbf{t})$ with respect to $\mathbf{t} \in \boldsymbol{\Theta}$, i.e.,

$$\mathbf{M}_n = \arg\min\{\sum_{i=1}^n \rho(X_i, \mathbf{t}) : \mathbf{t} \in \boldsymbol{\Theta}\}. \tag{5.2}$$

Of natural interest are the following questions:

1. Under what regularity conditions (on $F, \boldsymbol{\theta}, \rho$) exists a solution of (5.2) that is a $\sqrt{n}$-consistent estimator of $\boldsymbol{\theta}_0$?
2. Under what regularity conditions the asymptotic representations of the first-order, and in some cases those of the second-order hold?

Some asymptotic results are considered in this generality, albeit under comparatively more stringent regularity conditions on $\rho(.)$. The situation is simpler for $p = 1$ (i.e. for a single parameter), where a second-order asymptotic distributional representation result holds under quite general regularity conditions. This will be followed by some further simplifications for the location model, to be treated in Section 5.3. For the sake of simplicity, we will denote $\mathbb{E}_{\boldsymbol{\theta}_0}(.)$ and $P_{\boldsymbol{\theta}_0}(.)$ by $\mathbb{E}(.)$ and $P(.)$, respectively.

Let us first consider the general vector parameter. In this case we will impose the following regularity conditions on ρ and on F:

A1 *First-order derivatives.*

The functions $\psi_j(x, \mathbf{t}) = (\partial/\partial t_j)\rho(x, \mathbf{t})$ are assumed to be absolutely continuous in t_k with the derivatives $\dot{\psi}_{jk}(x, \mathbf{t}) = (\partial/\partial t_k)\psi_j(x, \mathbf{t})$, such that $\mathbb{E}[\dot{\psi}_{jk}(X_1, \boldsymbol{\theta}_0)]^2 < \infty$, $(j, k = 1, \ldots, p)$. Further we assume that the matrices $\boldsymbol{\Gamma}(\boldsymbol{\theta}_0) = [\gamma_{jk}(\boldsymbol{\theta}_0)]_{j,k=1}^p$ and $\mathbf{B}(\boldsymbol{\theta}_0) = [b_{jk}(\boldsymbol{\theta}_0)]_{j,k=1}^p$ are positive definite, where

$$\gamma_{jk}(\boldsymbol{\theta}) = \mathbb{E}_{\boldsymbol{\theta}} \dot{\psi}_{jk}(\boldsymbol{\theta}),$$

and

$$b_{jk}(\boldsymbol{\theta}) = \text{Cov}_{\boldsymbol{\theta}}(\psi_j(X_1, \boldsymbol{\theta}), \psi_k(X_1, \boldsymbol{\theta})), \quad j, k = 1, \ldots, p.$$

A2 *Second- and third-order derivatives.*

$\dot{\psi}_{jk}(x, \mathbf{t})$ are absolutely continuous in the components of $\mathbf{t}$ and there exist functions $M_{jkl}(x, \boldsymbol{\theta}_0)$ such that $m_{jkl} = \mathbb{E} M_{jkl}(X_1, \boldsymbol{\theta}_0) < \infty$ and

$$|\ddot{\psi}_{jkl}(x, \boldsymbol{\theta}_0 + \mathbf{t})| \leq M_{jkl}(x, \boldsymbol{\theta}_0), \quad x \in \boldsymbol{R}_1, \ \|\mathbf{t}\| \leq \delta, \ \delta > 0,$$

where

$$\ddot{\psi}_{jkl}(x, \mathbf{t}) = \frac{\partial^2 \psi_j(x, \mathbf{t})}{\partial t_k \partial t_l}, \quad j, k, l = 1, \ldots, p.$$

Under the conditions A1 and A2, the point of global minimum in (5.2) is one of the roots of the system of equations:

$$\sum_{i=1}^{n} \psi_j(X_i, \mathbf{t}) = 0, \quad j = 1, \ldots, p. \tag{5.3}$$

The following theorem states an existence of a solution of (5.3) that is $\sqrt{n}$-consistent estimator of $\boldsymbol{\theta}_0$ and admits an asymptotic representation.

THEOREM 5.1 *Let $X_1, X_2, \ldots$ be i.i.d. random variables with distribution function $F(x, \boldsymbol{\theta}_0)$, $\boldsymbol{\theta}_0 \in \boldsymbol{\Theta}$, $\boldsymbol{\Theta}$ being an open set of $\boldsymbol{R}_p$. Let $\rho(x, \mathbf{t}) : \ \boldsymbol{R}_1 \times \boldsymbol{\Theta} \mapsto \boldsymbol{R}_1$ be a function absolutely continuous in the components of $\mathbf{t}$ and such that the function $h(\mathbf{t})$ of (5.1) has a unique minimum at $\mathbf{t} = \boldsymbol{\theta}_0$. Then, under the conditions A1 and A2, there exists a sequence $\{\mathbf{M}_n\}$ of solutions of (5.3) such that*

$$n^{\frac{1}{2}} \|\mathbf{M}_n - \boldsymbol{\theta}_0\| = O_p(1) \ \text{as } n \to \infty, \tag{5.4}$$

and

$$\mathbf{M}_n = \boldsymbol{\theta}_0 - n^{-1}(\boldsymbol{\Gamma}(\boldsymbol{\theta}_0))^{-1} \sum_{i=1}^{n} \boldsymbol{\psi}(X_i, \boldsymbol{\theta}_0) + \mathbf{O}_p(n^{-1}), \tag{5.5}$$

where $\boldsymbol{\psi}(x, \boldsymbol{\theta}) = (\psi_1(x, \boldsymbol{\theta}), \ldots, \psi_p(x, \boldsymbol{\theta}))^\top$.

The representation (5.5) further implies the asymptotic normality of $\mathbf{M}_n$ which we will state as a corollary:

Corollary 5.1 *Under the conditions of Theorem 5.1, $n^{\frac{1}{2}}(\mathbf{M}_n - \boldsymbol{\theta}_0)$ has asymptotically p-dimensional normal distribution $\mathcal{N}_p(\mathbf{0}, \mathbf{A}(\boldsymbol{\theta}_0))$ with*

$$\mathbf{A}(\boldsymbol{\theta}_o) = (\boldsymbol{\Gamma}(\boldsymbol{\theta}_0))^{-1} \mathbf{B}(\boldsymbol{\theta}_0).$$

PROOF of Theorem 5.1 will go in several steps. First we prove the following uniform second-order asymptotic linearity for the vector of partial derivatives of ρ:

$$\sup_{\|\mathbf{t}\| \leq C} \left\| n^{-\frac{1}{2}} \sum_{i=1}^{n} \left[\boldsymbol{\psi}(X_i, \boldsymbol{\theta}_0 + n^{-\frac{1}{2}} \mathbf{t}) - \boldsymbol{\psi}(X_i, \boldsymbol{\theta}_0) \right] - \boldsymbol{\Gamma}(\boldsymbol{\theta}_0) \mathbf{t} \right\| = \mathrm{O}_p(n^{-\frac{1}{2}}) \tag{5.6}$$

for fixed C, $0 < C < \infty$, as $n \to \infty$. When proved, (5.6) will imply the consistency and asymptotic representation of the estimator in the following way: First, given an $\varepsilon > 0$, there exists $K > 0$ and an integer n_0 so that, for $n \geq n_0$,

$$P(\|n^{-\frac{1}{2}} \sum_{i=1}^{n} \boldsymbol{\psi}(X_i, \boldsymbol{\theta}_0)\| > K) < \varepsilon/2. \tag{5.7}$$

Take $C > (\varepsilon + K)/\lambda_1(\boldsymbol{\theta}_0)$ where $\lambda_1(\boldsymbol{\theta}_0)$ is the minimum eigenvalue of $\boldsymbol{\Gamma}(\boldsymbol{\theta}_0)$. Then for $n \geq n_0$,

$$P(\sup_{\|\mathbf{t}\| = C} \|\mathbf{S}_n(\mathbf{t})\| > \varepsilon) < \varepsilon/2, \tag{5.8}$$

where $\mathbf{S}_n(\mathbf{t}) = n^{-\frac{1}{2}} \sum_{i=1}^{n} \left[\boldsymbol{\psi}(X_i, \boldsymbol{\theta}_0 + n^{-\frac{1}{2}}\mathbf{t}) - \boldsymbol{\psi}(X_i, \boldsymbol{\theta}_0)\right] + \boldsymbol{\Gamma}(\boldsymbol{\theta}_0)\mathbf{t}$.
By (5.7) and (5.8),

$$\begin{aligned}
&\inf_{\|\mathbf{t}\|=C} \left\{n^{-\frac{1}{2}}\mathbf{t}^\top \sum_{i=1}^{n} \boldsymbol{\psi}(X_i, \boldsymbol{\theta}_0 + n^{-\frac{1}{2}}\mathbf{t})\right\} \\
&\geq \inf_{\|\mathbf{t}\|=C}\{\mathbf{t}^\top \mathbf{S}_n(\mathbf{t})\} + \inf_{\|\mathbf{t}\|=C}\left\{-\mathbf{t}^\top n^{-\frac{1}{2}} \sum_{i=1}^{n} \boldsymbol{\psi}(X_i, \boldsymbol{\theta}_0) + \mathbf{t}^\top \boldsymbol{\Gamma}(\boldsymbol{\theta}_0)\mathbf{t}\right\} \\
&\geq -C \inf_{\|\mathbf{t}\|=C} \|\mathbf{S}_n(\mathbf{t})\| + C^2\lambda_1(\boldsymbol{\theta}_0) - C\|n^{-\frac{1}{2}} \sum_{i=1}^{n} \boldsymbol{\psi}(X_i, \boldsymbol{\theta}_0)\|.
\end{aligned}$$

Hence

$$\begin{aligned}
&P\Big(\inf_{\|\mathbf{t}\|=C}\left\{\mathbf{t}^\top n^{-\frac{1}{2}} \sum_{i=1}^{n} \boldsymbol{\psi}(X_i, \boldsymbol{\theta}_0 + n^{-\frac{1}{2}}\mathbf{t})\right\} > 0\Big) \qquad (5.9) \\
&\geq P\Big\{\|n^{-\frac{1}{2}} \sum_{i=1}^{n} \boldsymbol{\psi}(X_i, \boldsymbol{\theta}_0)\| + \inf_{\|\mathbf{t}\|=C} \|\mathbf{S}_n(\mathbf{t})\| < C\lambda_1(\boldsymbol{\theta}_0)\Big\} \\
&\geq 1 - \varepsilon \quad \text{for } n \geq n_0.
\end{aligned}$$

By Theorem 6.4.3 of Ortega and Rheinboldt (1970), we conclude from (5.9) that for $n \geq n_0$ the system of equations

$$\sum_{i=1}^{n} \psi_j(X_i, \boldsymbol{\theta}_0 + n^{-\frac{1}{2}}\mathbf{t}) = 0, \quad j = 1, \ldots, p,$$

has a root $\mathbf{T}_n$ that lies in the sphere $\|\mathbf{t}\| \leq C$ with probability exceeding $1 - \varepsilon$ for $n \geq n_0$. Then $\mathbf{M}_n = \boldsymbol{\theta}_0 + n^{-\frac{1}{2}}\mathbf{T}_n$ is a solution of (5.3) satisfying $P(\|n^{\frac{1}{2}}(\mathbf{M}_n - \boldsymbol{\theta}_0)\| \leq C) \geq 1 - \varepsilon$ for $n \geq n_0$. Inserting $\mathbf{t} \to n^{\frac{1}{2}}(\mathbf{M}_n - \boldsymbol{\theta}_0)$ in (5.6), we obtain the representation (5.5).

To prove the uniform asymptotic linearity (5.6), we refer to the Appendix and note that by **A1** and **A2**,

$$\begin{aligned}
&\sup_{\|\mathbf{t}\|\leq C} \left| n^{-\frac{1}{2}} \sum_{i=1}^{n} \left[\psi_j(X_i, \boldsymbol{\theta}_0 + n^{-\frac{1}{2}}\mathbf{t}) - \psi_j(X_i, \boldsymbol{\theta}_0) - n^{-\frac{1}{2}} \sum_{k=1}^{p} t_k \dot{\psi}_{jk}(X_i, \boldsymbol{\theta}_0)\right]\right| \\
&\leq C^2 n^{-\frac{3}{2}} \sum_{i=1}^{n} \sum_{k,l=1}^{p} M_{jkl}(X_i, \boldsymbol{\theta}_0) = \mathrm{O}_p(n^{-\frac{1}{2}}), \quad j = 1, \ldots, p.
\end{aligned}$$

Moreover, uniformly for $\|\mathbf{t}\| \leq C$,

$$\begin{aligned}
&I\!\!E\Big[n^{-1} \sum_{i=1}^{n} \sum_{k=1}^{p} t_k(\dot{\psi}_{jk}(X_i, \boldsymbol{\theta}_0) - \gamma_{jk}(\boldsymbol{\theta}_o))\Big]^2 \\
&= n^{-1}\mathrm{Var}\Big\{\sum_{k=1}^{p} t_k \dot{\psi}_{jk}(X_1, \boldsymbol{\theta}_0)\Big\} \leq K/n
\end{aligned}$$

for some $K > 0,\ j = 1, \ldots, p$. Combining the last two statements, we arrive at (5.6), and this completes the proof of the theorem. □

If $\rho(x, \mathbf{t})$ is convex in $\mathbf{t}$, then the solution of (5.2) is uniquely determined. However, since we did not assume the convexity of ρ, there may exist more roots of the system (5.3) satisfying (5.4). Let $\mathbf{M}_n$, $\mathbf{M}_n^\star$ be two such sequences of roots. Then $\mathbf{M}_n$ and $\mathbf{M}_n^\star$ are asymptotically equivalent up to $O_p(n^{-1})$; indeed, we have the following corollary of Theorem 5.1:

Corollary 5.2 *Let $\mathbf{M}_n$ and $\mathbf{M}_n^\star$ be two sequences of roots of (5.3) satisfying (5.4). Then, under the conditions of Theorem 5.1,*

$$\|\mathbf{M}_n - \mathbf{M}_n^\star\| = O_p(n^{-1}) \text{ as } n \to \infty.$$

In the rest of this section we shall deal with the case $p = 1$, the case of a single one-dimensional parameter. Theorem 5.1 and its corollaries apply naturally to this special case, but in this case we can supplement the representation (5.5) by its second-order, i.e., by the asymptotic (nonnormal) distribution of the remainder term.

In particular, assume that $\theta_0 \in \Theta$, Θ being an open interval in $\boldsymbol{R}_1$. Let $\rho(x,t) : \boldsymbol{R}_1 \times \Theta \mapsto \boldsymbol{R}_1$ be a function such that the function

$$h(t) = \mathbb{E}_{\theta_0} \rho(X_1, t) \tag{5.10}$$

exists for $t \in \Theta$ and has a unique minimum at $t = \theta_0$. Assume that $\rho(x,t)$ is absolutely continuous in t with the derivatives $\psi(x,t)$, $\dot{\psi}(x,t)$, and $\ddot{\psi}(x,t)$ with respect to t. As in the general case, the M-estimator M_n of θ_0 is defined as a solution of the minimization (5.2) in $t \in \Theta \subset \boldsymbol{R}_1$. Then M_n is a root of the equation

$$\sum_{i=1}^{n} \psi(X_i, t) = 0, \quad t \in \Theta. \tag{5.11}$$

We shall prove the second-order asymptotic representation of M_n; for that we need the following conditions on ρ and on F:

B1 *Moments of derivatives.* There exists $K > 0$ and $\delta > 0$ such that

$$\mathbb{E}[\psi(X_1, \theta_0 + t)]^2 \le K, \quad \mathbb{E}[\dot{\psi}(X_1, \theta_0 + t)]^2 \le K,$$

and

$$\mathbb{E}[\ddot{\psi}(X_1, \theta_0 + t)]^2 \le K, \text{ for } |t| \le \delta.$$

B2 *Fisher's consistency.* $0 < \gamma_1(\theta_0) = \mathbb{E}_\theta \dot{\psi}(X_1, \theta) < \infty$.

B3 *Uniform continuity in the mean.* There exist $\alpha > 0,\ \delta > 0$, and a function $M(x, \theta_0)$ such that $m = \mathbb{E} M(X_1, \theta_0) < \infty$ and

$$|\ddot{\psi}(x, \theta_0 + t) - \ddot{\psi}(x, \theta_0)| \le |t|^\alpha M(x, \theta_0) \text{ for } |t| \le \delta, \text{ a.s. } [F(x, \theta_0)].$$

THEOREM 5.2 ***[second-order asymptotic distribution representation].*** *Let $\{X_i, i \geq 1\}$ be a sequence of i.i.d. random variables with the distribution function $F(x, \theta_0)$, $\theta_0 \in \Theta$, Θ being an open interval in $\boldsymbol{R}_1$. Let $\rho(x,t) : \boldsymbol{R}_1 \times \Theta \mapsto \boldsymbol{R}_1$ be a function such that $h(t)$ of (5.10) has a unique minimum at $t = \theta_0$. Then, under the conditions B1–B3, there exists a sequence $\{M_n\}$ of roots of the equation (5.11) such that, as $n \to \infty$,*

$$n^{\frac{1}{2}}(M_n - \theta_0) = O_p(1), \tag{5.12}$$

$$M_n = \theta_0 - (n\gamma_1(\theta_0))^{-1} \sum_{i=1}^{n} \psi(X_i, \theta_0) + R_n, \tag{5.13}$$

and

$$nR_n \xrightarrow{\mathcal{D}} \left(\xi_1 - \xi_2 \frac{\gamma_2(\theta_0)}{2\gamma_1(\theta_0)}\right) \cdot \xi_2, \tag{5.14}$$

where

$$\gamma_2(\theta) = \mathbb{E}_\theta \ddot{\psi}(X_1, \theta),$$

and $(\xi_1, \xi_2)^\top$ is a random vector with normal $\mathcal{N}_2(\mathbf{0}, \mathbf{S})$ distribution, with $\mathbf{S} = [s_{ij}]_{i,j=1,2}$ and

$$s_{11} = (\gamma_1(\theta_0))^{-2} \mathrm{Var}_{\theta_0} \dot{\psi}(X_1, \theta_0), \quad s_{22} = (\gamma_1(\theta_0))^{-2} \mathbb{E}_{\theta_0} \psi^2(X_1, \theta_0),$$

$$s_{12} = s_{21} = (\gamma_1(\theta_0))^{-2} \mathrm{Cov}_{\theta_0}(\dot{\psi}(X_1, \theta_0), \psi(X_1, \theta_0)).$$

PROOF For notational simplicity, denote $\gamma_j(\theta_0)$ by γ_j, $j = 1, 2$. Consider the random process $Y_n = \{Y_n(t),\ t \in [-B, B]\}$ defined by

$$Y_n(t) = \gamma_1^{-1} \sum_{i=1}^{n} [\psi(X_i, \theta_0 + n^{-\frac{1}{2}} t) - \psi(X_i, \theta_0)] - n^{\frac{1}{2}} t, \ \ |t| \leq B,\ 0 < B < \infty.$$

The realizations of Y_n belong to the space $D[-B, B]$. We shall first show that Y_n converges to some Gaussian process in the Skorokhod J_1-topology on $D[-B, B]$ as $n \to \infty$, and then return to the proof of the theorem.

LEMMA 5.1 *Under the conditions of Theorem 5.2, Y_n converges weakly to the Gaussian process $Y = \{Y(t),\ t \in [-B, B]\}$ in the Skorokhod J_1-topology on $D[-B, B]$, where*

$$Y(t) = t\xi_1 - t^2 \frac{\gamma_2}{2\gamma_1}, \quad t \in [-B, B],$$

for any fixed B, $0 < B < \infty$, and where ξ_1 is a random variable with normal $\mathcal{N}(0, s_{11})$ distribution.

PROOF Denote

$$Z_n(t) = \gamma_1^{-1} \sum_{i=1}^{n} A_n(X_i, t), \ \ Z_n^0(t) = Z_n(t) - \mathbb{E} Z_n(t),$$

where

$$A_n(X_i, t) = \psi(X_i, \theta_0 + n^{-\frac{1}{2}}t) - \psi(X_i, \theta_0), \quad t \in \boldsymbol{R}_1, \ \ i = 1, \ldots, n.$$

We shall first prove that $Z_n^0 \Longrightarrow Y$ as $n \to \infty$. Following Section 2.5.10, we should prove the weak convergence of finite dimensional distributions of Z_n^0 to those of Y and the tightness of the sequence of distributions of Z_n^0, $n = 1, 2, \ldots$. To prove the former, we show that, for any $\boldsymbol{\lambda} \in \boldsymbol{R}_p$, $\mathbf{t} \in \boldsymbol{R}_p$, $p \geq 1$,

$$\mathrm{Var}\Big\{\sum_{j=1}^{p} \lambda_j Z_n(t_j)\Big\} \to s_{11}(\boldsymbol{\lambda}^\top \mathbf{t})^2. \quad \text{as } n \to \infty, \tag{5.15}$$

The weak convergence then follows by the classical central limit theorem. To prove (5.15), we write

$$\mathrm{Var}\Big\{\sum_{j=1}^{p} \lambda_j Z_n(t_j)\Big\} = \gamma_1^{-2} \sum_{i=1}^{n} \sum_{j,k=1}^{m} \lambda_j \lambda_k \mathrm{Cov}(A_n(X_i, t_j), A_n(X_i, t_k)).$$

It suffices to study $\mathrm{Cov}(A_n(X_1, t_1), A_n(X_1, t_2))$, $t_1, t_2 \in [0, B]$ (the considerations are analogous when (t_1, t_2) belong to another quadrant). Then we have, for $n \geq n_0$,

$$\begin{aligned}
&|\mathbb{E}\{A_n(X_1, t_1)A_n(X_1, t_2) - n^{-1}t_1 t_2(\dot{\psi}(X_1, \theta_0))^2\}| \\
&\leq |\mathbb{E}\{[A_n(X_1, t_1) - n^{-\frac{1}{2}}t_1(\dot{\psi}(X_1, \theta_0)]A_n(X_1, t_2)\}| \\
&+|\mathbb{E} n^{-\frac{1}{2}}t_1\dot{\psi}(X_1, \theta_0)[A_n(X_1, t_2) - n^{-\frac{1}{2}}t_2(\dot{\psi}(X_1, \theta_0)]\}| \\
&\leq \Big|\mathbb{E}\Big\{\int_0^{\frac{t_1}{\sqrt{n}}} \int_0^u \ddot{\psi}(X_1, \theta_0 + v)dv \int_0^{\frac{t_2}{\sqrt{n}}} \dot{\psi}(X_1, \theta_0 + w)dw\Big\}\Big| \\
&+\Big|\mathbb{E}\Big\{n^{-\frac{1}{2}}t_1\dot{\psi}(X_1, \theta_0) \int_0^{\frac{t_2}{\sqrt{n}}} \int_0^u \ddot{\psi}(X_1, \theta_0 + v)dvdu\Big\}\Big| \\
&\leq (K/2)n^{-\frac{3}{2}}|t_1 t_2|(|t_1| + |t_2|).
\end{aligned}$$

Similarly,

$$\begin{aligned}
&|\mathbb{E}A_n(X_1, t_1)\mathbb{E}A_n(X_1, t_2) - n^{-1}t_1 t_2(\mathbb{E}\dot{\psi}(X_1, \theta_0))^2| \\
\leq \ &|t_1 t_2|(|t_1| + |t_2|) \cdot \mathrm{O}(n^{-\frac{3}{2}})
\end{aligned}$$

and thus, we obtain (5.15). This further implies

$$\begin{aligned}
&\mathbb{E}\Big(|Z_n^0(t) - Z_n^0(t_1)| \cdot |Z_n^0(t_2) - Z_n^0(t)|\Big) \\
\leq \ &\frac{1}{2}\Big\{\mathbb{E}(Z_n^0(t) - Z_n^0(t_1))^2 + \mathbb{E}(Z_n^0(t_2) - Z_n^0(t))^2\Big\} \\
\to \ &\frac{s_{11}}{2}\Big\{(t - t_1)^2 + (t_2 - t)^2\Big\} \leq \frac{s_{11}}{2}(t_2 - t_1)^2,
\end{aligned}$$

as $n \to \infty$ for every t_1, t, t_2, $\ -B \leq t \leq t_2 \leq B$. Consequently, by (2.101), we conclude that $\{Z_n^0\}$ is tight, and this completes the proof of the weak convergence of Z_n^0 to Y. To prove the weak convergence of the process Y_n, it remains to show that

$$I\!\!E Z_n(t) - n^{-\frac{1}{2}}t - t^2 \frac{\gamma_2}{2n\gamma_1} \to 0 \text{ as } n \to \infty,$$

uniformly in $t \in [-B, B]$. For that, it suffices to prove the uniform convergence

$$n \Big| I\!\!E \Big\{ A_n(X_1, t) - n^{-\frac{1}{2}} t \dot{\psi}(X_1, \theta_0) - (t^2/(2n)) \ddot{\psi}(X_1, \theta_0) \Big\} \Big| \to 0.$$

But the left-hand side is majorized by $m \cdot |n^{-\frac{1}{2}}t|^{\alpha}(t^2/2)$ by **B.3**; this tends to 0 uniformly over $[-B, B]$ and thus, the lemma is proved. □

The propositions (5.12) and (5.13) of Theorem 5.2 with $R_n = O_p(n^{-1})$ follow from the general case in Theorem 5.1. hence, it remains to prove that nR_n has the asymptotic distribution as described in (5.14). The main idea is to make a random change of time $t \to n^{1/2}(M_n - \theta_0)$ in the process Y_n. This will be accomplished in several steps. First, we shall show that the two-dimensional process

$$\mathbf{Y}_n^\star = \Big\{ \mathbf{Y}_n^\star(t) = \Big(Y_n(t), n^{\frac{1}{2}}(M_n - \theta_0) \Big)^\top, \quad t \in [-B, B] \Big\},$$

which belongs to $(D[-B, B])^2$, converges weakly to the Gaussian process

$$\mathbf{Y}^\star = \Big\{ \Big(t\xi_1 + t^2 \frac{\gamma_2}{2\gamma_1}, \xi_2 \Big)^\top, \quad t \in [-B, B] \Big\}. \tag{5.16}$$

Notice that the second component of $\mathbf{Y}_n^\star(t)$ does not depend on t. It follows from (5.13) that $\mathbf{Y}_n^\star$ is asymptotically equivalent [up to the order of $O_p(n^{-1/2})$] to the process

$$\mathbf{Y}_n^{0\star} = \Big\{ \mathbf{Y}_n^{0\star}(t) = \Big(Y_n(t), -n^{-\frac{1}{2}} \gamma_1^{-1} \sum_{i=1}^{n} \psi(X_i, \theta_0) \Big)^\top, \quad |t| \le B \Big\},$$

and this process converges to $\mathbf{Y}^\star$ by Lemma 5.1 and by the classical central limit theorem.

Second, denoting $[a]_B = a \cdot I[-B \le a \le B]$, $a \in \boldsymbol{R}_1$, $B > 0$, we consider the process

$$\Big[\mathbf{Y}_n^\star\Big]_B = \Big\{ \Big[\mathbf{Y}_n^\star(t)\Big]_B = \Big(Y_n(t), \Big[n^{\frac{1}{2}}(M_n - \theta_0) \Big]_B \Big)^\top, \quad t \in [-B, B] \Big\}.$$

Then, by Lemma 5.1,

$$\Big[\mathbf{Y}_n^\star\Big]_B \Longrightarrow \mathbf{Y}^\star \tag{5.17}$$

for every fixed $B > 0$, where $\mathbf{Y}^\star$ is the Gaussian process given in (5.16), which has continuous sample paths. Hence, as in Section 2.5.10, we can make the random change of time $t \to \left[n^{1/2}(M_n - \theta_0)\right]_B$ in the process $\left[\mathbf{Y}_n^{\star}\right]_B$. By (5.17), and referring to Billingsley (1968), Section 17, pp. 144–145, we conclude that

$$Y_n\left(\left[n^{1/2}(M_n - \theta_0)\right]_B\right) \to \xi_1\left[\xi_2\right]_B - \frac{\gamma_2\left(\left[\xi_2\right]_B\right)^2}{2\gamma_1}. \tag{5.18}$$

for every fixed $B > 0$, as $n \to \infty$. Now, given $\varepsilon > 0$, there exists $B_0 > 0$ such that, for every $B \geq B_0$,

$$I\!P\left\{\left[\xi_2\right]_B \neq \xi_2\right\} < \varepsilon \quad \text{and} \quad I\!P\left\{\xi_1\xi_2 \neq \xi_1\left[\xi_2\right]_B\right\} < \varepsilon.$$

Moreover, by (5.12), there exist B_1 ($\geq B_0$) and an integer n_0 such that

$$I\!P(n^{\frac{1}{2}}|M_n - \theta_0)| > B) < \varepsilon \quad \text{for } B \geq B_1 \text{ and } n \geq n_0. \tag{5.19}$$

Combining (5.18)–(5.19), we obtain

$$\begin{aligned}
&\limsup_{n\to\infty} I\!P\left\{Y_n(n^{\frac{1}{2}}(M_n - \theta_0)) \leq y\right\} \\
&\leq \limsup_{n\to\infty} I\!P\left\{Y_n\left(\left[n^{\frac{1}{2}}(M_n - \theta_0)\right]_B\right) \leq y\right\} + \varepsilon \\
&= I\!P\left\{\xi_1\left[\xi_2\right]_B - (2\gamma_1)^{-1}\gamma_2(\left[\xi_2\right]_B)^2 \leq y\right\} + \varepsilon \\
&\leq I\!P\left\{\xi_1\xi_2 - (2\gamma_1)^{-1}\gamma_2\xi_2^2 \leq y\right\} + 3\varepsilon.
\end{aligned}$$

Similarly,

$$\begin{aligned}
&\limsup_{n\to\infty} I\!P\left\{Y_n(n^{\frac{1}{2}}(M_n - \theta_0)) > y\right\} \\
&\leq \limsup_{n\to\infty} I\!P\left\{Y_n\left(\left[n^{\frac{1}{2}}(M_n - \theta_0)\right]_B\right) > y\right\} + \varepsilon \\
&\leq I\!P\left\{\xi_1\xi_2 - (2\gamma_1)^{-1}\gamma_2\xi_2^2 > y\right\} + 3\varepsilon,
\end{aligned}$$

and this completes the proof of (5.14) and hence of the theorem. □

5.3 M-Estimation of Location: Fixed Scale

In the simple location model we can obtain deeper results under simpler regularity conditions. Hence, it is convenient to consider the location model separately, mainly for the following reasons:

1. The M-estimators were originally developed just for the classical location model.
2. The consistency, asymptotic normality and Bahadur-type representations for M-estimators of location can be proved under weaker regularity conditions on ρ and F than in the general case. In fact, Bahadur (1966) developed

the asymptotic representation for the sample quantiles which coincide with M-estimators generated by step-functions ψ.

3. The M-estimators of location may not be scale-equivariant, although they are translation-equivariant. To obtain a scale-equivariant M-estimator of location, one may studentize M_n by an appropriate scale-statistic S_n and define M_n as a solution of the minimization

$$\sum_{i=1}^{n} \rho((X_i - t)/S_n) := \min \quad (\text{with respect to } t \in \boldsymbol{R}_1). \tag{5.20}$$

The present section considers M-estimators with fixed scale not involving the studentization; the studentized M-estimators will be studied in the next section.

Let $X_1, X_2, \ldots$ be i.i.d. random variables with the distribution function $F(x - \theta)$. Let $\rho: \boldsymbol{R}_1 \mapsto \boldsymbol{R}_1$ be an absolutely continuous function with the derivative ψ and such that the function

$$h(t) = \int \rho(x - t) dF(x) \tag{5.21}$$

has a unique minimum at $t = 0$. Let us first consider the case when ψ is an absolutely continuous function that could be decomposed as

$$\psi(t) = \psi_1(t) + \psi_2(t), \quad t \in \boldsymbol{R}_1, \tag{5.22}$$

where ψ_1 has an absolutely continuous derivative ψ_1', and ψ_2 is a piecewise linear continuous function, constant outside a bounded interval. More precisely, we shall impose the following conditions on ψ_1, ψ_2 and F:

A1 *Smooth component* ψ_1:

ψ_1 is absolutely continuous with an absolutely continuous derivative ψ_1' such that

$$\int (\psi_1'(x + t))^2 dF(x) < K_1 \text{ for } |t| \leq \delta,$$

and ψ_1' is absolutely continuous and $\int |\psi_1''(x+t)| dF(x) < K_2$ for $|t| \leq \delta$, where δ, K_1 and K_2 are positive constants.

A2 *Piecewise linear component* ψ_2:

ψ_2 is absolutely continuous with the derivate

$$\psi_2'(x) = \alpha_\nu \quad \text{for } r_\nu < x \leq r_{\nu+1}, \ \nu = 1, \ldots, k,$$

where $\alpha_0, \alpha_1, \ldots \alpha_k$ are real numbers, $\alpha_0 = \alpha_k = 0$, and $-\infty = r_0 < r_1 < \ldots < r_k < r_{k+1} = \infty$.

A3 *F smooth around the difficult points:*

F has two bounded derivatives f and f' and $f > 0$ in a neighborhood of $r_1, \ldots r_k$.

A4 *Fisher's consistency:*

$\gamma = \gamma_1 + \gamma_2 > 0$, where $\gamma_i = \int \psi_i'(x)dF(x)$ for $i = 1, 2$, and $\int \psi^2(x)dF(x) < \infty$.

The M-estimator M_n of θ is then defined as a solution of the minimization

$$\sum_{i=1}^{n} \rho(X_i - t) := \min \text{ with respect to } t \in \boldsymbol{R}_1. \tag{5.23}$$

Under the conditions A1–A4, M_n is a root of the equation

$$\sum_{i=1}^{n} \psi(X_i - t) = 0. \tag{5.24}$$

If ρ is convex, and hence ψ nondecreasing, M_n can be defined uniquely in the form

$$M_n = \frac{1}{2}(M_n^+ + M_n^-), \tag{5.25}$$

where

$$M_n^- = \sup\{t : \sum_{i=1}^{n} \psi(X_i - t) > 0\}, \quad M_n^+ = \inf\{t : \sum_{i=1}^{n} \psi(X_i - t) < 0\}. \tag{5.26}$$

If ρ is not convex, then the Equation (5.24) may have more roots. The conditions A1–A4 guarantee that there exists at least one root of (5.24) that is $\sqrt{n}$-consistent estimator of θ and that admits an asymptotic representation. This is formally expressed in the following theorem:

THEOREM 5.3 *Let $X_1, X_2, \ldots$ be i.i.d. random variables with distribution function $F(x - \theta)$. Let $\rho : \boldsymbol{R}_1 \mapsto \boldsymbol{R}_1$ be an absolutely continuous function whose derivative ψ can be decomposed as (5.22) and such that the function $h(t)$ of (5.21) has a unique minimum at $t = 0$. Then under the conditions A1–A4, there exists a sequence $\{M_n\}$ of roots of the equation (5.24) such that*

$$n^{\frac{1}{2}}(M_n - \theta) = O_p(1) \tag{5.27}$$

and

$$M_n = \theta + (n\gamma)^{-1} \sum_{i=1}^{n} \psi(X_i - \theta) + R_n, \quad \textit{where } R_n = O_p(n^{-1}). \tag{5.28}$$

The asymptotic representation (5.28) immediately implies that the sequence $\{n^{1/2}(M_n - \theta)\}$ is asymptotically normally distributed as $n \to \infty$; hence, we have the following corollary:

Corollary 5.3 *Under the conditions of Theorem 5.3, there exists a sequence $\{M_n\}$ of solutions of the equation (5.24) such that $n^{1/2}(M_n - \theta)$ has asymptotically normal distribution $\mathcal{N}(0, \sigma^2(\psi, F))$ with $\sigma^2(\psi, F) = \gamma^{-2} \int \psi^2(x)dF(x)$.*

PROOF of Theorem 5.3. We shall first prove the following uniform asymptotic linearity:

$$\sup_{|t|\leq C}\left|\sum_{i=1}^{n}[\psi_j(X_i-\theta-n^{-\frac{1}{2}}t)-\psi_j(X_i-\theta)]+n^{\frac{1}{2}}t\gamma_j\right|=\mathrm{O}_p(1), \tag{5.29}$$

as $n\to\infty$, for any fixed $C>0$ and $j=1,2$. If $j=1$ and hence, $\psi=\psi_1$ is the smooth component of ψ, we may refer to the proof of Lemma 5.1: Denoting

$$\begin{aligned} Z_{nj}(t) &= \sum_{i=1}^{n}[\psi_j(X_i-n^{-\frac{1}{2}}t)-\psi_j(X_i)] \\ Z_{nj}^0(t) &= Z_{nj}(t)-\mathbb{E}_0 Z_{nj}(t), \quad j=1,2; \end{aligned} \tag{5.30}$$

(we can put $\theta=0$ without loss of generality), we get from the mentioned proof that for $t_1\leq t\leq t_2$,

$$\limsup_{n\to\infty}\mathbb{E}_0(|Z_{n1}^0(t)-Z_{n1}^0(t_1)||Z_{n1}^0(t_2)-Z_{n1}^0(t)|)\leq K(t_2-t_1)^2, \tag{5.31}$$

where $K>0$ is a constant. Hence, by (2.101), we have

$$\sup_{|t|\leq C}|Z_{n1}^0|=\mathrm{O}_p(1). \tag{5.32}$$

Moreover, by A1,

$$\left|n\mathbb{E}_0[\psi_1(X_1-n^{-\frac{1}{2}}t)-\psi_1(X_1)+n^{-\frac{1}{2}}t\psi_1'(X_1)]\right|\leq K^\star \tag{5.33}$$

for $n\geq n_0$ and uniformly for $|t|\leq C$; $K^\star>0$ is a constant. Combining (5.32) and (5.33) gives (5.29) for the case $j=1$.

Let now ψ_2 be the piecewise linear continuous function of A2. Without loss of generality, we may consider the particular case when for some $-\infty<r_1<r_2<\infty$,

$$\psi_2(x)=r_1, x \text{ or } r_2 \text{ according as } x<r_1,\ r_1\leq x\leq r_2 \text{ or } >r_2. \tag{5.34}$$

Note that ψ_2 is first-order Lipschitz, so that for any $t_1\leq t_2$,

$$\begin{aligned} &\mathrm{Var}_0[Z_{n2}(t_2)-Z_{n2}(t_1)] \\ &\leq \sum_{i=1}^{n}\mathbb{E}_0[\psi_2(X_i-n^{-\frac{1}{2}}t_2)-\psi_2(X_i-n^{-\frac{1}{2}}t_1)]^2 \\ &\leq K(t_2-t_1)^2, \quad 0<K<\infty. \end{aligned} \tag{5.35}$$

Moreover, by A3,

$$\begin{aligned} &\left|n\mathbb{E}_0[\psi_2(X_i-n^{-\frac{1}{2}}t_2)-\psi_2(X_i-n^{-\frac{1}{2}}t_1)+n^{-\frac{1}{2}}\psi_2'(X_1)(t_2-t_1)]\right| \\ &\leq n\mathbb{E}_0\int_{n^{-\frac{1}{2}}t_1}^{n^{-\frac{1}{2}}t_2}\int_0^u\int_{\mathbf{R}_1}|\psi_2'(x)f'(x+v)|dxdvdu \\ &\leq K(t_2-t_1) \quad \text{for } n\geq n_0,\ 0<K<\infty. \end{aligned} \tag{5.36}$$

Combining (5.35) and (5.36), we arrive at (5.29) for the case $j = 2$. hence, we conclude that

$$\sup_{|t|\le C}\left|\sum_{i=1}^{n}[\psi(X_i-\theta-n^{-\frac{1}{2}}t)-\psi(X_i-\theta)+n^{\frac{1}{2}}\gamma t\right| = O_p(1). \tag{5.37}$$

Following the arguments in Equations (5.7)–(5.9), we further conclude that given $\varepsilon > 0$, there exist $C > 0$ and an integer n_0 so that the equation $\sum_{i=1}^{n}\psi(X_i-\theta-n^{-\frac{1}{2}}t) = 0$ has a root T_n in the interval $[-C, C]$ with probability exceeding $1-\varepsilon$ for $n \ge n_0$. Then $M_n = \theta+n^{-\frac{1}{2}}T_n$ is a root of (5.24), and it satisfies (5.27). Finally, it is sufficient to insert $t \to n^{\frac{1}{2}}(M_n - \theta)$ in (5.37), in order to obtain the asymptotic representation (5.28). This completes the proof of the theorem. □

Theorem 5.3 guarantees an existence of a $\sqrt{n}$-consistent M-estimator even in the case where ρ is not convex and hence, ψ is not monotone. In such a case there may exist more roots of Equation (5.24) that are $\sqrt{n}$-consistent estimators of θ. Incidentally, all of them admit the asymptotic representation (5.28) and are mutually asymptotically equivalent up to $O_p(n^{-1})$, as it follows from (5.37). hence, we have the following corollary:

Corollary 5.4 *Let M_n and $M_n^\star$ be two roots of (5.24) satisfying (5.27). Then, under the conditions of Theorem 5.3,*

$$|M_n - M_n^\star| = O_p(n^{-1}) \quad \textit{as } n \to \infty.$$

There still may exist other roots that are not consistent estimators of θ. This begs the question which roots to choose as an estimator. If we know some $\sqrt{n}$-consistent estimator T_n (e.g., the sample median is $\sqrt{n}$-consistent in the symmetric model under mild conditions), then we may look for the root nearest to T_n. Another possibility is to look for a one-step or k-step version of M_n starting again with a $\sqrt{n}$-consistent initial estimator of θ. These estimators deserve a special attention and will be considered in Chapter 7. However, we may remark that all results hold under the crucial assumption that the function $h(t)$ of (5.21) has a unique minimum at $t = 0$. This condition deserves a careful examination not only in the location model but also in the general case. Without this condition the M-estimators may be inconsistent, even for the maximum likelihood estimation, as it was demonstrated by Freedman and Diaconis (1982).

5.3.1 Possibly Discontinuous but Monotone ψ

We now turn to the important case that while ρ is absolutely continuous, its derivative ψ may have jump discontinuities. This case can typically appear in the location and regression models where it covers the estimation based on

L_1-norm. More precisely, we shall assume that ρ is absolutely continuous with the derivative ψ which can be written as the sum

$$\psi = \psi_c + \psi_s,$$

where ψ_c is an absolutely continuous function satisfying the conditions A1–A4 of the present section and ψ_s is a step-function

$$\psi_s = \beta_j \quad \text{for } q_j < x < q_{j+1},\ j = 0, 1, \ldots, m, \tag{5.38}$$

where $\beta_0 < \beta_1 < \ldots < \beta_m$ are real numbers and $-\infty = q_0 < q_1 < \ldots < q_m < q_{m+1} = \infty$, m being a positive integer. We assume that the distribution function F is absolutely continuous with the density f which has a bounded derivative f' in a neighborhood of $q_1, \ldots . q_m$. We denote

$$\gamma_s = \sum_{j=1}^{m} (\beta_j - \beta_{j-1}) f(q_j). \tag{5.39}$$

The M-estimator M_n of θ is defined as a solution of the minimization (5.23). We assume that $h(t)$ of (5.21) has a unique minimum at $t = 0$. Notice that the equation (5.24) may not have any roots.

Following the proof of Theorem 4.9 (for the case $x_{i1} = 1,\ x_{ij} = 0,\ i = 1, \ldots, n;\ j = 2, \ldots, p$), we obtain

$$\sup_{|t| \le C} \left| n^{-\frac{1}{2}} \sum_{i=1}^{n} [\psi_s(X_i - \theta - n^{-\frac{1}{2}} t) - \psi_s(X_i - \theta)] + \gamma_s t \right| = O_p(n^{-\frac{1}{4}}) \tag{5.40}$$

for any fixed $C > 0$. Combined with (5.37), this gives

$$\sup_{|t| \le C} \left| n^{-\frac{1}{2}} \sum_{i=1}^{n} [\psi(X_i - \theta - n^{-\frac{1}{2}} t) - \psi(X_i - \theta)] + \gamma t \right| = O_p(n^{-\frac{1}{4}}), \tag{5.41}$$

with $\gamma = \gamma_c + \gamma_s$ and

$$\gamma_c = \int_{\boldsymbol{R}_1} \psi_c'(x) dF(x). \tag{5.42}$$

If ρ is convex and thus, ψ nondecreasing and if $\gamma > 0$, then M_n can be uniquely determined as in (5.25) and (5.26). In this case, following the proof of (4.99), we can prove that

$$n^{-\frac{1}{2}} \sum_{i=1}^{n} \psi(X_i - M_n) = O_p(n^{-\frac{1}{2}}); \tag{5.43}$$

Moreover, following the proof of (4.102), we obtain that M_n is $\sqrt{n}$-consistent estimator of θ:

$$n^{\frac{1}{2}} (M_n - \theta) = O_p(1). \tag{5.44}$$

Substituting $n^{1/2}(M_n - \theta)$ for t in (5.41) and using (5.43), we come to the asymptotic representation

$$M_n = \theta + (n\gamma)^{-1} \sum_{i=1}^{n} \psi(X_i - \theta) + O_p(n^{-\frac{3}{4}}). \tag{5.45}$$

The results are summarized in the following theorem:

THEOREM 5.4 *Let $X_1, X_2, \dots$ be i.i.d. random variables with the distribution function $F(x-\theta)$. Let $\rho : \boldsymbol{R}_1 \mapsto \boldsymbol{R}_1$ be an absolutely continuous function such that $h(t)$ of (5.21) has a unique minimum at $t = 0$. Assume that $\psi = \rho' = \psi_c + \psi_s$, where ψ_c is absolutely continuous, nondecreasing function satisfying A1–A4, and ψ_s is a nondecreasing step function of (5.38). Let F have the second derivative f' in a neighborhood of the jump points of ψ_s, and let γ of (5.42) be positive. Then the M-estimator M_n defined in (5.25) and (5.26) is a $\sqrt{n}$-consistent estimator of θ that admits the asymptotic representation (5.45).*

5.3.2 Possibly Discontinuous and Nonmonotone ψ

If ρ is absolutely continuous but not convex, then the consistency of M_n is guaranteed by methods of Section 5.3, provided ψ is smooth. However, the estimators related to discontinuous and non-monotone functions ψ also appear in the literature (e.g., Andrews et al., 1972). Some of these estimators even have a special name, like the *skipped mean* generated by

$$\rho(x) = \begin{cases} \frac{x^2}{2}, & \text{if } |x| \le k, \\ \frac{k^2}{2}, & \text{if } |x| > k, \end{cases} \qquad \psi(x) = \begin{cases} x, & \text{if } |x| < k, \\ 0, & \text{if } |x| > k, \end{cases}$$

or the *skipped median* corresponding to

$$\rho(x) = \begin{cases} |x|, & \text{if } |x| \le k, \\ k, & \text{if } |x| > k, \end{cases} \qquad \psi(x) = \begin{cases} \text{sign}x, & \text{if } |x| < k, \\ 0, & \text{if } |x| > k. \end{cases}$$

Then there is a question of consistency of M_n, defined as the point of global minimum in (5.23). Let us study the class of M_n generated by a bounded ψ, vanishing outside the interval $[-C, C]$. This covers both above estimators. We shall impose the following conditions on the model:

B1 *Shape of ρ :*

ρ is a nonnegative, absolutely continuous, even function, nondecreasing on $[0, \infty)$.

B2 *Shape of the derivative:*

The derivative ψ of ρ is bounded and $\psi(x) = 0$ for $x \notin [-C, C]$, $0 < C < \infty$.

B3 *Moments of ρ :*

The function $h(t)$ defined in (5.21) is locally convex in a neighborhood of 0 and has a unique minimum at $t = 0$.

B4 *Smoothness of* F :

F has an absolutely continuous symmetric density f with two derivatives f', f'' bounded a.e. and namely at $-C, C$.

B5 *Fisher's consistency:*

$$\gamma = \int_{\boldsymbol{R}} f(x) d\psi(x) > 0.$$

THEOREM 5.5 *Let $X_1, X_2, \ldots$ be i.i.d. random variables with the distribution function $F(x - \theta)$. Assume the conditions B1–B5. Then the point M_n of global minimum of (5.23) is a $\sqrt{n}$-consistent estimator of θ, which admits the asymptotic representation (5.45).*

PROOF Put $\theta = 0$ without loss of generality. We shall first show that

$$\sup_{|t| \leq T} \left| \frac{1}{n} \sum_{i=1}^{n} [\rho(X_i - n^{-\frac{1}{2}} t - \rho(X_i) + n^{-\frac{1}{2}} t \psi(X_i)] - \frac{\gamma t^2}{2} \right| = \mathrm{O}_p(n^{-\frac{3}{2}}) \quad (5.46)$$

as $n \to \infty$, for any $0 < T < \infty$. Indeed, for $-T \leq t \leq u \leq T$,

$$\operatorname{Var}\left[\rho(X_i - n^{-\frac{1}{2}} u) - \rho(X_i - n^{-\frac{1}{2}} t) + n^{-\frac{1}{2}} (u - t) \psi(X_i)\right] \leq K n^{-\frac{3}{2}} (u - t)^2, \quad (5.47)$$

$0 < K < \infty$, because, regarding **B2**,

$$\begin{aligned}
& \mathbb{E}\{\rho(X - n^{-\frac{1}{2}} t) - \rho(X - n^{-\frac{1}{2}} u) + n^{-\frac{1}{2}} (t - u) \psi(X_i)\}^2 \quad (5.48) \\
&= \int_{\boldsymbol{R}} \left[\int_{n^{-\frac{1}{2}} t}^{n^{-\frac{1}{2}} u} (-\psi(x - v) + \psi(x)) dv \right]^2 dF(x) \\
&\leq n^{-\frac{1}{2}} (u - t) \int_{n^{-\frac{1}{2}} t}^{n^{-\frac{1}{2}} u} \int_{\boldsymbol{R}} (\psi(x - v) - \psi(x))^2 dF(x) \\
&\leq K \cdot n^{-\frac{1}{2}} (u - t) \int_{n^{-\frac{1}{2}} t}^{n^{-\frac{1}{2}} u} \int_{\boldsymbol{R}} |(\psi(x - v) - \psi(x)) f(x)| dx dv \\
&= K \cdot n^{-\frac{1}{2}} (u - t) \int_{\boldsymbol{R}} \int_{n^{-\frac{1}{2}} t}^{n^{-\frac{1}{2}} u} \int_0^v |\psi(x)(f(x + w) - f(x))| dw dv dx \\
&\leq K_1 n^{-\frac{3}{2}} (u - t)^2. \quad (5.49)
\end{aligned}$$

hence, (5.47) implies that $n^{\frac{5}{4}} \sum_{i=1}^{n} [\rho(X_i - n^{-\frac{1}{2}} t) - \rho(X_i) + n^{-\frac{1}{2}} t \psi(X_i) - h(n^{-\frac{1}{2}} t) + h(0)]$, considered as a sequence of random processes in $\mathcal{C}[-T, T]$, is tight. Moreover, for $0 < t \leq \varepsilon$ (and similarly, for $-\varepsilon \leq t < 0$)

$$\begin{aligned}
h(t) - h(0) &= t h'(0) - \int_{\mathbb{R}} \psi(x) \int_0^t \int_0^u f'(x + v) dv du dx \\
&= \gamma \frac{t^2}{2} + \mathrm{O}(|t|^3) \quad (5.50)
\end{aligned}$$

and this together with (5.47) further implies (5.46).

Let M_n be the point of global minimum of $\sum_{i=1}^n \rho(X_i - t)$. It minimizes the function

$$G_n(t) = \frac{1}{n}\sum_{i=1}^n [\rho(X_i - t) - \rho(X_i)]$$

with the expected value $\Gamma(t) = h(t) - h(0)$. By B3, $\Gamma(t)$ has a unique minimum at $t = 0$ and $\Gamma(0) = 0$, hence, $\Gamma(t) > 0\ \forall t \neq 0$. Moreover, Γ is convex for $|t| \leq \varepsilon$, $\varepsilon > 0$. Because of the condition B2 is $\rho(X_{n:i} - t) = \rho(C)$ for $|X_{n:i} - t| \geq C$ or $\leq -C$, hence, M_n lies between two central order statistics, say $X_{n:i}$ and $X_{n:j}$; thus, $M_n = \mathrm{O}_p(1)$. Moreover, $\frac{1}{n}\sum_{i=1}^n [\rho(X_i - M_n) - \rho(X_i) \leq 0 < \Gamma(M_n)$; then (5.46) implies that $M_n = o_p(1)$. We further obtain from (5.46), (5.47), (5.50) and B1–B5

$$\left|\frac{1}{n}\sum_{i=1}^n [\rho(X_i - t) - \rho(X_i) + t\psi(X_i)] - \gamma\frac{t^2}{2}\right| = \mathrm{O}_p\left(n^{-\frac{1}{2}}|t| + |t|^2\right)$$

uniformly for $|t| \leq r_n$ for every sequence $\{r_n\}$ of positive numbers such that $r_n \downarrow 0$ as $n \to \infty$. Denote

$$U_n = \arg\min_t \left\{\frac{t^2}{2} - \frac{t}{\gamma}\cdot\frac{1}{n}\sum_{i=1}^n \psi(X_i)\right\} = \frac{1}{n\gamma}\sum_{i=1}^n \psi(X_i).$$

We immediately see that $n^{\frac{1}{2}}(U_n - \theta) = \mathrm{O}_p(1)$ is asymptotically normal $\mathcal{N}\left(0, \gamma^{-2}\int_{\mathbf{R}} \psi^2(x)dF(x)\right)$. We want to prove $n^{\frac{1}{2}}|M_n - U_n| = o_p(1)$. By (5.46), (5.48), (5.50) we obtain

$$\begin{aligned} G_n(M_n) &= -\gamma M_n U_n + \frac{1}{2}M_n^2 + \mathrm{O}_p\left(n^{-\frac{1}{2}}|M_n| + |M_n|^2\right) \quad (5.51)\\ &\leq -\frac{1}{2}U_n^2 + \mathrm{O}_p(n^{-1}) \leq -\gamma M_n U_n + \frac{1}{2}M_n^2 + \mathrm{O}_p(n^{-1}), \end{aligned}$$

hence, $M_n = \mathrm{O}_p(n^{-\frac{1}{2}})$. Finally,

$$0 \geq G_n(M_n) - G_n(U_n) = 2\gamma(U_n - T_n)^2 + \mathrm{O}_p(n^{-\frac{3}{2}}),$$

hence,

$$T_n - U_n = o_p(n^{-\frac{1}{2}}).$$

□

5.3.3 Second-Order Distributional Representations

The representations in (5.28) and (5.45) convey a qualitative difference: Jump discontinuities in the score function ψ lead to a slower rate of convergence for the remainder term R_n. Besides that, the distributions of the normalized R_n also have different functional forms. This reveals that M-estimators based on

ψ-functions having jump discontinuities may entail more elaborate iterative solutions compared to smooth ψ-functions.

The second-order asymptotic distributional representation in the location model for a smooth ψ follows from Theorem 5.2. However, the symmetric case with $F(x) + F(-x) = 1$ and $\rho(-x) = \rho(x)$, $x \in \boldsymbol{R}_1$ deserves a separate formulation, because it leads to simpler and intuitive form.

THEOREM 5.6 *Let $\{X_i, i \geq 1\}$ be a sequence of i.i.d. random variables with distribution function $F(x - \theta)$, F symmetric about origin. Let $\rho : \boldsymbol{R}_1 \mapsto \boldsymbol{R}_1$ be an absolutely continuous symmetric function. Assume the conditions A1–A4 of Section 5.3. Then there exists a sequence $\{M_n\}$ of roots of the Equation (5.24) that admits representation (5.28), wherein*

$$nR_n \xrightarrow{\mathcal{D}} \xi_1^\star \xi_2^\star \quad \text{as } n \to \infty;$$

the random vector $(\xi_1^\star, \xi_2^\star)^\top$ has bivariate normal distribution $\mathcal{N}_2(\mathbf{0}, \mathbf{S}^\star)$ with $\mathbf{S}^\star = [s_{i,j}]_{i,j=1,2}$ and

$$s_{11}^\star = \gamma^{-2} \int (\psi'(x))^2 dF(x) - 1, \quad s_{12}^\star = s_{21}^\star = 0$$

$$s_{22}^\star = \gamma^{-2} \int \psi^2(x) dF(x).$$

Let us now consider the situation described in Theorem 5.4, where $\psi = \psi_c + \psi_s$, a sum of absolutely continuous and step function components. If ψ_s does not vanish (this practically means that γ_s in (5.39) is positive) then the asymptotic distribution of $n^{3/4} R_n$ is considerably different from that of nR_n in Theorem 5.6.

THEOREM 5.7 *Assume the conditions of Theorem 5.4. Moreover, assume that F satisfies the following conditions:*

- *F has an absolutely continuous density f and finite Fisher's information $I(f) = \int_{\boldsymbol{R}_1} \left(\frac{f'(x)}{f(x)}\right)^2 dF(x) < \infty$.*
- *f' is bounded and continuous in a neighborhood of jump points $q_1, \ldots, q_s$ of ψ_s, where ψ_s is nondecreasing and $\gamma_s = \sum_{j=1}^m (\beta_j - \beta_{j-1}) f(q_j) > 0$.*

Denote

$$\nu_s^2 = \sum_{j=1}^{m} (\beta_j - \beta_{j-1})^2 f(q_j) \ (> 0). \tag{5.52}$$

Then M_n defined in (5.25) and (5.26) admits the asymptotic representation (5.45) and

$$n^{\frac{3}{4}} R_n \xrightarrow{\mathcal{D}} \xi \text{ as } n \to \infty,$$

where

$$\mathbb{P}(\xi \le x) = 2\int_0^\infty \Phi\Big(\frac{x}{w\sqrt{t}}\Big)d\Phi(t), \quad x \in \boldsymbol{R}_1,$$

Φ being the standard normal distribution function and

$$w = \gamma^{-\frac{3}{2}}\Big(\int_{\boldsymbol{R}_1} \psi^2(x)dF(x)\Big)^{\frac{1}{2}}\nu_s.$$

Remark 5.1 *Fix p, $0 < p < 1$ such that $f(F^{-1}(p)) \neq 0$ and put $\psi_s(x) = p - I[x \le 0]$ and $\psi_c \equiv 0$. Then $\int \psi(x-t)dF(x) = 0$ for $t = F^{-1}(p)$. Replacing $\psi(x)$ by $\psi^\star(x) = \psi(x - F^{-1}(p))$, we have $\gamma = \gamma_s = \nu_s^2 = f(F^{-1}(p))$. The M-estimator generated by $\psi^\star$ is then the residual of the sample quantile $X_{n:[np]} - F^{-1}(p)$ from $F^{-1}(p)$. Theorem 5.7 has the following corollary, which coincides with the result of Kiefer (1967).*

Corollary 5.5 *Assume that F is twice differentiable at $F^{-1}(p)$ and $f(F^{-1}(p)) > 0$. Then*

$$n^{\frac{1}{2}}\left(X_{n:[np]} - F^{-1}(p)\right) = \frac{1}{n^{\frac{1}{2}}f(F^{-1}(p))}\sum_{i=1}^n \Big(p - I[X_i \le F^{-1}(p)]\Big) + R_n(p)$$

and

$$\lim_{n\to\infty} \mathbb{P}\Big(n^{\frac{1}{4}}f(F^{-1}(p))R_n(p) \le x\Big) = 2\int_0^p \Phi\left(\frac{x}{(p(1-p))^{1/4}\sqrt{t}}\right)d\Phi(t).$$

PROOF of Theorem 5.7. Denote

$$Z_n(t) = n^{-\frac{1}{4}}\sum_{i=1}^n A_n(X_i, t), \tag{5.53}$$

$$Z_n^0(t) = Z_n(t) + n^{\frac{1}{4}}\gamma_s t, \quad t \in \boldsymbol{R}_1$$

where $A_n(X_i, t) = \psi_s(X_i - \theta - n^{-\frac{1}{2}}t) - \psi_s(X_i - \theta)$, $\quad i = 1, \ldots, n$. We shall prove that

$$\nu_s^{-1}Z_n^0 \xrightarrow{\mathcal{D}} W \tag{5.54}$$

in the Skorokhod J_1-topology on $D[-B, B]$ for any fixed $B > 0$, where $W = \{W(t) : t \in [-B, B]\}$ is a centered Gaussian process with

$$\mathbb{E}W(s)W(t) = \begin{cases} |s| \wedge |t|, & \text{if } s \cdot t > 0, \\ 0, & \text{otherwise.} \end{cases}$$

The function ψ can be alternatively rewritten as $\psi = \beta_0 + \sum_{j=1}^m(\beta_j - \beta_{j-1})\psi_j$ where

$$\psi_j(x) = \begin{cases} 0, & \text{if } x \le q_j, \\ 1, & \text{if } x > q_j, \end{cases}$$

for each $j(= 1, \ldots, m)$. Hence, we can prove (5.54) only for a single function ψ_j, say, for the function $\psi_s = \psi$ with $q_j = q$. Moreover, we can put $\theta = 0$.

Then, $\nu_s^2 = \gamma_s = f(q) > 0$, and for $-B \leq t_1 < \ldots t_a < 0 \leq t_{a+1} < \ldots < t_b \leq B$ $(0 \leq a \leq b,\ a, b$ integers). We can easily show that for $\boldsymbol{\lambda} \in \boldsymbol{R}_b$

$$\text{Var} \sum_{j=1}^{b} \lambda_j Z_n(t_j) \rightarrow \nu_s^2 \boldsymbol{\lambda}^\top \mathbf{C}_b \boldsymbol{\lambda} \ \text{ as } n \rightarrow \infty$$

where $\mathbf{C}_b = (c_{jk})_{j,k=1}^{b}$ and for each $j, k(= 1, \ldots, b)$,

$$c_{jk} = \begin{cases} |t_j| \wedge |t_k|, & \text{if } t_j t_k > 0, \\ 0, & \text{otherwise.} \end{cases}$$

By the central limit theorem we conclude that the finite-dimensional distributions of $Z_n - I\!\!E Z_n$ converge to those of W.

To prove the tightness of $Z_n - I\!\!E Z_n$, let us write the inequalities for $t \leq u$

$$\begin{aligned} &\limsup_{n \rightarrow \infty} I\!\!E[Z_n(u) - Z_n(t) - I\!\!E(Z_n(u) - Z_n(t))]^4 \\ &\leq \ \limsup_{n \rightarrow \infty} \Big\{ 11 I\!\!E[\psi_s(X_1 - n^{-\frac{1}{2}} u) - \psi_s(X_1 - n^{-\frac{1}{2}} t)]^4 \\ &\quad + (n-1)(I\!\!E[\psi_s(X_1 - n^{-\frac{1}{2}} u) - \psi_s(X_1 - n^{-\frac{1}{2}} t)]^2)^2 \Big\} \\ &= \ \limsup_{n \rightarrow \infty} \Big\{ 11 |F(q + n^{-\frac{1}{2}} u) - F(q + n^{-\frac{1}{2}} t)| \\ &\quad + (n-1) |F(q + n^{-\frac{1}{2}} t) - F(q + n^{-\frac{1}{2}} u)|^2 \Big\} = \nu_s (u - t)^2. \end{aligned}$$

This implies the tightness by (2.101) and altogether it gives the weak convergence of $\nu_s^{-1}(Z_n - I\!\!E Z_n)$ to W. To prove (5.54), we shall approximate $I\!\!E Z_n(t)$ by $-n^{-\frac{1}{2}} \gamma_s t$ uniformly in $[-B, B]$. But

$$\begin{aligned} &\sup_{0 \leq t \leq B} |I\!\!E Z_n(t) + n^{\frac{1}{4}} \gamma_s t| \leq n^{\frac{3}{4}} \sup_{0 \leq t \leq B} |I\!\!E A_n(X_1, t) + n^{-\frac{1}{2}} t f(q)| \\ &= n^{\frac{3}{4}} \sup_{0 \leq t \leq B} |F(q + n^{-\frac{1}{2}} t) - F(q) - n^{-\frac{1}{2}} t f(q)| \\ &\leq \ n^{\frac{3}{4}} \sup_{0 \leq t \leq B} \int_0^{n^{-\frac{1}{2}} t} \int_0^u |f'(q + v)| dv du \leq K B^2 n^{-\frac{1}{4}}, \end{aligned}$$

and we get analogous inequalities for $\sup_{-B \leq t \leq 0}$. This proves the weak convergence (5.54).

Now we consider the process $Y_n = \{Y_n(t) : \ |t| \leq B\}$ with

$$Y_n(t) = n^{-\frac{1}{4}} \sum_{i=1}^{n} [\psi_c(X_i - \theta - n^{-\frac{1}{2}} t) - \psi_c(X_i - \theta)] + n^{\frac{1}{4}} t \gamma_c. \tag{5.55}$$

By Lemma 5.1,

$$\sup\{|Y_n(t)| : \ |t| \leq B\} = o_p(n^{-\frac{1}{4}}). \tag{5.56}$$

Combining (5.54) and (5.56), we obtain that $W_n \xrightarrow{\mathcal{D}} W$ as $n \rightarrow \infty$ in Skorokhod J_1-topology on $D[-B, B]$ for any fixed $B > 0$, where

$$W_n(t) = \nu_s^{-1} \left\{ n^{-\frac{1}{4}} \sum_{i=1}^{n} [\psi(X_i - \theta - n^{-\frac{1}{2}}t) - \psi(X_i - \theta)] + n^{\frac{1}{4}}\gamma t \right\}, \; t \in \boldsymbol{R}_1.$$

We shall make a random change of time $t \to \left[n^{\frac{1}{2}}(M_n - \theta)) \right]_B$ in the process (5.53), similarly as in the proof of Lemma 5.1. By (5.45),

$$n^{\frac{1}{2}}(M_n - \theta) = \sum_{i=1}^{n} U_{ni} + o_p(1)$$

where $U_{ni} = n^{-\frac{1}{2}}\gamma^{-1} \sum_{i=1}^{n} \psi(X_i - \theta)$, $\mathbb{E}U_{ni} = 0$, $\operatorname{Var} U_{ni} = n^{-1}\sigma^2\gamma^{-2}$, $\sigma^2 = \int \psi^2(x) dF(x)$, $i = 1, \ldots, n$. By (5.53) and (5.54),

$$\operatorname{Cov}(Z_n(t), \sum_{i=1}^{n} U_{ni}) = O(n^{-\frac{1}{4}}|t|), \quad \operatorname{Cov}(Y_n(t), \sum_{i=1}^{n} U_{ni}) = O(n^{-\frac{1}{4}}|t|),$$

hence, the finite dimensional distributions of the two-dimensional process

$$\left\{ (W_n(t), n^{\frac{1}{2}}(M_n - \theta))) : \; |t| \le B \right\} \tag{5.57}$$

converge to those of the process

$$\left\{ (W(t), \xi^0) : \; |t| \le B \right\} \tag{5.58}$$

where ξ^0 is a random variable with normal $\mathcal{N}(0, \sigma^2/\gamma^2)$ distribution, independent of W. Also, $n^{1/2}(M_n - \theta)$ is relatively compact, while the weak convergence (5.54) ensures the relative compactness of (5.55). This implies the weak convergence of (5.57) to (5.58). Then, applying the random change of time $t \to \left[n^{\frac{1}{2}}(M_n - \theta) \right]_B$ and then letting $B \to \infty$, we obtain

$$\begin{aligned} & W_n(n^{\frac{1}{2}}(M_n - \theta)) \\ & = \nu_s^{-1} \left\{ n^{-\frac{1}{4}} \sum_{i=1}^{n} [\psi(X_i - M_n) - \psi(X_i - \theta)] + n^{\frac{3}{4}}\gamma(M_n - \theta) \right\} \\ & \to W(\xi^0) = I[\xi^0 > 0] W_1(\xi^0) + I[\xi^0 < 0] W_2(|\xi^0|), \end{aligned} \tag{5.59}$$

where W_1 and W_2 are independent copies of the standard Wiener process. By (5.45) and (5.59),

$$W_n(n^{\frac{1}{2}}(M_n - \theta)) = \gamma \nu_s^{-1} n^{\frac{3}{4}} R_n + \nu_s^{-1} n^{-\frac{1}{4}} \sum_{i=1}^{n} \psi(X_i - M_n), \tag{5.60}$$

and the last term is negligible by (5.43). Combining (5.59) and (5.60), we come to the proposition of the theorem. □

5.4 Studentized M-Estimators of Location

Let $X_1, \ldots, X_n$ be independent observations with a joint distribution function $F(x-\theta)$. Consider some scale statistic $S_n = S_n(X_1, \ldots, X_n)$ such that

$$\begin{aligned} &S_n(\mathbf{x}) > 0 \text{ a.e. } \mathbf{x} \in \boldsymbol{R}_n && (5.61) \\ &S_n(\mathbf{x}+c) = S_n(\mathbf{x}), \quad c \in \boldsymbol{R}_1, \ \mathbf{x} \in \boldsymbol{R}_n \text{ (translation invariance)}, \\ &S_n(c\mathbf{x}) = cS_n(\mathbf{x}), \quad c > 0, \ \mathbf{x} \in \boldsymbol{R}_n, \text{ (scale equivariance)}. \end{aligned}$$

Assume that there exists a functional $S = S(F) > 0$ such that

$$n^{\frac{1}{2}}(S_n - S) = \mathrm{O}_p(1) \quad \text{as } n \to \infty. \tag{5.62}$$

The M-estimator of θ studentized by S_n is defined as a solution M_n of the minimization

$$\sum_{i=1}^{n} \rho\Big(\frac{X_i - t}{S_n}\Big) := \min \tag{5.63}$$

with respect to $t \in \boldsymbol{R}_1$. We assume that ρ is absolutely continuous with the derivative ψ. Moreover, we shall assume throughout that the function

$$h(t) = \int_{-\infty}^{\infty} \rho\Big(\frac{x-t}{S(F)}\Big) dF(x) \tag{5.64}$$

has the only minimum at $t = 0$.

The studentized estimator is equivariant both in location and scale. We shall study the asymptotic distribution and the asymptotic representation of M_n. Its asymptotic behavior typically depends on the studentizing statistic S_n, with exception of the case when both F and ρ are symmetric. The other properties are similar as in the fixed-scale case: The remainder term in the asymptotic representation is typically of order $\mathrm{O}_p(n^{-1})$ or $\mathrm{O}_p(n^{-3/4})$, depending on whether $\psi = \rho'$ is smooth or whether it has jump discontinuities. We should require F to be smooth in a neighborhood of S-multiples of "difficult points" of ψ. The continuous component of ψ does not need to be monotone; however, for the sake of simplicity, we shall consider only nondecreasing step functions ψ.

Summarizing, we distinguish three sets of conditions on ψ and on F:

[A1] $\psi : \boldsymbol{R}_1 \mapsto \boldsymbol{R}_1$ is a step function,

$$\psi(x) = \beta_j \quad \text{for } x \in (q_j, q_{j+1}], \ j = 0, 1, \ldots, m,$$

where
$-\infty < \beta_0 \le \beta_1 \le \ldots \le \beta_m < \infty$ and
$-\infty = q_0 < \ldots < q_m < q_{m+1} = \infty$.

[A2] F has two bounded derivatives f and f' in a neighborhood of $Sq_1, \ldots, Sq_m$ and $f(Sq_j) > 0, \quad j = 1, \ldots, m$.

[B1] ψ is absolutely continuous with derivative ψ', which is a step function,

$$\psi'(x) = \alpha_\nu \quad \text{for } \mu_\nu < x < \mu_{\nu+1}, \ \nu = 0, 1, \ldots, k,$$

where $\alpha_0, \alpha_1, \ldots, \alpha_k \in \boldsymbol{R}_1, \quad \alpha_0 = \alpha_k = 0,$
and $-\infty = \mu_0 < \mu_1 < \ldots < \mu_k < \mu_{k+1} = \infty.$

[B2] F has a bounded derivative f in a neighborhood of $S\mu_1, \ldots, S\mu_k$.

[C1] ψ is an absolutely continuous function with an absolutely continuous derivative ψ', and there exists a $\delta > 0$ such that

$$\mathbb{E}_0 \sup \Big\{ \Big|\psi''\Big(\frac{X_1 - t}{Se^u}\Big)\Big|(X_1 - t)^2 : \ |t|, |u| \leq \delta \Big\} < \infty,$$

$$\mathbb{E}_0 \sup \Big\{ \Big|\psi'\Big(\frac{X_1 - t}{Se^u}\Big)(X_1 - t)\Big| : \ |t|, |u| \leq \delta \Big\} < \infty,$$

and

$$\mathbb{E}_0 \Big(X_1 \psi'\Big(\frac{X_1}{S}\Big)\Big)^2 < \infty.$$

The asymptotic representation of M_n is formulated in the following main theorem of the section.

THEOREM 5.8 *Let M_n be a solution of the minimization (5.63) such that*

$$\sqrt{n}(M_n - \theta) = O_p(1) \ \text{ as } n \to \infty. \tag{5.65}$$

Let S_n satisfy (5.61) and (5.62), and let $h(t)$ in (5.64) have a unique minimum at $t = 0$. Then,

(i) Under conditions [A1] and [A2],

$$M_n = \frac{M_n^+ + M_n^-}{2}, \ \text{ where}$$
$$M_n^- = \sup\Big\{t : \sum_{i=1}^n \psi\Big(\frac{X_i - t}{S_n}\Big) > 0\Big\}, \tag{5.66}$$
$$M_n^+ = \inf\Big\{t : \sum_{i=1}^n \psi\Big(\frac{X_i - t}{S_n}\Big) < 0\Big\},$$

and it admits the representation

$$M_n = \theta + (n\gamma_1)^{-1} \sum_{i=1}^n \psi\Big(\frac{X_i - \theta}{S}\Big) - \frac{\gamma_2}{\gamma_1}\Big(\frac{S_n}{S} - 1\Big) + O_p(n^{-\frac{3}{4}}), \tag{5.67}$$

where

$$\gamma_1 = \sum_{\nu=1}^m (\beta_\nu - \beta_{\nu-1} f(Sq_\nu), \ \gamma_2 = \sum_{\nu=1}^m Sq_\nu(\beta_\nu - \beta_{\nu-1} f(Sq_\nu). \tag{5.68}$$

(ii) Under the conditions [B1] and [B2] or [C1],

$$M_n = \theta + (n\gamma_1)^{-1} \sum_{i=1}^{n} \psi\Big(\frac{X_i - \theta}{S}\Big) - \frac{\gamma_2}{\gamma_1}\Big(\frac{S_n}{S} - 1\Big) + O_p(n^{-1}), \tag{5.69}$$

where

$$\gamma_1 = S^{-1} \int_{-\infty}^{\infty} \psi'\Big(\frac{x}{S}\Big) dF(x), \;\; \gamma_2 = S^{-1} \int_{-\infty}^{\infty} x\psi'\Big(\frac{x}{S}\Big) dF(x). \tag{5.70}$$

Theorem 5.8 has several interesting corollaries. First, when ρ is not convex (and ψ is not monotone), the minimization (5.63) can have more solutions even under [**B1**] and [**B2**] or under [**C1**]. However, any pair $M_n^{(1)}$ and $M_n^{(2)}$ of consistent solutions are close to each other in the following sense:

Corollary 5.6 *Let $M_n^{(1)}$ and $M_n^{(2)}$ be any pair of solutions of (5.63). If both $M_n^{(1)}$ and $M_n^{(2)}$ satisfy (5.65) then, under* [**B1**] *and* [**B2**] *or* [**C1**], *respectively,*

$$M_n^{(1)} - M_n^{(2)} = O_p(n^{-1}). \tag{5.71}$$

Second, while the representations (5.67) and (5.69) do not directly imply the asymptotic normality of M_n, they do imply the asymptotic normality of a linear combination of M_n and S_n:

Corollary 5.7 *Assume that*

$$\sigma^2 = \int_{-\infty}^{\infty} \psi^2\Big(\frac{x}{S}\Big) dF(x) < \infty. \tag{5.72}$$

Then, under the conditions of Theorem 5.8,

$$n^{\frac{1}{2}}\Big\{\gamma_1(M_n - \theta) + \gamma_2\Big(\frac{S_n}{S}\Big)\Big\} \tag{5.73}$$

has asymptotically normal distribution $\mathcal{N}(0, \sigma^2)$.

However, γ_2 vanishes under the symmetry of F and ρ; this considerably simplifies the asymptotic representation and distribution of M_n:

Corollary 5.8 *If $F(x) + F(-x) = 1$ and $\rho(-x) = \rho(x)$, $x \in \boldsymbol{R}_1$, then*

$$M_n = \theta + (n\gamma_1)^{-1} \sum_{i=1}^{n} \psi\Big(\frac{X_i - \theta}{S}\Big) + R_n, \tag{5.74}$$

where $R_n = O_p(n^{-3/4})$ under [A1] and [A2], and $R_n = O_p(n^{-1})$ under [B1] and [B2] or [C1], respectively. Moreover, $\sqrt{n}(M_n - \theta)$ is then asymptotically normally distributed $\mathcal{N}(0, \sigma^2/\gamma_1^2)$.

Without the symmetry conditions, but assuming that S_n itself admits an asymptotic representation, we can still get a representation and the asymptotic normality for M_n itself:

Corollary 5.9 *Assume that S_n admits the asymptotic representation:*

$$\frac{S_n}{S} - 1 = n^{-1}\sum_{i=1}^{n}\phi(X_i - \theta) + o_p(n^{-\frac{1}{2}}), \tag{5.75}$$

where $\int_{-\infty}^{\infty}\phi(x)dF(x) = 0$, $\int_{-\infty}^{\infty}\phi^2(x)dF(x) < \infty$. Then

$$M_n = \theta + (n\gamma_1)^{-1}\sum_{i=1}^{n}\Big\{\psi\Big(\frac{X_i - \theta}{S}\Big) - \gamma_2\phi(X_i - \theta)\Big\} + o_p(n^{-\frac{1}{2}}), \tag{5.76}$$

and $n^{1/2}(M_n - \theta)$ is asymptotically normally distributed with expectation 0 and with the variance

$$\gamma_1^{-2}\int_{-\infty}^{\infty}\Big(\psi\Big(\frac{x}{S}\Big) - \gamma_2\phi(x)\Big)^2 dF(x). \tag{5.77}$$

The proof of Theorem 5.8 is based on the following asymptotic linearity lemma:

LEMMA 5.2 *For every $t, u: \ |t|, |u| < C$, let*

$$Z_n(t, u) = n^{-\frac{1}{2}}\sum_{i=1}^{n}\Big\{\psi\Big(\frac{X_i - \theta - tn^{-\frac{1}{2}}}{Se^{un^{-\frac{1}{2}}}}\Big) - \psi\Big(\frac{X_i - \theta}{S}\Big)\Big\}. \tag{5.78}$$

(i) Under **[A1]** *and* **[A2]**, *for any fixed $C > 0$,*

$$\sup\Big\{|Z_n(t, u) + t\gamma_1 + u\gamma_2| : \ |t|, |u| \le C\Big\} = O_p(n^{-\frac{1}{4}}). \tag{5.79}$$

(ii) Under **[B1]** *and* **[B2]** *or* **[C1]** $\ \forall C > 0$,

$$\sup\Big\{|Z_n(t, u) + t\gamma_1 + u\gamma_2| : \ |t|, |u| \le C\Big\} = O_p(n^{-\frac{1}{2}}). \tag{5.80}$$

PROOF of Lemma 5.2. Without loss of generality, we assume $\theta = 0$.
(i) We assume, without loss of generality, that $\psi(x) = 0$ or 1 according to $x \le q$ or $x > q$, respectively. Then

$$Z_n(t, u) - Z_n(0, u) = -n^{\frac{1}{2}}[F_n(Sqe^{un^{-\frac{1}{2}}} + n^{-\frac{1}{2}}t) - F_n(Sq)], \tag{5.81}$$

where $F_n(x) = n^{-1}\sum_{i=1}^{n} I[X_i \le x]$ is the empirical distribution function corresponding to $X_1, \ldots, X_n$. Then, by Komlós, Májor and Tusnády (1975),

$$\begin{aligned} Z_n(t, u) &= n^{-\frac{1}{2}}[F(Sqe^{n^{-\frac{1}{2}}u} + n^{-\frac{1}{2}}t) - F(Sq)] + \mathrm{O}_p(n^{-\frac{1}{4}}) \\ &= -t\gamma_1 - u\gamma_2 + \mathrm{O}_p(n^{-\frac{1}{4}}) \end{aligned} \tag{5.82}$$

uniformly in $|t|, |u| \le C$.
(ii) Assume the conditions [B1] and [B2]. If we denote

$$\psi_\nu(x) = \begin{cases} \mu_\nu, & \text{for } x < \mu_\nu, \\ x, & \text{for } \mu_\nu \le x \le \mu_{\nu+1}, \\ \mu_{\nu+1}, & \text{for } x > \mu_{\nu+1}, \quad \text{for } \nu = 1, \ldots, k, \end{cases} \tag{5.83}$$

then $\psi \equiv \sum_{\nu=1}^{k} \alpha_\nu \psi_\nu$. hence, we can consider only $\psi \equiv \psi_1$. Then,

$$\gamma_1 = S^{-1}(F(\mu_2 S) - F(\mu_1 S)), \;\; \gamma_2 = S^{-1} \int_{\mu_1 S}^{\mu_2 S} x dF(x), \tag{5.84}$$

and, if $t > 0$ (the case $t < 0$ is treated analogously),

$$\begin{aligned}
& \Big| Z_n(t,u) - Z_n(0,u) + \frac{t}{n} \sum_{i=1}^{n} \frac{1}{Se^{un^{-\frac{1}{2}}}} \psi'\Big(\frac{X_i}{Se^{un^{-\frac{1}{2}}}}\Big) \Big| \\
= & \Big| \frac{1}{\sqrt{n} Se^{un^{-\frac{1}{2}}}} \sum_{i=1}^{n} \int_0^{\frac{t}{\sqrt{n}}} \Big[-\psi'\Big(\frac{X_i - T}{Se^{un^{-\frac{1}{2}}}}\Big) + \psi'\Big(\frac{X_i}{Se^{un^{-\frac{1}{2}}}}\Big)\Big] dT \Big| \\
= & \Big| \frac{\sqrt{n}}{Se^{un^{-\frac{1}{2}}}} \int_0^{\frac{t}{\sqrt{n}}} \{ -F_n(S\mu_2 e^{un^{-\frac{1}{2}}} + T) + F_n(S\mu_2 e^{un^{-\frac{1}{2}}}) \\
& + F_n(S\mu_1 e^{un^{-\frac{1}{2}}} + T) - F_n(S\mu_1 e^{un^{-\frac{1}{2}}}) \} dT \Big| \qquad (5.85) \\
\leq & \Big| \frac{\sqrt{n}}{Se^{un^{-\frac{1}{2}}}} \int_0^{\frac{t}{\sqrt{n}}} \int_0^{T} \{ -f(S\mu_2 e^{un^{-\frac{1}{2}}} + V) + f(S\mu_1 e^{un^{-\frac{1}{2}}} + V)\} dV dT \Big| \\
& + \Big| \frac{1}{Se^{un^{-\frac{1}{2}}}} \int_0^{\frac{t}{\sqrt{n}}} \{ -q_n(S\mu_2 e^{un^{-\frac{1}{2}}} + T) + q_n(S\mu_2 e^{un^{-\frac{1}{2}}}) \\
& + q_n(S\mu_1 e^{un^{-1/2}} + T) - q_n(S\mu_1 e^{un^{-1/2}}) \} dT \Big|
\end{aligned}$$

where

$$q_n(x) = n^{1/2}(F_n(x) - F(x)), \; x \in \boldsymbol{R}_1.$$

Again, by Komlós, Májor, and Tusnády (1975),

$$q_n(x) = B_n(F(x)) + \mathrm{O}_p(n^{-1/2} \log n) \;\text{ uniformly in } x \in \boldsymbol{R}_1, \tag{5.86}$$

where the Brownian bridge $B_n(.)$ depends on $X_1, \ldots, X_n$. Hence, by (5.85) and (5.86),

$$\begin{aligned}
& \Big| Z_n(t,u) - Z_n(0,u) + \frac{t}{nSe^{un^{-\frac{1}{2}}}} \sum_{i=1}^{n} \psi'\Big(\frac{X_i}{Se^{un^{-\frac{1}{2}}}}\Big) \Big| \\
= & \; C^2 \mathrm{O}_p(n^{-\frac{1}{2}}) + \mathrm{O}_p(n^{-\frac{3}{4}}) \qquad (5.87)
\end{aligned}$$

uniformly in $|t|, |u| \leq C$. On the other hand,

$$\frac{t}{nSe^{un^{-\frac{1}{2}}}} \sum_{i=1}^{n} \psi'\Big(\frac{X_i}{Se^{un^{-\frac{1}{2}}}}\Big) - t\gamma_1 = \mathrm{O}_p(n^{-\frac{1}{2}}) \tag{5.88}$$

uniformly in $|t|, |u| \leq C$.

Similarly, for $u > 0$ (the case $u < 0$ is analogous),

$$\Big| Z_n(0,u) + \frac{u}{n} \sum_{i=1}^{n} \frac{X_i}{S} \psi'\Big(\frac{X_i}{S}\Big) \Big|$$

$$= \Big|n^{-\frac{1}{2}}\sum_{i=1}^{n}\frac{X_i}{S}\int_0^{\frac{t}{\sqrt{n}}}\Big\{-e^{-U}I[S\mu_1 e^U < X_i < S\mu_2 e^U] + I[S\mu_1 < X_i < S\mu_2]\Big\}dU\Big| \quad (5.89)$$

$$= \Big|\frac{n^{\frac{1}{2}}}{S}\int_0^{\frac{t}{\sqrt{n}}} x\Big\{I[S\mu_1 < X_i < S\mu_2] - e^{-U}I[S\mu_1 e^U < X_i < S\mu_2 e^U]\Big\} d(F_n - F)\Big| + \frac{1}{S}\mathrm{O}_p(n^{-\frac{1}{2}}) = \mathrm{O}_p(n^{-\frac{1}{2}}) \text{ uniformly in } |u| \le C.$$

Moreover,

$$\Big|\frac{u}{n}\sum_{i=1}^{n}\frac{X_i}{S}\psi'\Big(\frac{X_i}{S}\Big) - u\gamma_2\Big| = \Big|u\int_{\mu_1 S}^{\mu_2 S} x d(F_n - F)\Big| = \mathrm{O}_p(n^{-\frac{1}{2}}) \quad (5.90)$$

uniformly in $|u| \le C$. Combining (5.85)–(5.90), we arrive at (5.80) under conditions [B1] and [B2].

Assume now the condition [C1] and denote

$$Z_n^{\star}(t,u) = \sum_{\varepsilon_1=0,1}\sum_{\varepsilon_2=0,1}(-1)^{\varepsilon_1+\varepsilon_2}Z_n(t-\varepsilon_1 t, u-\varepsilon_2 u). \quad (5.91)$$

Then, obviously,

$$Z_n(t,u) = Z_n^{\star}(t,u) + Z_n^{\star}(t,0) + Z_n^{\star}(0,u), \quad (5.92)$$

and for $t > 0$ and $u > 0$ (other quadrants are treated analogously),

$$Z_n^{\star}(t,u) = n^{-\frac{1}{2}}\sum_{i=1}^{n}\int_0^{\frac{t}{\sqrt{n}}}\int_0^{\frac{u}{\sqrt{n}}}\Big\{\frac{X_i - T}{(Se^U)^2}\psi''\Big(\frac{X_i - T}{Se^U}\Big) + \frac{1}{Se^U}\psi'\Big(\frac{X_i - T}{Se^U}\Big)\Big\}dU\,dT.$$

thus, under [C1],

$$\mathbb{E}_0 \sup\Big\{|Z_n^{\star}(t,u)| : \ |t|, |u| \le C\Big\} \le C^2 K n^{-\frac{1}{2}} \quad (5.93)$$

for some K, $0 < K < \infty$, and $n \ge n_0$.

Similarly, for $t > 0$ (the case $t < 0$ is analogous),

$$Z_n^{\star}(t,u) + t\gamma_1 = \frac{1}{S\sqrt{n}}\sum_{i=1}^{n}\int_0^{\frac{t}{\sqrt{n}}}\Big\{-\psi'\Big(\frac{X_i - T}{S}\Big) + \mathbb{E}_0\psi'\Big(\frac{X_i}{S}\Big)\Big\}dT \quad (5.94)$$

$$= \frac{1}{S\sqrt{n}}\sum_{i=1}^{n}\int_0^{\frac{t}{\sqrt{n}}}\int_0^{T}\psi''\Big(\frac{X_i - V}{S}\Big)dV\,dT - \frac{t}{Sn}\sum_{i=1}^{n}\Big[\psi'\Big(\frac{X_i}{S}\Big) - \mathbb{E}_0\psi'\Big(\frac{X_i}{S}\Big)\Big].$$

By [**C1**],

$$\mathbb{E}_0\Big\{\sup\Big|\frac{1}{S\sqrt{n}}\sum_{i=1}^{n}\int_0^{\frac{t}{\sqrt{n}}}\int_0^{T}\psi''\Big(\frac{X_i-V}{S}\Big)dV\,dT\Big| : |t|\le C\Big\}$$
$$\le \frac{C^2K}{S^2}n^{-\frac{1}{2}}$$

for some K, $0<K<\infty$, and $n\ge n_0$. On the other hand,

$$\mathbb{E}_0\sup\Big\{\Big|\frac{t}{nS}\sum_{i=1}^{n}\Big[\psi'\Big(\frac{X_i-V}{S}\Big)-\mathbb{E}_0\psi'\Big(\frac{X_i-V}{S}\Big)\Big]\Big| : |t|\le C\Big\}^2$$
$$\le \frac{C^2}{nS^2}\mathbb{E}_0\Big(\psi'\Big(\frac{X_1}{S}\Big)\Big)^2.$$

Finally, for $u>0$ (the case $u<0$ is analogous),

$$\begin{aligned}
&Z_n^\star(0,u)+u\gamma_2\\
&= n^{-\frac{1}{2}}\sum_{i=1}^{n}\int_0^{\frac{u}{\sqrt{n}}}\Big\{-\frac{X_1}{Se^U}\psi'\Big(\frac{X_1}{Se^U}\Big)+\mathbb{E}_0\Big[\frac{X_1}{Se^U}\psi'\Big(\frac{X_1}{Se^U}\Big)\Big]\Big\}dU\\
&= \frac{1}{S\sqrt{n}}\sum_{i=1}^{n}\int_0^{\frac{u}{\sqrt{n}}}\int_0^{U}\Big\{\Big(\frac{X_i}{Se^V}\Big)^2\psi''\Big(\frac{X_i}{Se^V}\Big)+\frac{X_i}{Se^V}\psi'\Big(\frac{X_i}{Se^V}\Big)\Big\}dV\,dU\\
&\quad -\frac{u}{n}\sum_{i=1}^{n}\Big\{\frac{X_i}{S}\psi'\Big(\frac{X_i}{S}\Big)-\mathbb{E}_0\Big(\frac{X_i}{S}\psi'\Big(\frac{X_i}{S}\Big)\Big)\Big\}.
\end{aligned}\tag{5.95}$$

By [**C1**],

$$\mathbb{E}_0\sup\Big\{\Big|\tfrac{1}{S\sqrt{n}}\textstyle\sum_{i=1}^{n}\int_0^{\frac{u}{\sqrt{n}}}\int_0^{U}\Big\{\Big(\frac{X_i}{Se^V}\Big)^2\psi''\Big(\frac{X_i}{Se^V}\Big)+\frac{X_i}{Se^V}\psi'\Big(\frac{X_i}{Se^V}\Big)\Big\}dV\,dU\Big|$$
$$: |u|\le C\Big\}\le C^2Kn^{-\frac{1}{2}}$$

for some K, $0<K<\infty$, and $n\ge n_0$, and

$$\mathbb{E}_0\sup\Big\{\frac{u}{n}\sum_{i=1}^{n}\Big[\frac{X_i}{S}\psi'\Big(\frac{X_i}{S}\Big)-\mathbb{E}_0\Big(\frac{X_i}{S}\psi'\Big(\frac{X_i}{S}\Big)\Big)\Big] : |u|\le C\Big\}^2$$
$$\le \frac{C^2}{nS^2}\mathbb{E}_0\Big[X_1\psi'\Big(\frac{X_1}{S}\Big)\Big]^2. \tag{5.96}$$

Combining (5.92)–(5.96), with the aid of the Markov inequality, we arrive at (5.80) under the condition [**C1**]. The lemma is proved. □

PROOF of Theorem 5.8. Inserting $u\to n^{1/2}\ln(S_n/S)=\mathrm{O}_p(1))$ in (5.79) and (5.80) and recalling (5.62), we obtain

$$\sup_{|t|\le C}\Big\{\Big|n^{-\frac{1}{2}}\sum_{i=1}^{n}\Big[\psi\Big(\frac{X_i-\theta-n^{-\frac{1}{2}}}{S_n}\Big)-\psi\Big(\frac{X_i-\theta}{S}\Big)\Big]$$

$$+t\gamma_1 + n^{\frac{1}{2}}\Big(\frac{S_n}{S} - 1\Big)\gamma_2\Big|\Big\} \tag{5.97}$$

$$= \begin{cases} O_p(n^{-\frac{1}{4}}), & \text{under } [\mathbf{A1}] - [\mathbf{A2}], \\ O_p(n^{-\frac{1}{2}}), & \text{under } [\mathbf{B1}] - [\mathbf{B2}] \text{ or } [\mathbf{C1}]. \end{cases}$$

(i) Under [A1] and [A2], where ψ is a nondecreasing step function, we can follow the proof of (4.102) and obtain the $\sqrt{n}$-consistency of M_n. Moreover, analogously, as in the proof of (4.99), we can conclude that

$$n^{-\frac{1}{2}} \sum_{i=1}^{n} \psi\Big(\frac{X_i - M_n}{S_n}\Big) = O_p(n^{-\frac{1}{2}}).$$

Inserting $u \to n^{1/2}(M_n - \theta)$ in (5.97), we obtain the first proposition of the theorem.

(ii) Under conditions [B1] and [B2] and under [C1], we assume $n^{1/2}(M_n - \theta) = O_p(1)$. Inserting $u \to n^{1/2}(M_n - \theta)$ in (5.97), we obtain the second proposition.

□

The corollaries of Theorem 5.8 follow immediately.

One open question was tacitly avoided in Theorem 5.8: If ρ is not convex then the existence of a solution of (5.63) satisfying (5.65) may be doubtful. The following theorem deals with this problem.

THEOREM 5.9 *Let S_n satisfy (5.61) and (5.62), and let $h(t)$ in (5.64) have a unique minimum at $t = 0$ and $\gamma_1 > 0$.*

(i) Under [B1] and [B2] or under [C1], there exists a solution M_n of the equation

$$\sum_{i=1}^{n} \psi\Big(\frac{X_i - t}{S_n}\Big) = 0 \tag{5.98}$$

such that $n^{1/2}(M_n - \theta) = O_p(1)$, and M_n admits the representation (5.69). The Corollaries 5.6–5.9 apply to M_n.

(ii) Moreover, if [B1] and [B2] or [C1], respectively, apply also to ψ' in the role of ψ, then there exists a local minimum $M_n^\star$ of $\sum_{i=1}^{n} \rho((X_i - t)/S_n)$ satisfying $n^{1/2}(M_n - \theta) = O_p(1)$.

PROOF (i) Under the conditions of the theorem, $\gamma_1 > 0$ and

$$n^{-\frac{1}{2}} \sum_{i=1}^{n} \psi\Big(\frac{X_i - \theta}{S_n}\Big) = O_p(1), \quad n^{\frac{1}{2}}\Big(\frac{S_n}{S} - 1\Big) = O_p(1),$$

Hence, given $\varepsilon > 0$, there exist $C > 0$ and n_0 so that for $n \geq n_0$,

$$\mathbb{P}\Big\{\Big|n^{-\frac{1}{2}}\sum_{i=1}^{n}\psi\Big(\frac{X_i-\theta}{S_n}\Big)\Big| > \frac{C\gamma_1}{3}\Big\} < \frac{\varepsilon}{3},$$

$$\mathbb{P}\Big\{\Big|n^{\frac{1}{2}}\Big(\frac{S_n}{S}-1\Big)\gamma_2\Big| > \frac{C\gamma_1}{3}\Big\} < \frac{\varepsilon}{3}.$$

By (5.97),

$$\begin{aligned} & n^{-\frac{1}{2}}\sum_{i=1}^{n}\psi\Big(\frac{X_i-\theta-n^{-\frac{1}{2}}C}{S_n}\Big) \qquad (5.99)\\ = \;& n^{-\frac{1}{2}}\sum_{i=1}^{n}\psi\Big(\frac{X_i-\theta}{S}\Big) - C\gamma_1 - n^{\frac{1}{2}}\Big(\frac{S_n}{S}-1\Big)\gamma_2 + \mathrm{O}_p(n^{-\frac{1}{4}})\\ < \;& 0 \text{ for } n \geq n_0 \text{ with probability } \geq 1-\varepsilon. \end{aligned}$$

Analogously, for $n \geq n_0$,

$$n^{-\frac{1}{2}}\sum_{i=1}^{n}\psi\Big(\frac{X_i-\theta+n^{-\frac{1}{2}}C}{S_n}\Big) > 0 \text{ with probability } \geq 1-\varepsilon. \qquad (5.100)$$

Due to the continuity of ψ, (5.99) and (5.100) imply that there exists T_n such that

$$\sum_{i=1}^{n}\psi\Big(\frac{X_i-\theta-n^{-\frac{1}{2}}T_n}{S_n}\Big) = 0$$

and $\mathbb{P}(|T_n| < C) \geq 1-2\varepsilon$ for $n \geq n_0$. Put $M_n = n^{-\frac{1}{2}}T_n + \theta$. Then $\sum_{i=1}^{n}\psi\Big(\frac{X_i-M_n}{S_n}\Big) = 0$ and $n^{1/2}(M_n-\theta) = \mathrm{O}_p(1)$. (ii) Let M_n be the solution of (5.98) from part (i). We can virtually repeat the proof of Lemma 5.2 for

$$n^{-\frac{1}{2}}\sum_{i=1}^{n}\psi'\left(\frac{X_i-tn^{-\frac{1}{2}}}{Se^{un^{-\frac{1}{2}}}}\right)$$

and obtain for some γ_1', γ_2' ($|\gamma_1'|, |\gamma_2'| < \infty$)

$$\begin{aligned} & n^{-1}\sum_{i=1}^{n}\psi'\Big(\frac{X_i-M_n}{S_n}\Big)\\ = \;& n^{-1}\sum_{i=1}^{n}\psi'\Big(\frac{X_i-\theta}{S_n}\Big) - \gamma_1'(M_n-\theta) - \gamma_2'\Big(\frac{S_n}{S}-1\Big) + \mathrm{O}_p(n^{-1})\\ = \;& \gamma_1 S + \mathrm{o}_p(1). \end{aligned}$$

Hence, given $\varepsilon > 0$, there exists n_0, and $\mathbb{P}(A_n) \geq 1-\varepsilon$ for $n \geq n_0$, where $A_n = \Big\{\omega \in \Omega:\; n^{-1}\sum_{i=1}^{n}\psi'\Big(\frac{X_i-M_n}{S_n}\Big) > 0\Big\}$. Put $M_n^\star = M_n$ for $\omega \in A_n$; let $M_n^\star$ be any point of local minimum of $\sum_{i=1}^{n}\rho\big(\frac{X_i-t}{S_n}\big)$ at the other points. Then $M_n^\star$ is a local minimum of the above function which is a $\sqrt{n}$-consistent estimator of θ. $\square$

5.5 M-Estimation in Linear Regression Model

Consider the linear model

$$\mathbf{Y} = \mathbf{X}\boldsymbol{\beta} + \mathbf{E}, \tag{5.101}$$

where $\mathbf{Y} = (Y_1, \ldots, Y_n)^\top$ is the vector of observations, $\mathbf{X} = \mathbf{X}_n$ is a (known or observable) design matrix of order $(n \times p)$, $\boldsymbol{\beta} = (\beta_1, \ldots, \beta_p)^\top$ is an unknown parameter, and $\mathbf{E} = (E_1, \ldots, E_n)^\top$ is a vector of i.i.d. errors with a distribution function F.

The M-estimator of location parameter extends to the model (5.101) in a straightforward way: Given an absolutely continuous $\rho: \ \boldsymbol{R}_1 \mapsto \boldsymbol{R}_1$ with derivative ψ, we define an M-estimator of $\boldsymbol{\beta}$ as a solution of the minimization

$$\sum_{i=1}^{n} \rho(Y_i - \mathbf{x}_i^\top \mathbf{t}) := \min \tag{5.102}$$

with respect to $\mathbf{t} \in \boldsymbol{R}_p$, where $\mathbf{x}_i^\top$ is the ith row of $\mathbf{X}_n$, $i = 1, \ldots, n$. Such M-estimator $\mathbf{M}_n$ is regression equivariant:

$$\mathbf{M}_n(\mathbf{Y} + \mathbf{X}\mathbf{b}) = \mathbf{M}_n(\mathbf{Y}) + \mathbf{b} \ \text{ for } \ \mathbf{b} \in \boldsymbol{R}_p, \tag{5.103}$$

but $\mathbf{M}_n$ is generally not scale equivariant: It does not satisfy

$$\mathbf{M}_n(c\mathbf{Y}) = c\mathbf{M}_n(\mathbf{Y}) \ \text{ for } \ c > 0. \tag{5.104}$$

On the other hand, the studentization leads to estimators that are scale as well as regression equivariant. The studentized M-estimator is defined as a solution of the minimization

$$\sum_{i=1}^{n} \rho\Big(\frac{Y_i - \mathbf{x}_i^\top \mathbf{t}}{S_n}\Big) := \min, \tag{5.105}$$

where $S_n = S_n(\mathbf{Y}) \geq 0$ is an appropriate scale statistic. For the best results S_n should be regression invariant and scale equivariant:

$$S_n(c(\mathbf{Y} + \mathbf{X}\mathbf{b})) = cS_n(\mathbf{Y}) \ \text{ for } \ \mathbf{b} \in \boldsymbol{R}_p \ \text{ and } \ c > 0. \tag{5.106}$$

The minimization (5.105) should be supplemented by a rule how to define $\mathbf{M}_n$ if $S_n(\mathbf{Y}) = 0$. However, in typical cases it appears with probability zero, and the specific rule does not effect the asymptotic properties of $\mathbf{M}_n$.

To find a regression invariant and scale equivariant S_n is not straightforward. For example, the square root of the residual sum of squares

$$S_n(\mathbf{Y}) = (\mathbf{Y}^\top(\mathbf{I}_n - \mathbf{H})\mathbf{Y})^{\frac{1}{2}} \tag{5.107}$$

with the projection matrix $\mathbf{H} = \mathbf{X}(\mathbf{X}^\top\mathbf{X})^{-1}\mathbf{X}^\top$ and the identity matrix $\mathbf{I}_n$ satisfies (5.106). However, as nonrobust, it is not used in robust procedures. We shall rather use the scale statistics based on regression quantiles. Another possibility are statistics of the type $\|\mathbf{Y} - \mathbf{X}\widehat{\boldsymbol{\beta}}(\frac{1}{2})\|$ where $\widehat{\boldsymbol{\beta}}(\alpha)$ is α-regression quantile (L_1-estimator of $\boldsymbol{\beta}$ (and $\|\cdot\|$ is an appropriate norm.

We shall deal only with the studentized M-estimators. The nonstudentized M-estimators are covered as a special case (though of course they typically need weaker regularity conditions). Similarly, as in the location model, we shall assume that $\psi = \rho^{\top}$ can be decomposed into the sum

$$\psi = \psi_a + \psi_c + \psi_s, \tag{5.108}$$

where ψ_a is absolutely continuous function with absolutely continuous derivative, ψ_c is a continuous, piecewise linear function that is constant in a neighborhood of $\pm\infty$, and ψ_s is a nondecreasing step function.

More precisely, we shall impose the following conditions on (5.105):

[M1] $S_n(\mathbf{Y})$ is regression invariant and scale equivariant, $S_n > 0$ a.s. and

$$n^{\frac{1}{2}}(S_n - S) = \mathrm{O}_p(1)$$

for some functional $S = S(F) > 0$.

[M2] The function $h(t) = \int \rho\Big(\frac{z-t}{S}\Big)dF(z)$ has the unique minimum at $t = 0$.

[M3] For some $\delta > 0$ and $\eta > 1$,

$$\int_{-\infty}^{\infty} \Big\{|z| \sup_{|u|\leq\delta} \sup_{|v|\leq\delta} |\psi_a''(e^{-v}(z+u)/S)|\Big\}^{\eta} dF(z) < \infty$$

and

$$\int_{-\infty}^{\infty} \Big\{|z|^2 \sup_{|u|\leq\delta} |\psi_a''(z+u)/S)|\Big\}^{\eta} dF(z) < \infty,$$

where $\psi_a'(z) = (d/dz)\psi_a(z)$, and $\psi_a''(z) = (d^2/dz^2)\psi_a(z)$.

[M4] ψ_c is a continuous, piecewise linear function with knots at $\mu_1, \ldots, \mu_k$, which is constant in a neighborhood of $\pm\infty$. hence, the derivative ψ_c' of ψ_c is a step function

$$\psi_c'(z) = \alpha_\nu \;\text{ for }\; \mu_\nu < z < \mu_{\nu+1},\; \nu = 0, 1, \ldots, k,$$

where $\alpha_0, \alpha_1, \ldots, \alpha_k \in \boldsymbol{R}_1$, $\alpha_0 = \alpha_k = 0$ and $-\infty = \mu_0 < \mu_1 < \ldots < \mu_k < \mu_{k+1} = \infty$. We assume that $f(z) = \frac{dF(z)}{dz}$ is positive and bounded in neighborhoods of $S\mu_1, \ldots S\mu_k$.

[M5] $\psi_s(z) = \lambda_\nu$ for $q_\nu < z \leq q_{\nu+1}$, $\nu = 0, \ldots, m$ where $-\infty = q_0 < q_1 < \ldots q_m < q_{m+1} = \infty$, $-\infty < \lambda_0 < \lambda_1 < \ldots < \lambda_m < \infty$.
We assume that $0 < f(z) = \frac{dF(z)}{dz}$ and $f'(z) = \frac{d^2F(z)}{dz^2}$ are bounded in neighborhoods of $Sq_1, \ldots, Sq_m$.

The asymptotic representation for $\mathbf{M}_n$ will involve the functionals

$$\gamma_1 = S^{-1}\int_{-\infty}^{\infty} \Big(\psi_a'\Big(\frac{z}{S}\Big) + \psi_c'\Big(\frac{z}{S}\Big)\Big)dF(z), \tag{5.109}$$

$$\gamma_2 = S^{-1} \int_{-\infty}^{\infty} z\Big(\psi_a'\Big(\frac{z}{S}\Big) + \psi_c'\Big(\frac{z}{S}\Big)\Big) dF(z),$$

$$\gamma_1^\star = \sum_{\nu=1}^{m} (\lambda_\nu - \lambda_{\nu-1}) f(Sq_\nu), \tag{5.110}$$

and

$$\gamma_2^\star = S \sum_{\nu=1}^{m} (\lambda_\nu - \lambda_{\nu-1}) q_\nu f(Sq_\nu).$$

Condition [M3] is essentially a moment condition. It holds, for example, if ψ_a'' is bounded and either

(1) $\psi_a''(z) = 0$ for $z < a$ or $z > b$, $-\infty < a < b < \infty$, or
(2) $\int_{-\infty}^{\infty} |z|^{2+\varepsilon} dF(z) < \infty$ for some $\varepsilon > 0$.

[M1] can be omitted if either S is known or if the considered M-estimator non-studentized; then [M3] can be replaced by

[M3]′

$$\int_{-\infty}^{\infty} \Big\{ \sup_{|u| \le \delta} |\psi_a''((z+a)/S)| \Big\}^\eta dF(z) < \infty$$

for some $\delta > 0$ and $\eta > 1$.

Conditions [M4] and [M5] depict explicitly the trade-off between the smoothness of ψ and smoothness of F. The class of functions ψ_c covers the usual Huber's and Hampel's proposals.

Moreover, the following conditions will be imposed on the matrix $\mathbf{X}_n$:

[X1] $x_{i1} = 1$, $i = 1, \ldots, n$.
[X2] $n^{-1} \sum_{i=1}^{n} \|\mathbf{x}_i\|^4 = O_p(1)$.
[X3] $\lim_{n\to\infty} \mathbf{Q}_n = \mathbf{Q}$, where $\mathbf{Q}_n = n^{-1}\mathbf{X}_n^\top \mathbf{X}_n$ and $\mathbf{Q}$ is a positive definite $p \times p$ matrix.

Let $\mathbf{M}_n$ be a solution of the minimization (5.105). If $\psi = \rho'$ is continuous (i.e., $\psi_s \equiv 0$), then $\mathbf{M}_n$ is a solution of the system of equations

$$\sum_{i=1}^{n} \mathbf{x}_i \psi\Big(\frac{Y_i - \mathbf{x}_i^\top \mathbf{t}}{S_n}\Big) = \mathbf{0}. \tag{5.111}$$

However, this system may have more roots, while only one of them leads to a global minimum of (5.105). We shall prove that there exists at least one root of (5.111) which is a $\sqrt{n}$-consistent estimator of $\boldsymbol{\beta}$.

If ψ is a nondecreasing step function, $\psi_a = \psi_c \equiv 0$, then $\mathbf{M}_n$ is a point of minimum of the convex function $\sum_{i=1}^{n} \rho((Y_i - \mathbf{x}_i^\top t)/S_n)$ of $\mathbf{t} \in \boldsymbol{R}_p$, and its consistency and asymptotic representation may be proved using a different argument. The basic results on studentized M-estimators of regression are summarized in the following three theorems.

THEOREM 5.10 *Consider the model (5.101) and assume the conditions [M1]– [M4], [X1]–[X3], and that γ_1 defined in (5.109) is different from zero. Then, provided $\psi_s \equiv 0$, there exists a root $\mathbf{M}_n$ of the system (5.111) such that*

$$n^{\frac{1}{2}}\|\mathbf{M}_n - \boldsymbol{\beta}\| = O_p(1) \quad \text{as } n \to \infty. \tag{5.112}$$

Moreover, any root $\mathbf{M}_n$ of (5.111) satisfying (5.112) admits the representation

$$\mathbf{M}_n - \boldsymbol{\beta} = (n\gamma_1)^{-1}\mathbf{Q}_n^{-1}\sum_{i=1}^{n}\mathbf{x}_i\psi\Big(\frac{E_i}{S}\Big) - \frac{\gamma_2}{\gamma_1}\Big(\frac{S_n}{S} - 1\Big)\mathbf{e}_1 + \mathbf{R}_n, \tag{5.113}$$

where $\|\mathbf{R}_n\| = O_p(n^{-1})$ and $\mathbf{e}_1 = (1, 0, \ldots, 0)^\top \in \boldsymbol{R}_p$.

THEOREM 5.11 *Consider the linear model (5.101) and assume the conditions [M1], [M2], [M5], and [X1]–[X3]. Let $\mathbf{M}_n$ be the point of global minimum of (5.105). Then, provided that $\psi_a = \psi_c \equiv 0$,*

$$n^{\frac{1}{2}}\|\mathbf{M}_n - \boldsymbol{\beta}\| = O_p(1) \quad \text{as } n \to \infty,$$

and $\mathbf{M}_n$ admits the representation

$$\mathbf{M}_n - \boldsymbol{\beta} = (n\gamma_1^\star)^{-1}\mathbf{Q}_n^{-1}\sum_{i=1}^{n}\mathbf{x}_i\psi\Big(\frac{E_i}{S}\Big) - \frac{\gamma_2^\star}{\gamma_1^\star}\Big(\frac{S_n}{S} - 1\Big)\mathbf{e}_1 + \mathbf{R}_n, \tag{5.114}$$

where $\|\mathbf{R}_n\| = O_p(n^{-3/4})$ and $\mathbf{e}_1 = (1, 0, \ldots, 0)^\top \in \boldsymbol{R}_p$.

Remark 5.2 *Notice that only the first (intercept) components of the second terms in representations (5.113) and (5.114) are different from zero; the slope components of $\mathbf{M}_n$ are not affected by S_n.*

Combining the above results, we immediately obtain the following theorem for the general class of M-estimators:

THEOREM 5.12 *Consider the model (5.101) and assume the conditions [M1]– [M4] and [X1]–[X3]. Let ψ be either continuous or monotone, and let $\gamma_1 + \gamma_1^\star \neq 0$. Then, for any M-estimator $\mathbf{M}_n$ satisfying $\|n^{1/2}(\mathbf{M}_n - \boldsymbol{\beta})\| = O_p(1)$,*

$$\mathbf{M}_n - \boldsymbol{\beta} = \frac{1}{n(\gamma_1 + \gamma_1^\star)}\mathbf{Q}_n^{-1}\sum_{i=1}^{n}\mathbf{x}_i\psi\Big(\frac{E_i}{S}\Big) - \frac{\gamma_2 + \gamma_2^\star}{\gamma_1 + \gamma_1^\star}\Big(\frac{S_n}{S} - 1\Big)\mathbf{e}_1 + \mathbf{R}_n$$

where

$$\|\mathbf{R}_n\| = \begin{cases} O_p(n^{-1}), & \text{if } \psi_s \equiv 0, \\ O_p(n^{-3/4}), & \text{otherwise.} \end{cases}$$

Theorems 5.10–5.12 have several interesting corollaries, parallel to those in the location model.

Corollary 5.10 *Under the conditions of Theorem 5.10, let* $\mathbf{M}_n^{(1)}$ *and* $\mathbf{M}_n^{(2)}$ *be any pair of roots of the system of equations (5.111), both satisfying (5.112). Then*

$$\|\mathbf{M}_n^{(1)} - \mathbf{M}_n^{(2)}\| = O_p(n^{-1}).$$

Corollary 5.11 *Assume that*

$$\sigma^2 = \int_{-\infty}^{\infty} \psi^2\Big(\frac{z}{S}\Big)dF(z) < \infty. \tag{5.115}$$

Then, under the conditions of Theorems 5.10–5.12, respectively, the sequence

$$n^{\frac{1}{2}}\Big\{\tilde{\gamma}_1(\mathbf{M}_n - \boldsymbol{\beta}) + \tilde{\gamma}_2\Big(\frac{S_n}{S} - 1\Big)\mathbf{e}_1\Big\}$$

has the asymptotic p-dimensional normal distribution $\mathcal{N}_p(\mathbf{0}, \sigma^2\mathbf{Q}^{-1})$*; here* $\tilde{\gamma}_i$ *stands for* $\gamma_i, \gamma_i^\star$ *or* $\gamma_i + \gamma_i^\star$*, respectively,* $i = 1, 2$.

Corollary 5.12 *Let* $F(z) + F(-z) = 1,\ \rho(-z) = \rho(z), z \in \boldsymbol{R}_1$*. Then, under the conditions of either of Theorems 5.3–5.5, respectively,*

$$\mathbf{M}_n - \boldsymbol{\beta} = (n\tilde{\gamma}_1)^{-1}\mathbf{Q}_n^{-1}\sum_{i=1}^{n}\mathbf{x}_i\psi\Big(E_iS\Big) + \mathbf{R}_n$$

where $\|\mathbf{R}_n\| = O_p(n^{-1})$ *provided that* $\psi_s \equiv 0$ *and* $\|\mathbf{R}_n\| = O_p(n^{-3/4})$ *otherwise. Moreover, if* $\sigma^2 < \infty$ *for* σ^2 *in (5.115),* $n^{1/2}(\mathbf{M}_n - \boldsymbol{\beta})$ *is asymptotically normally distributed*

$$\mathcal{N}_p\Big(\mathbf{0}, \frac{\sigma^2}{\tilde{\gamma}_1^2}\,\mathbf{Q}^{-1}\Big).$$

In some situations is convenient to consider the *restricted M-estimator* $\mathbf{M}_n$ of $\boldsymbol{\beta}$ defined as a solution of the minimization (5.105) under some constraint, e.g. under an hypothesis H_0 of interest. Let us illustrate the restricted M-estimator under the linear constraint:

$$\mathbf{A}\boldsymbol{\beta} = \mathbf{c}, \tag{5.116}$$

where $\mathbf{A}$ is a $q \times p$ matrix of the full rank and $c \in \mathbf{R}_1$. Similarly as in the unrestricted case, we get the following representation for restricted M-estimator:

Corollary 5.13 *Let* $\mathbf{M}_n$ *be the restricted M-estimator of* $\boldsymbol{\beta}$ *under the constraint (5.116). Then, under the conditions of Theorem 5.12,* $\mathbf{M}_n$ *admits the asymptotic representation*

$$\begin{aligned} &\mathbf{M}_n - \boldsymbol{\beta} \qquad (5.117)\\ &= \frac{1}{\tilde{\gamma}_1}\Big[\mathbf{I}_p - \mathbf{Q}^{-1}\mathbf{A}^\top\left(\mathbf{A}\mathbf{Q}^{-1}\mathbf{A}^\top\right)^{-1}\mathbf{A}\Big]\cdot\Big[\mathbf{Q}^{-1}\zeta_n - \tilde{\gamma}_2\mathbf{e}_1\Big(\frac{S_n}{S} - 1\Big)\Big] + \mathbf{R}_n \end{aligned}$$

where $\zeta_n = \frac{1}{n}\sum_{i=1}^{n}\mathbf{x}_i\psi\Big(\frac{E_i}{S}\Big)$*, and* $\tilde{\gamma}_\nu = \gamma_\nu + \gamma_\nu^*$*,* $\nu = 1, 2$.

To prove Theorems 5.10–5.12, we need the following asymptotic linearity lemma:

LEMMA 5.3 *Consider the model (5.101) with $\mathbf{X}_n$ satisfying [X1]–[X3]. Let $\psi : \boldsymbol{R}_1 \mapsto \boldsymbol{R}_1$ be a function of the form $\psi = \psi_a + \psi_c + \psi_s$.*
(i) If $\psi_s \equiv 0$, then, under the conditions [M3]–[M4],

$$\sup_{\|\mathbf{t}\| \le C, |u| \le C} \left\| n^{-\frac{1}{2}} \sum_{i=1}^{n} \mathbf{x}_i \left\{ \psi\left(\frac{E_i - n^{-\frac{1}{2}} \mathbf{x}_i^\top \mathbf{t}}{S e^{n^{-\frac{1}{2}} u}} \right) - \psi\left(\frac{E_i}{S} \right) + n^{-\frac{1}{2}} (\gamma_1 \mathbf{x}_i^\top \mathbf{t} + \gamma_2 u) \right\} \right\|$$

$$= O_p(n^{-\frac{1}{2}}), \tag{5.118}$$

for any fixed $C > 0$ as $n \to \infty$.

(ii) If $\psi_a = \psi_c \equiv 0$, then under the condition **[M5]**,

$$\sup_{\|\mathbf{t}\| \le C, |u| \le C} \left\| n^{-\frac{1}{2}} \sum_{i=1}^{n} \mathbf{x}_i \left\{ \psi\left(\frac{E_i - n^{-\frac{1}{2}} \mathbf{x}_i^\top \mathbf{t}}{S e^{n^{-\frac{1}{2}} u}} \right) - \psi\left(\frac{E_i}{S} \right) + n^{-\frac{1}{2}} (\gamma_1^\star \mathbf{x}_i^\top \mathbf{t} + \gamma_2^\star u) \right\} \right\|$$

$$= O_p(n^{-\frac{1}{4}}) \tag{5.119}$$

for any fixed $C > 0$ as $n \to \infty$.

The proof of Lemma 5.3 is relegated to the Appendix (see Section A.2).

PROOF of Theorem 5.10. Notice that $\gamma_1 \ge 0$ by condition [M2]; hence $\gamma_1 > 0$ under the conditions of the theorem. Put

$$\mathbf{E}_n(\mathbf{t}, s) = n^{-\frac{1}{2}} \sum_{i=1}^{n} \mathbf{x}_i \psi\left(\frac{E_i - n^{-\frac{1}{2}} \mathbf{x}_i^\top \mathbf{t}}{s} \right), \quad \mathbf{t} \in \boldsymbol{R}_p, \; s > 0.$$

Inserting $u \to n^{1/2} \ln\left(\frac{S_n}{S}\right)$ $(= \mathrm{O}_p(1))$ in (5.118), we obtain (in view of [M1])

$$\sup_{\|t\| \le C} \left\| \mathbf{E}_n(\mathbf{t}, S_n) - \mathbf{E}_n(\mathbf{0}, S) + \gamma_1 \mathbf{Q} \mathbf{t} + \gamma_2 \mathbf{q}_1 n^{\frac{1}{2}} \left(\frac{S_n}{S} - 1 \right) \right\| = \mathrm{O}_p(n^{-\frac{1}{2}}), \tag{5.120}$$

where $\mathbf{q}_1$ is the first column of the matrix $\mathbf{Q}$ from the condition **[X3]**. Notice that

$$\mathbf{E}_n(n^{\frac{1}{2}}(\mathbf{M}_n - \boldsymbol{\beta}), S_n) = n^{-\frac{1}{2}} \sum_{i=1}^{n} \mathbf{x}_i \psi\left(\frac{Y_i - \mathbf{x}_i^\top \mathbf{M}_n}{S_n} \right)$$

and thus, $n^{1/2}(\mathbf{M}_n - \boldsymbol{\beta})$ is a root of $\mathbf{E}_n(\mathbf{t}, S_n) = \mathbf{0}$ if and only if $\mathbf{M}_n$ is a root of (5.111). The $\sqrt{n}$-consistency of $\mathbf{M}_n$ will follow from Proposition 6.3.4 in Ortega and Rheinboldt (1973, p. 163) if we can show that

$$\mathbf{t}^\top \mathbf{E}_n(\mathbf{t}, S_n) < 0$$

for all $\mathbf{t}$, $\|\mathbf{t}\| = C$ in probability as $n \to \infty$ for sufficiently large $C > 0$. To prove this, notice that (5.120) enables to write that, given $\eta > 0$ and $\varepsilon > 0$,

$$\begin{aligned} &\mathbb{P}\Big\{\sup_{\|\mathbf{t}\|=C} \mathbf{t}^\top \mathbf{E}_n(\mathbf{t}, S_n) \geq 0\Big\} \\ &\quad \leq \mathbb{P}\Big\{\sup_{\|\mathbf{t}\|=C} \mathbf{t}^\top \mathbf{E}_n(\mathbf{t}, S_n) \geq 0,\ \sup_{\|\mathbf{t}\|\leq C} \Big\|\mathbf{E}_n(\mathbf{t}, S_n) - \mathbf{E}_n(\mathbf{0}, S) \\ &\quad + \gamma_1 \mathbf{Q}\mathbf{t} + \gamma_2 \mathbf{q}_1 n^{\frac{1}{2}}\Big(\frac{S_n}{S} - 1\Big)\Big\| < \eta\Big\} + \frac{\varepsilon}{2} \end{aligned}$$

for $n \geq n_0$ and for any fixed $C > 0$. Being the first column hence,

$$\begin{aligned} &\mathbb{P}\Big\{\sup_{\|\mathbf{t}\|=C} \mathbf{t}^\top \mathbf{E}_n(\mathbf{t}, S_n) \geq 0\Big\} \\ &\leq \mathbb{P}\Big\{\sup_{\|\mathbf{t}\|=C} \Big[\mathbf{t}^\top \mathbf{E}_n(\mathbf{0}, S) - \gamma_1 \mathbf{t}^\top \mathbf{Q}\mathbf{t} - \gamma_2 \mathbf{t}^\top \mathbf{q}_1 n^{\frac{1}{2}}\Big(\frac{S_n}{S} - 1\Big)\Big] \geq -C\eta\Big\} + \frac{\varepsilon}{2} \\ &\leq \mathbb{P}\Big\{C\|\mathbf{E}_n(\mathbf{0}, S_n)\| + |\gamma_2| C \|\mathbf{q}_1\| n^{\frac{1}{2}}\Big|\frac{S_n}{S} - 1\Big| \geq \gamma_1 C^2 \lambda_{\min} - C\eta\Big\} + \frac{\varepsilon}{2} \leq \varepsilon \end{aligned}$$

for $n \geq n_1$, and for sufficiently large $C > 0$ because both $\|\mathbf{E}_n(\mathbf{0}, S)\|$ and $n^{1/2}|S_n/S - 1|$ are $O_p(1)$ as $n \to \infty$; $\lambda_{\min}$ is the minimal eigenvalue of $\mathbf{Q}$. Hence, given $\varepsilon > 0$, there exist $C > 0$ and n_1 such that

$$\mathbb{P}\Big\{\sup_{\|\mathbf{t}\|=C} \mathbf{t}^\top \mathbf{E}_n(\mathbf{t}, S_n) < 0\Big\} > 1 - \varepsilon \quad \text{for } n \geq n_1,$$

and this in turn implies that, with probability $> 1 - \varepsilon$ and for $n \geq n_1$, there exists a root of the system $\mathbf{E}_n(\mathbf{t}, S_n) = \mathbf{0}$ inside the ball $\|\mathbf{t}\| \leq C$. This completes the proof of (5.112).

Inserting $n^{1/2}(\mathbf{M}_n - \boldsymbol{\beta})$ for $\mathbf{t}$ in (5.120), we obtain

$$\begin{aligned} &\Big\|\gamma_1 \mathbf{Q}_n n^{\frac{1}{2}}(\mathbf{M}_n - \boldsymbol{\beta}) + \gamma_2 \mathbf{q}_1^{(n)} n^{\frac{1}{2}}\Big(\frac{S_n}{S} - 1\Big) - n^{-\frac{1}{2}} \sum_{i=1}^n \mathbf{x}_i \psi(E_i/S)\Big\| \\ &= O_p(n^{-\frac{1}{4}}), \end{aligned}$$

$\mathbf{q}_1^{(n)}$ being the first column of $\mathbf{Q}_n$. This gives the representation (5.113). □

PROOF of Theorem 5.11. Inserting $n^{1/2}\ln\Big(\frac{S_n}{S} - 1\Big)$ for u in (5.119), we obtain

$$\begin{aligned} &\sup_{\|\mathbf{t}\|\leq C} \Big\|n^{-\frac{1}{2}} \sum_{i=1}^n \mathbf{x}_i \Big\{\psi\Big(\frac{E_i - n^{-\frac{1}{2}}\mathbf{x}_i^\top \mathbf{t}}{S_n}\Big) - \psi\Big(\frac{E_i}{S}\Big)\Big\} + \gamma_1^\star \mathbf{Q}\mathbf{t} + \gamma_2^\star \mathbf{q}_1 n^{\frac{1}{2}}\Big(\frac{S_n}{S} - 1\Big)\Big\| \\ &\quad = O_p(n^{-\frac{1}{4}}). \end{aligned}$$

This in turn implies, through the integration over $\mathbf{t}$ and using [M1], that

$$\begin{aligned} &\sup_{\|\mathbf{t}\|\leq C} \Big|\sum_{i=1}^n \Big[\rho\Big(\frac{E_i - n^{-\frac{1}{2}}\mathbf{x}_i^\top \mathbf{t}}{S_n}\Big) - \rho\Big(\frac{E_i}{S}\Big)\Big] + \mathbf{t}^\top \mathbf{Z}_n - a\mathbf{t}^\top \mathbf{Q}\mathbf{t}\Big| \\ &\qquad = o_p(1) \quad \text{as } n \to \infty, \end{aligned} \tag{5.121}$$

where $a = \frac{\gamma_1^\star}{2S} > 0$ by (5.110) and $[M5]$ and where

$$\mathbf{Z}_n = \frac{1}{S}\Big\{n^{-\frac{1}{2}}\sum_{i=1}^{n}\mathbf{x}_i\psi\Big(\frac{E_i}{S}\Big) - \gamma_2^\star\mathbf{q}_1 n^{\frac{1}{2}}\Big(\frac{S_n}{S}-1\Big)\Big\};$$

obviously $\|\mathbf{Z}_n\| = O_p(1)$. If $\mathbf{M}_n$ minimizes (5.105), then $n^{1/2}(\mathbf{M}_n - \boldsymbol{\beta})$ minimizes the convex function

$$G_n(\mathbf{t}) = \sup_{\|\mathbf{t}\|\le C}\Big|\sum_{i=1}^{n}\Big[\rho\Big(\frac{E_i - n^{-\frac{1}{2}}\mathbf{x}_i^\top\mathbf{t}}{S_n}\Big) - \rho\Big(\frac{E_i}{S}\Big)\Big]\Big|$$

with respect to $\mathbf{t}\in\boldsymbol{R}_p$. By (5.121),

$$\min_{\|\mathbf{t}\|\le C} G_n(\mathbf{t}) = \min_{\|\mathbf{t}\|\le C}\Big\{-\mathbf{t}^\top\mathbf{Z}_n + a\mathbf{t}^\top\mathbf{Q}\mathbf{t}\Big\} + o_p(1)$$

for any fixed $C > 0$. Denoting $\mathbf{U}_n = \arg\min_{\mathbf{t}\in\boldsymbol{R}_p}\Big\{-\mathbf{t}^\top\mathbf{Z}_n + a\mathbf{t}^\top\mathbf{Q}\mathbf{t}\Big\}$, we get $\mathbf{U}_n = \frac{a}{2}\mathbf{Q}^{-1}\mathbf{Z}_n$ $(= O_p(1))$ and

$$\min_{\mathbf{t}\in\boldsymbol{R}_p}\Big\{-\mathbf{t}^\top\mathbf{Z}_n + a\mathbf{t}^\top\mathbf{Q}\mathbf{t}\Big\} = \frac{1}{4a}\mathbf{Z}_n^\top\mathbf{Q}\mathbf{Z}_n.$$

hence, we can write

$$-\mathbf{t}^\top\mathbf{Z}_n + a\mathbf{t}^\top\mathbf{Q}\mathbf{t} = a\Big\{(\mathbf{t}-\mathbf{U}_n)^\top\mathbf{Q}(\mathbf{t}-\mathbf{U}_n) - \mathbf{U}_n^\top\mathbf{Q}\mathbf{U}_n)\Big\}$$

and then rewrite (5.121) in the form

$$\sup_{\|\mathbf{t}\|\le C}\Big|G_n(\mathbf{t}) - a\Big[(\mathbf{t}-\mathbf{U}_n)^\top\mathbf{Q}(\mathbf{t}-\mathbf{U}_n) - \mathbf{U}_n\mathbf{Q}\mathbf{U}_n)\Big]\Big| \xrightarrow{P} \text{ as } n\to\infty. \quad (5.122)$$

Substituting $\mathbf{U}_n$ for $\mathbf{t}$, we further obtain

$$G_n(\mathbf{U}_n) + a\mathbf{U}_n^\top\mathbf{Q}\mathbf{U}_n = o_p(1). \quad (5.123)$$

We want to show that

$$\|n^{\frac{1}{2}}(\mathbf{M}_n - \boldsymbol{\beta}) - \mathbf{U}_n\| = o_p(1). \quad (5.124)$$

Consider the ball $\mathcal{B}_n$ with center $\mathbf{U}_n$ and radius $\delta > 0$. This ball lies in a compact set with probability exceeding $(1-\varepsilon)$ for $n \ge n_0$ because $\|\mathbf{t}\| \le \|\mathbf{t}-\mathbf{U}_n\| + \|\mathbf{U}_n\| \le \delta + C$ for $\mathbf{t}\in\mathcal{B}_n$ and for some $C > 0$ (with probability exceeding $1-\varepsilon$ for $n \ge n_0$). Hence, by (5.121),

$$\Delta_n = \sup_{\mathbf{t}\in\mathcal{B}_n}\Big|\sum_{i=1}^{n}\Big[\rho\Big(\frac{E_i - n^{-\frac{1}{2}}\mathbf{x}_i^\top\mathbf{t}}{S_n}\Big) - \rho\Big(E_i S\Big)\Big] + \mathbf{t}^\top\mathbf{Z}_n - a\mathbf{t}^\top\mathbf{Q}\mathbf{t}\Big| = o_p(1).$$

Let $\mathbf{t}\notin\mathcal{B}_n$; then $\mathbf{t} = \mathbf{U}_n + k\mathbf{v}$ where $k > \delta$ and $\|\mathbf{v}\| = 1$. Let $\mathbf{t}^\star$ be the boundary point of $\mathcal{B}_n$ that lies on the line from $\mathbf{U}_n$ to $\mathbf{t}$ (i.e., $\mathbf{t}^\star = \mathbf{U}_n + \delta\mathbf{v}$). Then $\mathbf{t}^\star = (1-(\delta/k))\mathbf{U}_n + (\delta/k)\mathbf{t}$, and, by (5.122)–(5.123),

$$a\delta^2\lambda_{\min} + G_n(\mathbf{U}_n) - 2\Delta_n \le G_n(\mathbf{t}^\star) \le \frac{\delta}{k}G_n(\mathbf{t}) + (1-\frac{\delta}{k})G_n(\mathbf{U}_n),$$

where $\lambda_{\min}$ is the minimal eigenvalue of $\mathbf{Q}$. hence,

$$\inf_{\|\boldsymbol{t}-\boldsymbol{U}_n\|\geq\delta} G_n(\mathbf{t}) \geq G_n(\mathbf{U}_n) + \frac{k}{\delta}(a\delta^2\lambda_{\min} - 2\Delta_n)$$

and this implies that, given $\delta > 0$ and $\varepsilon > 0$, there exist n_0 and $\eta > 0$ such that for $n \geq n_0$

$$I\!P\Big\{ \inf_{\|\boldsymbol{t}-\boldsymbol{U}_n\|\geq\delta} G_n(\mathbf{t}) - G_n(\mathbf{U}_n) > \eta \Big\} > 1 - \varepsilon.$$

thus,

$$I\!P(\|n^{\frac{1}{2}}(\mathbf{M}_n - \boldsymbol{\beta}) - \mathbf{U}_n)\| \leq \delta) \to 1$$

for any fixed $\delta > 0$, as $n \to \infty$, and this is (5.124). We obtain the asymptotic representation (5.114) after inserting $t \to n^{1/2}(\mathbf{M}_n - \boldsymbol{\beta})$ and $u \to n^{1/2}\ln\left(\frac{S_n}{S}\right)$ into (5.119). □

PROOF of Theorem 5.12. If ψ is continuous, then $\psi_s \equiv 0$, and the proposition follows from Theorem 5.10.

Hence, let $\psi_s \not\equiv 0$ but let ψ be nondecreasing. First, combining (i) and (ii) of Lemma 5.1, we obtain

$$\sup_{\|\mathbf{t}\|\leq C, |u|\leq C} \Big\| n^{-\frac{1}{2}} \sum_{i=1}^{n} \mathbf{x}_i \Big\{ \psi\Big(\frac{E_i - n^{-\frac{1}{2}}\mathbf{x}_i^\top \mathbf{t}}{Se^{n^{-\frac{1}{2}}u}}\Big) - \psi\Big(\frac{E_i}{S}\Big) + n^{-\frac{1}{2}}(\tilde{\gamma}_1 \mathbf{x}_i^\top \mathbf{t} + \tilde{\gamma}_2 u) \Big\} \Big\|$$

$$= O_p(n^{-\frac{1}{4}})$$

for any fixed $C > 0$ where $\tilde{\gamma}_i = \gamma_i + \gamma_i^\star$, $i = 1, 2$.

Let $\mathbf{M}_n$ be the global solution of the minimization of $\sum_{i=1}^n \rho\left(\frac{Y_i - \mathbf{x}_i^\top \mathbf{t}}{S_n}\right)$, what is a function convex in $\mathbf{t}$ (because ψ is nondecreasing). Then, following step by step the proof of Theorem 5.4, we arrive at $n^{1/2}\|\mathbf{M}_n - \boldsymbol{\beta}\| = O_p(1)$. Inserting $\mathbf{t} \to n^{1/2}(\mathbf{M}_n - \boldsymbol{\beta})$ and $u \to n^{1/2}\ln\left(\frac{S_n}{S}\right)$ in (5.52), we receive the representation of $n^{1/2}(\mathbf{M}_n - \boldsymbol{\beta})$. □

5.6 Some Studentizing Scale Statistics for Linear Models

In Section 5.4 we considered some studentizing scale statistics for the simple location model, that were translation invariant and scale equivariant [see (5.62)]. In the linear models the concept of translation invariance extends to that of the affine invariance. This can be stated as follows: Consider the linear model $\mathbf{Y} = \mathbf{X}\boldsymbol{\beta} + \mathbf{E}$, as introduced in (5.101) and the transformation

$$\mathbf{Y}^\star(\mathbf{b}) = \mathbf{Y} + \mathbf{X}\mathbf{b}, \ \mathbf{b} \in \boldsymbol{R}_p.$$

Let $S_n = S_n(\mathbf{Y})$ be a scale statistic such that

$$S_n(\mathbf{Y} + \mathbf{X}\mathbf{b}) = S_n(\mathbf{Y}) \ \forall \mathbf{b} \in \boldsymbol{R}_p, \mathbf{Y} \in \boldsymbol{R}_n. \tag{5.125}$$

Then S_n is said to be affine invariant. On the other hand, the scale equivariance means that $S_n(c\mathbf{Y}) = cS_n(\mathbf{Y}) \ \forall c \in \boldsymbol{R}^+$, $\mathbf{Y} \in \boldsymbol{R}_n$, analogously as in (5.62). Combining this with (5.125), we say that S_n is affine invariant and scale equivariant if

$$S_n(c(\mathbf{Y} + \mathbf{X}\mathbf{b})) = cS_n(\mathbf{Y}) \ \forall \mathbf{b} \in \boldsymbol{R}_p, \ c > 0, \ \mathbf{Y} \in \boldsymbol{R}_n. \tag{5.126}$$

If we consider the usual least square estimator $\widehat{\boldsymbol{\beta}}$ of $\boldsymbol{\beta}$ (see (5.101)), then the square root of the residual sum of squares

$$\|\mathbf{Y} - \mathbf{X}\widehat{\boldsymbol{\beta}}\| = [(\mathbf{Y} - \mathbf{X}\widehat{\boldsymbol{\beta}})^\top (\mathbf{Y} - \mathbf{X}\widehat{\boldsymbol{\beta}})]^{\frac{1}{2}},$$

defined in (5.107), possesses the affine equivariance and scale invariance properties in (5.126). But this is a highly nonrobust statistic, while in the robust inference there is a genuine need to explore alternative scale statistics that are basically robust and satisfy (5.126). Some of such scale statistics in linear models we present in this section.

1. *Median absolute deviation from the median* (MAD). An extension of MAD to model (5.101) was proposed by Welsh (1986): Let $\widehat{\boldsymbol{\beta}}^0$ be an initial $\sqrt{n}$-consistent and affine and scale equivariant estimator of $\boldsymbol{\beta}$, and denote by

$$Y_i(\widehat{\boldsymbol{\beta}}^0) = Y_i - \mathbf{x}_i^\top \widehat{\boldsymbol{\beta}}^0, \ \ i = 1, \ldots, n,$$

$$\xi_{\frac{1}{2}}(\widehat{\boldsymbol{\beta}}^0) = \text{med}_{1 \le i \le n} Y_i(\widehat{\boldsymbol{\beta}}^0),$$

and

$$S_n = \text{med}_{1 \le i \le n} |Y_i(\widehat{\boldsymbol{\beta}}^0) - \xi_{\frac{1}{2}}(\widehat{\boldsymbol{\beta}}^0)|.$$

Then S_n satisfies (5.125) and (5.126) provided that $\widehat{\boldsymbol{\beta}}^0$ is affine and scale equivariant. Moreover, under some regularity conditions on F, S_n is $\sqrt{n}$-consistent estimator of the population median deviation. Welsh (1986) also derived its asymptotic representation of the type (5.72), hence, Theorem 5.11 applies.

2. *L-statistics based on regression quantiles.* The α-regression quantile $\widehat{\boldsymbol{\beta}}(\alpha)$ for model (5.101) was introduced by Koenker and Bassett (1978) as a solution of the minimization problem

$$\sum_{i=1}^{n} \rho_\alpha(Y_i - \mathbf{x}_i \mathbf{t}) := \min \quad \text{with respect to } \mathbf{t} \in \boldsymbol{R}_p,$$

where

$$\rho_\alpha(z) = |z|\{\alpha I(z > 0] + (1 - \alpha) I[z < 0]\}, \ \ z \in \boldsymbol{R}_1.$$

The Euclidean distance of two regression quantiles

$$S_n = \|\widehat{\boldsymbol{\beta}}_n(\alpha_2) - \widehat{\boldsymbol{\beta}}_n(\alpha_1)\|, \ \ 0 < \alpha_1 < \alpha_2 < 1,$$

satisfies (5.125) and (5.126) and $S_n \xrightarrow{P} S(F) = F^{-1}(\alpha_2) - F^{-1}(\alpha_1)$. Its asymptotic representation follows, for example, from that for $\widehat{\boldsymbol{\beta}}_n(\alpha)$ derived by Ruppert and Carroll (1980). The Euclidean norm may be replaced by L_p-norm or by another appropriate norm. An alternative statistic is the deviation of the first components of regression quantiles, $S_n = \widehat{\beta}_{n1}(\alpha_2) - \widehat{\beta}_{n2}(\alpha_1)$, $0 < \alpha_1 < \alpha_2 < 1$, with the same population counterpart as above. More generally, Bickel and Lehmann (1979) proposed various measures of spread of the distribution F, that can also serve as the scale functional $S(F)$. The corresponding scale statistic is then an estimator of $S(F)$ based on regression quantiles. As an example, we can take

$$S(F) = \left\{ \int_{\frac{1}{2}}^{1} [F^{-1}(u) - F^{-1}(1-u)]^2 d\Lambda(u) \right\}^{\frac{1}{2}}$$

where Λ is the uniform distribution on $(1/2, 1-\delta)$, $0 < \delta < 1/2$; then

$$S(F) = \left\{ \int_{\frac{1}{2}}^{1} \|\widehat{\boldsymbol{\beta}}(u) - \widehat{\boldsymbol{\beta}}(1-u)\|^2 \, d\Lambda(u) \right\}^{\frac{1}{2}}.$$

3. Falk (1986) proposed a histogram and kernel type estimators of the value $1/f(F^{-1}(\alpha))$, $0 < \alpha < 1$, in the location model, $f(x) = \frac{dF(x)}{dx}$. Dodge and Jurečková (1993) extended Falk's estimators to the linear model in the following way: First, a histogram type $(n\nu_n)^{\frac{1}{2}}$-consistent estimator of $1/f(F^{-1}(\alpha))$ satisfying (5.125) and (5.126) has the form

$$H_n^{(\alpha)} = \frac{\widehat{\boldsymbol{\beta}}_{n1}(\alpha + \nu_n) - \widehat{\boldsymbol{\beta}}_{n1}(\alpha - \nu_n)}{2\nu_n} \tag{5.127}$$

where

$$\nu_n = o(n^{-\frac{1}{3}}) \text{ and } n\nu_n \to \infty \text{ as } n \to \infty.$$

Second, the kernel type estimator of $1/f(F^{-1}(\alpha))$ has the form

$$\chi_n^{(\alpha)} = \nu_n^{-2} \int_0^1 \widehat{\beta}_{n1}(u) k\Big(\frac{\alpha - u}{\nu_n}\Big) du,$$

where

$$\nu_n \to 0, \ n\nu_n^2 \to \infty \ \text{ and } \ n\nu_n^3 \to 0 \ \text{ as } n \to \infty$$

and where $k : \boldsymbol{R}_1 \mapsto \boldsymbol{R}_1$ is the kernel function with a compact support, which is continuous on its support and satisfies

$$\int k(x)dx = 0 \text{ and } \int xk(x)dx = -1.$$

Again, $\chi_n^{(\alpha)}$ is a $(n\nu_n)^{1/2}$-consistent estimator of $1/f(F^{-1}(\alpha))$, whose asymptotic variance may be less than that of (5.127) for some kernels. Due to their lower rates of consistency, we shall not use the above estimators for a simple studentization but rather in an inference on the population quantiles based on the regression data.

4. Jurečková and Sen (1994) constructed a scale statistics based on *regression rank scores*, which are dual to the regression quantiles in the linear programming sense and represent an extension of the rank scores to the linear model (see Gutenbrunner and Jurečková [1992] for a detailed account of this concept). More precisely, regression rank scores $\widehat{\mathbf{a}}_n(\alpha) = (\hat{a}_{n1}(\alpha), \ldots, \hat{a}_{nn}(\alpha))^\top$, $0 < \alpha < 1$, for the model (5.101) are defined as a solution of the maximization
$$\mathbf{Y}^\top \widehat{\mathbf{a}}_n(\alpha) := \max$$
under the restriction
$$\mathbf{X}^\top(\widehat{\mathbf{a}}_n(\alpha) - \mathbf{1}_n^\top(1-\alpha)) = \mathbf{0}, \;\; \widehat{\mathbf{a}}_n(\alpha) \in [0,1]^n, \;\; 0 < \alpha < 1.$$
In the location model, they reduce to the rank scores considered in Hájek and Šidák (1967) and Hájek et al. (1999). The proposed scale statistic is of the form
$$S_n = n^{-1} \sum_{i=1}^{n} Y_i \hat{b}_{ni},$$
where
$$\hat{b}_{ni} = -\int_{-\alpha_0}^{1-\alpha_0} \phi(\alpha) d\widehat{\mathbf{a}}_{ni}(\alpha), \;\; 0 < \alpha_0 < 1/2,$$
and the score-generating function $\phi : [0,1] \mapsto \boldsymbol{R}_1$ is nondecreasing, square integrable and skew-symmetric, standardized so that $\int_{\alpha_0}^{1-\alpha_0} \phi^2(\alpha) d\alpha = 1$. Then S_n satisfies (5.126) and is a $\sqrt{n}$-consistent estimator of $S(F) = \int_{\alpha_0}^{1-\alpha_0} \phi(\alpha) F^{-1}(\alpha) d\alpha$. In the location model, S_n reduces to the Jaeckel (1972) measure of dispersion of ranks of residuals. Some of these details are provided in the last section of Chapter 6, which deals with regression rank scores in depth.
5. Víšek (2010) constructed a scale-equivariant and regression-invariant scale statistic based on (down)weighting order statistics of the squared residuals from a suitable robust estimate.

5.7 Hadamard Differentiability Approaches for M-Estimators in Linear Models

In this section we briefly describe an alternative approach based on Hadamard (or compact) differentiability, mostly generalizations of results in Sections 4.5 dealing with i.i.d. observations. This approach has been studied at length by Ren and Sen (1991, 1994, 1995a,b), among others.

The M-estimation procedure based on the norm $\rho(.) : \;\; \boldsymbol{R} \mapsto \boldsymbol{R}$, given in (5.102), can be alternatively formulated in terms of the estimating equation $\sum_{i=1}^{n} \mathbf{x}_i \psi(\mathbf{Y}_i - \mathbf{x}_i^\top \boldsymbol{\beta}) = \mathbf{0}$, whenever $\rho(.)$ has a derivative ψ (a.e.); we have designated ψ as the score function. The basic statistic (empirical process) considered in the preceding section is

$$\mathbf{M}_n(\mathbf{u}) = \sum_{i=1}^{n} \mathbf{x}_i \psi(e_i - \mathbf{x}_i^\top \mathbf{u}), \quad \mathbf{u} \in \boldsymbol{R}_p. \tag{5.128}$$

The approach of Section 4.5 clearly extends to the M-estimators; however, the M-estimators are defined implicitly, and hence their asymptotics should be treated more carefully. Notice that the Y_i are not identically distributed, though the errors e_i are i.i.d. random variables even in the linear model. We shall use the following notation:

$$\mathbf{C}_n = \mathbf{X}^\top \mathbf{X} = \sum_{i=1}^{n} \mathbf{x}_i \mathbf{x}_i^\top = ((r_{nij}))_{i,j=1,\ldots,p}$$

$$\mathbf{C}_n^0 = \mathrm{Diag}(\sqrt{r_{n11}}, \ldots, \sqrt{r_{npp}})^\top, \quad 1 \le i \le n,$$

$$\mathbf{c}_{ni} = (\mathbf{C}_n^0)^{-1} \mathbf{x}_i = (c_{ni1}, \ldots, c_{nip})^\top, \quad 1 \le i \le n,$$

and

$$\mathbf{Q}_n = \sum_{i=1}^{n} \mathbf{c}_{ni} \mathbf{c}_{ni}^\top = (\mathbf{C}_n^0)^{-1} \mathbf{C}_n (\mathbf{C}_n^0)^{-1}.$$

We modify $\mathbf{M}_n(\mathbf{u})$ in (5.128) as

$$\mathbf{M}_n(\mathbf{u}) = \sum_{i=1}^{n} \mathbf{c}_{ni} \psi(e_i - \mathbf{c}_{ni}^\top \mathbf{u}), \quad \mathbf{u} \in \boldsymbol{R}_p. \tag{5.129}$$

Our goal is to verify whether

$$\sup_{\|\mathbf{u}\| \le K} \|\mathbf{M}_n(\mathbf{u}) - \mathbf{M}_n(\mathbf{0}) + \mathbf{Q}_n \mathbf{u} \gamma\| \xrightarrow{P} 0 \text{ as } n \to \infty \tag{5.130}$$

for every $K: \; 0 < K < \infty$, where $\gamma = \int_{\boldsymbol{R}} f(x) d\psi(x)$. Whenever $I(f) < \infty$, we can write

$$\gamma = \int_{\boldsymbol{R}} \psi(x) \Big\{ -\frac{f'(x)}{f(x)} \Big\} dF(x).$$

Introduce the vector empirical function

$$\mathbf{S}_n^\star(t, \mathbf{u}) = \sum_{i=1}^{n} \mathbf{c}_{ni} I[e_i \le F^{-1}(t) + \mathbf{c}_{ni}^\top \mathbf{u}], \quad \text{for } t \in [0,1], \quad \mathbf{u} \in \boldsymbol{R}_p. \tag{5.131}$$

Then we have from (5.129) and (5.131)

$$\mathbf{M}_n(\mathbf{u}) = \int_{\boldsymbol{R}} \psi(F^{-1}(t)) d\mathbf{S}_n^\star(t, \mathbf{u}), \quad \mathbf{u} \in \boldsymbol{R}_p,$$

which is a linear function of $\mathbf{S}_n^\star(.)$.

Similarly as in Section 4.5, note that if a functional $\tau(F)$ of F is Hadamard (or compact) differentiable, then its Hadamard derivative is a linear functional [see (4.48)–(4.49)]. This provides an intuitive justification for adopting a plausible Hadamard differentiability approach for the study of (5.130).

Let us keep the notations and definitions of the first- (and second-) order Hadamard derivatives, as introduced in Section 4.5 [see (4.48)–(4.52)]. For simplicity of presentation, take first the case of $p = 1$, when the c_{ni} and u are both real numbers. We assume that

$$\sum_{i=1}^{n} c_{ni}^2 = 1 \quad \text{and} \quad \max_{1 \le i \le n} c_{ni}^2 \to 0 \quad \text{as } n \to \infty.$$

Moreover, assume that F is absolutely continuous and has a positive and continuous derivative $F' = f$ with limits at $\pm\infty$. Define $\mathbf{S}_n^\star(t, u)$ as before (for $p = 1$), and replacing c_{ni} by c_{ni}^+ and c_{ni}^- with

$$c_{ni} = c_{ni}^+ - c_{ni}^-; \;\; c_{ni}^+ = \max(0, c_{ni}), \;\; c_{ni}^- = -\min(0, c_{ni}),$$

we obtain parallel expressions for $S_n^{+\star}(t, u)$ and $S_n^{-\star}(t, u)$.

Denote as U_n the empirical distribution function of the $F(e_i)$, $1 \le i \le n$, and let U be the classical uniform (0,1) distribution function. Let $\tau(.)$ be a right continuous function having left-hand limits. Then

THEOREM 5.13 *(Ren and Sen 1991). Suppose that $\tau : D[0,1] \mapsto \boldsymbol{R}$ is a functional Hadamard differentiable at U and that the c_{ni} and F satisfy the conditions mentioned before. Then, for any $K > 0$, as $n \to \infty$,*

$$\sup_{|u|<K} \Big| \sum_{i=1}^{n} c_{ni}^+ \tau\Big(\frac{S_n^{+\star}(., u)}{\sum_{i=1}^n c_{ni}^+}\Big) - \sum_{i=1}^{n} c_{ni}^- \tau\Big(\frac{S_n^{+\star}(.u)}{\sum_{i=1}^n c_{ni}^-}\Big) - \tau(U) \sum_{i=1}^{n} c_{ni}$$
$$-\tau_U'\Big(S_n^\star(., u) - U(.) \sum_{i=1}^{n} c_{ni}\Big)\Big| \to 0, \;\; \textit{in probability.} \tag{5.132}$$

If $\tau(.)$ is such that $M_n(u)$ may be its Hadamard derivative, it is possible to show that for any $0 < K < \infty$, as $n \to \infty$,

$$\sup_{|u| \le K} \Big| \sum_{i=1}^{n} c_{ni} \Big\{ \tau\Big(\frac{S_n^\star(., u)}{\sum_{i=1}^n c_{ni}}\Big) - \tau\Big(\frac{S_n^\star(., 0)}{\sum_{i=1}^n c_{ni}}\Big) \Big\}$$
$$-[M_n(u) - M_n(0)] \Big| \xrightarrow{P} 0 \tag{5.133}$$

and

$$\sup_{|u| \le K} \Big| \sum_{i=1}^{n} c_{ni} \Big\{ \tau\Big(\frac{S_n^\star(., u)}{\sum_{i=1}^n c_{ni}}\Big) - \tau\Big(\frac{S_n^\star(., 0)}{\sum_{i=1}^n c_{ni}}\Big) \Big\} + u\gamma \Big| \xrightarrow{P} 0.$$

what implies the asymptotic linearity.

Indeed, define

$$\mathbf{D}_p = \text{Diag}\,\Big(\sum_{i=1}^{n} c_{ni1}, \ldots, \sum_{i=1}^{n} c_{nip}\Big)$$

$$\boldsymbol{\tau}(\mathbf{S}_n^\star(.\mathbf{u})) = \Big(\tau\Big(\frac{S_{n1}^\star(.,\mathbf{u})}{\sum_{i=1}^n c_{ni1}}\Big),\ldots,\tau\Big(\frac{S_{np}^\star(.,\mathbf{u})}{\sum_{i=1}^n c_{nip}}\Big)\Big)^\top. \tag{5.134}$$

Our goal is to show that

$$\sup_{\|\mathbf{u}\|\leq K} \|\mathbf{D}_p\Big\{\tau(S_n^\star(.,\mathbf{u}) - \tau(S_n^\star(.,\mathbf{0}))\Big\} - [M_n(u) - M_n(0)]\| \xrightarrow{P} 0,$$

as $n \to \infty$, under appropriate regularity conditions, and

$$\sup_{\|\mathbf{u}\|<K} \|\mathbf{D}_p\Big\{\tau(S_n^\star(.,\mathbf{u}) - \tau(S_n^\star(.,\mathbf{0}))\Big\} + \mathbf{Q}_n\mathbf{u}\gamma\| \xrightarrow{P} 0, \tag{5.135}$$

as $n \to \infty$. These two propositions in turn imply that

$$\sup_{\|\mathbf{u}\|<K} \|\mathbf{M}_n(\mathbf{u}) - \mathbf{M}_n(\mathbf{0}) + \mathbf{Q}_n\mathbf{u}\gamma\| \xrightarrow{P} 0,$$

and hence, the uniform asymptotic linearity.

The Hadamard differentiability approach does not depend on the convexity, which was used by Pollard (1991) to derive the uniform asymptotic linearity, with the aid of pointwise linearity results. The Hadamard differentiability approach allows to derive the asymptotic approximation of the estimating equations [i.e., $\mathbf{M}_n(\mathbf{u})$] uniformly over a compact set, and so it provides a good method for the study of the asymptotic properties of M-estimators of regression through asymptotic linear approximations. The regularity conditions appear to be less stringent.

General second-order asymptotic representations for a second-order Hadamard differentiable functional, studied in Section 4.5 for i.i.d. random variables, also holds for M-estimators of regression parameters. The relevant methodology has been developed by Ren and Sen (1995b).

5.8 Further Developments

1. Portnoy (1984b, 1985) studied the consistency and asymptotic normality of the M-estimators in the linear model in the situation when p is permitted to grow with n, more precisely when $p \to \infty$ but $\frac{1}{n}(p\log n)^{3/2} \to 0$. Similar problem was studied by Gallant (1989).

2. A properly modified one-step version of M-estimator in linear model based on Newton-Raphson iteration inherits the breakdown point of the preliminary estimator and yet has the same asymptotic distribution as the fully iterated version, provided the initial estimate is n^α-consistent with $\alpha > 1/4$. This phenomenon was noticed by Jurečková and Portnoy (1987) and further studied by Coakley and Hettmansperger (1993), Welsh and Ronchetti (2002), among others. Simpson et. al. (1992) extended this result to the GM-estimators in linear model.

3. The strong representations for location and regression M-estimators were derived by Carroll (1978) and Martinsek (1989). Liese and Vajda (1994) studied general conditions for strong consistency of M-estimates. He and Wang (1995) derived strong Bahadur representations for multivariate location and scatter estimation. Arcones (1996) obtained the strong Bahadur representation with the exact error rate for the minimum L_p distance estimator with random regressors. He and Shao (1996) derived strong Bahadur representations for a general class of M-estimators based on independent but not necessarily identically distributed random variables. The results they applied to M-estimators of regression with nonstochastic regressors.
4. Omelka (2010) derived the second-order asymptotic representations for the fixed-scale and studentized M-estimators in linear model.

5.9 Problems

5.2.1 Consider a generalized Laplace density

$$f(x;\theta) = \text{constant} \cdot \exp\{-|x-\theta|^{\alpha}\}, \quad x \in \boldsymbol{R},$$

where $\alpha \in (0,1)$ and $\theta \in \theta \subset \boldsymbol{R}$. Show that

$$\frac{\partial}{\partial \theta} \log f(x;\theta) = -\alpha|x-\theta|^{\alpha-1}\text{sign}(x-\theta)$$

so that

$$\int_{\boldsymbol{R}} \left|\frac{\partial}{\partial \theta} \log f(x;\theta)\right| dF(x;\theta) < \infty, \quad \forall \alpha \in (0,1),$$

but that for $\alpha \leq 1/2$,

$$\int_{\boldsymbol{R}} \left\{\frac{\partial}{\partial \theta} \log f(x;\theta)\right\}^2 dF(x;\theta) \not< \infty.$$

Comment on the asymptotic properties of the MLE of θ (when α is known and $0 < \alpha \leq 1/2$).

5.2.2 (Continuation). Consider the same model when $\alpha \in [1,2]$. Describe the asymptotic properties of the MLE of θ. What are the main differences in the two cases: (a) $0 < \alpha \leq 1/2$ and (b) $\alpha \geq 1$?

5.2.3 (Continuation). For the same model, let $\psi(x;\theta) = -\frac{f'_x(x;\theta)}{f(x;\theta)}$, $x \in \boldsymbol{R}$. Are the regularity conditions [A1]–[A2] in Section 5.2 satisfied? Consider the three cases separately: (a) $0 < \alpha < 1$, (b) $\alpha = 1$, (c) $1 < \alpha < 2$. What happens when $\alpha = 2$ or $\alpha > 2$ (but, not an integer)?

5.2.4 Let $\psi(x,t) = g(x-t)$, $x \in \boldsymbol{R}, t \in \boldsymbol{R}$ where $g(x)$ is an absolutely continuous, bounded, and skew-symmetric function. Show that (5.3) can be incorporated for the estimation of θ in Problem **5.2.1**. What happens when $g(-x) = -g(x)$ for all x, where $g(0+) \neq 0$? Can **[A1]**–**[A2]** be modified to include such a jump discontinuity of $\psi(.)$?

5.2.5 Consider the two-parameter Laplace density

$$f(x, \boldsymbol{\theta}) = \frac{1}{2\theta_2} \exp\left\{ -\frac{|x-\theta_1|}{\theta_2} \right\}, \quad x \in \boldsymbol{R},\ \theta_1 \in \boldsymbol{R},\ \theta_2 \in \boldsymbol{R}^+.$$

Let $\boldsymbol{\psi}(x, \mathbf{t}) = \left(\psi_1(x; \mathbf{t}), \psi_2(x; \mathbf{t})\right)^\top = \left(\text{sign}(x - t_1), \frac{|x-t_1|}{t_2} - 1\right)^\top$. Show that **[A1]**–**[A2]** may not hold here, although if one uses the median-estimator of θ_1 for the estimation of θ_2, then $\psi_2(.)$ can be tuned to satisfy both **[A1]** and **[A2]**.

5.3.1 Consider the function $\rho(y) = \log(1 + y^2)$, $y \in \boldsymbol{R}$, and define $h(t)$ as in (5.21). Assume that the distribution function F has a density f, symmetric about 0, such that f has heavy tails [i.e., $f(x)$ converges to 0 as $x \to \infty$ at a rate slower than the normal or exponential pdf.] Show then that $h(t) = 0$ may not have a unique minimum at $t = 0$.

5.3.2 (Continuation). Verify the last problem when $f(x)$ is a Cauchy density. Is $\rho(.)$ in Problem **5.3.1** convex?

5.3.3 Consider the following redescending score functions:

(a)

$$\psi_1(x) = \begin{cases} \sin(\frac{x}{a}) & \text{for} \quad -a\pi \le x \le a\pi \\ 0 & \text{otherwise.} \end{cases}$$

(b)

$$\psi_2(x) = \begin{cases} x(k^2 - x^2)^p & \text{for} \quad -k \le x \le k \\ 0 & \text{otherwise} \end{cases}$$

where k and p are positive numbers.

(c)

$$\psi_3(x) = \begin{cases} x & \text{for} \quad -k \le x \le k \\ k \cdot \text{sign} x & \text{for} \quad k \le |x| \le k + a,\ a > 0 \\ \left(k e^{\mathrm{k+a-|x|}}\right) \cdot \text{sign} x & \text{for} \quad x > k + a. \end{cases}$$

where $k > 0$.

In each case, find $h(t)$, defined by (5.21), and verify whether a unique root for $h(t) = 0$ exists or not.

5.4.1 For $h(t)$ defined by (5.64), show that when ψ is nonmonotone (as in Problem 5.3.3), there may not be a unique solution of $h(t) = 0$.

CHAPTER 6

Asymptotic Representations for R-Estimators

6.1 Introduction

Like the M-estimators in Chapter 5, the R-estimators are implicitly defined statistical functionals. Although for R-estimators a differentiable statistical functional approach can be formulated, it may call for bounded score functions that do not include the important cases of normal scores, log-rank scores, and other unbounded ones. Moreover the ranks are integer valued random variables, so that even for smooth score functions, the rank scores may not possess smoothness to a very refined extent. For this reason the treatment of Chapter 5 may not go through entirely for R-estimators of location or regression parameters. hence, to encompass a larger class of rank-based estimators, we will follow the tracks based on the asymptotic linearity of rank statistics in shift or regression parameters: This approach has been popularized in the past two decades, and some accounts of the related developments are also available in some other contemporary text books on nonparametrics. However, our aim is to go beyond these reported developments onto the second-order representations and to relax some of the currently used regularity conditions. In this way we aim to provide an up-to-date and unifying treatment of the asymptotic representations for R-estimators of location and regression parameters.

In Section 6.2 we start with the usual signed rank statistics and present the first-order asymptotic representations for allied R-estimators of location of a symmetric distribution. The next section deals with second-order representations for such R-estimators of location (under additional regularity conditions). Asymptotic representations for R-estimators of the slope in a simple linear regression model are studied in Section 6.4, and the second-order results are then presented in Section 6.5. Section 6.6 deals with the general multiple regression models and contains the first-order asymptotic representation results for the usual R-estimators of the regression and intercept parameters. The last two sections are devoted to regression rank scores estimators, which have been developed only in the recent past.

6.2 Asymptotic Representations for R-Estimators of Location

The R-estimator of location parameter, which is considered as the center of symmetry of a continuous (unknown) distribution function, was defined in Section 3.4, and was based on appropriate signed rank statistic (see [3.78]). To derive an asymptotic representation for such R-estimator, we should start with the asymptotic representation for signed-rank statistic

$$S_n^0(b) = n^{-1/2}\sum_{i=1}^{n} \operatorname{sign}(X_i - b)a_n^0(R_{ni}^+(b)), \quad b \in \boldsymbol{R}_1,$$

where $R_{ni}^+(b) =$ rank of $|X_i - b|$ among $|X_1 - b|, \ldots, |X_n - b|$, for $i = 1, \ldots, n$, with the *scores*

$$a_n^0(i) = I\!\!E[\varphi(U_{n:i})], \quad 1 \le i \le n;\ n \ge 1, \tag{6.1}$$

where $U_{n:1} < \ldots < U_{n:n}$ are the order statistics of a sample of size n from the uniform (0,1) distribution. The score function $\varphi = \{\varphi(u),\ u \in (0,1)\}$ is taken to be nondecreasing. We let

$$\begin{aligned} \varphi(u) &= \varphi^\star\left(\frac{1+u}{2}\right),\ u \in (0,1),\ \varphi^\star \nearrow \text{ in } u, \\ \varphi^\star(u) &+ \varphi^\star(1-u) = 0 \ \forall u \in [0,1] \end{aligned} \tag{6.2}$$

so that $\varphi^\star$ is skew-symmetric about 1/2. We assume that $X_1, \ldots, X_n$ have symmetric distribution function $F(x - \theta)$ where $F(x) + F(-x) = 1\ \forall x \in \boldsymbol{R}_1$. We may put $\theta = 0$ without loss of generality. For notational simplicity, we let $R_{ni}^+(0) = R_{ni}^+$, $\forall i, n$, and $F^+(x) = F(x) - F(-x) = 2F(x) - 1$, $x \ge 0$. Our main object is the difference

$$\begin{aligned} Q_n^0 &= S_n^0(0) - n^{-\frac{1}{2}}\sum_{i=1}^{n}\varphi^\star(F(X_i)) \\ &= n^{-\frac{1}{2}}\sum_{i=1}^{n}\operatorname{sign}X_i\{a_n^0(R_{ni}^+) - \varphi(F^+(|X_i|))\}. \end{aligned} \tag{6.3}$$

The convergence $Q_n^0 \xrightarrow{p} 0$, given in the following theorem, in fact brings the asymptotic representation of $S_n^0(0)$ by a sum of independent random variables:

THEOREM 6.1 *Let X_i, $i \ge 1$ be i.i.d. random variables with a continuous and symmetric distribution function F, defined on $\boldsymbol{R}_1$. Let $\varphi : (0,1) \mapsto \boldsymbol{R}_1^+$ be $\nearrow$ such that $\varphi(0) = 0$ and $\int_0^1 \varphi^2(u)du = A_\varphi^2 < \infty$. Then*

$$Q_n^0 \to 0, \quad \textit{in probability, as } n \to \infty. \tag{6.4}$$

PROOF Let $\mathcal{B}_n$ be the sigma-field generated by the vector of signs and the vector of the ranks [i.e., $\mathcal{B}_n = \mathcal{B}(\operatorname{sign}X_1, \ldots, \operatorname{sign}X_n; R_{n1}^+, \ldots, R_{nn}^+)$] for $n \ge 1$. Note that $\varphi^\star(F(X_i)) = (\operatorname{sign}X_i)\varphi(F^+(|X_i|))$, for every $i = 1, \ldots, n$ by (6.2) and that the vectors (sign $X_1, \ldots,$ sign X_n) and $(R_{n1}^+, \ldots, R_{nn}^+)$ are stochastically independent. Moreover, the R^+_{ni} are independent with the ordered variables $\varphi(F^+(|X_i|))$, $i = 1, \ldots, n$. Then it follows that

$$\mathbb{E}\Big[n^{-\frac{1}{2}}\sum_{i=1}^{n}\varphi^{\star}(F(X_i))|\mathcal{B}_n\Big] = \mathbb{E}\Big[n^{-\frac{1}{2}}\sum_{i=1}^{n}(\text{sign}X_i)\varphi(F(X_i))|\mathcal{B}_n\Big]$$
$$= n^{-\frac{1}{2}}\sum_{i=1}^{n}(\text{sign}X_i)\mathbb{E}\Big[\varphi(F(X_i))|R^+_{ni}\Big] = n^{-\frac{1}{2}}\sum_{i=1}^{n}(\text{sign}X_i)a^0_n(R^+_{ni})$$
$$= S^0_n(0), \qquad n \geq 1. \tag{6.5}$$

Therefore, we obtain by (6.3) and (6.5) that

$$\begin{aligned} \mathbb{E}[Q^0_n]^2 &= \mathbb{E}\Big[\varphi^2(F^+(|X_i|)\Big] - \mathbb{E}\Big[(S^0_n(0))^2\Big] \\ &= \int_0^1 \varphi^2(u)du - n^{-1}\sum_{i=1}^{n}\{a^0_n(i)\}^2, \end{aligned} \tag{6.6}$$

where, by (6.1), for $n \geq 1$

$$n^{-1}\sum_{i=1}^{n}\{a^0_n(i)\}^2 = n^{-1}\sum_{i=1}^{n}\Big\{\mathbb{E}[\varphi(U_{n:i})]\Big\}^2 \tag{6.7}$$
$$\leq n^{-1}\sum_{i=1}^{n}\mathbb{E}[\varphi^2(U_{n:i})] = n^{-1}\sum_{i=1}^{n}\mathbb{E}[\varphi^2(U_i)] = \int_0^1 \varphi^2(u)du = A^2_{\varphi}.$$

Using (6.2), we obtain that

$$\lim_{n\to\infty}\varphi^0_n(u) = \varphi(u), \quad \text{for every (fixed) } u \in [0,1)$$

where

$$\varphi^0_n(u) = a^0_n(i) \quad \text{for } (i-1)/n < u \leq i/n,\ i = 1, \ldots, n, \tag{6.8}$$

Notice that

$$n^{-1}\sum_{i=1}^{n}\{a^0_n(i)\}^2 = \int_0^1 \{\varphi^0_n(u)\}^2 du; \tag{6.9}$$

then it follows from (6.7), (6.8) and from the Fatou lemma (Section 2.5) that (6.6) converges to 0 as $n \to \infty$. Then, by the Chebyshev inequality, $Q_n \to 0$ in probability, as $n \to \infty$. □

It is not necessary to define the scores by (6.1). The convergence (6.4) holds for all suitably defined scores $a_n(i)$, $i = 1, \ldots, n$, satisfying

$$n^{-1}\sum_{i=1}^{n}[a_n(i) - a^0_n(i)]^2 \to 0 \quad \text{as} \quad n \to \infty. \tag{6.10}$$

This further ensures the stochastic convergence of $S_n(0) - S^0_n(0)$ to 0; more precisely, regarding the independence of the signs and the ranks,

$$\mathbb{E}\Big[(S_n(0) - S^0_n(0))^2\Big] = n^{-1}\mathbb{E}\Big[\sum_{i=1}^{n}\text{sign}X_i\{a_n(R^+_{ni}) - a^0_n(R^+_{ni})\}\Big]^2$$

$$= n^{-1}\sum_{i=1}^{n} \mathbb{E}\Big[a_n(R_{ni}^+) - a_n^0(R_{ni}^+)\Big]^2 = n^{-1}\sum_{i=1}^{n}\Big[a_n(i) - a_n^0(i)\Big]^2.$$

This and (6.10) ensure that the proposition of Theorem 6.1 also holds for $S_n^0(0)$ being replaced by $S_n(0)$. Moreover, we can write

$$\begin{aligned} n^{-1}\sum_{i=1}^{n} a_n^2(i) &= n^{-1}\sum_{i=1}^{n}\{a_n^0(i)\}^2 + n^{-1}\sum_{i=1}^{n}\Big[a_n(i) - a_n^0(i)\Big]^2 \\ &\quad -2n^{-1}\sum_{i=1}^{n} a_n^0(i)\Big[a_n^0(i) - a_n(i)\Big]; \end{aligned}$$

and this with (6.10) and with the Cauchy-Schwarz inequality imply that

$$n^{-1}\sum_{i=1}^{n} a_n^2(i) \to A_\varphi^2 = \int_0^1 \varphi^2(u)du, \quad \text{as } n \to \infty.$$

Alternatively, the following equivalent form of (6.10) is often more convenient to verify in practice:

$$\int_0^1 \Big\{\varphi_n(u) - \varphi(u)\Big\}^2 du \to 0 \quad \text{as } n \to \infty \tag{6.11}$$

where

$$\varphi_n(u) = a_n(i), \quad \text{for } (i-1)/n < u \leq i/n,\ i = 1, \ldots, n;\ n \geq 1; \tag{6.12}$$

this we obtain by the simple triangular inequality. We can use the "approximate" scores as

$$a_n(i) = \varphi(i/(n+1)), \ \ i = 1, \ldots, n;\ n \geq 1, \tag{6.13}$$

with the score-generating function $\varphi(.)$ satisfying (6.2). Using basic theorems of integral calculus, we obtain

$$\lim_{n\to\infty}\Big\{\int_0^1 \varphi_n^2(u)du\Big\} = \lim_{n\to\infty}\Big\{n^{-1}\sum_{i=1}^{n}\varphi^2\Big(\frac{i}{n+1}\Big)\Big\} = A_\varphi^2 = \int_0^1 \varphi^2(u)du.$$

Let us consider the empirical process

$$Z_n(t) = S_n(n^{-\frac{1}{2}}t) - S_n(0), \ \ t \in [-C, C] \tag{6.14}$$

with the scores defined in (6.13), $C > 0$ being an arbitrary fixed constant. The following theorem characterizes the asymptotic linearity of $Z_n(t)$, uniform in $|t| \leq C$.

THEOREM 6.2 *Let $X_1, X_2, \ldots$ be a sequence of i.i.d. random variables with a symmetric distribution function F. Assume that F possesses an absolutely continuous density f and finite Fisher's information*

$$0 < I(f) = \int_{\boldsymbol{R}_1} \Big(f'(x)/f(x)\Big)^2 dF(x) < \infty. \tag{6.15}$$

Let $\varphi : [0,1] \mapsto \boldsymbol{R}^+$ be a nondecreasing and square integrable function such that $\varphi(0) = 0$. Then, for any fixed $C > 0$, as $n \to \infty$,

$$\sup\Big\{|Z_n(t) + \gamma t| : |t| \le C\Big\} \xrightarrow{P} 0, \tag{6.16}$$

where

$$\gamma = -\int_{\boldsymbol{R}_1} \varphi^\star(F(x))f'(x)dx = \int_{\boldsymbol{R}_1} \varphi^\star(u)\psi_f(u)du = \langle \varphi^\star, \psi_f \rangle, \tag{6.17}$$

where

$$\psi_f(u) = -f'(F^{-1}(u))/f(F^{-1}(u)),\ \ 0 < u < 1, \tag{6.18}$$

is the Fisher score function and $\varphi^\star$ is related to φ by (6.2).
[Note that $\langle \psi_f, \psi_f \rangle = \int_0^1 \psi_f^2(u)du = \int_{\boldsymbol{R}_1} (f'(x)/f(x))^2 dF(x) = I(f)$].

PROOF of Theorem 6.2. We carry out the proof in several steps. First, we prove that for every $n(\ge 1)$ and every nondecreasing $\varphi(.)$, $S_n(b)$ is nonincreasing in $b \in \boldsymbol{R}_1$; this result has independent interest of its own.

LEMMA 6.1 *Under the hypothesis of Theorem 6.2, the statistic $S_n(b)$ is nonincreasing in b ($\in \boldsymbol{R}_1$) with probability one.*

PROOF Note that the sign$(X_i - b)$ are nonincreasing in b ($\in \boldsymbol{R}_1$), and assume the values $+1, 0$, or -1. On the other hand, for $X_i > b$, $|X_i - b|$ is nonincreasing in b, while for $X_i < b$, $|X_i - b|$ is nondecreasing. Hence, for the $X_i > b$, $R_{ni}^+(b)$ are nonincreasing in b, and for the $X_i < b$, $R_{ni}^+(b)$ are nondecreasing in b. Since sign$(X_i - b)$ is $+1$ or -1 according as X_i is $> b$ or $< b$, from the above monotonicity results we immediately conclude that $S_n(b)$ is nonincreasing in $b \in \boldsymbol{R}_1$. □

Returning to the proof of Theorem 6.2, in the second step we prove a pointwise convergence result.

LEMMA 6.2 *Let $t \in \boldsymbol{R}_1$ be an arbitrary fixed number. Then, under the hypothesis of Theorem 6.2, $Z_n(t) + \gamma t \to 0$ in probability as $n \to \infty$.*

PROOF Let us denote by

$$V_n(t) = n^{-\frac{1}{2}} \sum_{i=1}^{n} \text{sign}(X_i - t)\varphi(F^+(|X_i - t|)),\ \ t \in \boldsymbol{R}_1. \tag{6.19}$$

Then, by Theorem 6.1, (6.10) and (6.11),

$$S_n(0) - V_n(0) \to 0, \quad \text{in probability, as } n \to \infty. \tag{6.20}$$

Let $\{P_n\}$ and $\{P_n^\star\}$ denote the probability distributions with the densities $P_n = \prod_{i=1}^n f(X_i)$ and $P_n^\star = \prod_{i=1}^n f(X_i - n^{-1/2}t)$, respectively, for $n \ge 1$.

Then, by (2.94), the sequence $\{P_n^\star\}$ is *contiguous* to $\{P_n\}$ so that (6.20) implies that

$$S_n(0) - V_n(0) \to 0, \quad \text{under } \{P_n^\star\} \text{ as well, as } n \to \infty. \tag{6.21}$$

We rewrite (6.21) as

$$S_n(n^{-\frac{1}{2}}t) - V_n(n^{-\frac{1}{2}}t) \to 0, \quad \text{under } \{P_n\} \text{ measure, as } n \to \infty. \tag{6.22}$$

hence, it suffices to show that for any fixed $t \in \boldsymbol{R}_1$

$$V_n(n^{-\frac{1}{2}}t) - V_n(0) + \gamma t \to 0, \quad \text{in probability, as } n \to \infty. \tag{6.23}$$

Note that apart from the monotonicity and square integrability of $\varphi^\star$, we are not assuming any other regularity conditions (and in general $\varphi^\star$ is unbounded at the two extremities). Hence, for our subsequent manipulations, we consider the following approximation to $\varphi^\star$: For each k (≥ 1), let

$$\begin{aligned} \varphi_k^\star(u) &= \varphi^\star\Big(\frac{1}{k+1}\Big)I\Big(u < \frac{1}{k}\Big) + \varphi^\star(u)I\Big(\frac{1}{k} \leq u \leq \frac{k-1}{k}\Big) \\ &+\varphi^\star\Big(\frac{k}{k+1}\Big)I\Big(u > \frac{k-1}{k}\Big). \end{aligned} \tag{6.24}$$

Then we have

$$\lim_{k\to\infty} \Big\{ \int_0^1 \{\varphi_k^\star(u) - \varphi^\star(u)\}^2 du \Big\} = 0. \tag{6.25}$$

Noting that $(\text{sign } y)\varphi(F^+(|y|) = \varphi^\star(F(y))$, $\forall y \in \boldsymbol{R}_1$, we obtain from (6.19) and (6.24) that the parallel version for $V_n(t)$ corresponding to $\varphi_k^\star$ is

$$V_{nk}(t) = n^{-\frac{1}{2}} \sum_{i=1}^n \varphi_k^\star(F(X_i - t)), \ t \in \boldsymbol{R}_1.$$

As in (6.17), we let $\gamma_k = \langle \varphi_k^\star, \psi_f \rangle$, $k \geq 1$, so that by (6.25),

$$\begin{aligned} (\gamma_k - \gamma)^2 &= \langle (\varphi_k^\star - \varphi^\star), \psi_f \rangle^2 \leq \|\varphi_k^\star - \varphi^\star\|^2 \|\psi_f\|^2 \\ &= I(f)\|\varphi_k^\star - \varphi^\star\|^2 \to 0, \quad \text{as } k \to \infty. \end{aligned} \tag{6.26}$$

Further, note that by (6.25),

$$\text{Var}[V_{nk}(0) - V_n(0)] = I\!E[V_{nk}(0) - V_n(0)]^2 \tag{6.27}$$

$$= I\!E\Big[\varphi^\star(F(X_1)) - \varphi_k^\star(F(X_1))\Big]^2 = \int_0^1 \Big[\varphi^\star(u) - \varphi_k^\star(u)\Big]^2 du \to 0,$$

as $k \to \infty$,

so that $V_{nk}(0) - V_n(0)$ converges to 0 in probability, as $k \to \infty$ uniformly in $n = 1, 2, \ldots$. Using the contiguity of $\{P_n^\star\}$ to $\{P_n\}$, it follows as in (6.21) and (6.22) that for any fixed t,

$$V_{nk}(n^{-\frac{1}{2}}t) - V_n(n^{-\frac{1}{2}}t) \xrightarrow{P} 0, \quad \text{as } n \to \infty, \ k \to \infty. \tag{6.28}$$

Thus, regarding (6.23), (6.26), (6.27), and (6.28), we have to show that for every (fixed) k and any (fixed) $t \in [-C, C]$,

$$V_{nk}(n^{-\frac{1}{2}}t) - V_{nk}(0) + t\gamma_k \xrightarrow{P} 0. \tag{6.29}$$

as $n \to \infty$. Note that

$$\begin{aligned} &\mathbb{E}\Big[V_{nk}(n^{-\frac{1}{2}}t) - V_{nk}(0) + t\gamma_k\Big] \\ &= n^{\frac{1}{2}}\Big[\int_{\boldsymbol{R}_1} \varphi_k^\star(F(x))d\Big[F(x+n^{-\frac{1}{2}}t) - F(x)\Big] \\ &\quad -n^{-\frac{1}{2}}t\Big\{\int_{\boldsymbol{R}_1} \varphi_k^\star(F(x))f'(x)dx\Big\}\Big] \quad (6.30) \\ &= n^{\frac{1}{2}}\Big[\int_{\boldsymbol{R}_1} \varphi_k^\star(F(x))d\Big\{F(x+n^{-\frac{1}{2}}t) - F(x) - n^{-\frac{1}{2}}tf(x)\Big\}\Big]. \end{aligned}$$

Since $\varphi_k^\star$ is monotone skew-symmetric for every fixed k and is bounded from below and above by $\varphi^\star(1/(k+1))$ and $\varphi^\star(k/(k+1))$, respectively, integrating the right hand side of (6.30) by parts and noting that

$$F(x+n^{-\frac{1}{2}}t) - F(x) - n^{-\frac{1}{2}}tf(x) = \mathrm{O}(n^{-1})$$

a.e. for all $x: \ |\varphi_k^\star(F(x))| \le |\varphi^\star(k/(k+1))|)$, we obtain that (6.30) converges to 0 as $n \to \infty$. Finally, note that

$$\mathrm{Var}\Big[V_{nk}(n^{-\frac{1}{2}}t) - V_{nk}(0)\Big] = \mathrm{Var}\Big[\varphi_k^\star(F(X_1 - n^{-\frac{1}{2}}t)) - \varphi_k^\star(F(X_1))\Big]. \tag{6.31}$$

We can prove that (6.31) converges to 0 as $n \to \infty$ by similar arguments. Then (6.29) follows from the Chebyshev inequality and from (6.30). □

Let us now return to the third phase of the proof of Theorem 6.2. We want to prove that the stochastic convergence in Lemma 6.2 is uniform in $t \in [-C, C]$ for every fixed C ($0 < C < \infty$). Let C, ε, and η be arbitrary positive numbers. Consider a partition of the interval $[-C, C]: \ -C = t_0 < t_1 < \ldots < t_p = C$ such that

$$t_i - t_{i-1} \le \frac{\eta}{2|\gamma|}, \quad \text{for every } i = 1, \ldots, p. \tag{6.32}$$

Then, by Lemma 6.2, for every positive ε and η, there exists a positive integer n_0, such that for every i ($= 0, 1, \ldots, p$),

$$\mathbb{P}\Big\{|Z_n(t_i) + \gamma t_i| \ge \eta/4\Big\} \le \frac{\varepsilon}{p+1}, \quad \forall n \ge n_0.$$

On the other hand, by Lemma 6.1,

$$|Z_n(t) + \gamma t| \le |Z_n(t_{i-1}) + \gamma t_{i-1}| + |Z_n(t_i) + \gamma t_i| + |\gamma(t_i - t_{i-1})|$$

for every $t \in [t_{i-1}, t_i], i = 1, 2, \ldots, p$. hence,

$$\begin{aligned} &\mathbb{P}\Big\{\sup\{|Z_n(t) + t\gamma| : \ t \in [-C, C]\} \ge \eta\Big\} \\ &\le \sum_{i=0}^{p} \mathbb{P}\Big\{|Z_n(t_i) + \gamma t_i| \ge \eta/4\Big\} < \varepsilon \quad \text{for every } n \ge n_0. \end{aligned} \tag{6.33}$$

This completes the proof of Theorem 6.2. □

In order to prove the asymptotic representation for the R-estimator, we should first prove that a general R-estimator of location is $\sqrt{n}$-consistent.

LEMMA 6.3 *Let $X_1, X_2, \ldots$ be a sequence of i.i.d. random variables with a distribution function $F(x-\theta)$ with unknown θ and such that F has an absolutely continuous symmetric density f having a finite Fisher information $I(f)$. Let $\varphi: [0,1) \mapsto \boldsymbol{R}^+$ be a nondecreasing square integrable score function such that $\varphi(0)=0$. Let γ be defined as (6.17), and assume that $\gamma \neq 0$.*

Then the R-estimator T_n of θ based on $S_n(.)$ by means of the relation (3.78) satisfies

$$n^{\frac{1}{2}}|T_n - \theta| = O_p(1), \quad \text{as } n \to \infty. \tag{6.34}$$

PROOF Note that by definition of T_n, for any (fixed) $C > 0$,

$$\begin{aligned}
I\!P_\theta\Big\{n^{\frac{1}{2}}|T_n-\theta| > C\Big\} &= I\!P_0\Big\{T_n > n^{-\frac{1}{2}}C\Big\} + I\!P_0\Big\{T_n < -n^{-\frac{1}{2}}C\Big\} \\
&\leq I\!P_0\Big\{S_n(n^{-\frac{1}{2}}C) \geq 0\Big\} + I\!P_0\Big\{S_n(-n^{-\frac{1}{2}}C) \leq 0\Big\}.
\end{aligned} \tag{6.35}$$

Further, as φ is assumed to be nondecreasing, $\gamma = \int_0^1 f(F^{-1}(u))d\varphi^\star(u) > 0$, while $S_n(0) \stackrel{P}{\sim} n^{-\frac{1}{2}} \sum_{i=1}^n \varphi^\star(F(X_i)) \stackrel{\mathcal{D}}{\sim} \mathcal{N}(0, \langle\varphi^\star, \varphi^\star\rangle^2)$ by Theorem 6.1. thus, there exists a positive $C^\star$ and an integer n_0 such that for $n \geq n_0$,

$$I\!P_0\Big\{|S_n(0)| > C^\star\Big\} < \varepsilon/2 \tag{6.36}$$

For the first term on the right-hand side of (6.35), we have for any $\eta > 0$,

$$\begin{aligned}
I\!P_0\Big\{S_n(n^{-\frac{1}{2}}C) \geq 0\Big\} &= I\!P_0\Big\{S_n(n^{-\frac{1}{2}}C) - S_n(0) + \gamma C \geq -S_n(0) + \gamma C\Big\} \\
&\leq I\!P_0\Big\{Z_n(C) + \gamma C \geq \eta\Big\} + I\!P_0\Big\{S_n(0) > \gamma C - \eta\Big\}
\end{aligned} \tag{6.37}$$

Thus, choosing C so large that $\gamma C > C^\star + \eta$, we obtain by using Lemma 6.2 and (6.36) that (6.37) can be made smaller than ε for every $n \geq n_0$. A similar case holds for the second term on the right-hand side of (6.35). □

Now we are in a position to formulate the main theorem of this section.

THEOREM 6.3 *Let $\{X_i;\ i \geq 1\}$ be a sequence of i.i.d. random variables with a distribution function $F(x-\theta)$ where F has an absolutely continuous, symmetric density function f having a finite Fisher information $I(f)$. Let $\varphi: [0,1) \mapsto \boldsymbol{R}^+$ be a nondecreasing, square integrable score function such that $\varphi(0) = 0$, and assume that γ, defined by (6.17), is different from 0. Then, for the R-estimator T_n, defined by (3.76)–(3.78),*

$$T_n - \theta = (n\gamma)^{-1} \sum_{i=1}^n \varphi^\star(F(X_i - \theta)) + o_p(n^{-\frac{1}{2}}) \quad \text{as } n \to \infty. \tag{6.38}$$

PROOF By Theorem 6.2 and Lemma 6.3, we conclude that when θ is the true parameter value,

$$n^{\frac{1}{2}}(T_n - \theta) - \gamma^{-1}S_n(\theta) \xrightarrow{P} 0, \quad \text{as } n \to \infty. \tag{6.39}$$

On the other hand, by Theorem 6.1,

$$S_n(\theta) - n^{-\frac{1}{2}} \sum_{i=1}^{n} \varphi^\star(F(X_i - \theta)) \xrightarrow{P} 0, \quad \text{as } n \to \infty, \tag{6.40}$$

so (6.38) is a direct consequence of (6.39) and (6.40). □

6.3 Representations for R-Estimators in Linear Model

Let us consider the univariate linear model

$$Y_i = \theta + \boldsymbol{\beta}^\top \mathbf{x}_i + e_i, \ \ i = 1, \ldots, n. \tag{6.41}$$

Here, $\mathbf{Y} = (Y_1, \ldots, Y_n)^\top$ is an n-vector of observable random variables, $\mathbf{x}_i = (x_{i1}, \ldots, x_{ip})^\top$, $1 \leq i \leq n$ are vectors of given regression constants (not all equal), θ is an intercept parameter, $\boldsymbol{\beta} = (\beta_1, \ldots, \beta_p)^\top$ is the unknown vector of regression parameters, $p \geq 1$, and the e_i are i.i.d. errors having a continuous (but unknown) distribution function F defined on $\boldsymbol{R}_1$. We do not need to assume that F is symmetric about 0, although it is necessary for the estimation of θ. The estimation of $\boldsymbol{\beta}$ will be based on suitable linear rank statistics while the estimation of θ on an aligned signed-rank statistic.

First, consider the estimation of $\boldsymbol{\beta}$. For every $\mathbf{b} \in \boldsymbol{R}_p$ let

$$R_{ni}(\mathbf{b}) = \text{ Rank of } (Y_i - \mathbf{b}^\top \mathbf{x}_i) \tag{6.42}$$

among the $Y_j - \mathbf{b}^\top \mathbf{x}_j$ $(1 \leq j \leq n)$, for $i = 1, \ldots, n$. For each n (≥ 1) we consider a set of scores $a_n(1) \leq \ldots \leq a_n(n)$ (not all equal) as in Section 6.4. Define a vector of aligned linear rank statistics

$$\begin{aligned} \mathbf{L}_n(\mathbf{b}) &= (L_{n1}(\mathbf{b}), \ldots, L_{np}(\mathbf{b}))^\top \\ &= \sum_{i=1}^{n} (\mathbf{x}_i - \bar{\mathbf{x}}_n) a_n(R_{ni}(\mathbf{b})), \ \ \mathbf{b} \in \boldsymbol{R}_p, \end{aligned} \tag{6.43}$$

where $\bar{\mathbf{x}}_n = n^{-1}\sum_{i=1}^{n} \mathbf{x}_i$. Note that the ranks $R_{ni}(\mathbf{b})$ are translation invariant, and hence, there is no need for adjustment for θ in (6.43). Further, under $\boldsymbol{\beta} = \mathbf{0}$, $R_{n1}(\mathbf{0}), \ldots R_{nn}(\mathbf{0})$ are interchangeable random variables [assuming each permutation of $1, \ldots, n$ with the common probability $(n!)^{-1}$], so we have

$$I\!\!E_{\boldsymbol{\beta}=0}(\mathbf{L}_n(\mathbf{0})) = \mathbf{0}, \ \ \forall n \geq 1.$$

We may be tempted to "equate" $\mathbf{L}_n(\mathbf{b})$ to $\mathbf{0}$ to obtain suitable estimators of $\boldsymbol{\beta}$. However, this system of equations may not have a solution, because the statistics are step functions. Instead of that, we may define the R-estimator of $\boldsymbol{\beta}$ by the minimization of the L_1-norm

$$\|\mathbf{L}_n(\mathbf{b})\| = \|\mathbf{L}_n(\mathbf{b}_1)\|_1 = \sum_{j=1}^{p} |L_{nj}(\mathbf{b})|, \ \mathbf{b} \in \boldsymbol{R}_p$$

and define

$$\mathbf{b}^0 : \| \ \mathbf{L}_n(\mathbf{b}^0)\| = \inf_{\mathbf{b} \in \boldsymbol{R}_p} \|\mathbf{L}_n(\mathbf{b})\|. \tag{6.44}$$

Such $\mathbf{b}^0$ may not be uniquely determined. Let

$$\mathcal{D}_n = \text{ set of all } \mathbf{b}^0 \text{ satisfying (6.44).} \tag{6.45}$$

The justification of $\mathcal{D}_n$ in estimation of $\boldsymbol{\beta}$ rests in establishing that $\mathcal{D}_n$ represents a sufficiently shrinking neighborhood of $\boldsymbol{\beta}$, in probability, as $n \to \infty$. The arbitrariness in (6.44) we can eliminate by letting

$$\widehat{\boldsymbol{\beta}}_n = \text{ center of gravity of } \mathcal{D}_n. \tag{6.46}$$

The L_1-norm in (6.44) can be replaced by L_2 or, in general, an L_r norm for an arbitrary $r > 0$; we shall see later that the choice of r in an L_r-norm is not that crucial. A slightly different formulation used Jaeckel (1972) who eliminated this arbitrariness by using the measure of *rank dispersion*

$$D_n(\mathbf{b}) = \sum_{i=1}^{n} (Y_i - \mathbf{b}^\top \mathbf{x}_i) a_n(R_{ni}(\mathbf{b})), \ \ \mathbf{b} \in \boldsymbol{R}_p \tag{6.47}$$

where $a_n(1), \ldots, a_n(n)$ are again nondecreasing scores, and proposed to estimate $\boldsymbol{\beta}$ by minimizing $D_n(\mathbf{b})$ with respect to $\mathbf{b} \in \boldsymbol{R}_p$. If we set $\bar{a}_n = n^{-1} \sum_{i=1}^{n} a_n(i) = 0$, without loss of generality, we can show that $D_n(\mathbf{b})$ is translation-invariant. Indeed, if we denote $D_n(\mathbf{b}, k)$ the measure (6.47) calculated for the pseudo-observations $Y_i + k - \mathbf{b}^\top \mathbf{x}_i, \ i = 1, \ldots, n$, we can write

$$\begin{aligned} D_n(\mathbf{b}, k) &= \sum_{i=1}^{n} (Y_i + k - \mathbf{b}^\top \mathbf{x}_i)(a_n(R_{ni}(\mathbf{b})) - \bar{a}_n) \qquad (6.48) \\ &= D_n(\mathbf{b}) + k \sum_{i=1}^{n} (a_n(R_{ni}(\mathbf{b})) - \bar{a}_n) = D_n(\mathbf{b}) \end{aligned}$$

because $a_n(R_{ni}(\mathbf{b}))$ are translation-invariant. Moreover, Jaeckel (1972) showed that

$$\begin{aligned} &D_n(\mathbf{b}) \text{ is nonnegative, continuous, piecewise linear,} \\ &\qquad \text{and convex function of } \mathbf{b} \in \boldsymbol{R}_p. \end{aligned} \tag{6.49}$$

These properties ensure that $D_n(\mathbf{b})$ is differentiable in $\mathbf{b}$ almost everywhere and

$$(\partial/\partial \mathbf{b}) D_n(\mathbf{b}) \Big|_{\mathbf{b}^0} = -L_n(\mathbf{b}^0) \tag{6.50}$$

at any point $\mathbf{b}^0$ of differentiability of D_n. If D_n is not differentiable in $\mathbf{b}^0$, we can work with the subgradient $\nabla D_n(\mathbf{b}_0)$ of D_n at $\mathbf{b}_0$ defined as the operation satisfying

$$D_n(\mathbf{b}) - D_n(\mathbf{b}^0) \geq (\mathbf{b} - \mathbf{b}^0)\nabla D_n(\mathbf{b}_0) \tag{6.51}$$

for all $\mathbf{b} \in \boldsymbol{R}_p$. The following representation of R-estimate, which is an extension of the original construction, will be proved under the following assumptions:

1. *Smoothness of* F: The errors e_i in (6.41) are i.i.d. random variables with a distribution function F (not depending on n) which possesses an absolutely continuous density f with a positive and finite Fisher information $I(f)$ [defined by (6.15)].
2. *Score-generating function*: The function $\varphi: (0,1) \mapsto \boldsymbol{R}_1$ is assumed to be nonconstant, nondecreasing and square integrable on $(0,1)$, so
$$0 < A_\varphi^2 = \int_0^1 \varphi^2(u)du < \infty. \tag{6.52}$$
The scores $a_n(i)$ are defined either by
$$a_n(i) = \mathbb{E}[\varphi(U_{n:i})], \; i = 1, \ldots, n, \tag{6.53}$$
or by
$$a_n(i) = \varphi(\mathbb{E}U_{n:i}) = \varphi(i/(n+1)), \; i = 1, \ldots, n.$$
3. *Generalized Noether condition*: Assume that
$$\lim_{n\to\infty} \max_{1\leq i\leq n} (\mathbf{x}_i - \bar{\mathbf{x}}_n)^\top \mathbf{V}_n^{-1}(\mathbf{x}_i - \bar{\mathbf{x}}_n) = 0, \tag{6.54}$$
where
$$\mathbf{V}_n = \sum_{i=1}^n (\mathbf{x}_i - \bar{\mathbf{x}}_n)(\mathbf{x}_i - \bar{\mathbf{x}}_n)^\top, \; n > p. \tag{6.55}$$

For convenience, apply the following orthonormal transformation on the model (6.41): Put

$$\mathbf{x}_i^\star = \mathbf{V}_n^{-\frac{1}{2}}\mathbf{x}_i, \tag{6.56}$$

and introduce the reparametrization

$$\boldsymbol{\beta}_n^\star = \boldsymbol{\beta}^\star = \mathbf{V}_n^{\frac{1}{2}}\boldsymbol{\beta}, \; n = 1, 2, \ldots \tag{6.57}$$

Then,

$$\mathbf{V}_n^\star = \sum_{i=1}^n (\mathbf{x}_i^\star - \bar{\mathbf{x}}_n^\star)(\mathbf{x}_i^\star - \bar{\mathbf{x}}_n^\star)^\top = \mathbf{I}_p,$$

$$\boldsymbol{\beta}^\top \mathbf{x}_i = \boldsymbol{\beta}^{\star\top}\mathbf{x}_i^\star, \; 1 \leq i \leq n. \tag{6.58}$$

The reparametrization (6.57) accompanied by a conjugate transformation (6.56) leave the estimation problem invariant and this simplification does not reduce the generality of the model. Consider the reduced linear rank statistics $\mathbf{L}_n^\star(\mathbf{b}) = \left(L_{n1}^\star(\mathbf{b}), \ldots, L_{np}^\star(\mathbf{b})\right)^\top$,

$$\mathbf{L}_n^\star(\mathbf{b}) = \sum_{i=1}^{n} (\mathbf{x}_i^\star - \bar{\mathbf{x}}_n^\star) a_n(R_{ni}^\star(\mathbf{b})), \ \ \mathbf{b} \in \boldsymbol{R}_p,$$

where $R_{ni}^\star(\mathbf{b})$ are defined as in (6.42) with the $\mathbf{x}_i$ replaced by $\mathbf{x}_i^\star$. Moreover, consider the random vectors $\mathbf{S}_n^\star(\mathbf{b}) = \left(S_{n1}^\star(\mathbf{b}), \ldots, S_{nn}^\star(\mathbf{b})\right)^\top$,

$$\mathbf{S}_n^\star(\mathbf{b}) = \sum_{i=1}^{n} (\mathbf{x}_i^\star - \bar{\mathbf{x}}_n^\star) \varphi(F(Y_i - \theta - \mathbf{b}^\top \mathbf{x}_n^\star)), \ \ \mathbf{b} \in \boldsymbol{R}_p. \tag{6.59}$$

We shall show that $\mathbf{L}_n^\star(\mathbf{b})$ and $\mathbf{S}_n^\star(\mathbf{b})$ are close to each other for large n.

THEOREM 6.4 *(Heiler and Willers). Let $\{Y_i;\ i \geq 1\}$ follow the model (6.41) where the i.i.d. errors e_i have a continuous distribution function F. Let φ be a nondecreasing function such that $\int_0^1 \varphi^2(u)du < \infty$. Then, under assumption 2,*

$$\|\mathbf{L}_n^\star(\mathbf{0}) - \mathbf{S}_n^\star(\mathbf{0})\| \xrightarrow{P} 0 \quad \text{as } n \to \infty. \tag{6.60}$$

PROOF Let $R_{ni} = R_{ni}(\mathbf{0})$ and let $\mathcal{B}_n = \mathcal{B}(R_{n1}(0), \ldots, R_{nn}(0))$ be the sigma-field generated by the vector of ranks at the nth stage, $n \geq 1$. When $\boldsymbol{\beta} = \mathbf{0}$, then $F(Y_i)$ are i.i.d. random variables having the uniform (0, 1) distribution function, $i = 1, \ldots, n$. Moreover, the vector of ranks and order statistics are stochastically independent. thus, $\mathbb{E}[\varphi(F(Y_i))|\mathcal{B}_n] = \mathbb{E}[\varphi(U_{n:R_{ni}})|R_{ni}$ given$] = a_n(R_{ni})$ for every $i = 1, \ldots, n$. hence, we obtain

$$\mathbb{E}[S_{nj}^*(\mathbf{0})|\mathcal{B}_n] = \sum_{i=1}^{n} (x_{ij}^* - \bar{x}_{nj}^*) \mathbb{E}[\varphi(F(Y_i))|\mathcal{B}_n] \tag{6.61}$$

$$= \sum_{i=1}^{n} (x_{ij}^* - \bar{x}_{nj}^*) a_n(R_{ni}(0)) = L_{nj}^*(\mathbf{0}) \text{ for every } n \geq 2,\ j = 1, \ldots, p.$$

Now we have

$$\mathbb{E}[(S_{nj}^*(\mathbf{0}) - L_{nj}^*(\mathbf{0})]^2 = \mathbb{E}[(S_{nj}^*(\mathbf{0})]^2 - \mathbb{E}[(L_{nj}^*(\mathbf{0})]^2$$

$$= \int_0^1 \varphi^2(u)du - \frac{n}{n-1}\Big[n^{-1}\sum_{i=1}^{n} a_n^2(i)\Big],\ j = 1, \ldots, p. \tag{6.62}$$

Proceeding as in the proof of Theorem 6.1, we find that the right-hand side of (6.62) converges to 0 as $n \to \infty$, and hence, (6.60) holds. □

Let P_n and $P_n^\star$ be the probability distributions with the respective densities $p_n = \prod_{i=1}^{n} f(Y_i - \theta)$ and $p_n^\star = \prod_{i=1}^{n} f(Y_i - \theta - \boldsymbol{\lambda}^\top \mathbf{x}_i^\star)$, $\boldsymbol{\lambda} \in \boldsymbol{R}_p$ fixed. Then $\{P_n^\star\}$ are contiguous to $\{P_n\}$. Proceeding similarly as in (6.21)–(6.22), we conclude from (6.60) that

$$\|\mathbf{L}_n^\star(\boldsymbol{\lambda}) - \mathbf{S}_n^\star(\boldsymbol{\lambda})\| \xrightarrow{P} 0 \quad \text{as } n \to \infty \tag{6.63}$$

under $\boldsymbol{\beta}^\star = \mathbf{0}$ and for any fixed $\boldsymbol{\lambda} \in \boldsymbol{R}_p$. In the following step, we shall prove the uniform asymptotic linearity of $\mathbf{L}_n^\star(\boldsymbol{\lambda})$ in $\boldsymbol{\lambda}$.

THEOREM 6.5 *Under assumptions 1–3, for monotone score generating function* φ,

$$\sup\{\|\mathbf{L}_n^*(\boldsymbol{\lambda}) - \mathbf{L}_n^*(\mathbf{0}) + \gamma\boldsymbol{\lambda}\| : \ \|\boldsymbol{\lambda}\| \leq C\} \xrightarrow{P} 0 \tag{6.64}$$

as $n \to \infty$ *for any fixed* C, $0 < C < \infty$, *where*

$$\gamma = \int_{-\infty}^{\infty} \varphi(F(x))\{-f'(x)\}dx = \int_0^1 \varphi(u)\psi_f(u)du = \langle \varphi, \psi_f \rangle, \tag{6.65}$$

and $\psi_f(.)$ *is defined by (6.18).*

PROOF For a fixed $\boldsymbol{\lambda}$, we have by (6.59)

$$\begin{aligned} & \mathbf{S}_n^\star(\boldsymbol{\lambda}) - \mathbf{S}_n^\star(\mathbf{0}) + \gamma\boldsymbol{\lambda} \\ = \ & \sum_{i=1}^n (\mathbf{x}_i^\star - \bar{\mathbf{x}}_n^\star)\big[\varphi(F(Y_i - \theta - \boldsymbol{\lambda}^\top \mathbf{x}_i^\star)) - \varphi(F(Y_i - \theta)) \\ & + \gamma(\mathbf{x}_i^\star - \bar{\mathbf{x}}_n^\star)^\top \boldsymbol{\lambda}\big]. \end{aligned} \tag{6.66}$$

The right-hand side of (6.66) is a sum of independent summands and its stochastic convergence to $\mathbf{0}$ follows from the multivariate central limit theorem, in its degenerate form. Thus, under $\boldsymbol{\beta}^\star = \mathbf{0}$ and under all three assumptions,

$$\|\mathbf{S}_n^\star(\boldsymbol{\lambda}) - \mathbf{S}_n^\star(\mathbf{0}) + \gamma\boldsymbol{\lambda}\| \xrightarrow{P} 0 \quad \text{as } n \to \infty,$$

for any fixed $\boldsymbol{\lambda} \in \boldsymbol{R}_p$. hence,

$$\begin{aligned} & \|\mathbf{L}_n^\star(\boldsymbol{\lambda}) - \mathbf{L}_n^\star(\mathbf{0}) + \gamma\boldsymbol{\lambda}\| \\ & \leq \|\mathbf{L}_n^\star(\boldsymbol{\lambda}) - \mathbf{S}_n^\star(\boldsymbol{\lambda})\| + \|\mathbf{S}_n^\star(\boldsymbol{\lambda}) - \mathbf{S}_n^\star(\mathbf{0}) + \gamma\boldsymbol{\lambda}\| \\ & + \|\mathbf{L}_n^\star(\mathbf{0}) - \mathbf{S}_n^\star(\mathbf{0})\| \xrightarrow{P} 0 \quad \text{as } n \to \infty. \end{aligned} \tag{6.67}$$

under all three assumptions and for fixed $\boldsymbol{\lambda} \in \boldsymbol{R}_p$.

It remains to prove that the convergence in (6.21) is uniform over any compact set $I\!\!K = \{\boldsymbol{\lambda} : \ \|\boldsymbol{\lambda}\| \leq C\}$, $0 < C < \infty$. In the case $p = 1$ the uniformity result follows from the monotonicity of $L_n^\star(b)$ in b. For $p \geq 2$, the proof of (6.64) rests on (6.50), (6.51), on the convexity of $D_n(\mathbf{b})$ in (6.49) and a convexity argument, included in (2.103)–(2.104). □

The following theorem gives the asymptotic representation for $\widehat{\boldsymbol{\beta}}$.

THEOREM 6.6 *Let* $\widehat{\boldsymbol{\beta}}_n$ *be the R-estimator of* $\boldsymbol{\beta}$ *in the linear model (6.41), defined by (6.46). Then, under assumptions 1, 2, and 3,* $\widehat{\boldsymbol{\beta}}_n$ *admits the asymptotic representation*

$$\widehat{\boldsymbol{\beta}}_n - \boldsymbol{\beta} = \gamma^{-1}\mathbf{V}_n^{-1} \sum_{i=1}^n (\mathbf{x}_i - \bar{\mathbf{x}}_n)\varphi(F(e_i)) + o_p\left(\|\mathbf{V}_n^{-\frac{1}{2}}\|\right). \tag{6.68}$$

Moreover, $\mathbf{V}_n^{\frac{1}{2}}(\widehat{\boldsymbol{\beta}}_n - \boldsymbol{\beta})$ *is asymptotically normally distributed*

$$\mathbf{V}_n^{\frac{1}{2}}(\widehat{\boldsymbol{\beta}}_n - \boldsymbol{\beta}) \sim \mathcal{N}_p(\mathbf{0}, \gamma^{-2}A_\varphi^2\mathbf{I}_p). \tag{6.69}$$

PROOF By virtue of (6.56)–(6.58), we have

$$R_{ni}^\star(\boldsymbol{\lambda}) = R_{ni}(\mathbf{V}_n^{-\frac{1}{2}}\boldsymbol{\lambda}),\ \mathbf{x}_i^\star = \mathbf{V}_n^{-\frac{1}{2}}\mathbf{x}_i,\ \ i = 1, \ldots, n,$$

so that

$$\mathbf{L}_n^\star(\boldsymbol{\lambda}) = \mathbf{V}_n^{-\frac{1}{2}}\mathbf{L}_n(\mathbf{V}_n^{-\frac{1}{2}}\boldsymbol{\lambda}),\ \ \forall \boldsymbol{\lambda} \in \boldsymbol{R}_p.$$

thus, we may rewrite (6.64) in the form

$$\sup\left(\left\|\mathbf{V}_n^{-\frac{1}{2}}\left\{\mathbf{L}_n(\mathbf{V}_n^{-\frac{1}{2}}\boldsymbol{\lambda}) - \mathbf{L}_n(\mathbf{0}) + \mathbf{V}_n^{\frac{1}{2}}\boldsymbol{\lambda}\gamma\right\}\right\| :\ \|\boldsymbol{\lambda}\| \leq C\right) \xrightarrow{P} 0$$

as $n \to \infty$, under assumptions 1 to 3 and under $\boldsymbol{\beta} = \mathbf{0}$. This can be further rewritten as

$$\sup\left(\left\|\mathbf{V}_n^{-\frac{1}{2}}\left\{\mathbf{L}_n(\mathbf{b}) - \mathbf{L}_n(\mathbf{0}) + \gamma\mathbf{V}_n\mathbf{b}\right\}\right\| :\ \|\mathbf{V}_n^{\frac{1}{2}}\mathbf{b}\| \leq C\right) \xrightarrow{P} 0 \tag{6.70}$$

as $n \to \infty$. Finally, translating (6.70) to the case of true $\boldsymbol{\beta}$, we have under the three assumptions

$$\sup_{\|\mathbf{V}_n^{\frac{1}{2}}(\mathbf{b}-\boldsymbol{\beta})\| \leq C}\left(\left\|\mathbf{V}_n^{-\frac{1}{2}}\left\{\mathbf{L}_n(\mathbf{b}) - \mathbf{L}_n(\boldsymbol{\beta}) + \gamma\mathbf{V}_n(\mathbf{b} - \boldsymbol{\beta})\right\}\right\|\right) \xrightarrow{P} 0 \tag{6.71}$$

under $\boldsymbol{\beta}$ and for any fixed C, $0 < C < \infty$.

The uniform asymptotic linearity in (6.71), along with the monotonicity of φ, enables us to conclude that the diameter of $\mathcal{D}_n$, defined in (6.44)–(6.46), is $O_p(\|\mathbf{V}_n^{-\frac{1}{2}}\|)$, and that under the three assumptions,

$$\sup\left(\left\|\mathbf{V}_n^{-\frac{1}{2}}\left\{\mathbf{L}_n(\boldsymbol{\beta}) - \gamma\mathbf{V}_n(\mathbf{b} - \boldsymbol{\beta})\right\}\right\| :\ \mathbf{b} \in \mathcal{D}_n\right) \xrightarrow{P} 0, \quad \text{as } n \to \infty.$$

For every $\widehat{\boldsymbol{\beta}}_n$ belonging to $\mathcal{D}_n$, we have that

$$\begin{aligned} \gamma\mathbf{V}_n^{\frac{1}{2}}(\widehat{\boldsymbol{\beta}}_n - \boldsymbol{\beta}) &= \mathbf{V}_n^{-\frac{1}{2}}\mathbf{L}_n(\boldsymbol{\beta}) + \mathbf{o}_p(1) \\ &= \mathbf{L}_n^\star(\boldsymbol{\beta}^\star) + \mathbf{o}_p(1) = \mathbf{S}_n^\star(\boldsymbol{\beta}^\star) + \mathbf{o}_p(1). \end{aligned} \tag{6.72}$$

Moreover, $\mathbf{S}_n^\star(\boldsymbol{\beta}^\star)$ under $\boldsymbol{\beta}^\star$ has the same distribution as $\mathbf{S}_n^\star(\mathbf{0})$ under $\boldsymbol{\beta}^\star = \mathbf{0}$, by (6.59). Using the multivariate central limit theorem, we conclude that under $\boldsymbol{\beta}^\star = \mathbf{0}$ and under assumptions 2 and 3,

$$\mathbf{S}_n^\star(\mathbf{0}) \sim \mathcal{N}_p(\mathbf{0}, A_\varphi^2\mathbf{I}_p), \tag{6.73}$$

where A_φ^2 is defined by (6.54). By (6.72) and (6.73), we arrive in the proposition of the theorem. □

The propositions apply also to other forms of the scores, mentioned above. An analogous argument may be based on the convexity of the Jaeckel measure of dispersion (6.47). By the asymptotic linearity, (6.64) and (6.50), it can be approximated by a quadratic function of $\mathbf{b}$ uniformly over $\|\mathbf{b}\| \leq C$. We have used similar arguments already in connection with regression quantiles (Chapter 4) and with M-estimators generated by a step function (Chapter 5).

Let us consider a representation for an R-estimator of the intercept parameter θ. We should assume that F is symmetric around 0 (otherwise $\hat{\theta}_n$ is generally biased) and that

$$\lim_{n\to\infty} n^{\frac{1}{2}} \bar{\mathbf{x}}_n^\top \mathbf{V}_n^{-\frac{1}{2}} = \boldsymbol{\xi}^\top \quad \text{exists.}$$

Rewrite (6.41) as

$$\begin{aligned} Y_i &= \theta + \boldsymbol{\beta}^\top \mathbf{x}_i + e_i \\ &= (\theta - \boldsymbol{\beta}^\top \bar{\mathbf{x}}_n) + \boldsymbol{\beta}^\top (\mathbf{x}_i - \bar{\mathbf{x}}_n) + e_i \\ &= \theta_n^0 + \boldsymbol{\beta}^\top (\mathbf{x}_i - \bar{\mathbf{x}}_n) + e_i, \quad i = 1, \ldots, n, \end{aligned}$$

where $\theta_n^0 = \theta + \boldsymbol{\beta}^\top \bar{\mathbf{x}}_n$. Replace $\boldsymbol{\beta}$ by its R-estimator $\widehat{\boldsymbol{\beta}}_n$, and consider the residuals

$$\begin{aligned} \hat{X}_i &= Y_i - \widehat{\boldsymbol{\beta}}_n^\top (\mathbf{x}_i - \bar{\mathbf{x}}_n) - \widehat{\boldsymbol{\beta}}_n^\top \bar{\mathbf{x}}_n \\ &= \theta + e_i + (\boldsymbol{\beta} - \widehat{\boldsymbol{\beta}}_n)^\top (\bar{\mathbf{x}}_n + (\mathbf{x}_i - \bar{\mathbf{x}}_n)) \\ &= \theta + e_i + (\boldsymbol{\beta} - \widehat{\boldsymbol{\beta}}_n)^\top \mathbf{V}_n^{\frac{1}{2}} \Big\{ \mathbf{V}_n^{-\frac{1}{2}} (\bar{\mathbf{x}}_n + (\mathbf{x}_i - \bar{\mathbf{x}}_n)) \Big\}, \end{aligned} \tag{6.74}$$

$i = 1, \ldots, n$. Note that by (6.69),

$$\|(\boldsymbol{\beta} - \widehat{\boldsymbol{\beta}}_n)^\top \mathbf{V}_n^{\frac{1}{2}}\| = O_p(1),$$

while by (6.54),

$$\max_{1\leq i\leq n} \|\mathbf{V}_n^{-\frac{1}{2}} (\mathbf{x}_i - \bar{\mathbf{x}}_n)\| = o_p(1).$$

hence, the last term on the right-hand side of (6.74) is negligible in probability uniformly in $\{i : 1 \leq i \leq n\}$, while the results of Section 6.2 apply for $\theta + e_i$, $i = 1, \ldots, n$. An R-estimator $\hat{\theta}_n$ of θ, based on the residuals $\hat{X}_i$, $1 \leq i \leq n$, may be defined as in (3.78).

The estimator will be based on the signed-rank statistic with the score function $\varphi(u) = \varphi^\star((u+1)/2)$, $0 \leq u \leq 1$ where $\varphi^\star$ is skew-symmetric about 1/2 [i.e., $\varphi^\star(u) + \varphi^\star(1-u) = 0$, $0 < u < 1$]. The unknown $\boldsymbol{\beta}$ in this signed-rank statistic is replaced by $\widehat{\boldsymbol{\beta}}_n$ (aligned signed-rank statistic based on the $\hat{X}_i$), and we can prove the uniform asymptotic linearity in α, $\boldsymbol{\lambda}$, parallel to (6.64). As such, we have

$$\begin{aligned} & n^{\frac{1}{2}} (\hat{\theta}_n - \theta) \\ &= \frac{1}{\gamma\sqrt{n}} \Big\{ \sum_{i=1}^n \varphi^\star(F(e_i)) - \boldsymbol{\xi}^\top \sum_{i=1}^n \mathbf{V}_n^{-\frac{1}{2}} (\mathbf{x}_i - \bar{\mathbf{x}}_n) \varphi^\star(F(e_i)) \Big\} + o_p(1) \\ &= \frac{1}{\gamma\sqrt{n}} \sum_{i=1}^n \Big[1 - \boldsymbol{\xi}^\top \mathbf{V}_n^{-\frac{1}{2}} (\mathbf{x}_i - \bar{\mathbf{x}}_n) \Big] \varphi^\star(F(e_i)) + o_p(1). \end{aligned} \tag{6.75}$$

This asymptotic representation implies the asymptotic normality of $n^{\frac{1}{2}}(\hat{\theta}_n - \theta)$. Combining (6.68) and (6.75), we conclude that, under the regularity conditions,

$$n^{\frac{1}{2}} \begin{bmatrix} \hat{\theta}_n - \theta \\ \widehat{\boldsymbol{\beta}}_n - \boldsymbol{\beta} \end{bmatrix} = \frac{1}{\gamma\sqrt{n}} \sum_{i=1}^{n} \begin{bmatrix} 1 - \boldsymbol{\xi}^\top \mathbf{d}_{ni} \\ n^{\frac{1}{2}} \mathbf{d}_{ni} \end{bmatrix} \varphi^\star(F(e_i)) + o_p(1) \qquad (6.76)$$

as $n \to \infty$, where

$$\mathbf{d}_{ni} = \mathbf{V}_n^{-\frac{1}{2}}(\mathbf{x}_i^\star - \bar{\mathbf{x}}_n^\star) = \mathbf{V}_n^{-1}(\mathbf{x}_i - \bar{\mathbf{x}}_n), \;\; 1 \leq i \leq n.$$

6.4 Regression Rank Scores

Consider the linear model

$$Y_i = \boldsymbol{\beta}^\top \mathbf{x}_i + e_i, \;\; i = 1, \ldots, n, \qquad (6.77)$$

where the Y_i are the observable random variables, $\mathbf{x}_i = (x_{i1}, \ldots, x_{ip})^\top, \; i \geq 1$ are given p-vectors of known regression constants, $x_{i1} = 1 \; \forall i \geq 1$, $\boldsymbol{\beta} = (\beta_1, \ldots, \beta_p)^\top$ is a p-vector of unknown parameters, and the e_i are i.i.d. random variables having a continuous distribution function F defined on $\boldsymbol{R}_1$. Assume that $p \geq 2$, and note that β_1 is an intercept, while $\beta_2, \ldots, \beta_p$ are slopes. The case of $p = 1$ relates to the location model treated in Section 6.2.

In Section 4.7, the α-regression quantile $\widehat{\boldsymbol{\beta}}(\alpha)$ (RQ) has been introduced and its possible characterization as an optimal solution of the linear program (4.89) has been shown. The optimal solution of the dual linear program (4.91) extends, rather surprisingly, the concept of rank scores from the location model to the linear regression model; hence, we will name it the *regression rank scores* (RR). The duality of RQs to RRs is not only in the linear programming sense, but it also extends the duality of order statistics and ranks from the location to the linear regression model.

The vector of regression rank scores, denoted by the vector $\hat{\mathbf{a}}_n(\alpha) = (\hat{a}_{n1}(\alpha), \ldots, \hat{a}_{nn}(\alpha))^\top$, $\alpha \in (0, 1)$, is defined as the optimal solution of the linear programming problem:

$$\sum_{i=1}^{n} Y_i \hat{a}_{ni}(\alpha) = \max,$$

$$\sum_{i=1}^{n} x_{ij} \hat{a}_{ni}(\alpha) = (1 - \alpha) \sum_{i=1}^{n} x_{ij}, \;\; j = 1, \ldots, p;$$

$$\hat{a}_{ni}(\alpha) \in [0, 1], \;\; \forall 1 \leq i \leq n, \;\; 0 \leq \alpha \leq 1. \qquad (6.78)$$

In the location model, where $p = 1$ and $x_i = 1, \; i = 1, \ldots, n$, we have $\hat{a}_{ni}(\alpha) =$

$a_n^\star(R_i, \alpha)$, where R_i is the rank of Y_i among $Y_1, \ldots, Y_n$ and

$$a_n^\star(R_i, \alpha) = \begin{cases} 0, & \text{if } \frac{R_i}{n} < \alpha, \\ R_i - n\alpha, & \text{if } \frac{R_i - 1}{n} \leq \alpha \leq \frac{R_i}{n}, \\ 1, & \text{if } \alpha < \frac{R_i - 1}{n}, \end{cases} \tag{6.79}$$

$i = 1, \ldots, n$, are scores considered by Hájek (1965), as a starting point for an extension of the Kolmogorov–Smirnov test to regression alternatives.

While the regression quantiles are suitable mainly for estimation, the regression rank scores are used for testing hypotheses in the linear model, particularly when the hypothesis concerns only some components of $\boldsymbol{\beta}$ and other components are considered as nuisance. In turn, one may construct estimators (RR-estimators) of sub-vectors of $\boldsymbol{\beta}$ by inverting such tests.

Let us illustrate some finite-sample properties of regression rank scores, in order to see how far the analogy with the ordinary ranks goes and where the deviations start. In the model (6.77), denote $\mathbf{X} = \mathbf{X}_n$ the $n \times p$ matrix with components x_{ij}, $1 \leq i \leq n$ and $1 \leq j \leq p$, assume that $\mathbf{X}$ is of rank p and satisfies

$$x_{i1} = 1, \quad i = 1, \ldots, n. \tag{6.80}$$

Let $\widehat{\boldsymbol{\beta}}(\alpha)$ be the α-regression quantile and let $\hat{\mathbf{a}}_n(\alpha)$ be the vector of regression rank scores, $0 < \alpha < 1$. Some finite-sample properties of these statistics are summarized in the following lemma:

LEMMA 6.4 *(1)* $\hat{\mathbf{a}}_n(\alpha, \mathbf{Y} + \mathbf{X}\mathbf{b}) = \hat{\mathbf{a}}_n(\alpha, \mathbf{Y})\ \forall \mathbf{b} \in \boldsymbol{R}_p$, *(i.e., the regression rank scores are invariant to the regression with matrix* $\mathbf{X}$ *).*

(2)

$$\bar{\mathbf{x}}^\top \hat{\beta}_1(\alpha_1) \leq x\bar{b}f^{top} \hat{\beta}_2(\alpha_2) \quad \textit{for any } \alpha_1 \leq \alpha_2$$

where $\bar{\mathbf{x}} = \frac{1}{n}\sum_{i=1}^n \mathbf{x}_i$.

(3) If, on the contrary, $\sum_{i=1}^n x_{ij} = 0$ *for* $j = 1, \ldots, p$, *then*

$$\hat{\beta}_1(\alpha_1) = \hat{\beta}_2(\alpha_2) \quad \textit{for any } 0 < \alpha_1, \alpha_2 < 1$$

(i.e., the regression quantiles corresponding to different α*'s are not distinguishable).*

Remark 6.1 *By proposition 1, every statistical procedure based on regression rank scores is invariant with respect to the* $\mathbf{X}$*-regression. This is important mainly in the models where we have a nuisance* $\mathbf{X}$*-regression. Moreover, the monotonicity of* $\bar{\mathbf{x}}^\top \hat{\beta}_1(\alpha)$ *claimed in proposition 2 is in correspondence with the fact that the first component of the regression quantile represents the location quantile, while the other components estimate the slope parameters.*

PROOF Proposition 1 follows directly from (6.78). Moreover, for every $\mathbf{t} \in \boldsymbol{R}_p$,

$$\sum_{i=1}^{n}[\rho_{\alpha_2}(Y_i - \mathbf{x}_i^\top \mathbf{t}) - \rho_{\alpha_1}(Y_i - \mathbf{x}_i^\top \mathbf{t})] = (\alpha_2 - \alpha_1)\sum_{i=1}^{n}(Y_i - \mathbf{x}_i^\top \mathbf{t}),$$

where $\rho_\alpha(x) = |x|\{\alpha I[x > 0] + (1-\alpha)I[x < 0]\}$. This immediately implies 3. This further implies under (6.80) that

$$\sum_{i=1}^{n}\rho_{\alpha_2}(Y_i - \mathbf{x}_i^\top \widehat{\boldsymbol{\beta}}(\alpha_2)) \geq \sum_{i=1}^{n}\rho_{\alpha_1}(Y_i - \mathbf{x}_i^\top \widehat{\boldsymbol{\beta}}(\alpha_1)) + n(\alpha_2 - \alpha_1)(\bar{Y} - \widehat{\boldsymbol{\beta}}(\alpha_2)), \tag{6.81}$$

where $\bar{Y} = n^{-1}\sum_{i=1}^{n} Y_i$. Similarly

$$\sum_{i=1}^{n}\rho_{\alpha_1}(Y_i - \mathbf{x}_i^\top \widehat{\boldsymbol{\beta}}(\alpha_1)) \geq \sum_{i=1}^{n}\rho_{\alpha_2}(Y_i - \mathbf{x}_i^\top \widehat{\boldsymbol{\beta}}(\alpha_2)) - n(\alpha_2 - \alpha_1)(\bar{Y} - \widehat{\boldsymbol{\beta}}(\alpha_1)), \tag{6.82}$$

Combining (6.81) and (6.82), we arrive at proposition 2. □

LEMMA 6.5 *Under (6.80) and for* $0 < \alpha_1 \leq \alpha_2 < 1$,

$$\begin{aligned}(\alpha_2 - \alpha_1)\bar{\mathbf{x}}^\top \widehat{\boldsymbol{\beta}}_1(\alpha_1) &\leq n^{-1}\sum_{i=1}^{n} Y_i[\hat{a}_i(\alpha_1) - \hat{a}_i(\alpha_2)] \\ &\leq (\alpha_2 - \alpha_1)\bar{\mathbf{x}}^\top \widehat{\boldsymbol{\beta}}(\alpha_2)\end{aligned} \tag{6.83}$$

Moreover, if α *is a point of continuity of* $\widehat{\boldsymbol{\beta}}(.)$, *then*

$$\bar{\mathbf{x}}^\top \widehat{\boldsymbol{\beta}}(\alpha) = -\frac{1}{n}\sum_{i=1}^{n} Y_i \hat{a}_i'(\alpha) \tag{6.84}$$

where $\hat{a}_i'(\alpha) = \frac{\partial\, a_i(\alpha)}{\partial \alpha}$.

Remark 6.2 *For a fixed* $\alpha \in (0,1)$, *let* $I(\alpha) = \{i : 0 < \hat{a}_i(\alpha) < 1\}$; *then* $Y_i = \mathbf{x}_i^\top \widehat{\boldsymbol{\beta}}(\alpha)$ *for* $i \in I(\alpha)$ *and* $\#I(\alpha) = p$. *The identity (6.84) means that while* $\bar{\mathbf{x}}^\top \widehat{\boldsymbol{\beta}}(\alpha)$ *coincides just with one order statistic (i.e., with one observation) in the location model, in the linear regression model it coincides with a linear combination of those* Y_i*'s which have an exact fit for* α. *The weight of* Y_i *in (6.84) is positive or negative according as whether* $\hat{a}_i(\alpha)$ *is decreasing or increasing in* α *for* $i \in I(\alpha)$. *The* $\hat{a}_i(\alpha)$, $i \in I(\alpha)$, *satisfy*

$$-\frac{1}{n}\sum_{i=1}^{n}\hat{a}_i'(\alpha) = 1, \ \sum_{i\in I(\alpha)} x_{ij}\hat{a}_i'(\alpha) = -\sum_{i=1}^{n} x_{ij}, \ j = 2, \ldots, p.$$

Remark 6.3 *The sets* $\{\widehat{\boldsymbol{\beta}}(\alpha): 0 < \alpha < 1\}$ *and* $\{I(\alpha): 0 < \alpha < 1\}$ *form a sufficient statistic for the model (6.77). In fact, with probability 1, every* i_0 *belongs to some* $I(\alpha_{i_0})$, $i_0 = 1, \ldots, n$ *by the continuity of* $\hat{a}_i(\alpha)$ *and due to the fact that* $\hat{a}_i(0) = 1$, $\hat{a}_i(1) = 0$, $i = 1, \ldots, n$. *Fixing the corresponding* α_0, *we have* $Y_{i_0} = \mathbf{x}_{i0}^\top \widehat{\boldsymbol{\beta}}(\alpha_{i_0})$ *with probability 1,* $i_0 = 1, \ldots, n$.

PROOF The duality of $\widehat{\boldsymbol{\beta}}(\alpha)$ and $\hat{\mathbf{a}}(\alpha)$ implies that

$$\sum_{i=1}^{n} \rho_\alpha(Y_i - \mathbf{x}_i^\top \widehat{\boldsymbol{\beta}}(\alpha)) = \sum_{i=1}^{n} Y_i(\hat{a}_i(\alpha) - (1-\alpha)).$$

hence, (6.83) follows from (6.81) and (6.82). Moreover, $\hat{\beta}_1(\alpha)$ is a step function of α, and $\hat{a}_i(\alpha)$ is a continuous, piecewise linear function of α, $0 < \alpha < 1$. The breakpoints of two functions coincide (e.g., see Gutenbrunner and Jurečková (1992)). thus, (6.84) follows from (6.83). □

6.5 Inference Based on Regression Rank Scores

The statistical inference based on RR's is typically based on the functionals of the weighted *regression rank scores process*

$$\{n^{\frac{1}{2}} \sum_{i=1}^{n} d_{ni}\hat{a}_{ni}(\alpha) : 0 \le \alpha \le 1\} \tag{6.85}$$

with appropriate coefficients $(d_{n1}, \ldots, d_{nn})$. In the location model such process was studied by Hájek (1965) who among others proved its weak convergence to Brownian Bridge in the uniform topology on $[0, 1]$. Analogously, the process (6.85) can be approximated by an empirical process as well as by the Hájek process, uniformly in the segment $[\alpha_n^\star, 1 - \alpha_n^\star] \subset (0, 1)$ with a specific $\alpha_n^\star \downarrow 0$ as $n \to \infty$.

The statistical tests and estimates are based on *linear regression rank score statistics* (linear RR statistics), which are constructed in the following way: Choose a *score-generating function* $\varphi : (0, 1) \mapsto \boldsymbol{R}_1$, nondecreasing and square integrable on $(0, 1)$ and put

$$\varphi_n(\alpha) = \begin{cases} \varphi(\alpha_n^\star), & \text{if } 0 \le \alpha < \alpha_n^\star, \\ \varphi(\alpha), & \text{if } \alpha_n^\star \le \alpha \le 1 - \alpha_n^\star, \\ \varphi(1 - \alpha_n^\star), & \text{if } 1 - \alpha_n^\star < \alpha \le 1. \end{cases}$$

Let $\hat{\mathbf{a}}_n(\alpha) = (\hat{a}_{n1}(\alpha), \ldots, \hat{a}_{nn}(\alpha))^\top$ be the RR's corresponding to the model (6.77); calculate the *scores* $\widehat{\mathbf{b}} = (\hat{b}_{n1}, \ldots, \hat{b}_{nn})^\top$ generated by φ in the following way

$$\hat{b}_{ni} = -\int_0^1 \varphi_n(\alpha) d\hat{a}_{ni}(\alpha), \ i = 1, \ldots, n. \tag{6.86}$$

The *linear RR-statistic* is then

$$S_{nn} = n^{-\frac{1}{2}} \sum_{i=1}^{n} d_{ni} \hat{b}_{ni}. \tag{6.87}$$

Note that (6.86) can be rewritten as

$$\hat{b}_{ni} = -\int_0^1 \varphi_n(\alpha) \hat{a}'_{ni}(\alpha) d\alpha, \;\; i = 1, \ldots, n. \tag{6.88}$$

and that the derivatives $\hat{a}'_{ni}(\alpha)$ are step-functions on $[0,1]$. In the location model this reduces to

$$\hat{b}_{ni} = n \int_{\frac{R_i - 1}{n}}^{\frac{R_i}{n}} \varphi_n(\alpha), \;\; i = 1, \ldots, n.$$

Typical choices of φ are in correspondence with the classical rank-tests theory (Wilcoxon, normal scores, and median scores).

The asymptotic representations of RR-process and linear RR-statistics were proven under regularity conditions on the distribution of the errors, on $\mathbf{X}_n$ and on $\mathbf{d}_n$: namely, the density f is supposed being absolutely continuous, positive and the derivative f' of f being bounded a.e. on its domain; the tails of f are exponentially decreasing to 0 in a rate faster than those of the t distribution with 4 degrees of freedom. The conditions on $\mathbf{X}_n$ and $\mathbf{d}_n$ concern mainly their rate of convergence.

Important is the relation of the linear RR-statistic to the ordinary R-statistic, which in turn demonstrates that the tests based on both are asymptotically equivalent in the following way [Gutenbrunner (1993)]:

$$\sup \left\{ |n^{-\frac{1}{2}} \sum_{j=1}^{n} d_{ni} [\hat{a}_{ni}(\alpha) - a_n^{\star}(R_i, \alpha)]| : \; \alpha_n^{\star} \leq \alpha \leq 1 - \alpha_n^{\star} \right\} \xrightarrow{P} 0. \tag{6.89}$$

for a suitable $\alpha_n^{\star} \downarrow 0$ as $n \to \infty$, where $R_{n1}, \ldots, R_{nn}$ denote the ranks of errors $e_1, \ldots, e_n$, and $a_n^{\star}(R_i, \alpha)$, $i = 1, \ldots, n$, are the Hájek scores defined in (6.79). Under a more stringent condition on the distribution tails, the uniform approximation of RR process holds even over the whole segment [0,1].

The asymptotic representation of simple linear rank score statistic (6.87) was proved for the class of score functions satisfying some Chernoff-Savage-type condition, which covers the normal scores RR-test. Namely, the score function φ should satisfy

$$|\varphi'(t)| \leq c[t(1-t)]^{-1-\delta^{\star}}$$

with $c > 0$ and $0 < \delta < \delta^{\star}$. Let S_{nn} be the linear RR-statistic (6.87) corresponding to φ. Then,

$$S_{nn} = T_n + o_p(1) \quad \text{as } n \to \infty,$$

where

$$T_n = n^{-\frac{1}{2}} \sum_{i=1}^{n} d_{ni}\varphi(F(e_i)).$$

This provides the asymptotic normal distributions of S_{nn} under the hypothesis and under contiguous alternatives, which coincide with those of simple linear rank statistics studied in Section 6.2.

6.5.1 RR-Tests

Tests based on regression rank scores are useful in the situation when the model (6.77) is partitioned in the following way:

$$\mathbf{Y} = \mathbf{X}_1\boldsymbol{\beta}_1 + \mathbf{X}_2\boldsymbol{\beta}_2 + \mathbf{e} \tag{6.90}$$

where $\mathbf{X} = \left[\mathbf{X}_1 \vdots \mathbf{X}_2\right]$ is a $n \times (p+q)$ matrix with rows $\mathbf{x}_{ni}^\top = \mathbf{x}_i^\top = (\mathbf{x}_{1i}^\top, \mathbf{x}_{2i}^\top) \in \mathbb{R}_{p+q}$ and with $x_{i1} = 1,\ i = 1, \ldots, n$. We want to test the hypothesis $\mathbf{H}_0 : \boldsymbol{\beta}_2 = \mathbf{0}$, treating $\boldsymbol{\beta}_1$ as a nuisance. Define

$$\mathbf{D}_n = n^{-1}\mathbf{X}_1^\top \mathbf{X}_1, \quad \widehat{\mathbf{H}}_1 = \mathbf{X}_1(\mathbf{X}_1^\top \mathbf{X}_1)^{-1}\mathbf{X}^\top,$$
$$\mathbf{Q}_n = n^{-1}(\mathbf{X}_2 - \widehat{\mathbf{X}}_2)^\top (\mathbf{X}_2 - \widehat{\mathbf{X}}_2) \tag{6.91}$$

where $\widehat{\mathbf{X}}_2 = \widehat{\mathbf{H}}_1\mathbf{X}_2$ is the projection of $\mathbf{X}_2$ on the space spanned by the columns of $\mathbf{X}_1$. We also assume

$$\lim_{n\to\infty} \mathbf{D}_n = \mathbf{D}, \quad \lim_{n\to\infty} \mathbf{Q} = \mathbf{Q} \tag{6.92}$$

where $\mathbf{D}$ and $\mathbf{Q}$ are positive definite $p \times p$ and $q \times q$ matrices, respectively. Defining

$$\widehat{\mathbf{S}}_n = n^{-1/2}(\mathbf{X}_2 - \widehat{\mathbf{X}}_2)^\top \widehat{\mathbf{b}}_n. \tag{6.93}$$

The proposed test criterion for $\mathbf{H}_0$ is

$$\mathcal{T}_n^2 = \frac{\widehat{\mathbf{S}}_n^\top \mathbf{Q}_n^{-1} \widehat{\mathbf{S}}_n}{A^2(\varphi)}, \quad A^2(\varphi) = \int_0^1 \varphi^2(u)du - \bar{\varphi}^2. \tag{6.94}$$

It is important that $\mathcal{T}_n^2$ needs no estimation of the nuisance parameter. Under $\mathbf{H}_0$ is $\mathcal{T}_n^2$ asymptotically χ^2 distributed with q degrees of freedom. Under the local alternative

$$\mathbf{H}_n : \ \boldsymbol{\beta}_{2n} = n^{-1/2}\boldsymbol{\beta}_0, \quad \boldsymbol{\beta}_0 \in \mathbb{R}_q \ \text{ fixed },$$

the asymptotic distribution of $\mathcal{T}_n^2$ is noncentral χ^2 with q degrees of freedom and with the noncentrality parameter

$$\eta^2 = \boldsymbol{\beta}_0^\top \mathbf{Q}\boldsymbol{\beta}_0 \cdot \frac{\gamma^2(\varphi, f)}{A^2(\varphi)}, \quad \gamma(\varphi, f) = -\int_0^1 \varphi(u)\mathrm{d}f(F^{-1}(u)).$$

The performance of the RR tests in the linear model is studied in Gutenbrunner et al. (1993). Analogous tests in the linear autoregressive model were developed by Hallin and Jurečková (1999).

6.5.2 RR-Estimators

Using the estimation based on regression rank scores can lead to estimating a component of $\boldsymbol{\beta}$ only, while the classical R-estimation of a component needs to solve the minimization for the entire vector, as we have seen in (6.44)–(6.48). To illustrate this point, write

$$\boldsymbol{\beta}^\top = (\beta_0, \boldsymbol{\beta}^{(1)\top}, \boldsymbol{\beta}^{(2)\top}), \tag{6.95}$$

where $\boldsymbol{\beta}^{(1)}$ is $(p-1)$-vector, and $\boldsymbol{\beta}^{(2)}$ is q-vector. Let

$$\mathbf{x}_i = \begin{pmatrix} x_{i1} \\ \mathbf{x}_i^{(1)} \\ \mathbf{x}_i^{(2)} \end{pmatrix}$$

be an analogous partition on $\mathbf{x}_i$, $i = 1, \ldots, n$. We also partition the design matrix $\mathbf{X}_n$ as $(\mathbf{1}_n, \mathbf{X}_{n1}, \mathbf{X}_{n2})$ where $\mathbf{X}_{n1}$ is a $n \times (p-1)$ matrix and $\mathbf{X}_{n2}$ is a $n \times q$ matrix. thus, $\mathbf{x}_i^{(2)\top}$ is the ith row of $\mathbf{X}_{n2}$. For $\mathbf{t} \in \boldsymbol{R}_q$, consider the pseudo-observations

$$Y_i - \mathbf{t}^\top \mathbf{x}_i^{(2)}, \quad i = 1, \ldots, n.$$

Let $\hat{a}_{ni}(\alpha, \mathbf{Y}_n - \mathbf{X}_{n2}\mathbf{t})$, $i = 1, \ldots, n$, be the regression rank scores derived from (6.78) when $\mathbf{Y}_n$ and $\mathbf{X}_n$ are replaced by $(\mathbf{Y}_n - \mathbf{X}_{n2}\mathbf{t})$ and $(\mathbf{1}_n, \mathbf{X}_{n1})$, respectively. This relates to the model with the parameter $(\beta_0, \boldsymbol{\beta}^{(1)\top})$. Define the scores $\hat{b}_{ni}$ as in (6.88), and denote them as

$$\hat{b}_{ni}(\mathbf{Y}_n - \mathbf{X}_{n2}\mathbf{t}), \quad i = 1, \ldots, n. \tag{6.96}$$

By (6.78)

$$\frac{1}{n} \sum_{i=1}^{n} \hat{b}_{ni}(\mathbf{Y}_n - \mathbf{X}_{n2}\mathbf{t}) = \bar{\varphi} = \int_0^1 \varphi(u) du. \tag{6.97}$$

for every $n \geq 1$. Denote

$$D_n(\mathbf{t}) = \sum_{i=1}^{n} (Y_i - \mathbf{t}^\top \mathbf{x}_i^{(2)})[\hat{b}_{ni}(\mathbf{Y}_n - \mathbf{X}_{n2}\mathbf{t}) - \bar{\varphi}], \quad \mathbf{t} \in \boldsymbol{R}_q \tag{6.98}$$

an extension of the Jaeckel (1972) measure of dispersion (see Section 6.3) based on regression rank scores. It is continuous, piecewise linear, and convex in $\mathbf{t} \in \boldsymbol{R}_q$. It is also differentiable with respect to $\mathbf{t}$ a.e., and the vector of derivatives, whenever it exists, coincides with a particular vector of linear rank scores statistics. Then the RR-estimator $\bar{\boldsymbol{\beta}}_n^{(2)}$ of $\boldsymbol{\beta}^{(2)}$ is defined as

$$\bar{\boldsymbol{\beta}}_n^{(2)} = \arg\min\{D_n(\mathbf{t}) : \mathbf{t} \in \boldsymbol{R}_q\}. \tag{6.99}$$

The uniform asymptotic linearity (in $\mathbf{t}$) of the corresponding RR statistics [proved in Jurečková (1992)] then leads to the asymptotic representation for RR-estimators in the same fashion as in Section 6.3:

Whenever

$$0 < \gamma = -\int_0^1 \varphi(\alpha) \mathrm{df}(\mathrm{F}^{-1}(\alpha)) < \infty,$$

then

$$n^{\frac{1}{2}}(\bar{\boldsymbol{\beta}}_n^{(2)} - \boldsymbol{\beta}^{(2)}) = \frac{1}{\gamma\sqrt{n}}(\mathbf{D}^{(2)})^{-1}\sum_{i=1}^{n}(\mathbf{x}_i^{(2)} - \hat{\mathbf{x}}_i^{(2)})\varphi(F(e_i)) + o_p(1), \quad (6.100)$$

where

$$\widehat{\mathbf{X}}_{n2} = \mathbf{X}_{n1}(\mathbf{X}_{n1}^{\top}\mathbf{X}_{n1})^{-1}\mathbf{X}_{n1}^{\top}\mathbf{X}_{n2}$$

is the projection of $\mathbf{X}_{n2}$ on the space spanned by the columns of $\mathbf{X}_{n1}$, and $\hat{\mathbf{x}}_i^{(2)\top}$ is the ith row of $\widehat{\mathbf{X}}_{n2}$ and

$$\mathbf{D}^{(2)} = \lim_{n\to\infty} \mathbf{D}_n^{(2)} = \lim_{n\to\infty} \tfrac{1}{n}(\mathbf{X}_{n2} - \widehat{\mathbf{X}}_{n2})^{\top}(\mathbf{X}_{n2} - \widehat{\mathbf{X}}_{n2}).$$

For $\boldsymbol{\beta}^{(1)} = \mathbf{0}$, we have $\mathbf{X}_{n1} = \mathbf{0}$, so $\widehat{\mathbf{X}}_{n2} = \mathbf{0}$. Hence, $\mathbf{D}^{(2)}$ reduces to $\mathbf{D}$, defined by **C2**. Looking back at (6.100), we note that the $\varphi(F(e_i))$ are i.i.d. random variables with 0 mean and finite variance A_φ^2, while the $\mathbf{x}_i^{(2)} - \hat{\mathbf{x}}_i^{(2)}$ satisfy the generalized Noether condition. Therefore, we can use the Hájek–Šidák Central Limit Theorem (see [2.85]), extended to the multivariate case, and obtain

$$n^{\frac{1}{2}}(\bar{\boldsymbol{\beta}}_n^{(2)} - \boldsymbol{\beta}^{(2)}) \xrightarrow{\mathcal{D}} \mathcal{N}_q\left(\mathbf{0}, \frac{A_\varphi^2}{\gamma^2}\cdot(\mathbf{D}^{(2)})^{-1}\right)$$

as $n \to \infty$. As a special case of $p = 1$, we have

$$n^{\frac{1}{2}}(\bar{\boldsymbol{\beta}}_n - \boldsymbol{\beta}) \xrightarrow{\mathcal{D}} \mathcal{N}_q\left(\mathbf{0}, \frac{A_\varphi^2}{\gamma^2}\cdot\mathbf{D}^{-1}\right).$$

Under suitable regularity conditions, the R-estimator and RR-estimator of $\boldsymbol{\beta}^{(2)}$ (or $\boldsymbol{\beta}$), based on a common score function φ, are asymptotically equivalent, and they share the same asymptotic relative efficiency (ARE) properties. thus, the choice between the estimators can be based mainly on computational and diagnostic aspects, because it is more convenient to compute the R-estimator in the case of low dimension of $\boldsymbol{\beta}^{(2)}$ (e.g., if $q = 1$). While the RR are calculated with the aid of linear programming techniques, the minimization in (6.99) is only with respect to a q-dimensional vector.

6.5.3 Studentizing Scale Statistics and Regression Rank Scores

In dealing with studentized M-estimators in linear models in Section 5.5, we looked for appropriate studentizing scale statistics. To obtain a regression-

and scale-equivariant M-estimator, we should studentize it by a scale statistic S_n, which is regression-invariant and scale-equivariant in the sense

$$S_n(\mathbf{Y}+\mathbf{X}\mathbf{b}) = S_n(\mathbf{Y}), \quad \mathbf{b} \in \boldsymbol{R}_p$$
$$S_n(a\mathbf{Y}) = aS_n(\mathbf{Y}), \quad a > 0. \tag{6.101}$$

Surprisingly, not many scale-statistics satisfying (6.101) can be found in the literature. Many studentizing scale statistics used in the literature are only translation and not regression invariant. The root of the *residual sum of squares,*

$$S_n = \{(\mathbf{Y}-\widehat{\mathbf{Y}})^\top(\mathbf{Y}-\widehat{\mathbf{Y}})\}^{\frac{1}{2}},$$

where $\widehat{\mathbf{Y}} = \mathbf{X}\widehat{\boldsymbol{\beta}}$ with $\widehat{\boldsymbol{\beta}}$ being the LSE, satisfies (6.101); however, such S_n is closely connected with the normal distribution and is quite nonrobust.

One possible scale statistic is an L-statistic based on regression quantiles, mentioned in Section 4.7. As an alternative, we propose another class of scale statistics, based on *regression rank scores.* These statistics are applicable in the studentization and in various other contexts.They represent an extension of Jaeckel's (1972) rank dispersion of the residuals. However, Jaeckel's dispersion itself is only translation- invariant and not regression-invariant, hence it does not satisfy (6.95).

The scale statistic is defined in the following way: Select a score function $\varphi: (0,1) \mapsto \boldsymbol{R}_1$, nondecreasing, nonconstant, square integrable on (0,1) and such that $\varphi(\alpha) = -\varphi(1-\alpha), \ 0 < \alpha < 1$. Fix a number $\alpha_0, \ 0 < \alpha_0 < 1/2$, and assume, without loss of generality, that φ is standardized so that

$$\int_{\alpha_0}^{1-\alpha_0} \varphi^2(\alpha) \mathrm{d}\alpha = 1.$$

Calculate the *regression scores* $\widehat{\mathbf{b}} = (\hat{b}_{n1}, \ldots, \hat{b}_{nn})^\top$ generated by φ for the linear model (6.41). Then

$$\hat{b}_{ni} = -\int_{\alpha_0}^{1-\alpha_0} \varphi(\alpha) \mathrm{d}\hat{a}_{ni}(\alpha), \quad i = 1, \ldots, n,$$

where the $\hat{a}_{ni}(\alpha)$ are obtained as in (6.78). The proposed scale statistic S_n has the form

$$S_n = S_n(\mathbf{Y}) = n^{-1}\sum_{i=1}^{n} Y_i \hat{b}_{ni} = n^{-1}\mathbf{Y}^\top \widehat{\mathbf{b}}_n. \tag{6.102}$$

S_n is regression-invariant because

$$S_n(\mathbf{Y}+\mathbf{X}\mathbf{d}) = n^{-1}(\mathbf{Y}+\mathbf{X}\mathbf{d})^\top \widehat{\mathbf{b}}_n(\mathbf{Y}+\mathbf{X}\mathbf{d}) = n^{-1}\mathbf{Y}^\top \widehat{\mathbf{b}}_n = S_n(\mathbf{Y}).$$

Since $\hat{\mathbf{a}}_n(\alpha)$ and $\widehat{\mathbf{b}}_n$ are scale-invariant, S_n is also scale-equivariant. Furthermore, because $n^{-1}\sum_{i=1}^n \hat{b}_{ni} = 0$ and S_n is regression-invariant, and hence, also translation-invariant, (6.102) implies that $S_n \geq 0$ and $S_n > 0$ with probability 1.

Under some regularity conditions on F, φ, and $\mathbf{X}$, the statistic S_n converges in probability to the functional $S(F) = \int_{\alpha_0}^{1-\alpha_0} \varphi(\alpha)F^{-1}(\alpha)d\alpha$, and $n^{1/2}(S_n - S(F)$ is asymptotically normal.

More precisely,

$$S_n \xrightarrow{P} S(F) \quad \text{as } n \to \infty, \tag{6.103}$$

where

$$S(F) = \int_{\alpha_0}^{1-\alpha_0} \varphi(\alpha)F^{-1}(\alpha)d\alpha. \tag{6.104}$$

Furthermore, $n^{\frac{1}{2}}(S_n - S(F)) \xrightarrow{\mathcal{D}} \mathcal{N}(0, \sigma^2)$, where

$$\begin{aligned} \sigma^2 &= \sigma^2(\varphi, F, \alpha_0) \qquad (6.105) \\ &= \int_{F^{-1}(\alpha_0)}^{F^{-1}(1-\alpha_0)} \{F(x \wedge y) - F(x)F(y)\}\varphi(F(x))\varphi(F(y))dxdy. \end{aligned}$$

Finally, S_n admits the asymptotic representation

$$\begin{aligned} S_n - S(F) &= n^{-1}\sum_{i=1}^{n} \psi(e_i) + o_p(n^{-\frac{1}{2}}) \qquad (6.106) \\ &= \int_{F^{-1}(\alpha_0)}^{F^{-1}(1-\alpha_0)} \varphi(F(z))\{F(z) - F_n(z)\}dz + o_p(n^{-\frac{1}{2}}) \end{aligned}$$

where

$$\psi(z) = \int_{F^{-1}(\alpha_0)}^{F^{-1}(1-\alpha_0)} \Big\{\frac{\alpha - I[F(z) \le \alpha]}{f(F^{-1}(\alpha))}\Big\}\varphi(\alpha)d\alpha, \quad z \in \boldsymbol{R}_1, \tag{6.107}$$

and $F_n(z)$ is the empirical distribution function of $e_1, \ldots, e_n$.

Remark 6.4 *The scale statistic S_n is robust in the sense that its influence function, and hence, its global sensitivity are bounded, as can be deduced from (6.106). We can also admit $\alpha_0 = 0$ in the definition of S_n; then the asymptotic properties are retained but S_n is not robust.*

6.6 Bibliographical Notes

- Tail-monotone densities were studied by Parzen (1970, 1979, 1980), and by Seneta (1976) and used by Csörgö and Révész (1978) and Csörgö et al. (1982). See also Feller (1966).
- Theorem 6.4 is an extension of the original construction of Jurečková (1971a), due to Heiler and Willers (1988).
- Linear regression rank scores tests were constructed by Gutenbrunner et al. (1993), and the R-tests of Kolmogorov-Smirnov type by Jurečková (1992a).

- The RR-estimators were defined and their properties were proved in Jurečková (1991). RR-estimators of sub-vectors of $\boldsymbol{\beta}$ were proposed in (Jurečková (1992c)). The relations of various estimates are studied in Jurečková and Sen (1993).
- The asymptotic representations of RR-process and of linear RR-statistics are proved by Jurečková (1992d); some results under more restrictive conditions (smooth density of errors with exponential tails) are proved in Gutenbrunner and Jurečková (1992) and in Gutenbrunner et al. (1993).
- The scale statistics based on regression rank scores, regression invariant and scale equivariant were proposed by Jurečková and Sen (1994) where the precise conditions for assertions (6.104)–(6.107) can be found. The proof of these propositions follows from Theorem 3.1 in Gutenbrunner and Jurečková (1992) who deal with a more general setup.

6.7 Problems

6.2.1 For every $i: \ 1 \le i \le n$ and $n \ge 1$,

$$a_n^{(0)}(i) = I\!\!E\varphi(U_{n:i}) = i\binom{n}{i}\int_0^1 \varphi(u)u^{i-1}(1-u)^{n-i}du.$$

Hence, for every (fixed) $u \in [0,1)$, $a_n^0([nu]) \to \varphi(u)$ as $n \to \infty$. For this pointwise convergence to hold, $\int_0^1 |\varphi(u)|du < \infty$ is sufficient but not necessary.

6.2.2 For every $n \ge 1$,

$$\frac{1}{n}\sum_{i=1}^{n}\{a_n(i) - a_n^0(i)\}^2 \le \int_0^1 \{\varphi_n(u) - \varphi_n^0(u)\}^2 du$$

$$\le 2\Big\{\int_0^1 \{\varphi_n(u) - \varphi(u)\}^2 du + \int_0^1 \{\varphi_n^0(u) - \varphi(u)\}^2 du\Big\},$$

where by (6.6) and (6.9), the second term on the right hand side converges to 0 as $n \to \infty$. hence, (6.11) entails (6.10).

6.2.3 For a nondecreasing φ,

$$\varphi_n(u)\varphi(u)du = n^{-1}\sum_{i=1}^{n}\varphi\Big(\frac{1}{n+1}\Big)\varphi(\xi_{ni}),$$

where $\xi_{ni} \in \left(\frac{i-1}{n}, \frac{i}{n}\right)$, $1 \le i \le n$. If φ is square integrable, then $\frac{1}{n}\varphi^2\left(\frac{n}{n+1}\right)$ $\to 0$ and $\frac{1}{n}\varphi(\xi_{nn}) = \mathrm{o}\Big(\frac{1}{\sqrt{n}}\Big)$ as $n \to \infty$.

6.2.4 Define $\{P_n\}$ and $\{P_n^\star\}$ as in after (6.20). Then under $\{P_n\}$

$$\log(P_n^\star/P_n) \xrightarrow{\mathcal{D}} \mathcal{N}\Big(-\frac{t^2}{2}I(f), I(f)\Big).$$

as $n \to \infty$. Hence, invoking (2.94), it follows that $\{P_n^\star\}$ is contiguous to $\{P_n\}$.

6.3.1 By repeated integration over $(0, u)$, $u \in [0, 1)]$, we can show that under $\mathbf{A_2}$, $\varphi(u) \leq c_1 - c_2 \log(1 - u) \quad \forall u \in [0, 1]$ where c_1, c_2 are nonnegative constants. hence, using Problem 6.2.1 we can verify that

$$\max\{a_n^0(i) : 1 \leq i \leq n\} = a_n^0(n) \leq C + c \log n.$$

6.4.1 The contiguity of $\{P_n^\star\}$ to $\{P_n\}$, as established in the location model, holds for the simple regression model too.

6.6.3 The generalized Noether condition in (6.54) ensures that the usual Noether condition holds for each coordinate.

CHAPTER 7

Asymptotic Interrelations of Estimators

7.1 Introduction

We have seen in the preceding chapters that a parameter can be estimated in various ways. This begs a natural question, how to compare the competing estimators. In the finite sample case, the estimators are compared by means of their *risks*, which are computed by reference to a suitable *loss function.* The choice of the loss function is a basic concept in this respect, but there is no unique choice of the loss which can be judged the most appropriate. The exact computation of a risk usually brings some technical difficulties, especially when the sample size is not so small. Fortunately, in an asymptotic setup, by allowing the sample size to increase indefinitely, it is often possible to induce various approximations and/or simplifications by which either the limit of a risk or the value of risk at the asymptotic distribution can be obtained in a closed form, what enables to compare the competing estimators asymptotically. The asymptotic normality considerations usually lead to the choice of squared error loss, and thus to the *mean square error* for a real-valued estimator and to *quadratic risks* in the multiparameter case. However, in the finite sample setup we can consider various loss criteria and their risk counterparts.

Let X_1, $X_2, \ldots$ be a sequence of independent identically distributed observations with a distribution function $F(x-\theta)$, $\theta \in \Theta$, and let $\{T_{1n}\}$ and $\{T_{2n}\}$ be two sequences of estimators of the function $g(\theta)$. If $n^{\frac{1}{2}}(T_{in}-g(\theta)) \xrightarrow{\mathcal{D}} \mathcal{N}(0,\sigma_i^2)$ as $n \to \infty$, $i = 1, 2$, then the asymptotic relative efficiency of $\{T_{2n}\}$ with respect to $\{T_{1n}\}$ is formulated as $e_{2,1} = \sigma_1^2/\sigma_2^2$.

If, alternatively, $\{T_{2n'}\}$ is based on n' observations, then $n^{\frac{1}{2}}(T_{2n'} - g(\theta))$ has the same asymptotic distribution $\mathcal{N}(0,\sigma_1^2)$ as $n^{\frac{1}{2}}(T_{1n} - g(\theta))$ if and only if $n' = n'(n)$ is chosen so that $\lim_{n\to\infty}[n/n'(n)]$ exists and

$$\lim_{n\to\infty} \frac{n}{n'(n)} = e_{2,1}^{\star} = \frac{\sigma_1^2}{\sigma_2^2} = e_{2,1}.$$

The equality $e_{2,1} = 1$ means that $\{T_{1n}\}$ and $\{T_{2n}\}$ are equally asymptotically efficient. A more refined comparison is based on the *deficiency* of $\{T_{2n}\}$ with respective to $\{T_{1n}\}$: Assume that

$$\mathbb{E}[n(T_{in} - g(\theta))^2] = \tau^2 + \frac{a_i}{n} + o(n^{-1}), \quad i = 1, 2:$$

then the deficiency of $\{T_{2n}\}$ with respect to $\{T_{1n}\}$ is defined as $d = \frac{a_2 - a_1}{\tau^2}$. On the other hand, d is also equal to $\lim_{n\to\infty} d(n) = \lim_{n\to\infty}[n'(n) - n]$, where $n'(n)$ is chosen in such a way that $\mathbb{E}[n(T_{2n'} - g(\theta))^2]$ coincides with $\mathbb{E}[n(T_{1n} - g(\theta))^2]$ up to terms of order n^{-1}. Similar considerations also apply when the (asymptotic) mean square error is replaced by a general (asymptotic) risk.

We often want not only to compare two or more estimators, but also to study whether there exist closer relations of their functionals and/or other characteristics. This is possible with the aid of asymptotic representations derived in the preceding chapters, considering the asymptotic representations of the difference $n^{\frac{1}{2}}(T_{2n} - T_{1n})$. If the leading terms in the respective representations of $n^{\frac{1}{2}}(T_{1n} - g(\theta))$ and $n^{\frac{1}{2}}(T_{2n} - g(\theta))$ differ from each other, then $n^{\frac{1}{2}}(T_{2n} - T_{1n})$ has a nondegenerate asymptotic normal distribution. However, it may be more interesting to enquire whether this asymptotic distribution is degenerate and under what conditions? This would mean that $\{T_{1n}\}$ and $\{T_{2n}\}$ are asymptotically equivalent; the order of this equivalence coincides with the order of the remainder terms of the respective representations. We can speak about the *equivalence of* $\{T_{1n}\}$ *and* $\{T_{2n}\}$ *of the first order* if we are able to prove

$$n^{\frac{1}{2}}(T_{2n} - T_{1n}) = o_p(1) \text{ as } n \to \infty, \tag{7.1}$$

while we speak about the asymptotic equivalence of the second order if we know the exact order of the right-hand side of (7.1). Typically, if $\|n^{\frac{1}{2}}(T_{2n} - T_{1n})\| \xrightarrow{P} 0$ as $n \to \infty$, then the order of this distance is either $O_p(n^{-\frac{1}{2}})$ or $O_p(n^{-\frac{1}{4}})$, respectively, depending on the smoothness of the score functions generating the estimators. Moreover, $n^{\frac{1}{2}}(T_{2n} - T_{1n})$ can again have a nondegenerate asymptotic (nonnormal) distribution as $n \to \infty$, after an suitable standardization by a positive power of n; this gives an additional information on the interrelation of the sequences $\{T_{1n}\}$ and $\{T_{2n}\}$. The asymptotic interrelations of estimators will be studied in the present chapter; although we shall concentrate on robust estimators, analogous results hold in the parametric setup. For instance, considering the classes of M-, R-, and L-estimators of location or regression parameters, we know that either of these classes contains an asymptotically efficient element for a given distribution. One naturally expects that the optimal elements are asymptotically equivalent; if it is the case, then what is the order of the equivalence? If we have, for instance, an M-estimator of θ, what is its asymptotically equivalent R-estimation counterpart, if any? We shall consider such and similar questions.

The approach based on asymptotic representations was used by Ibragimov and Hasminskii (1970, 1971, 1972, 1979, sec. 3.1 and 3.2), who proved an asymptotic equivalence of Bayesian and maximum likelihood estimators in a locally asymptotic normal family of distributions. Jaeckel (1971) established a close relation of M-, L-, and R-estimators of location. The asymptotic relations of M-, L-, and R-estimators in the location and in the linear regression models were studied in Jurečková (1977, 1983a, b, 1985, 1986), Hušková and Jurečková (1981, 1985), Rivest (1982) and by van Eeden (1983), among others. Hanousek (1988, 1990) established the asymptotic equivalence of M-estimators to P-estimators (Pitman-type estimators) and to B-estimator (Bayes-type estimators) in the location model. The (second order) asymptotic relations of estimators can lead to goodness-of-fit tests on the shape of the distribution; an excellent example of this type of tests is the Shapiro-Wilk test of normality (Shapiro and Wilk (1965)) based on the relation of two estimators of the variance. Other tests of this type, based on relations of two robust estimators, were studied by Jurečková (1995c), Jurečková and Sen (2000), Sen et al. (2003) and Jurečková and Picek (2007).

For many estimators, defined implicitly, we can construct one-step or k- step versions starting with an initial estimator. As we shall se, the one-step version is asymptotically equivalent to the noniterative estimator with a suitable consistent initial estimator, not necessary $\sqrt{n}$-consistent, and the order of this equivalence is increasing with k. Approximations of robust estimators by their one- or k-step versions were studied by Bickel (1973), Janssen et al. (1987), Jurečková (1983, 1986), Jurečková and Portnoy (1987), Jurečková and Welsh (1990), Simpson et al. (1992), Jurečková and Malý (1995), Bergesio and Yohai (2011), among others; the effect of the initial estimator on the quality of the one-step version was studied by Jurečková and Sen (1990).

7.2 Asymptotic interrelations of location estimators

Let $X_1, X_2, \ldots, X_n$ be independent observations with a distribution function $F(x-\theta)$, where F is an unknown member of a family $\mathcal{F}$ of symmetric distribution functions and θ is the parameter to be estimated. Among various estimators of θ, the M-, L-, and R-estimators play the most important role. Either of these three types is defined with the aid of some score (or weight) function that determines its robustness and efficiency properties. Selecting these score functions properly, we may obtain asymptotically equivalent estimators. However, we should emphasize that the relation between the score functions, leading to asymptotically equivalent estimators, depend on the underlying distribution function F which is considered as unknown. Hence we cannot calculate one estimator numerically once we have a value of another estimator. However, with the knowledge of such interrelations, we may select the estimator that better fits the specific situation. The asymptotic equivalence may also carry some specific features from one type of estimator to another. Each of these classes has some advantages as well as disadvantages: M-estimators have attractive minimax properties, but they are generally not scale-equivariant and have to be supplemented by an additional estimator of scale, or they have to be studentized, which may violate their minimax properties. L-estimators are computationally appealing, mainly in the location model. R-estimators retain the advantages and disadvantages of the rank tests on which they are based.

The close relation of M-, L-, and R-estimators of location was first noticed by Jaeckel (1971) who found sufficient conditions for their asymptotic equivalence. These results were then extended to the linear model and to the second-order asymptotic equivalence by the other authors referred to in Section 7.1. Let us present some of the most interesting results of this kind.

7.2.1 Asymptotic relations of location M- and L-estimators

Let $X_1, X_2, \ldots$ be independent random variables, identically distributed according to distribution function $F(x-\theta)$ such that $F(x)+F(-x)=1,\ x \in \boldsymbol{R}_1$; denote $X_{n:1} \leq \ldots X_{n:n}$ the order statistics corresponding to $X_1, \ldots, X_n$. Let M_n be the M-estimator of θ generated by a nondecreasing step-function ψ of the form

$$\psi(x) = \alpha_j, \quad \text{for} \quad s_j < x < s_{j+1}, \quad j = 1, \ldots, k, \tag{7.2}$$

where $-\infty = s_0 < s_1 < \ldots < s_k < s_{k+1} = \infty$, $-\infty < \alpha_0 \leq \alpha_1 \leq \ldots \leq \alpha_k < \infty$, $\alpha_j = -\alpha_{k-j+1}$, $s_j = -s_{k-j+1}$, $j = 1, \ldots, k$ (at least two α's different). It means that M_n is a solution of the minimization

$$\sum_{i=1}^{n} \rho(X_i - t) := \min \text{ with respect to } t \in \boldsymbol{R}_1, \tag{7.3}$$

where ρ is a continuous convex symmetric piecewise linear function with the derivative $\rho' = \psi$. The first theorem gives an L-estimator counterpart of M_n, which is a linear combination of several quantiles.

THEOREM 7.1 *Let M_n be an M-estimator of θ defined in (7.2.1)-(7.2.2). Assume that distribution function F has two bounded derivatives f and f', and that f is positive in neighborhoods of $s_1, \ldots, s_k$. Then*

$$M_n - L_n = O_p(n^{-\frac{3}{4}}) \tag{7.4}$$

where L_n is the L-estimator, $L_n = \sum_{i=1}^{k} a_j X_{n:[np_j]}$ with

$$p_j = F(s_j),\ a_j = \gamma^{-1}(\alpha_j - \alpha_{j-1}) f(s_j),\ j = 1, \ldots, k$$

$$\gamma = \sum_{j=1}^{k} (\alpha_j - \alpha_{j-1}) f(s_j) \ (> 0). \tag{7.5}$$

PROOF By Theorem 5.4, M_n admits the asymptotic representation (5.45).

On the other hand, let F_n be the empirical distribution function corresponding to $X_1, \ldots, X_n$:

$$F_n(x) = n^{-1} \sum_{i=1}^{n} I[X_i \leq x], \ \ x \in \boldsymbol{R}_1,$$

and let $Q_n(t)$ be the corresponding empirical quantile function, $0 < t < 1$:

$$Q_n(t) = \begin{cases} X_{n:i} & \text{if } (i-1)/n < t < i/n, \\ 0 & \text{if } t = 0; \quad i = 1, \ldots, n. \end{cases} \tag{7.6}$$

We may set $\theta = 0$ without loss of generality. Using the representation (5.44) and (7.4)–(7.6), we obtain that

$$\begin{aligned} n^{\frac{1}{2}}(M_n - L_n) &= n^{\frac{1}{2}} \gamma^{-1} \sum_{j=1}^{k} (\alpha_j - \alpha_{j-1})[F_n(s_j) - F(s_j)] \\ &\quad + f(s_j)(Q_n(F(s_j)) - s_j) + \mathrm{O}_p(n^{-\frac{1}{4}}) \end{aligned}$$

and this is of order $\mathrm{O}_p(n^{-\frac{1}{4}})$ by Kiefer (1967). □

Remark 7.1 *The condition that the first derivative of F is positive and finite in a neighborhood of each of $s_1, \ldots, s_k$ is crucial for (7.4). In a nonregular case where there are singular points at which the above assumption does not hold, M- and L– estimators are not asymptotically equivalent any more, and they may have different rates of consistency. The behavior of estimators in such nonregular cases was studied by Akahira (1975a,b) Akahira and Takeuchi (1981), Chanda (1975), de Haan and Taconis-Haantjes (1979), Ghosh and Sukhatme (1981), Ibragimov and Hasminskii (1979), Jurečková (1983c), among others.*

We will now consider the asymptotic interrelation of two important estimators: Huber's M-estimator and the trimmed mean. There has been some confusion in the history of this problem: One might intuitively expect that the Winsorized rather than the trimmed mean resembles Huber's estimator (see Huber 1964). Bickel (1965) was apparently the first who recognized the close connection between Huber's estimator and the trimmed mean. Jaeckel (1971) found the L-estimator counterpart to the given M-estimator such that the difference between the two is $\mathrm{O}_p(n^{-1})$, as $n \to \infty$, provided that the function ψ generating the M-estimator (and hence the function J generating the L-estimator) is sufficiently smooth. However, Huber's ψ-function is not smooth enough to be covered by Jaeckel's proof. This important special case, situated on the border of smooth and nonsmooth cases and yet attaining the order $\mathrm{O}_p(n^{-1})$, needs to be proved in a more delicate way.

THEOREM 7.2 *Let M_n be the Huber M-estimator of θ generated by the ψ-function*

$$\psi(x) = \begin{cases} x & \textit{if } |x| \leq c \\ c \operatorname{sign} x & \textit{if } |x| > c, \end{cases}$$

for some $c > 0$, *and let* L_n *be the* α*-trimmed mean,*

$$L_n = \frac{1}{n - 2[n\alpha]} \sum_{i=[n\alpha]+1}^{n-[n\alpha]} X_{n:i}.$$

Assume the following:

(i) $c = F^{-1}(1-\alpha)$.

(ii) F *has absolutely continuous symmetric density* f *and positive and finite Fisher information,*

$$0 < \mathcal{I}(f) = \int \left(\frac{f'(x)}{f(x)}\right)^2 dF(x) < \infty.$$

(iii) $f(x) > a > 0$ *for all* x *satisfying*

$$\alpha - \varepsilon \le F(x) \le 1 - \alpha + \varepsilon, \;\; 0 < \alpha < \frac{1}{2}, \;\; \varepsilon > 0.$$

(iv) $f'(x)$ *exists in interval* $(F^{-1}(\alpha - \varepsilon), \; F^{-1}(1-\alpha+\varepsilon))$.

Then

$$L_n - M_n = O_p(n^{-1}) \quad \textit{as } n \to \infty.$$

PROOF We may put $\theta = 0$ without loss of generality. It follows from Theorem 5.3 that

$$\begin{aligned} M_n &= \frac{1}{1-2\alpha}\Big\{\frac{1}{n}\sum_{i=1}^{n} X_{n:i} I[F^{-1}(\alpha) \le X_{n:i} \le F^{-1}(1-\alpha)] \\ &+ \frac{1}{n}F^{-1}(\alpha)\sum_{i=1}^{n} I[X_{n:i} \le F^{-1}(\alpha)] \\ &+ \frac{1}{n}F^{-1}(1-\alpha)\sum_{i=1}^{n} I[X_{n:i} > F^{-1}(1-\alpha)]\Big\} + O_p(n^{-1}) \\ &= \frac{1}{1-2\alpha}\Big\{\frac{1}{n}\sum_{i=i_n}^{j_n} X_{n:i} + F^{-1}(1-\alpha)\Big[1 - F_n(F^{-1}(1-\alpha)) \\ &\quad - F_n(F^{-1}(\alpha)-)\Big]\Big\} + O_p(n^{-1}), \end{aligned} \tag{7.7}$$

where

$$i_n = nF_n(F^{-1}(\alpha)-) + 1, \;\; j_n = nF_n(F^{-1}(1-\alpha)). \tag{7.8}$$

By (7.8),

$$X_{n:i} = X_{n:n-[n\alpha]} + O_p(n^{-\frac{1}{2}}) \tag{7.9}$$

for every integer i between $n - [n\alpha]$ and j_n; analogously

$$X_{n:i} = X_{n:[n\alpha]+1} + O_p(n^{-\frac{1}{2}}) \tag{7.10}$$

for every integer i between $[n\alpha]+1$ and i_n. Hence, combining (7.7)–(7.10), we obtain

$$\begin{aligned}
&(1-2\alpha)(M_n - L_n)\\
&= F^{-1}(1-\alpha)\Big[1 - F_n(F^{-1}(1-\alpha)) - F_n(F^{-1}(\alpha)-)\Big]\\
&-\Big[X_{n:n-[n\alpha]+1} + \mathrm{O}_p(n^{-\frac{1}{2}})\Big]\Big[F_n(F^{-1}(\alpha)-) - \alpha + \mathrm{O}_p(n^{-1})\Big]\\
&+\Big[X_{n:n-[n\alpha]} + \mathrm{O}_p(n^{-\frac{1}{2}})\Big]\Big[F_n(F^{-1}(1-\alpha)) - (1-\alpha) + \mathrm{O}_p(n^{-1})\Big]\\
&+\mathrm{O}_p(n^{-1}),
\end{aligned}$$

and this, by Bahadur's representation of sample quantiles, studied in Chapter 4, is equal to

$$\begin{aligned}
&F^{-1}(1-\alpha)\Big[1 - F_n(F^{-1}(1-\alpha)) - F_n(F^{-1}(\alpha)-)\Big]\\
&-\Big\{F^{-1}(\alpha) + \frac{\alpha - F_n(F^{-1}(\alpha)-)}{f(F^{-1}(\alpha))} + \mathrm{O}_p(n^{-\frac{1}{2}})\Big\}\\
&\quad\cdot\Big\{F_n(F^{-1}(\alpha)-) - \alpha + \mathrm{O}_p(n^{-\frac{1}{2}})\Big\}\\
&+\Big\{F^{-1}(1-\alpha) + \frac{1-\alpha - F_n(F^{-1}(1-\alpha))}{f(F^{-1}(\alpha))} + \mathrm{O}_p(n^{-\frac{1}{2}})\Big\}\\
&\cdot\Big\{F_n(F^{-1}(1-\alpha)) - (1-\alpha) + \mathrm{O}_p(n^{-1})\Big\} + \mathrm{O}_p(n^{-1}) = \mathrm{O}_p(n^{-1}).
\end{aligned}$$

This completes the proof. □

Combining Theorems 4.1 and 7.2, we immediately obtain the M-estimator counterpart of the α-Winsorized mean

$$L_n = \frac{1}{n}\Big\{[n\alpha]X_{n:[n\alpha]} + \sum_{i=[n\alpha]+1}^{n-[n\alpha]} X_{n:ii} + [n\alpha]X_{n:n-[n\alpha]+1}\Big\}. \tag{7.11}$$

Corollary 7.1 *Under the conditions of Theorem 7.2, to the α-Winsorized mean L_n given in (7.11) there exists an M-estimator M_n generated by the function*

$$\psi(x) = \begin{cases} F^{-1}(\alpha) - [\alpha/f(F^{-1}(\alpha))] & \text{if } x < F^{-1}(\alpha),\\ x & \text{if } F^{-1}(\alpha) \le x \le F^{-1}(1-\alpha),\\ F^{-1}(1-\alpha) + [\alpha/f(F^{-1}(\alpha))] & \text{if } x > F^{-1}(1-\alpha) \end{cases}$$

such that

$$L_n - M_n = O_p(n^{-\frac{3}{4}}).$$

Let us now consider a fully general L-estimator of the form

$$L_n = \sum_{i=1}^{n} c_{ni} X_{n:i}, \tag{7.12}$$

where the coefficients c_{ni}, $i = 1, \ldots, n$, are generated by a function $J : (0,1) \mapsto \boldsymbol{R}_1$, $J(u) = J(1-u)$, $0 < u < 1$, according to (4.26) or (4.27), respectively. We shall find an M-estimator counterpart M_n to L_n in two situations:

(1) J trimmed, that is, $J(u) = 0$ for $0 \le u < \alpha$ and $1 - \alpha < u \le 1$ and

(2) J possibly untrimmed but under more restrictive conditions on F. In either case the corresponding M-estimator is generated by the following continuous and skew-symmetric ψ-function:

$$\psi(x) = -\int_{\boldsymbol{R}_1} (I[y \ge x] - F(y))J(F(y))dy, \quad x \in \boldsymbol{R}_1. \tag{7.13}$$

THEOREM 7.3 *For an L-estimator L_n in (7.12) suppose that the c_{ni}, generated by a function $J : (0,1) \mapsto \boldsymbol{R}_1$, satisfy*

(i)

$$J(u) = J(1-u), \; 0 < u < 1, \quad \int_0^1 J(u)du = 1. \tag{7.14}$$

(ii) $J(u) = 0$ for $u \le \alpha$ and $u \ge 1-\alpha$ for some $\alpha \in (0, \frac{1}{2})$, J is continuous on (0,1) up to a finite number of points $s_1, \ldots, s_m$, $\alpha < s_1 < \ldots < s_m < 1-\alpha$ and J is Lipschitz of order $\nu \le 1$ in intervals (α, s_1), $(s_1, s_2), \ldots (s_m, 1-\alpha)$.

(iii) F is absolutely continuous with symmetric density f and $F^{-1}(u) = \inf\{x : F(x) \ge u\}$ satisfies the Lipschitz condition of order 1 in neighborhoods of $s_1, \ldots, s_m$.

(iv) $\int_{-A}^{A} f^2(x)dx < \infty$ where $A = F^{-1}(1-\alpha+\varepsilon)$, $\varepsilon > 0$.

Then

$$L_n - M_n = O_p(n^{-\nu}) \tag{7.15}$$

where M_n is the M-estimator generated by the function (7.13).

PROOF Assume that $\theta = 0$ without loss of generality. ψ-function (7.13) satisfies the conditions [**A1**]–[**A4**] of Section 5.3, for trimmed J satisfying (i)–(ii) and for F satisfying (iii)–(iv). Moreover, the derivative $h'(t)$ of the function h in (5.22) is decreasing at $t = 0$. Hence (7.15) follows by combining the asymptotic representation (4.28) of L_n (see Theorem 4.4) with the asymptotic representation (5.28) of M_n (Theorem 5.3). □

For an untrimmed J, more restrictive conditions on J and F are needed; better results will be obtained for the c_{ni} in (4.27):

$$c_{ni} = \int_{(i-1)/n}^{i/n} J(u)du, \quad i = 1, \ldots, n. \tag{7.16}$$

THEOREM 7.4 *Let L_n be an L-estimator (7.12) with the coefficients c_{ni} defined in (7.16) and satisfying the following conditions:*

(i) $J(u)$ is continuous up to a finite number of points $s_1, \ldots, s_m$, $0 < s_1 < \ldots < s_m < 1$ and satisfies the Lipschitz condition of order $\nu > 0$ in each of the intervals $(0, s_1)$, (s_1, s_2), $\ldots (s_m, 1)$;

(ii) $F^{-1}(s) = inf\{x : F(x) \geq s\}$ satisfies the Lipschitz condition of order 1 in a neighborhood of $s_1, \ldots, s_m$.

(iii) F is absolutely continuous with symmetric density f, and

$$\sup\{|x|^{\beta}F(x)(1 - F(x)) : x \in \boldsymbol{R}_1\} < \infty \tag{7.17}$$

for some $\beta > \frac{2}{\nu+1-\Delta}$, $0 < \Delta < 1$.

(iv) J and F satisfy

$$\int_{\boldsymbol{R}_1} J^2(F(x))f(x+t)dx \leq K_1 \quad \textit{and}$$

$$\int_{\boldsymbol{R}_1} |J'(F(x))|f(x)f(x+t)dx \leq K_2,$$

for $|t| \leq \delta$, $K_1, K_2, \delta > 0$.

Then

$$L_n - M_n = O_p(n^{-r}) \quad \textit{as } n \to \infty,$$

where M_n is the M-estimator generated by the ψ-function of (7.13) and

$$r = \min\left\{\frac{\nu+1}{2}, 1\right\}. \tag{7.18}$$

PROOF The proposition follows from the representation of L_n and M_n derived in Theorems 4.4 and 5.3; the conditions of these theorems are guaranteed by assumptions (i)–(iv) of Theorem 7.4. □

L-estimators in the location model are both location and scale equivariant, while M-estimators are generally only location equivariant. This shortcoming of M-estimators can be removed by a studentization using an appropriate scale statistic. Studentized M-estimators are studied in detail in Section 5.4, where their asymptotic representations are also derived. Hence starting with an L-estimator L_n with the weight function J, we are naturally interested in its M-estimator counterpart. Recall that the studentized M-estimator of θ is defined as a solution of the minimization (5.61) with the scale statistics $S_n = S_n(X_1, \ldots, X_n)$ being translation invariant and scale equivariant in the sense of (5.62). If $\psi = \rho'$ is continuous, then M_n is also a root of the equation

$$\sum_{i=1}^{n} \psi\Big(\frac{X_i - t}{S_n}\Big) = 0, \; t \in \boldsymbol{R}_1. \tag{7.19}$$

It is assumed [see (5.63)] that S_n is $\sqrt{n}$-consistent estimator of some positive functional $S(F)$. Then typically the studentized M-estimator M_n, asymptotically equivalent to L_n, will be a root of (7.19) with

$$\psi(x) = -\int_{\boldsymbol{R}_1} (I[y \geq xS(F)] - F(y))J(F(y))dy, \; x \in \boldsymbol{R}_1. \tag{7.20}$$

The question is which conditions are needed for this asymptotic equivalence? The following theorem gives sufficient conditions for the asymptotic equivalence of a trimmed L-estimator and its studentized M-estimation counterpart.

THEOREM 7.5 *Let L_n be the L-estimator generated by the function J; assume conditions (i)–(iii) of Theorem 7.3 and condition (iv) with A replaced by $A = a + bF^{-1}(1-\alpha)$, $a \in \boldsymbol{R}_1$, $b > 0$. Assume further that there exists a scale statistic S_n satisfying (5.62) and (5.63). Then*

$$L_n - M_n = O_p(n^{-\nu})$$

where M_n is the M-estimator generated by ψ of (7.20) and studentized by S_n.

PROOF The result follows from the asymptotic representations of L_n and M_n derived in Theorems 4.4 and 5.8, respectively, whose conditions are easily verified in the trimmed case. □

Finally, consider possibly untrimmed L-estimator L_n and a scale statistic S_n satisfying (5.62) and (5.63). We expect that L_n is asymptotically equivalent to an M-estimator M_n generated by ψ of (7.20). The following theorem gives one set of sufficient conditions for this asymptotic equivalence.

THEOREM 7.6 *Let L_n be an L-estimator of (7.12) with the coefficients c_{ni} defined in (7.17), generated by J satisfying (7.14) and conditions (i) and (ii) of Theorem 7.4; further assume that F satisfies condition (iii) of Theorem 7.4, and that together J and F satisfy the following conditions:*

$$\int_{\boldsymbol{R}_1} x^2 J^2(F(x)) f(xe^u + t)dx \leq K_1,$$

$$\int_{\boldsymbol{R}_1} |J'(F(x))| f(x) f(xe^u + t)dx \leq K_2,$$

$$\int_{\boldsymbol{R}_1} x^2 |J'(F(x))| f(x) f(xe^u + t)dx \leq K_3$$

for $|t|,\ |u| \leq \delta$, $K_1, K_2, K_3, \delta > 0$. Assume that there exists a scale statistic S_n satisfying (5.62) and (5.63). Then

$$L_n - M_n = O_p(n^{-r})$$

with r given in (7.18), where M_n is the estimator generated by ψ of (7.20) and studentized by S_n.

PROOF The proposition follows from the asymptotic representation of L_n and M_n given in Theorems 4.4 and 5.8. □

7.2.2 Asymptotic relations of location M- and R-estimators

Let $\{X_i;\ i \geq 1\}$ be a sequence of independent observations from a population with a distribution function $F(x-\theta)$, where $F(x)+F(-x)=1,\ x \in \boldsymbol{R}_1$, and F has an absolutely continuous density f and finite and positive Fisher's information $\mathcal{I}(f)$. Let $\varphi:\ [0,1) \mapsto \boldsymbol{R}^+$ be a nondecreasing score function such that

$$\varphi(0)=0 \ \text{ and } \int_0^1 \varphi^2(u)du < \infty, \tag{7.21}$$

and assume that γ, defined in (6.3), is nonzero. Let R^+_{ni} denote the rank of $|X_i - t|$ among $|X_1-t|,\ldots,|X_n-t|,\ i=1,\ldots,n,\ t \in \boldsymbol{R}_1$, and consider the linear signed-rank statistic

$$S_n(\mathbf{X}_n - t\mathbf{1}_n) = n^{-\frac{1}{2}} \sum_{i=1}^{n} \text{sign}(X_i - t)a_n(R^+_{ni}(t)) \tag{7.22}$$

with the scores $a_n(1) \leq \ldots a_n(n)$ satisfying (6.12) and (6.11). The R-estimator T_n of θ was defined in (3.77)–(3.78) as

$$T_n = \frac{1}{2}(T_n^- + T_n^+) \tag{7.23}$$

where

$$T_n^- = \sup\{t:\ S_n(t) > 0\},\ \ T_n^+ = \inf\{t:\ S_n(t) < 0\}. \tag{7.24}$$

The following theorem gives an M-estimator counterpart that is first order asymptotically equivalent to T_n.

THEOREM 7.7 *Let $\{X_i:\ i \geq 1\}$ be a sequence of independent identically distributed random variables with a distribution function $F(x-\theta)$ where F has an absolutely continuous, symmetric density f and positive and finite Fisher's information $I(f)$. Let T_n be an R-estimator of θ defined in (7.23) and (7.24), generated by a nondecreasing function $\phi:\ [0,1) \mapsto \boldsymbol{R}^+$ satisfying (7.21). Assume that*

$$0 < \gamma = -\int_{\boldsymbol{R}_1} \phi^\star(F(x))f'(x)dx < \infty, \tag{7.25}$$

where

$$\phi^\star(u) = \begin{cases} \phi(2u-1), & \frac{1}{2} \leq u < 1, \\ -\phi^\star(1-u), & 0 < u \leq \frac{1}{2}. \end{cases} \tag{7.26}$$

Then

$$n^{\frac{1}{2}}(T_n - M_n) = o_p(1) \ \text{ as } n \to \infty, \tag{7.27}$$

where M_n is the M-estimator of θ generated by the ψ-function

$$\psi(x) = c\phi^\star(F(x)),\ \ \forall x \in \boldsymbol{R}_1,\ \text{for some } c > 0. \tag{7.28}$$

PROOF By Theorem 6.3,

$$\begin{aligned} n^{\frac{1}{2}}(T_n - \theta) &= (n\gamma)^{-1}\sum_{i=1}^{n}\phi^\star(F(X_i-\theta)) + o_p(1) \\ &= (n\gamma)^{-1}\sum_{i=1}^{n}\psi(X_i-\theta) + o_p(1). \end{aligned} \tag{7.29}$$

Hence the theorem follows from Theorems 6.3 and 5.4. □

7.2.3 Asymptotic relations of location L- and R-estimators

The asymptotic interrelations of L- and R-estimators of location follow from the results in Sections 7.2.1 and 7.2.2. Indeed, given an L-estimator generated by a symmetric function J, the M-estimation counterpart is generated by the function ψ in (7.13). On the other hand, the R-estimation counterpart is generated by the score-generating function $\phi(u) = \phi^\star((u+1)/2),\ 0 < u < 1$, where $\phi^\star$ is determined by the relation $\phi^\star(u) = \psi(F^{-1}(u)),\ 0 < u < 1$. Hence we get the following relation between J and $\phi^\star$:

$$\phi^\star(u) = -\int_0^1 (I[v \geq u] - v)J(v)dv, \quad 0 < u < 1. \tag{7.30}$$

The respective L-, M-, and R-estimators, generated by this triple of functions, are then asymptotically equivalent to each other, but the conditions under which this asymptotic equivalence holds need to be formulated, combining the above results.

As mentioned earlier, such asymptotic relations do not enable us to calculate the estimators numerically from each other because the relations between the pertaining weight functions involve the unknown F. The relations are rather between the classes of estimators than between the individual members. On the other hand, the knowledge of these relations gives us a better picture of the related classes of estimators. For instance, the Wilcoxon score R-estimator has the score function $\phi^\star(u) = u - \frac{1}{2},\ 0 < u < 1$. Hence the related class of M-estimators is generated by the functions $\psi(x) = F(x) - \frac{1}{2},\ x \in \boldsymbol{R}_1, F \in \mathcal{F}$. Further the corresponding class of L-estimators is generated by functions $f(F^{-1}(u)/\int_R f^2(x)dx),\ 0 < u < 1,\ F \in \mathcal{F}$. These estimators have bounded influence functions, and it makes no sense to winsorize $\phi^\star$ and in this way obtain the trimmed Hodges-Lehmann estimator, as some authors recommend.

Another example is the normal scores R-estimator generated by $\phi^\star(u) = \Phi^{-1}(u)$ [Φ being the N(0,1) d.f.], $0 < u < 1$. Then

$$\psi(x) = \Phi^{-1}(F(x)),\ x \in \boldsymbol{R}_1 \tag{7.31}$$

and

$$J(u) = \frac{f(F^{-1}(u))}{\Phi'(\Phi^{-1}(u))} \cdot \left\{ \int_{\boldsymbol{R}_1} f(F^{-1}(\Phi(x)))dx \right\}^{-1}, \ 0 < u < 1. \tag{7.32}$$

Such estimators are considered as nonrobust because they have unbounded influence functions. However, looking at (7.31) and (7.32), we see that the normal scores estimator is sensitive to the tails of the underlying (unknown) distribution. Indeed, considering the Parzen (1979) classification of tails, we see that $J(0) = J(1) = 0$ for a heavy-tailed F, $J(0) = J(1) = 1$ for an exponentially tailed F and $\lim_{u \to 0,1} J(u) = \infty$ for a short tailed F.

R-estimation counterpart of the α-trimmed mean corresponds to the score function

$$\phi^\star(u) = \begin{cases} F^{-1}(\alpha), & 0 < u < F^{-1}(\alpha), \\ F^{-1}(u), & F^{-1}(\alpha) \le u \le F^{-1}(1-\alpha), \\ F^{-1}(1-\alpha), & F^{-1}(1-\alpha) < u < 1. \end{cases} \tag{7.33}$$

Taking into account the Huber M-estimator and its L- and R-estimator counterparts, we come to a conclusion that not only the class of M-estimators but also L-estimators and R-estimators contain an element, asymptotically minimax over a family of contaminated normal distributions. More about the asymptotic interrelations and their consequences may be found in Jurečková (1989).

7.3 Asymptotic relations in the linear model

7.3.1 M- and R-estimators

Using the asymptotic representations derived in Chapters 5 and 6, we may derive the asymptotic interrelations of M- and R-estimators in the linear regression model. Consider the linear model

$$\mathbf{Y} = \mathbf{X}\boldsymbol{\beta} + \mathbf{E}, \tag{7.34}$$

where $\mathbf{Y} = (Y_1, \ldots, Y_n)^\top$ is the vector of observations, $\mathbf{X} = \mathbf{X}_n$ is a known design matrix of order $n \times p$, $\boldsymbol{\beta} = (\beta_1, \ldots, \beta_p)^\top$ is a parameter and $\mathbf{E} = (E_1, \ldots, E_n)^\top$ is vector of independent identically distributed errors with a distribution function F. M-estimator of $\boldsymbol{\beta}$ is defined as a solution of the minimization (5.102) with the function ρ. The asymptotic representation of $\mathbf{M}_n$ was derived in Section 5.5 in the form

$$\mathbf{M}_n - \boldsymbol{\beta} = \gamma_M^{-1} \mathbf{Q}_n^{-1} \sum_{i=1}^{n} \mathbf{x}_i \psi(E_i) + o_p(\|\mathbf{Q}_n^{-\frac{1}{2}}\|), \tag{7.35}$$

with $\gamma_M = \int_{\boldsymbol{R}} f(x) d\psi(x)$ and $\mathbf{Q}_n = \sum_{i=1}^{n} \mathbf{x}_i \mathbf{x}_i^\top$.

The R-estimator $\mathbf{T}_n$ of $\boldsymbol{\beta}$ is defined as a solution of the minimization of the measure of *rank dispersion* [see (6.47)]

$$D_n(\mathbf{b}) = \sum_{i=1}^{n}(Y_i - \mathbf{x}_i^\top \mathbf{b})[a_n(R_{ni}(\mathbf{b})) - \bar{a}_n], \tag{7.36}$$

where $R_{ni}(\mathbf{b})$ is the rank of the residual $Y_i - \mathbf{x}_i^\top \mathbf{b}$ among $Y_1 - \mathbf{x}_1^\top \mathbf{b}, \ldots, Y_n - \mathbf{x}_n^\top \mathbf{b}$, $\mathbf{b} \in \boldsymbol{R}_p$, $i = 1, \ldots, n$; $a_n(1), \ldots, a_n(n)$ are scores generated by a nondecreasing, square-integrable score function $\phi: (0,1) \mapsto \boldsymbol{R}_1$ and $\bar{a}_n = n^{-1}\sum_{i=1}^{n} a_n(i)$. Notice that $D_n(\mathbf{b})$ is invariant to the translation; hence the above procedure is not able to estimate the intercept, which should be estimated by signed-rank statistics. Thus, for simplicity of comparison, we will consider the model without an intercept:

$$\sum_{i=1}^{n} x_{ij} = 0, \; j = 1, \ldots, p. \tag{7.37}$$

Under the regularity conditions of Section 6.3, $\mathbf{T}_n$ admits the asymptotic representation

$$\mathbf{T}_n - \boldsymbol{\beta} = \gamma_R^{-1} \sum_{i=1}^{n} \mathbf{Q}_n^{-1}\mathbf{x}_i \phi(F(E_i)) + o_p(\|\mathbf{Q}_n^{-\frac{1}{2}}\|) \tag{7.38}$$

with γ_R defined in (7.25).

Combining the representations (7.35) and (7.38), and regarding their regularity conditions, we obtain the following (first order) asymptotic relations of $\mathbf{M}_n$ and $\mathbf{T}_n$.

THEOREM 7.8 *(i) Assume that the errors E_i in the model (7.34) are independent identically distributed random variables with a distribution function F which has an absolutely continuous density f with a positive and finite Fisher's information $\mathcal{I}(f)$.*

(ii) The design matrix $\mathbf{X}_n$ let satisfy (7.37) and the generalized Noether condition

$$\lim_{n\to\infty} \max_{1\le i\le n} \mathbf{x}_i^\top \mathbf{Q}_n^{-1}\mathbf{x}_i = 0.$$

(iii) Let $\mathbf{T}_n$ be the R-estimator of $\boldsymbol{\beta}$ defined through the minimization (7.36), generated by the score function $\varphi: (0,1) \mapsto \boldsymbol{R}_1$, nonconstant, nondecreasing and square integrable, and such that $\varphi(1-u) = -\phi(u), \; 0 < u < 1$.

Then

$$\mathbf{M}_n - \mathbf{T}_n = o_p(\|\mathbf{Q}_n^{-\frac{1}{2}}\|),$$

where $\mathbf{M}_n$ is the M-estimator of $\boldsymbol{\beta}$ generated by the function $\psi(x) = a + b\varphi(F(x)), \; \forall x \in \boldsymbol{R}_1$, with $a \in \boldsymbol{R}_1, \; b > 0$ being arbitrary numbers.

7.3.2 Asymptotic relations of Huber's M-estimator and trimmed LSE

The trimmed least squares estimator, introduced in (4.96), is a straightforward extension of the trimmed mean to the linear regression model. Noting the close relation of the class of trimmed means and that of Huber's M-estimators of location, we expect an analogous phenomenon in the linear regression model. Actually, the relation of the trimmed LSE and of Huber's estimator is quite analogous; the only difference may appear in the first components when F is asymmetric because then the population functionals do not coincide, and both estimators may be biased in the first components. This bias appears in the asymmetric location model too.

Consider the model (7.34), but this time with an intercept, i.e.

$$x_{i1} = 1,\ i = 1, \ldots, n. \tag{7.39}$$

The (α_1, α_2)-trimmed LSE $\mathbf{T}_n(\alpha_1, \alpha_2)$ of $\boldsymbol{\beta}$, defined in 4.96) can be also described as a solution of the minimization

$$\sum_{i=1}^{n} a_i(Y_i - \mathbf{x}_i\mathbf{t})^2 := \min \quad \text{with respect to } \mathbf{t} \in \boldsymbol{R}_p,$$

with the same a_i as in (4.95).

The following theorem provides one set of regularity conditions leading to the asymptotic equivalence of trimmed LSE and of the Huber estimator of order $O_p(n^{-1})$, up to a bias that may appear in the first components in case of asymmetric F.

THEOREM 7.9 *Assume the following conditions in the linear model (7.34):*

(i) $\mathbf{X}_n$ *satisfies (7.39) and the following two conditions:*

  (a) $\lim_{n\to\infty} n^{-1}\mathbf{X}_n^\top \mathbf{X}_n = \mathbf{Q}$ *is positive definite,*

  (b) $\max\{|x_{ij}| : 1 \le i \le n;\ 1 \le j \le p\} = O(1)$ *as* $n \to \infty$.

(ii) F *is absolutely continuous with density* f *such that* $0 < f(x) < \infty$ *for* $F^{-1}(\alpha_1) - \varepsilon \le x \le F^{-1}(\alpha_2) + \varepsilon$ *with some* $\varepsilon > 0$ *and* f *has a bounded derivative* f' *in a neighborhood of* $F^{-1}(\alpha_1)$ *and* $F^{-1}(\alpha_2)$.

Then

$$\|\mathbf{T}_n(\alpha_1, \alpha_2) - \mathbf{M}_n - \mathbf{e}_1\eta(\alpha_2 - \alpha_1)^{-1}\| = O_p(n^{-1}),$$

where $\eta = (1-\alpha_2)F^{-1}(\alpha_2) + \alpha_1 F^{-1}(\alpha_1)$, $\mathbf{e}_1 = (1, 0, \ldots, 0)^\top \in \boldsymbol{R}_p$, *and* $\mathbf{M}_n$ *is the M-estimator generated by the* ψ*-function*

$$\psi(x) = \begin{cases} F^{-1}(\alpha_1), & x < F^{-1}(\alpha_1), \\ x, & F^{-1}(\alpha_1) \le x \le F^{-1}(\alpha_2), \\ F^{-1}(\alpha_2), & x > F^{-1}(\alpha_2). \end{cases}$$

PROOF Theorem follows from the asymptotic representations of $T_n(\alpha_1, \alpha_2)$ (Theorem 4.10) and of M_n (Theorem 5.10). □

The following corollary concerns the symmetric case.

Corollary 7.2 *If F is symmetric around 0 and $\alpha_1 = \alpha$, $\alpha_2 = 1-\alpha$, $0 < \alpha < \frac{1}{2}$, then, under the conditions of Theorem 7.9,* $\|\mathbf{T}_n(\alpha) - \mathbf{M}_n\| = O_p(n^{-1})$.

7.4 Approximation by One-Step Versions

Many estimators, like maximum likelihood, M- and R-estimators, are defined implicitly as a solution of a minimization problem or of a system of equations. Sometimes it may be difficult to obtain this solution algebraically, and there may exist more solutions among which one is consistent and/or efficient, and so on. We have already touched on this problem in the context of M-estimators generated by a nonconvex ρ-function. In the case of continuously differentiable ρ, there exists a root of the pertaining system of equations that is a $\sqrt{n}$-consistent estimator of the parameter; however, there is no criterion distinguishing the right one among the multiple roots.

The asymptotic equivalence of an M-estimator M_n with a one-step Newton-Raphson adjustment starting with an appropriate consistent initial estimate is well known in the context of likelihood estimation with a sufficiently smooth kernel (e.g., Lehmann 1983, Section 6.3). This smoothness precludes the standard Huber and Hampel kernels. Moreover, unlike in the MLE case, we can still estimate some functional of the unknown distribution function. Bickel (1975) introduced the one-step version of the M-estimator M_n in the linear model, which is asymptotically equivalent to M_n in probability as $n \to \infty$. Jurečková (1983) and Jurečková and Portnoy (1987) studied the order of approximation of M_n by its one-step version in the linear model; the latter paper considered also the effect of the initial estimator on the resulting breakdown point. Welsh (1987) constructed a one-step version of L-estimator in the linear model; its relation to the one-step M-estimator was studied by Jurečková and Welsh (1990). Janssen, Jurečková, and Veraverbeke (1985) studied the order of an approximation of M-estimator of a general parameter by its one- and two-step versions. Jurečková and Sen (1990) studied the effect of the initial estimator on the second order asymptotics of the one-step version of M-estimator of general parameter. Jurečková and Malý (1995) studied the orders of approximations of studentized M-estimators of location by their one-step versions.

In the context of M-estimators, we generally observe an interesting phenomenon relating to multiple (k-step) iterations: If the ρ-function is sufficiently smooth, then the order of approximation of M_n by its k-step version increases rapidly with k [more precisely, it is $O_p(n^{-(k-1)/2})$]. However, if $\psi = \rho'$ has jump discontinuities, this order is only $\mathrm{O}_p(n^{-1-2^{-k-1}})$ (see Jurečková and Malý 1995).

7.4.1 One-step version of general M-estimator

Let $X_1, X_2, \ldots$ be a sequence of independent identically distributed random variables with a distribution function $F(x, \theta_0)$, where $\theta_0 \in \Theta$, an open interval in $\boldsymbol{R}_1$. Let $\rho: \boldsymbol{R}_1 \times \Theta \mapsto \boldsymbol{R}_1$ be a function, absolutely continuous in θ with the derivative $\psi(x, \theta) = \partial\rho(x, \theta)/\partial\theta$. We assume that $\boldsymbol{E}_{\theta_0}\rho(X_1, \theta)$ exists for all $\theta \in \Theta$ and has a minimum at $\theta = \theta_0$.

The M-estimator M_n of θ_0 is defined as a solution of the minimization (5.2). Let us first consider the case that the function $\psi(x, \theta)$ is also absolutely continuous in θ. Then M_n is a solution of the equation

$$\sum_{i=1}^{n} \psi(X_i, t) = 0. \tag{7.40}$$

By Theorem 5.2, there exists a sequence $\{M_n\}$ of roots of (7.40) satisfying $n^{\frac{1}{2}}(M_n - \theta_0) = \mathrm{O}_p(1)$ as $n \to \infty$, which admits the asymptotic representation

$$M_n = \theta_0 - (n\gamma_1(\theta_0))^{-1} \sum_{i=1}^{n} \psi(X_i, \theta_0) + R_n, \tag{7.41}$$

where $R_n = \mathrm{O}_p(n^{-1})$ as $n \to \infty$, and

$$\gamma_1(\theta) = \boldsymbol{E}_\theta \dot{\psi}(X_1, \theta), \text{ and } \dot{\psi}(x, \theta) = (\partial/\partial\theta)\psi(x, \theta). \tag{7.42}$$

It is often difficult to find an explicit and consistent solution of (7.40). Therefore, a standard technique is to look at an iterative solution of this equation. Starting with an initial consistent estimator $M_n^{(0)}$, we may consider the successive estimators $\{M_n^{(k)}; k \geq 1\}$ defined recursively by $M_n^{(k)} = M_n^{(k-1)}$ if $\gamma_n^{(k-1)} = 0$, and if $\gamma_n^{(k-1)} \neq 0$, as

$$M_n^{(k)} = M_n^{(k-1)} - (n\hat{\gamma}_n^{(k-1)})^{-1} \sum_{i=1}^{n} \psi(X_i, M_n^{(k-1)}), \tag{7.43}$$

for $k = 1, 2, \ldots$, where we take

$$\hat{\gamma}_n^{(k)} = n^{-1} \sum_{i=1}^{n} \dot{\psi}(X_i, M_n^{(k)}), \quad k = 0, 1, 2, \ldots. \tag{7.44}$$

We shall show that

$$n(M_n^{(1)} - M_n) = \mathrm{O}_p(1), \quad n(M_n^{(k)} - M_n) = \mathrm{o}_p(1), \; k \geq 2$$

under some regularity conditions, provided that

$$\sqrt{n}(M_n^{(0)} - \theta_0) = \mathrm{O}_p(1), \quad \text{as } n \to \infty.$$

Together with (7.41) this implies that

$$M_n^{(1)} = \theta_0 - (n\gamma_1(\theta_0))^{-1} \sum_{i=1}^{n} \psi(X_i, \theta_0) + R_n^{(1)}, \tag{7.45}$$

where $R_n^{(1)} = O_p(n^{-1})$ as $n \to \infty$. A closer look at the behavior of the remainder $R_n^{(1)}$ reveals the role of the initial estimator $M_n^{(0)}$.

The *second order distributional representation* (SOADR) of M_n, which is the asymptotic representation (7.41) supplemented with the asymptotic (non-normal) distribution of nR_n, is given by Theorem 5.2. We shall now derive the asymptotic distribution of $nR_n^{(1)}$ and the second order distributional representation for the one-step version. Its most important corollary will be the asymptotic distribution of $n(M_n^{(1)} - M_n)$, a properly standardized difference of M_n and of its one-step version. Whenever this asymptotic distribution degenerates for some initial $M_n^{(0)}$, we can obtain an approximation of M_n of order $o_p(n^{-1})$.

We shall confine ourselves to the situations where the initial estimator $M_n^{(0)}$ itself is asymptotically linear in the sense that it admits the asymptotic representation

$$M_n^{(0)} = \theta_0 + n^{-1} \sum_{i=1}^{n} \phi(X_i, \theta_0) + o_p(n^{-\frac{1}{2}}) \tag{7.46}$$

with a suitable function $\varphi(x, \theta)$ on $\boldsymbol{R}_1 \times \Theta$ such that $0 < \mathbb{E}_\theta \varphi^2(X_1, \theta_0) < \infty$ in neighborhood of θ_0. We shall impose the following conditions on F and φ:

1. $h(\theta) = \mathbb{E}_{\theta_0} \psi(X_1, \theta)$ exists $\forall \theta \in \Theta$ and has a unique root at $\theta = \theta_0$.
2. $\dot{\psi}(x, \theta)$ is absolutely continuous in θ, and there exist $\delta > 0$, $K_1, K_2 > 0$, such that

$$\mathbb{E}_{\theta_0} |\dot{\psi}(X_1, \theta_0 + t)|^2 \leq K_1, \ \ \mathbb{E}_{\theta_0} |\ddot{\psi}(X_1, \theta_0 + t)|^2 \leq K_2 \ \forall |t| \leq \delta$$

where

$$\dot{\psi}(x, \theta) = \frac{\partial}{\partial \theta} \psi(x, \theta), \quad \ddot{\psi}(x, \theta) = \frac{\partial}{\partial \theta} \dot{\psi}(x, \theta).$$

3. $\gamma_1(\theta_0)$ as defined in (7.42) is nonzero and finite.
4. $0 < \mathbb{E}_{\theta_0} \psi^2(X_1, \theta) < \infty$ in a neighborhood of θ_0.
5. There exist $\alpha > 0$, $\delta > 0$, and a function $H(x, \theta_0)$ such that

$$\mathbb{E}_{\theta_0} H(X_1, \theta_0) < \infty$$

and

$$|\ddot{\psi}(x, \theta_0 + t) - \ddot{\psi}(x, \theta_0)| \leq |t|^\alpha H(x, \theta_0) \quad \text{a.e. } [F(x, \theta_0)]$$

for $|t| \leq \delta$.

Let

$$U_{n1} = n^{\frac{1}{2}}\Big(n^{-1}\sum_{i=1}^{n} \dot{\psi}(X_1,\theta_0) - \gamma_1(\theta_0)\Big),$$

$$U_{n2} = n^{-\frac{1}{2}}\sum_{i=1}^{n} \psi(X_i,\theta_0),$$

$$U_{n3} = n^{-\frac{1}{2}}\sum_{i=1}^{n} \phi(X_i,\theta_0),$$

and let $\mathbf{U}_n = (U_{n1}, U_{n2}, U_{n3})^\top$. Under the above conditions, $\mathbf{U}_n$ is asymptotically normally distributed,

$$\mathbf{U}_n \xrightarrow{\mathcal{D}} \mathcal{N}_3(\mathbf{0}, \mathbf{S}),$$

where $\mathbf{S}$ is a (3×3) matrix with the elements

$$\begin{aligned}
s_{11} &= \mathrm{Var}_{\theta_0}\dot{\psi}(X_1,\theta_0), \;\; s_{22} = \mathbb{E}_{\theta_0}\psi^2(X_1,\theta_0), \\
s_{33} &= \mathbb{E}_{\theta_0}\phi^2(X_1,\theta_0), \;\; s_{12} = \mathrm{Cov}_{\theta_0}(\dot{\psi}(X_1,\theta_0), \psi(X_1,\theta_0)), \\
s_{13} &= \mathrm{Cov}_{\theta_0}(\dot{\psi}(X_1,\theta_0), \phi(X_1,\theta_0)) \;\; s_{23} = \mathrm{Cov}_{\theta_0}(\psi(X_1,\theta_0), \phi(X_1,\theta_0)) \\
s_{21} &= s_{12}, \;\; s_{31} = s_{13} \;\; s_{32} = s_{23}.
\end{aligned}$$

THEOREM 7.10 *Assume that $\rho(x,\theta)$ and $\psi(x,\theta) = \partial\rho(x,\theta)/\partial\theta$ are absolutely continuous in θ and the above conditions **1. – 5.** are satisfied. Let $M_n^{(1)}$ be the one-step estimator defined in (7.43) and (7.44) with $M_n^{(0)}$ satisfying (7.46). Then $M_n^{(1)}$ admits the asymptotic representation (7.45) and*

$$nR_n^{(1)} \xrightarrow{\mathcal{D}} U^\star \;\; \textit{as } n \to \infty,$$

where

$$U^\star = \gamma_1^{-2}U_2\Big(U_1 - \frac{U_2\gamma_2}{2\gamma_1}\Big) + \gamma_2(2\gamma_1^3)^{-1}(U_2 + \gamma_1 U_3)^2 \tag{7.47}$$

with $\gamma_1 = \gamma_1(\theta_0)$ and $\gamma_2 = \gamma_2(\theta_0) = \mathbb{E}_{\theta_0}\ddot{\psi}(X_1,\theta_0)$.

Notice that, by Theorem 5.2, the distribution of the first term on the right-hand side of (7.47) coincides with the asymptotic distribution of nR_n, where R_n is the remainder term in (7.41). Hence the effect of the initial estimate $M_n^{(0)}$ appears only in the second term of the right-hand side of (7.47). In this way we get the following corollary:

Corollary 7.3 *Under the conditions of Theorem 7.10,*

$$n(M_n^{(1)} - M_n) \xrightarrow{\mathcal{D}} \frac{\gamma_2}{2\gamma_1^3}(U_2 + \gamma_1 U_3)^2 \;\; \textit{as } n \to \infty.$$

Consequently,

$$M_n^{(1)} - M_n = o_p(n^{-1}) \tag{7.48}$$

if and only if either

$$\phi(x,\theta) = -\gamma_1^{-1}\psi(x,\theta)\ \forall x \in \boldsymbol{R}_1,\ \theta \in \Theta$$

for the function φ *in the representation (7.46) or* $\gamma_2(\theta_0) = \mathbb{E}_{\theta_0}\ddot{\psi}(X_1,\theta_0) = 0$.

Remark 7.2 *(1) The formula (7.48) tells us that* M_n *and* $M_n^{(1)}$ *are asymptotically equivalent up to the order* n^{-1} *if the initial estimator* $M_n^{(0)}$ *has the same influence function as* M_n.

(2) The asymptotic distribution of $n(M_n^{(1)} - M_n)$ *is the central* χ^2 *with 1 d.f., up to the multiplicative factor* $\sigma^2\gamma_2/(2\gamma_1^3)$, *where*

$$\begin{aligned}\sigma^2 &= \mathbb{E}(U_2 + \gamma_1 U_3)^2 \\ &= \mathbb{E}\psi^2(X_1,\theta_0) + \gamma_1^2\mathbb{E}\varphi^2(X_1,\theta_0) + 2\gamma_1\mathbb{E}(\psi(X_1,\theta_0)\varphi(X_1,\theta_0)).\end{aligned} \tag{7.49}$$

The asymptotic distribution is confined to the positive or negative part of $\boldsymbol{R}_1$ *depending on whether or not* γ_2/γ_1 *is positive.*

(3) The asymptotic relative efficiency of $M_n^{(1)}$ *to* M_n *is equal to 1. On the other hand, the second moment of the asymptotic distribution of* $n(M_n^{(1)} - M_n)$ *in (7.13) may be considered as a measure of deficiency of* $M_n^{(1)}$ *with respect to* M_n*; hence*

$$d(M_n^{(1)}, M_n) = \frac{3}{16}\sigma^4(\gamma_2\gamma_1^{-3})^2$$

with σ *defined by (7.49).*

(4) If $k \geq 2$, *then* $M_n^{(k)} - M_n = o_p(n^{-1})$.

(5) In the location model $\psi(x,t) = \psi(x-t)$ *and* $F(x,\theta) = F(x-\theta)$. *In the symmetric case when* $F(x) + F(-x) = 1$, $\psi(-x) = -\psi(x)$, $x \in \boldsymbol{R}_1$ *is* $\gamma_2 = 0$, *and hence* $M_n^{(1)} - M_n = o_p(n^{-1})$.

(6) Unless $M_n^{(0)}$ *and* M_n *have the same influence functions, the ratio of the second moment of the limiting distribution of* $n^{\frac{1}{2}}(M_n^{(0)} - M_n)$ *and of the first absolute moment of that of* $n(M_n^{(1)} - M_n)$ *is equal to* $\gamma_2/(2\gamma_1)$ *and hence is independent of the choice of* $M_n^{(0)}$.

PROOF of Theorem 7.10. By Theorem 5.2 and by virtue of the conditions, we have

$$\begin{aligned}
&n^{-1}\sum_{i=1}^{n}\ddot{\psi}(X_i,\theta_0) = \gamma_2 + o_p(1),\\
&n^{\frac{1}{2}}(\hat{\gamma}_n - \gamma_1) = U_{n1} + \gamma_2 U_{n3} + o_p(1),\\
&n^{\frac{1}{2}}(\hat{\gamma}_n^{-1} - \gamma_1^{-1}) = -\gamma_1^{-2}U_{n1} + \gamma_2\gamma_1^{-2}U_{n3} + o_p(1),\\
&n^{-\frac{1}{2}}\sum_{i=1}^{n}\psi(X_i, M_n^{(0)}) = U_{n2} + \gamma_1 U_{n3} + O_p(n^{-\frac{1}{2}}).
\end{aligned}$$

Hence,

$$nR_n^{(1)} = \gamma_1^{-2}\Big\{\frac{1}{2}\gamma_1\gamma_2 U_{n3}^2 + \gamma_2 U_{n2}U_{n3} + U_{n1}U_{n2}\Big\} + o_p(1),$$

and this gives the desired result. □

7.4.2 One-step version of M-estimator of location

The location model can be considered as a special case of the general linear model. Theorem 7.10 provides the orders of approximations of estimators by their one-step versions also in the location model, at least under some conditions. We will leave the precise formulation of these results to the reader as an exercise and rather turn our attention to the cases not covered by the general model, namely to the studentized estimators and to less smooth ρ-functions.

Let $X_1, \ldots, X_n$ be independent observations with a common distribution function $F(x-\theta)$. The studentized M-estimator M_n of θ was defined in Section 5.4 as a solution of the minimization (5.61) with the studentizing scale statistic S_n satisfying the conditions (5.62) and (5.63). Theorem 5.8 provided the asymptotic representation of M_n which can be written in a unified form as

$$M_n = \theta + (n\gamma_1)^{-1}\sum_{i=1}^{n}\psi\Big(\frac{X_i-\theta}{S}\Big) - \frac{\gamma_2}{\gamma_1}\Big(\frac{S_n}{S}-1\Big) + O_p(n^{-\frac{1}{2}-r}), \qquad (7.50)$$

where γ_1 and γ_2 are suitable functionals of F, dependent on $S = S(F)$ but not on θ and $0 < r \le \frac{1}{2}$; specific values of r in Theorem 5.8 were $r = \frac{1}{4}$ or $r = \frac{1}{2}$ according to the smoothness of $\psi = \rho'$.

On the other hand, in Chapter 5 we proved the asymptotic linearity of the empirical process (Lemma 5.2)

$$Z_n(t,u) = n^{-\frac{1}{2}}\sum_{i=1}^{n}\Big\{\psi\Big(\frac{X_i-\theta-n^{-\frac{1}{2}}t}{Se^{n^{-\frac{1}{2}}u}}\Big) - \psi\Big(\frac{X_i-\theta}{S}\Big)\Big\}.$$

More precisely, for any $C > 0$, as $n \to \infty$,

$$\sup\Big\{|Z_n(t,u) + t\gamma_1 + u\gamma_2| : \ |t|, |u| \le C\Big\} = O_p(n^{-r}). \qquad (7.51)$$

Inserting $t \to n^{\frac{1}{2}}(M_n^{(0)} - \theta)$ $[= O_p(1)]$ and $u \to n^{\frac{1}{2}}(Z_n - Z)$, $Z_n = \ln S_n, Z = \ln S$ in (7.51), where $M_n^{(0)}$ is an initial estimator of θ, we express M_n as

$$M_n = M_n^{(0)} + \frac{1}{n\gamma_1}\sum_{i=1}^{n}\psi\Big(\frac{X_i - M_n^{(0)}}{S_n}\Big) + O_p(n^{-\frac{1}{2}-r}). \qquad (7.52)$$

Hence we can define the one-step version of the studentized M-estimator M_n

as

$$M_n^{(1)} = \begin{cases} M_n^{(0)} + \frac{1}{n\hat{\gamma}_1^{(1)}} \sum_{i=1}^n \psi\left(\frac{X_i - M_n^{(0)}}{S_n}\right), & \hat{\gamma}_1^{(1)} \neq 0, \\ M_n^{(0)}, & \hat{\gamma}_1^{(1)} = 0, \end{cases} \tag{7.53}$$

where $\hat{\gamma}_1^{(1)}$ is a consistent estimator of γ_1. The order of approximation of M_n by its one-step version (7.53) is given in the following theorem:

THEOREM 7.11 *Assume the conditions of Theorem 5.8. Let $M_n^{(1)}$ be the one-step version of M_n with an estimator $\hat{\gamma}_1^{(1)}$ of $\gamma_1 > 0$ such that*

$$\hat{\gamma}_1^{(1)} - \gamma_1 = O_p(n^{-r}), \tag{7.54}$$

where $r = \frac{1}{4}$ or $\frac{1}{2}$ under the conditions of parts (i) or (ii) of Theorem 5.8, respectively. Then

$$M_n - M_n^{(1)} = O_p(n^{-r-\frac{1}{2}}) \quad \text{as } n \to \infty.$$

Remark 7.3 *If ψ' exists and is continuous, we may take*

$$\hat{\gamma}_1^{(1)} = \frac{1}{nS_n} \sum_{i=1}^n \psi'\left(\frac{X_i - M_n^{(0)}}{S_n}\right). \tag{7.55}$$

Another possible choice is

$$\begin{aligned} \hat{\gamma}_1^{(1)} &= \frac{1}{n^{\frac{1}{2}}(t_2 - t_1)} \sum_{i=1}^n \Big\{ \psi\Big(\frac{X_i - M_n^{(0)} + n^{-\frac{1}{2}} t_2}{S_n}\Big) \\ &\quad - \psi\Big(\frac{X_i - M_n^{(0)} + n^{-\frac{1}{2}} t_1}{S_n}\Big) \Big\} \end{aligned} \tag{7.56}$$

for some $-\infty < t_1 < t_2 < \infty$. Actually, (7.54) applies to (7.56) as well.

PROOF of Theorem 7.11. Since $\gamma_1 > 0$, we have $\frac{\hat{\gamma}_1}{\gamma_1} = O_p(n^{-r})$ as $n \to \infty$. Combining (7.50) and (7.52), we obtain

$$\begin{aligned} n^{\frac{1}{2}}(M_n - M_n^{(1)}) &= -\frac{1}{\gamma_1 \sqrt{n}} \frac{\gamma_1}{\hat{\gamma}_1^{(1)}} \sum_{i=1}^n \Big[\psi\Big(\frac{X_i - M_n^{(0)}}{S_n}\Big) \\ &- \psi\Big(\frac{X_i - \theta}{S}\Big)\Big] + n^{\frac{1}{2}}(M_n^{(0)} - \theta) + n^{\frac{1}{2}} \frac{\gamma_2}{\gamma_1}\Big(\frac{S_n}{S} - 1\Big) \\ &- \Big(\frac{\gamma_1}{\hat{\gamma}_1^{(1)}} - 1\Big)\Big\{ \frac{1}{n^{\frac{1}{2}}\gamma_1} \sum_{i=1}^n \psi\Big(\frac{X_i - \theta}{S}\Big) - n^{\frac{1}{2}}(M_n^{(0)} - \theta) \\ &- n^{\frac{1}{2}} \frac{\gamma_2}{\gamma_1}\Big(\frac{S_n}{S} - 1\Big)\Big\} + O_p(n^{-r}) = O_p(n^{-r}). \end{aligned}$$

□

Similarly, we can define the k-step version of M_n $(k \geq 1)$,

$$M_n^{(k)} = \begin{cases} M_n^{(k-1)} + \frac{1}{n\hat{\gamma}_1^{(k)}} \sum_{i=1}^n \psi\left(\frac{X_i - M_n^{(k-1)}}{S_n}\right), & \text{if } \hat{\gamma}_1^{(k)} \neq 0, \\ M_n^{(k-1)} & \text{if } \hat{\gamma}_1^{(k)} = 0, \end{cases} \tag{7.57}$$

where $\hat{\gamma}_1^{(k)}$ is a consistent estimator of γ_1. For instance, we may take the estimators (7.55) and (7.56) with $M_n^{(0)}$ being replaced $M_n^{(k-1)}$, $k = 1, 2, \ldots$.

The asymptotic properties of $M_n^{(k)}$ were studied by Jurečková and Malý (1995); for the sake of simplicity, we shall restrict ourselves to the special case $S_n = 1$, namely to the non-studentized M-estimators.

The following asymptotic linearity lemma provides the key to our asymptotic considerations:

LEMMA 7.1 *Let $X_1, \ldots, X_n$ be independent identically distributed random variables with the distribution function F. Consider the following empirical process with time parameters $t, u \in \boldsymbol{R}_1$, and an index $\tau \geq \frac{1}{2}$:*

$$Q_n = Q_n(t, u) = \sum_{i=1}^n \left\{\psi(X_i - n^{-\frac{1}{2}}t - n^{-\tau}u) - \psi(X_i - n^{-\frac{1}{2}}t)\right\}.$$

(i) If either ψ and ψ' are absolutely continuous,

$$\begin{aligned} &\mathbb{E}(\psi'(X_1))^2 < \infty, \quad \textit{and} \\ &\mathbb{E}|\psi''_\delta(X_1)| < \infty \quad \textit{for } 0 < \delta \leq \delta_0, \quad \textit{where} \\ &\psi''_\delta(x) = \sup_{|y| \leq \delta} |\psi''(x + y)|, \; x \in \boldsymbol{R}_1 \end{aligned} \tag{7.58}$$

or ψ is absolutely continuous with the step-function derivative

$$\psi'(x) = \beta_j, \;\; p_j < x < p_{j+1}, \;\; j = 1, \ldots, \ell$$

with $\beta_0, \ldots, \beta_\ell \in \boldsymbol{R}_1$, $\beta_0 = \beta_\ell = 0$, $-\infty = p_0 < p_1 < \ldots < p_\ell < p_{\ell+1} = \infty$, and F has a bounded derivative $f(x)$ in a neighborhood of $p_1, \ldots, p_\ell$, then

$$\sup_{|t|, |u| \leq C} \left\{|n^{-\frac{1}{2}} Q_n(t, u) + n^{\frac{1}{2} - \tau}\gamma_1 u|\right\} = O_p(n^{-\tau})$$

for any $\tau \geq \frac{1}{2}$ and any $C > 0$, whenever $\gamma_1 = \mathbb{E}\psi'(X_1) > 0$.

(ii) If ψ is a step-function,

$$\psi(x) = \alpha_j \;\; \textit{if } q_j < x < q_{j+1}, \;\; j = 0, 1, \ldots, m,$$

and F is continuous with two bounded derivatives f and f', $f > 0$ in a neighborhood of $q_1, \ldots q_m$, then

$$\sup_{|t|, |u| \leq C} \left\{|n^{-\frac{1}{2}} Q_n(t, u) + n^{\frac{1}{2} - \tau}\gamma_1 u|\right\} = O_p(n^{-\tau/2})$$

for any $\tau \geq \frac{1}{2}$ with $\gamma_1 = \sum_{j=1}^m (\alpha_j - \alpha_{j-1}) f(q_j) > 0$.

PROOF Let first ψ and ψ' be absolutely continuous. Using the Taylor expansion, we get for some $0 < h_1,\ h_2 < 1$ and $K > 0$, due to (7.58),

$$\sup_{|t|,|u|\le C} |n^{-\frac{1}{2}} Q_n(t,u) + n^{\frac{1}{2}-\tau}\gamma_1 u|$$

$$= \sup_{|t|,|u|<C} \Big\{ | - n^{-\frac{1}{2}-\tau} u \sum_{i=1}^n [\psi'(X_i) - \mathbb{E}\psi'(X_i)]$$

$$+ n^{-1-\tau} ut \sum_{i=1}^n \psi''(X_i - h_2 n^{-\frac{1}{2}} t)$$

$$+ \frac{1}{2} n^{-\frac{1}{2}-2\tau} u^2 \sum_{i=1}^n \psi''(X_i - n^{-\frac{1}{2}} t - h_1 n^{-\tau} u)| \Big\}$$

$$\le n^{-\tau} C |n^{-\frac{1}{2}} \sum_{i=1}^n [\psi'(X_i) - \mathbb{E}\psi'(X_i)]|$$

$$+ n^{-\tau} C^2 K \sup_{|t|\le C} \Big\{ |n^{-1} \sum_{i=1}^n \psi''(X_i - h_2 n^{-\frac{1}{2}} t)| \Big\}$$

$$+ \frac{1}{2} n^{\frac{1}{2}-2\tau} C^2 K \sup_{|t|\le C} \Big\{ |n^{-1} \sum_{i=1}^n \psi''(X_i - h_1 n^{-\frac{1}{2}} t)| \Big\} = \mathrm{O}_p(n^{-\tau}).$$

Under the second condition of (i), it suffices to consider the function

$$\psi(x) = \begin{cases} p_1 & \text{for } x < p_1, \\ x & \text{for } p_1 \le x \le p_2, \\ p_2 & \text{for } x > p_2, \end{cases}$$

with $-\infty < p_1 < p_2 < \infty$, without loss of generality. Assume first that $u > 0$ (the case $u < 0$ is analogous), and denote by

$$a_n = n^{-\tau} u,$$
$$\mathcal{I}_i = I[X_i - n^{-\frac{1}{2}} t \in (r_1 - a_n, r_2 + a_n) \cup (r_2 - a_n, r_2 + a_n)],$$
$$\mathcal{J}_i = I[X_i - n^{-\frac{1}{2}} t \in (r_1 + a_n, r_2 - a_n)];$$
$$Q_n^{(i)} = \psi(X_i - n^{-\frac{1}{2}} t - a_n) - \psi(X_i - n^{-\frac{1}{2}} t),\ i = 1, \dots, n.$$

Then $Q_n = \sum_{i=1}^n Q_n^{(i)} \mathcal{I}_i - \sum_{i=1}^n a_n \mathcal{J}_i$, and we have

$$\sup \Big\{ \Big| n^{-\frac{1}{2}} \sum_{i=1}^n Q_n^{(i)} \mathcal{I}_i \Big| : |t| < C, |u| < C \Big\} \tag{7.59}$$

$$\le n^{\frac{1}{2}-\tau} K \sup \Big\{ \Big| \hat{F}_n(R_1^+) - \hat{F}_n(R_1^-) + \hat{F}_n(R_2^+) - \hat{F}_n(R_2^-) \Big| : |t| < C, |u| < C \Big\},$$

where $\hat{F}_n$ stands for the empirical distribution function based on $X_1, \dots, X_n$ and

$$R_j^- = p_j + n^{-\frac{1}{2}} t - a_n,\ \ R_j^+ = p_j + n^{-\frac{1}{2}} t + a_n,\ j = 1, 2.$$

Using the approximation of the empirical distribution function process by

Brownian bridge [see (2.116)–(2.117)], we come to the conclusion that the right-hand side of (7.59) is $O_p(n^{-\tau})$.

Similarly it can be shown that

$$\sup_{|t|,|u|\leq C}\left\{n^{-\frac{1}{2}}\sum_{i=1}^{n}\left|a_n\mathcal{I}_i + n^{\frac{1}{2}-\tau}u\gamma_1\right|\right\} = O_p(n^{-\tau}), \tag{7.60}$$

and this completes the proof of Part (i).

For Part (ii) we can consider the function

$$\psi(x) = \begin{cases} 0 & \text{if } x \leq q, \\ 1 & \text{if } x > q. \end{cases}$$

without loss of generality. Then

$$n^{-\frac{1}{2}}Q_n(t,u) = -n^{\frac{1}{2}}[\hat{F}_n(q + n^{-\frac{1}{2}}t + n^{-\tau}u) - \hat{F}_n(q + n^{-\frac{1}{2}}t)],$$

and the proposition follows analogously, using the approximation of the empirical distribution function process. □

Now we are able to derive the order of the approximation of M_n by its k-step version:

THEOREM 7.12 *Let $X_1, X_2, \ldots$ be independent identically distributed observations with the distribution function $F(x-\theta)$. Let M_n be the M-estimator of θ generated by an absolutely continuous function ρ with the derivative ψ. Let $M_n^{(k)}$ be the k-step version of M_n defined in (7.57), $k = 1, 2, \ldots$ with a $\sqrt{n}$-consistent initial estimator $M_n^{(0)}$. Then*

(i) Under the assumptions of part (i) of Lemma 7.1, provided $\hat{\gamma}_1^{(k)} - \gamma_1 = O_p(n^{-\frac{1}{2}})$, $k = 1, 2, \ldots$, we have

$$n^{\frac{1}{2}}(M_n^{(k)} - M_n) = O_p(n^{-\frac{k}{2}}),\ k = 1, 2, \ldots. \tag{7.61}$$

(ii) Under the assumptions of part (ii) of Lemma 7.1, provided $\hat{\gamma}_1^{(k)} - \gamma_1 = O_p(n^{-\frac{1}{4}})$, $k = 1, 2, \ldots$,

$$n^{\frac{1}{2}}(M_n^{(k)} - M_n) = O_p(n^{2^{-k-1}-\frac{1}{2}}),\ k = 1, 2, \ldots.$$

PROOF (i) By Theorem 7.11, the proposition (7.61) holds for $k = 1$. Assume that (7.61) holds for $k-1$. Inserting $t \to n^{\frac{1}{2}}(M_n - \theta)$ $[= O_p(1)]$ and $u \to n^{\tau}(M_n^{(k-1)} - M_n)$ $[O_p(1)$ for $\tau = k/2]$ in $Q_n(t,u)$ we obtain

$$n^{-\frac{1}{2}}\sum_{i=1}^{n}\psi(X_i - M_n^{(k-1)}) + n^{\frac{1}{2}}(M_n^{(k-1)} - M_n)\gamma_1 = O_p(n^{-\frac{k}{2}}).$$

Moreover, by definition of $M_n^{(k)}$, we have under $\hat{\gamma}_1^{(k)} \neq 0$

$$\begin{aligned} & n^{\frac{1}{2}}(M_n^{(k)} - M_n) \\ &= \frac{1}{n^{\frac{1}{2}}\hat{\gamma}_1^{(k)}} \sum_{i=1}^{n} \psi(X_i - M_n^{(k-1)}) + n^{\frac{1}{2}}(M_n^{(k-1)} - M_n) \end{aligned}$$

and this leads to the desired result by the induction hypothesis.

The proof of part (ii) is quite analogous. □

Remark 7.4 *Notice that while the approximation of M_n by $M_n^{(k)}$ really improves with increasing k in the case of continuous ψ, the order of approximation of M_n by its k-step version increases very slowly with k if ψ has jump discontinuities. It is never better than $O_p(n^{-1})$ in this case. This surprising conclusion also concerns the estimators based on L_1-norm.*

7.4.3 One-step versions in linear models

The one- and k-step iterations of estimators have even more applicability in the linear models than in the one-parameter ones. The ideas and properties of one- and k-step versions in the linear regression model are analogous to those in the location model. However, their proof are more technical. As a compromise, in order to bring the forms and properties of these estimators as well as the basic references and yet not to consume much space on the technicalities, we will give a brief description of one and k-step versions in the linear regression model along with some basic properties, with only either hints or references to proofs.

Consider the linear regression model (7.34) with the design matrix $\mathbf{X}$ of order $n \times p$. For simplicity, we assume that $x_{i1} = 1,\ i = 1, \ldots, n$. The *one-step version of M-estimator* of $\boldsymbol{\beta}$ was first proposed by Bickel (1975) and later investigated by Jurečková (1983), Jurečková and Portnoy (1987), and Jurečková and Welsh (1990). Keeping the notation of Section 7.3, we can define the one-step version $\mathbf{M}_n^{(1)}$ of the M-estimator $\mathbf{M}_n$, generated by an absolutely continuous function ρ with $\psi = \rho'$ and studentized by a scale statistics $S_n = S_n(\mathbf{Y})$, as

$$\mathbf{M}_n^{(1)} = \begin{cases} \mathbf{M}_n^{(0)} + \frac{1}{\hat{\gamma}_n}\mathbf{W}_n & \text{if } \hat{\gamma}_n \neq 0, \\ \mathbf{M}_n^{(0)} & \text{if } \hat{\gamma}_n = 0, \end{cases} \tag{7.62}$$

where

$$\mathbf{W}_n = \mathbf{Q}_n^{-1} \sum_{i=1}^{n} \mathbf{x}_i \psi\Big(\frac{Y_i - \mathbf{x}_i^{\top}\mathbf{M}_n^{(0)}}{S_n}\Big) \tag{7.63}$$

and $\hat{\gamma}_n$ is the estimator of γ_1 based on $Y_1, \ldots, Y_n$. For instance, we can consider

$$\hat{\gamma}_n = \frac{1}{2t\sqrt{n}} \sum_{i=1}^{n} \Big\{\psi\Big(\frac{Y_i - \mathbf{x}_i^{\top}\mathbf{M}_n^{(0)} + n^{-\frac{1}{2}}t}{S_n}\Big) - \psi\Big(\frac{Y_i - \mathbf{x}_i^{\top}\mathbf{M}_n^{(0)} - n^{-\frac{1}{2}}t}{S_n}\Big)\Big\}.$$

By the asymptotic linearity results of Chapter 5, $\hat{\gamma}_n$ is a $n^{\frac{1}{2}}$- or $n^{\frac{1}{4}}$-consistent estimator of γ_1, depending on whether ψ is smooth or has jump discontinuities. If ψ' exists, we may also consider

$$\hat{\gamma}_n = \frac{1}{nS_n}\sum_{i=1}^{n}\psi'\Big(\frac{Y_i - \mathbf{x}_i^\top \mathbf{M}_n^{(0)}}{S_n}\Big).$$

Then, under the conditions of Theorem 5.10 (for smooth ψ), provided that $n^{\frac{1}{2}}\|\mathbf{M}_n^{(0)} - \boldsymbol{\beta}\| = O_p(1)$, we have

$$\|\mathbf{M}_n - \mathbf{M}_n^{(1)}\| = O_p(n^{-1}),$$

while under the conditions of Theorem 5.11 (step-function ψ),

$$\|\mathbf{M}_n - \mathbf{M}_n^{(1)}\| = O_p(n^{-3/4}).$$

We expect that also the behavior of the k-step version will be analogous as in the location case.

The possible existence of leverage points in the matrix $\mathbf{X}_n$ leads to the fact that the breakdown point of the M-estimator does not exceed $1/(p+1)$. The situation is not better with L- and R-estimators in the linear model (see Donoho and Huber 1983). However, the following simple modification of the one-step estimator in (7.62)–(7.63) yields an estimator $\mathbf{M}_n^\star$, satisfying (7.63), which has the same breakdown point as the initial estimator $\mathbf{M}_n^{(0)}$:

$$\mathbf{M}_n^\star = \begin{cases} \mathbf{M}_n^{(0)} + \frac{1}{\hat{\gamma}_n}\mathbf{W}_n & \text{if } \frac{1}{\hat{\gamma}_n}\|\mathbf{W}_n\| \le a, \\ \mathbf{M}_n^{(0)} & \text{otherwise,} \end{cases}$$

for some fixed $a > 0$. We refer to Jurečková and Portnoy (1987) for the proof in case of a smooth ψ. Even if the initial estimator $\mathbf{M}_n^{(0)}$ has a lower rate of consistency, say $n^{\tau}\|\mathbf{M}_n^{(0)} - \boldsymbol{\beta}\| = O_p(1)$ for some $\tau \in (\frac{1}{4}, \frac{1}{2}]$, $\mathbf{M}_n^\star$ inherits the breakdown point from $\mathbf{M}_n^{(0)}$ and

$$\|\mathbf{M}_n^\star - \mathbf{M}_n\| = O_p(n^{-2\tau}).$$

This approach was later extended by Simpson et al. (1992) to the one-step GM-estimators in the linear model and by Bergesio and Yohai (2011) to the generalized linear model. However, if the initial estimator has a low rate of consistency, the convergence is slow and to get a reasonable approximation of an estimator by is one-step version needs many observations.

The *one-step version* $\mathbf{T}_n^{(1)}$ of R-estimator $\mathbf{T}_n$ generated by the score-generating function φ is defined analogously. Let $\varphi: (0,1) \mapsto \boldsymbol{R}_1$ be a nondecreasing, square-integrable score function, and let $R_{ni}(\mathbf{t})$ denote the rank of the residual $Y_i - \mathbf{x}_i^\top\mathbf{t}$, $\mathbf{t} \in \boldsymbol{R}_p$, $i = 1, \ldots, n$. Consider the vector of *linear rank statistics* $\mathbf{S}_n(\mathbf{t}) = (S_{n1}(\mathbf{t}), \ldots, S_{np}(\mathbf{t}))^\top$ where

$$S_{nj}(\mathbf{t}) = \sum_{i=1}^{n}(x_{ij} - \bar{x}_j)a_n(R_{ni}(\mathbf{t})), \;\; j = 1, \ldots, p,$$

and $a_n(i)$, $1 \le i \le n$, are the *scores* generated by φ as $a_n(i) = \varphi\left(\frac{i}{n+1}\right)$, $i = 1, \ldots, n$. The R-estimate $\mathbf{T}_n^*$ of $\boldsymbol{\beta}$ based on $\mathbf{S}_n(\mathbf{t})$ is obtained by the minimization of the Jaeckel measure of rank dispersion (6.47). However, this minimization means a new ordering in every iteration and the calculation can by long. Hence, starting with an initial $\sqrt{n}$-consistent estimate $\mathbf{T}_n^{(0)}$ (not necessary R-estimate), we define the one-step version of $\mathbf{T}_n^*$ as follows:

$$\mathbf{T}_n^{(1)} = \begin{cases} \mathbf{T}_n^{(0)} + \frac{1}{n\hat{\gamma}_n}\mathbf{Q}_n^{-1}\mathbf{S}_n(\mathbf{T}_n^{(0)}) & \text{if } \hat{\gamma}_n \neq 0, \\ \mathbf{T}_n^{(0)} & \text{if } \hat{\gamma}_n = 0, \end{cases},$$

where

$$\mathbf{Q}_n = \left[q_{jk}^{(n)}\right]_{j,k=1,\ldots,p},$$

$$q_{jk}^{(n)} = \frac{1}{n}\sum_{i=1}^{n}(x_{ij} - \bar{x}_j)(x_{ik} - \bar{x}_k), \quad \text{for } j,k = 1,\ldots,p$$

and $\hat{\gamma}_n$ is an estimator of $\gamma = \int_0^1 \varphi(u)\varphi(u,f)du$. The asymptotic linearity of the linear rank statistics offers the following possible estimator of γ :

$$\hat{\gamma}_n = \left\| \frac{1}{2\sqrt{n}}\mathbf{Q}_n^{-1}[\mathbf{S}_n(\mathbf{T}_n^{(0)} - n^{-\frac{1}{2}}\mathbf{e}_j) - \mathbf{S}_n(\mathbf{T}_n^{(0)} + n^{-\frac{1}{2}}\mathbf{e}_j)] \right\|$$

with $\mathbf{e}_j$ standing for the jth unit vector, $1 \le j \le p$. If $\mathbf{T}_n^{(0)}$ is a $\sqrt{n}$-consistent estimator of $\boldsymbol{\beta}$ and the conditions of Chapter 6 leading to the asymptotic linearity of $\mathbf{S}_n(\cdot)$ are satisfied, then

$$n^{\frac{1}{2}}\|\mathbf{T}_n^{(1)} - \mathbf{T}_n^*\| = o_p(1).$$

7.5 Further developments

1. Kraft and van Eeden (1972) proposed the *linearized rank estimator* in the linear regression model (7.34) based on the vector of linear rank statistics

$$\mathbf{S}_n(\mathbf{t}) = (S_{n1}(\mathbf{t}), \ldots, S_{np}(\mathbf{t}))^\top$$

$$S_{nj}(\mathbf{t}) = \sum_{i=1}^{n}(x_{ij} - \bar{x}_j)a_n(R_{ni}(\mathbf{t})), \;\; j = 1, \ldots, p.$$

The linearized rank estimator $\mathbf{T}_n^{(\ell)}$ is defined with the aid of an initial estimator $\mathbf{T}_n^{(0)}$ as

$$\mathbf{T}_n^{(\ell)} = \mathbf{T}_n^{(0)} + \frac{1}{nA^2(\varphi)}\mathbf{Q}_n^{-1}\mathbf{S}_n(\mathbf{T}_n^{(0)}),$$

where $A^2(\varphi) = \int_0^1(\varphi(u) - \bar{\varphi})^2 du$, $\;\bar{\varphi} = \int_0^1 \varphi(u)du$. If $\mathbf{T}_n^{(0)}$ is $\sqrt{n}$-consistent estimator of $\boldsymbol{\beta}$ and

$$\varphi(u) - \bar{\varphi} \equiv \varphi(u,f) = -\frac{f'(F^{-1}(u))}{f(F^{-1}(u))}, \; 0 < u < 1,$$

then $T_n^{(\ell)}$ is an asymptotically efficient estimator of $\boldsymbol{\beta}$. Under a general φ the asymptotic (normal) distribution of $T_n^{(\ell)}$ depends on $T_n^{(0)}$ and φ. If $\mathbf{T}_n$ is the ordinary R-estimator of $\boldsymbol{\beta}$ based on $\mathbf{S}_n(\mathbf{t})$, then

$$n^{\frac{1}{2}}(\mathbf{T}_n - \mathbf{T}_n^{(\ell)}) = \Big(1 - \frac{\gamma}{A^2(\varphi)}\Big)\mathbf{S}_n(\boldsymbol{\beta}) + o_p(1), \quad \gamma = \int_0^1 \varphi(u)\varphi(u, f)du$$

(see Kraft and van Eeden (1972) and Jurečková (1989) for detailed proofs).

2. Dodge and Jurečková (2000) constructed a family of adaptive convex combinations of two M-estimators (or others) of $\boldsymbol{\beta}$ in linear model (7.34), adaptive over the family of symmetric densities

$$\mathcal{F} = \Big\{f(e) = \frac{1}{s}f_0\Big(\frac{e}{s}\Big)\Big|\, s = \frac{1}{f(0)} > 0\Big\}$$

where f_0 is a fixed (but unknown) symmetric density with the finite second moment, $f_0(0) = 1$. The adaptive convex combination of the LSE and LAD estimators is an important example, when $\rho(z) = (1-\delta)z^2 + \delta|z|$, and the optimal $\hat{\delta}$, asymptotically minimizing the variance of the resulting estimator, is estimated from the data.

7.6 Problems

7.2.1 Are the regularity conditions of both Theorems 4.4 and 5.3 fulfilled under assumptions (i)–(iv) in Theorem 7.4?

7.2.2 Let θ be a center of symmetry of a symmetric distribution function F. Consider the following

(a) An L-estimator based on a score function $J(.)$.

(b) An M-estimator based on a score function $\psi(.)$.

(c) A R-estimator based on a score function $\varphi(.)$.

Which regularity conditions should distribution function F and the respective score functions satisfy in each case so that a first-order asymptotic distributional representation holds? Specify the regularity conditions under which the above estimators are (pairwise) asymptotically first-order equivalent.

7.2.3 Let Φ be the standard normal distribution function, and let $\mathcal{P}$ be the set of all symmetric probability measures arising from Φ through ε-contamination: $\mathcal{P} = \{F : F = (1-\varepsilon)\varphi + \varepsilon H,\ H \in \mathcal{M}\}$, where $\mathcal{M}$ is the set of all symmetric probability measures on $\boldsymbol{R}_1$. Then the Fisher information $\mathcal{I}(f)$ is minimized by the probability density

$$f_0(x) = \frac{1-\varepsilon}{\sqrt{2\pi}}\Big\{\mathrm{e}^{-\frac{1}{2}\mathrm{x}^2}\mathrm{I}[|\mathrm{x}| \le \mathrm{k}] + \mathrm{e}^{\frac{1}{2}\mathrm{k}^2 - \mathrm{k}|\mathrm{x}|}\mathrm{I}[|\mathrm{x}| > \mathrm{k}]\Big\},$$

where $2\frac{\phi'(k)}{k-2\Phi(-k)} = \frac{\varepsilon}{1-\varepsilon}$. Then the minimax M-estimator of θ corresponds to the score function

$$\psi_0(x) = -\frac{\partial \ln f_0(x)}{\partial x}$$

[Huber 1981].

7.2.4 Using the results in Problem 7.2.2, we can show that for the ε-error contaminated model in Problem 7.2.3, besides the M-estimator considered in Problem 7.2.3, the following two estimates are asymptotically minimax:

(a) The α-trimmed mean with

$$\alpha = F_0(-k) = (1-\varepsilon)\Phi(-k) + \frac{\varepsilon}{2},$$

and

(b) the R-estimator of θ based on the score function

$$\varphi_0(u) = \begin{cases} -k, & u \leq \alpha, \\ \Phi^{-1}\left(\frac{u-\frac{\varepsilon}{20}}{1-\varepsilon}\right), & \alpha \leq u \leq 1-\alpha, \\ k, & u \geq 1-\alpha, \end{cases}$$

with α defined by (a). [Huber 1981].

7.3.1 Characterize the second order asymptotic relations of M- and R- estimators of the slope of the regression line.

7.4.1 Using Theorem 7.10, show a second order asymptotic equivalence for the simple location model when $\psi = \rho'$ is smooth. State the needed regularity conditions.

7.4.2 Show that (7.54) holds for $\hat{\gamma}_1^{(1)}$ defined by (7.56).

CHAPTER 8

Robust Estimation: Multivariate Perspectives

8.1 Introduction

Chapters 3 to 7 mostly dealt with the methodological tools of robust estimation of location, regression and scale parameters, in univariate setups, assessing global robustness of R-estimators (shared to a certain extent by L-estimators), along with local robustness of M-estimators. Due emphasis has been laid down on their translation-scale (regression-scale) equivariance and invariance properties. The theoretical foundation of multivariate statistical estimation has been fortified by the notion of affine-equivariance and its dual affine-invariance, which are the natural generalizations of univariate translation-scale equivariance and invariance notion (Eaton 1983). This is due to the fact that multivariate parametric estimation theory has primarily evolved in the domain of *elliptically symmetric distributions*, which have their genesis in general multivariate normal distributions. However, general multivariate distributions, even symmetric in some sense, may not belong to the elliptically symmetric distribution family, and for which affine-equivariance (invariance) may not have a pioneering role, even under appropriate moment conditions. Also, it may not imply spherical symmetry.

The *Mahalanobis distance* based on elliptically symmetric distributions have witnessed an affine equivariance/invariance evolution in multivariate nonparametrics (Oja 2010; Serfling 2010, and the references cited therein). Yet, there are other norms of the multivariate symmetry, no less general than elliptical symmetry, where the affine equivariance/invariance may lack a complete statistical rationality. In addition, the Mahalanobis norm, being based on the inverse of the dispersion matrix, indicts a higher level of nonrobustness than in the univariate setup. In this vein, *diagonal symmetry*, *spherical symmetry*, *total symmetry*, *marginal symmetry*, and *symmetry in interchangeability* relate to different subclasses of symmetric multivariate distributions where affine equivariance/invariance may not transpire. For this reason, we shall devote more space to this broader conception of multivariate robust estimation than other contemporary work; it definitely needs a more elaborate treatise.

The diverse pattern of multivariate symmetry is assessed in Section 8.2. It is further incorporated in subsequent sections of this chapter. Robust estimation of multivariate location from the marginal symmetry point of view is presented in Section 8.3, and its linear model counterpart is considered in Section 8.4. The multivariate estimation based on the *spatial median*, *spatial ranks,* and *spatial quantile function* is treated in Section 8.5.

There is no universal optimality criterion in the multivariate data model; not only that, even the notion of robustness is rather imprecise. Then the robustness and optimality trade-off in multivariate estimation is a much more delicate task than in the univariate case. Moreover, the discovery of the *Stein-rule* or *shrinkage estimation* (James and Stein 1961) has rocked the foundation of the affine equivariance/invariance, especially from the decision-theoretic point of view. The *minimum risk estimation* and robust *Stein-rule estimation* are partially expounded in Sections 8.6 and 8.7. Section 8.8 is devoted to an outline of robust estimation of multivariate dispersion. The concluding section deals with some general remarks and observations.

8.2 The Notion of Multivariate Symmetry

A random variable X with a distribution function F, defined on $\mathbb{R}$, is said to be distributed symmetrically around $\theta \in \mathbb{R}$, if $F(\theta + x) + F(\theta - x-) = 1$, where $F(u-) = \mathbb{P}(X < u)$ and $F(u) = \mathbb{P}(X \leq u)$. We can write $F(x) = F_0(x - \theta)$, $x \in \mathbb{R}$ and say that F_0 is symmetric about 0. In other words, F is symmetric about θ if $X - \theta$ and $\theta - X$ both have a common distribution function F_0, symmetric about 0.

Without invoking symmetry, we say that F belongs to a location-scale family of distribution functions if there exists an F_0, free from location and scale parameters, such that $F(x) = F_0\left(\frac{x-\theta}{\lambda}\right)$, $\theta \in \mathbb{R}, \lambda \in \mathbb{R}^+, x \in \mathbb{R}$. Here θ need not be the mean and λ the standard deviation of X, but in most cases θ is a location, λ a scale parameter, and F_0 is symmetric about 0.

If $\mathbf{X} = (X_1, \ldots, X_p)^\top$ is a random vector, the marginal location/scale transformation

$$\mathbf{X} \mapsto \mathbf{Y} = (Y_1, \ldots, Y_p)^\top \quad \text{where } Y_j = \frac{X_j - \theta_j}{\lambda_j},\ j = 1, \ldots, p.$$

fails to capture a stochastic association or interdependence of the coordinates of $\mathbf{X}$ or $\mathbf{Y}$. Hence, it is deemed to use a more general transformation $\mathbf{X} \mapsto \mathbf{Y}$ which attempts to depict the interdependence of the coordinates in a canonical form, as the *affine transformation*

$$\mathbf{X} \mapsto \mathbf{Y} = \mathbf{\Lambda X} + \mathbf{a},\ \mathbf{a} \in \mathbb{R}_p \tag{8.1}$$

where $\mathbf{\Lambda}$ is a positive definite $p \times p$ matrix. Ideally, we wish to choose $\mathbf{\Lambda}$ so that the coordinates of $\mathbf{Y}$ are stochastically independent. This is possible when $\mathbf{X}$

has a multi-normal distribution with mean vector $\boldsymbol{\theta}$ and (positive definite) dispersion matrix $\boldsymbol{\Sigma}$, by letting $\boldsymbol{\Sigma}^{-1} = \boldsymbol{\Lambda}\boldsymbol{\Lambda}^\top$, so that $I\!\!E\mathbf{Y} = \boldsymbol{\xi} = \boldsymbol{\Lambda}\boldsymbol{\theta} + \mathbf{a}$ and dispersion matrix $\boldsymbol{\mathcal{V}}(\mathbf{Y}) = \boldsymbol{\Lambda}\boldsymbol{\Sigma}\boldsymbol{\Lambda}^\top = \mathbf{I}_p$. For non-normal distributions this can at best give the uncorrelation of the components of $\mathbf{Y}$, without any provision for their stochastic independence. The simplest example of an *affine-equivariant* estimator of $\boldsymbol{\theta}$ is the sample mean $\bar{\mathbf{X}}_n = \frac{1}{n}\sum_{i=1}^n \mathbf{X}_i$, leading to $\bar{\mathbf{Y}}_n = \boldsymbol{\Lambda}\bar{\mathbf{X}} + \mathbf{a}$ estimating $\boldsymbol{\xi}$.

An example of the *affine invariant* norm is the Mahalanobis norm. Indeed, the squared Mahalanobis distance of $\mathbf{X}$ from $\boldsymbol{\theta}$ is

$$\Delta^2 = \|\mathbf{X} - \boldsymbol{\theta}\|^2_\Sigma = (\mathbf{X} - \boldsymbol{\theta})^\top \boldsymbol{\Sigma}^{-1}(\mathbf{X} - \boldsymbol{\theta}) \tag{8.2}$$

while the corresponding squared Mahalanobis distance for $\mathbf{Y}$ from $\boldsymbol{\xi}$ takes the form

$$\begin{aligned}\Delta^{*2} &= (\mathbf{Y} - \boldsymbol{\xi})^\top [\boldsymbol{\mathcal{V}}(\mathbf{Y})]^{-1}(\mathbf{Y} - \boldsymbol{\xi}) \\ &= (\mathbf{X} - \boldsymbol{\theta})^\top \boldsymbol{\Lambda}^\top [\boldsymbol{\Lambda}\boldsymbol{\Sigma}\boldsymbol{\Lambda}^\top]^{-1} \boldsymbol{\Lambda}(\mathbf{X} - \boldsymbol{\theta}) \\ &= (\mathbf{X} - \boldsymbol{\theta})^\top \boldsymbol{\Sigma}^{-1}(\mathbf{X} - \boldsymbol{\theta}) = \Delta^2\end{aligned} \tag{8.3}$$

for all positive definite $\boldsymbol{\Lambda}$.

Let $\mathbf{X} \in \mathbb{R}_p$) have a distribution function F possessing a density $f(\mathbf{x}; \boldsymbol{\theta}, \boldsymbol{\Sigma})$ which can be expressed as

$$f(\mathbf{x}; \boldsymbol{\theta}, \boldsymbol{\Sigma}) = c(\boldsymbol{\Sigma}) h_0(\|\mathbf{x} - \boldsymbol{\theta}\|_\Sigma), \quad \mathbf{x} \in \mathbb{R}_p \tag{8.4}$$

where the function $h_0(y)$, $y \in \mathbb{R}^+$, does not depend on $\boldsymbol{\theta}, \boldsymbol{\Sigma}$. Since $\|\mathbf{x} - \boldsymbol{\theta}\|^2_\Sigma =$ const (> 0) depicts an ellipsoidal surface, the *equi-probability contours* of density (8.4) are ellipsoidal with center $\boldsymbol{\theta}$ and orientation matrix $\boldsymbol{\Sigma}$. Such a density is termed *elliptically symmetric* and the corresponding F belongs to the class $\mathcal{F}_E$ of the elliptically symmetric distributions. By the construction in (8.3)–(8.4), elliptically symmetric distributions admit the finite second-order moments and have the affine equivariant/invariant structure. This is true for the multivariate normal distribution. However, not all symmetric multivariate distribution functions are elliptically symmetric. An example is a mixture model with the density

$$f(\mathbf{x}; \boldsymbol{\theta}) = (1 - \pi)\, c_1(\boldsymbol{\Sigma}_1)\, h_1(\|\mathbf{x} - \boldsymbol{\theta}\|_{\Sigma_1}) + \pi\, c_2(\boldsymbol{\Sigma}_2)\, h_2(\|\mathbf{x} - \boldsymbol{\theta}\|_{\Sigma_2}) \tag{8.5}$$

where $\pi \in (0, 1)$, $\boldsymbol{\Sigma}_1$ and $\boldsymbol{\Sigma}_2$ are nonproportional, and $h_1(\cdot)$, $h_2(\cdot)$ are not necessarily of the same form. Thus, the equi-probability contours are not ellipsoidal, so that f does not belong to the elliptically symmetric family.

The density (8.5), though not elliptically symmetric, is still *diagonally symmetric* in the sense that $\mathbf{X} - \boldsymbol{\theta}$ and $\boldsymbol{\theta} - \mathbf{X}$ have the same distribution function F_0, symmetric about $\mathbf{0}$. The diagonal symmetry is also termed as *reflexive* or *antipodal* symmetry (Serfling 2010). A diagonally symmetric density $f(\mathbf{x}, \boldsymbol{\theta})$, whenever exists (a.e.), satisfies $f(\mathbf{x} - \boldsymbol{\theta}) = f(\boldsymbol{\theta} - \mathbf{x})$, $\forall \mathbf{x} \in \mathbb{R}_p$. It need not have the finite second moment or elliptically symmetric. Therefore, $\mathcal{F}_E \subseteq \mathcal{F}_D$, where $\mathcal{F}_E$, $\mathcal{F}_D$ are the respective families of elliptically and diagonally symmetric distributions.

Unlike the Mahalanobis norm in (8.2), the Euclidean norm $\{\Delta_0 : \Delta_0^2 = \|\mathbf{x} - \boldsymbol{\theta}\|^2\}$ is not affine-invariant but is *rotational invariant*, i.e., for $\mathbf{X} \in \mathbb{R}_p$

$$\mathbf{X} \mapsto \mathbf{Y} = \mathbf{A}\mathbf{X} + \mathbf{a} \Rightarrow \Delta_0^2(\mathbf{X}) = \Delta_0^2(\mathbf{Y})$$

for any orthogonal matrix $\mathbf{A}$. If $h_0(\cdot)$ in (8.4) depends on $\mathbf{x}, \boldsymbol{\theta}$ through $\|\mathbf{x} - \boldsymbol{\theta}\|$ alone, we have a *spherically symmetric* density. The terminology is based on the interpretation that $\|\mathbf{x} - \boldsymbol{\theta}\|^2 = c^2$ depicts a spherical surface whose center is $\boldsymbol{\theta}$ and radius c. The class $\mathcal{F}_S$ of spherically symmetric distributions is a subclass of $\mathcal{F}_D$. Note that

$$\mathbf{X} - \boldsymbol{\theta} = \|\mathbf{X} - \boldsymbol{\theta}\| \cdot \frac{\mathbf{X} - \boldsymbol{\theta}}{\|\mathbf{X} - \boldsymbol{\theta}\|} = U \cdot \mathbf{V} \quad \text{(say)}$$

where $\mathbf{V}$ has the domain $\mathcal{S}_{p-1} = \{\mathbf{v} \in \mathbb{R}_p : \mathbf{v}^\top \mathbf{v} = 1\}$, the unit spherical surface. $\mathbf{V}$ has a uniform distribution on $\mathcal{S}_{p-1}$, and U and $\mathbf{V}$ are stochastically independent for spherically symmetric distributions.

Without loss of generality, we shall take $\boldsymbol{\theta} = \mathbf{0}$ in the sequel; thus, we shall work with $\mathbf{X}$ instead of $\mathbf{X} - \boldsymbol{\theta}$. Let $\mathbf{r} = (r_1, \ldots, r_p)^\top$ be a vector with each r_i taking only 0 or 1 values, and let $\mathcal{I}$ be the totality of 2^p possible realizations of $\mathbf{r}$. Further, let $\mathbf{x} \circ \mathbf{r} = ((-1)^{r_1} x_1, \ldots, (-1)^{r_p} x_p)^\top$, $\mathbf{r} \in \mathcal{I}$. If

$$\mathbf{X} \circ \mathbf{r} \stackrel{\mathcal{D}}{=} \mathbf{X} \quad \forall \mathbf{r} \in \mathcal{I},$$

then the distribution function F_0 of $\mathbf{X}$ is said to be *totally symmetric* about $\mathbf{0}$. If $X_1, \ldots, X_p$ are independent and each marginal distribution function is symmetric about 0, then the total symmetry holds, but the converse may not be true. As such, the spherical symmetry implies total symmetry, but the total symmetry may neither imply coordinatewise independence nor spherical symmetry. Thus,

$$\mathcal{F}_S \subseteq \mathcal{F}_T \subseteq \mathcal{F}_D \subseteq \mathcal{F}$$

where $\mathcal{F}$ is the family of all distributions and $\mathcal{F}_T$ the class of totally symmetric distributions.

Another concept is the *marginal symmetry* of F : The distribution function F on $\mathbb{R}_p$ is called *marginally symmetric*, if all its marginal distribution functions $F_1, \ldots, F_p$ are symmetric about 0. The marginal symmetry does not preclude the stochastic independence of the coordinates. While each of elliptical symmetry, diagonal symmetry, spherical symmetry or total symmetry implies marginal symmetry, marginal symmetry may not imply any of them. Therefore, we obtain

$$\mathcal{F}_E \subseteq \mathcal{F}_M \qquad \text{and} \quad \mathcal{F}_S \subseteq \mathcal{F}_T \subseteq \mathcal{F}_M \subseteq \mathcal{F}$$

where $\mathcal{F}_M$ is the class of marginally symmetric distributions.

Finally, let $\mathbf{j} = (j_1, \ldots, j_p)^\top$ be a permutation of $(1, \ldots, p)^\top$ and let $\mathcal{J}$ be the set of $p!$ possible realizations of $\mathbf{j}$. Let $\mathbf{x} \circ \mathbf{j} = (x_{j_1}, \ldots, x_{j_p})^\top$, $\mathbf{j} \in \mathcal{J}$. If $\mathbf{X} \circ \mathbf{j} \stackrel{\mathcal{D}}{=} \mathbf{X} \quad \forall \mathbf{j} \in \mathcal{J}$, the coordinate variables of $\mathbf{X}$ are *exchangeable* or *interchangeable* in the conventional sense (note that $X_1, \ldots, X_p$ need not be independent). If, in addition, $\mathbf{X}$ is diagonally symmetric, then $\mathbf{X} \circ \mathbf{j}$ and $(-1)\mathbf{X} \circ \mathbf{j}$ have the same distributions as $\mathbf{X}$ for all $\mathbf{j} \in \mathcal{J}$. In this case, the distribution function F of $\mathbf{X}$ is said to be *interchangeably symmetric.* The interchangeable symmetry is close to elliptical symmetry with the structural restraints on $\boldsymbol{\Sigma}$ (such as $\boldsymbol{\Sigma} = \sigma^2[(1-\rho)\mathbf{I} + \rho \mathbf{1}\,\mathbf{1}^\top]$, $0 < \rho < 1$), though the interchangeable symmetry does not require the ellipsoidal contours. Spherical symmetry is a particular case of interchangeable symmetry (with $\rho = 0$), while total symmetry does not entail the interchangeability restraint. The interchangeability and the diagonal symmetry, compounded together, yield the interchangeable symmetry property. Some clarification of basic notion of multivariate symmetry can be found in problems at the end of the chapter.

This spectrum of multivariate symmetries is incorporated in the rest of this chapter; some other related notions used in the robust estimation theory will be introduced in later sections.

8.3 Multivariate Location Estimation

Let $\mathbf{X}_1, \ldots, \mathbf{X}_n$ be n *i.i.d.* random p-vectors with a continuous distribution function F, which has a density $f(\cdot)$ with a suitably defined location $\boldsymbol{\theta} \in \mathbb{R}_p$ and possibly with other nuisance parameters $\boldsymbol{\Gamma}$, belonging to a parametric space Ω. Thus, $f(\mathbf{x}; \boldsymbol{\theta}, \boldsymbol{\Gamma}) = f_0(\mathbf{x} - \boldsymbol{\theta}; \boldsymbol{\Gamma})$, $\mathbf{x} \in \mathbb{R}_p, \boldsymbol{\theta} \in \mathbb{R}_p, \boldsymbol{\Gamma} \in \Omega$. The likelihood function $L_n(\mathbf{X}_1, \ldots, \mathbf{X}_n; \boldsymbol{\theta}, \boldsymbol{\Gamma}) = \prod_{i=1}^n f_0(\mathbf{X}_i - \boldsymbol{\theta}; \boldsymbol{\Gamma})$ leads to the maximal likelihood estimators (MLE) of $\boldsymbol{\theta}$ and $\boldsymbol{\Gamma}$ as the solution of *estimating equations*

$$\sum_{i=1}^n \frac{\partial}{\partial \boldsymbol{\theta}} \ln f_0(\mathbf{x}_i - \boldsymbol{\theta}, \boldsymbol{\Gamma}) = \mathbf{0}, \tag{8.6}$$

$$\sum_{i=1}^n \frac{\partial}{\partial \boldsymbol{\Gamma}} \ln f_0(\mathbf{x}_i - \boldsymbol{\theta}, \boldsymbol{\Gamma}) = \mathbf{0}.$$

A closed expression for the MLE exists rather in the special cases only (especially, it does not exist when $\boldsymbol{\theta}$ and $\boldsymbol{\Gamma}$ are not orthogonal and/or f does not belong to the exponential family). If F is multi-normal $(\boldsymbol{\theta}, \boldsymbol{\Sigma})$, then (8.6) yields $\boldsymbol{\Sigma}^{-1}(\bar{\mathbf{X}}_n - \boldsymbol{\theta}) = \mathbf{0}$ and $\bar{\mathbf{X}}_n = \frac{1}{n}\sum_{i=1}^n \mathbf{X}_i$ is the MLE of $\boldsymbol{\theta}$, while the MLE of $\boldsymbol{\Sigma}$ is the sample covariance matrix $\mathbf{S}_n$, due to the parametric orthogonality. $\bar{\mathbf{X}}$ is affine equivariant and $\bar{X}_{nj} = \frac{1}{n}\sum_{i=1}^n X_{ij}$ is the marginal MLE of θ_j, as can be derived from the marginal likelihood function $L_n^{(j)}(\theta_j, \gamma_j) = \prod_{i=1}^n f_j(X_{ij} - \theta_j; \gamma_j)$, $1 \le j \le p$ (and the γ_j are appropriate subsets of $\boldsymbol{\Gamma}$). This is an ideal situation, which rests on a basic property of the multivariate normal density:

$$\frac{\partial}{\partial \boldsymbol{\theta}} \ln f_0(\mathbf{x}-\boldsymbol{\theta};\boldsymbol{\Gamma}) = \boldsymbol{\Lambda}\mathbf{U}$$

$$U_j = \frac{\partial}{\partial \theta_j} \ln f_j(x_j-\theta_j;\boldsymbol{\gamma}_j),\ \mathbf{U}=(U_1,\ldots,U_p)^\top$$

and $\boldsymbol{\Lambda}$ is a $p\times p$ matrix which depends only on the subsets $\boldsymbol{\gamma}_j,\ 1\le j\le p$ of $\boldsymbol{\Gamma}$. Thus, concerning the estimation of $\boldsymbol{\theta}$, the marginal maximum likelihood estimators capture all information and coincide with the joint MLE of $\boldsymbol{\theta}$.

Let now $f(\mathbf{x}-\boldsymbol{\theta};\boldsymbol{\Gamma})$ belong to the elliptically symmetric family $\mathcal{F}_E$ in (8.4). Then

$$\frac{\partial}{\partial \boldsymbol{\theta}} \ln f(\mathbf{x}-\boldsymbol{\theta}) = \frac{\boldsymbol{\Sigma}^{-1}(\mathbf{x}-\boldsymbol{\theta})}{\|\mathbf{x}-\boldsymbol{\theta}\|_{\Sigma}} \cdot \Big(-\frac{h_0'(y)}{h_0(y)}\Big|_{y=\|\mathbf{x}-\boldsymbol{\theta}\|_{\Sigma}}\Big). \tag{8.7}$$

If f is multi-normal, then $-\frac{h_0'(y)}{h_0(y)} = y$ and (8.7) reduces to $\boldsymbol{\Sigma}^{-1}(\mathbf{x}-\boldsymbol{\theta})$. However, for the multivariate Laplace density where $h_0(y)=\text{const}\cdot \mathrm{e}^{-\mathrm{y}}$, (8.6) and (8.7) lead to the estimating equation $\sum_{i=1}^{n} \frac{\boldsymbol{\Sigma}^{-1}(\mathbf{X}_i-\boldsymbol{\theta})}{\|\mathbf{X}_i-\boldsymbol{\theta}\|_{\Sigma}} = \mathbf{0}$, and this further leads to the spatial median as the MLE of $\boldsymbol{\theta}$. This will be considered in detail in Section 8.5.

We see that the MLE of $\boldsymbol{\theta}$ heavily depends on the form of $f(\cdot)$ and may be highly nonrobust to the model departures. Comparing with the univariate case, the nonrobustness prospects may be further enhanced by the presence of the multidimensional nuisance parameter $\boldsymbol{\Gamma}$.

Let $\mathbf{X}_1,\ldots,\mathbf{X}_n$ be *i.i.d.* random p-vectors with a distribution function F defined on $\mathbb{R}_p$; let $F_1,\ldots,F_p$ denote the marginal distribution functions. Assuming that $\theta_j = F_j^{-1}(\frac{1}{2})$ is uniquely defined for $j=1,\ldots,p$, what happens if $F_j(x)$ is strictly monotone for x in a neighborhood of θ_j. For some asymptotic properties we may need $f_j(x)$ to be continuous in a neighborhood of θ_j and $f_j(\theta_j)>0,\ 1\le j\le p$. Denote $\boldsymbol{\theta}=(\theta_1,\ldots,\theta_p)^\top$ the vector of marginal medians.

Let $X_{j1},\ldots,X_{jn}$ be the n variables for the j-th coordinate, and let $X_{j:1} < \ldots < X_{j:n}$ be the associated order statistics (ties are neglected by virtue of the assumed continuity of $F_j,\ 1\le j\le p$). Let

$$\widehat{\boldsymbol{\theta}}_n = (\hat{\theta}_{n1},\ldots,\hat{\theta}_{np})^\top \quad \text{where} \tag{8.8}$$

$$\hat{\theta}_{nj} = \begin{cases} X_{j:\frac{n+1}{2}} & \text{for} \quad n \quad \text{odd} \\ \frac{1}{2}(X_{j:\frac{n}{2}} + X_{j:\frac{n}{2}+1}) & \text{for} \quad n \quad \text{even}\ ,\ j=1,\ldots,p \end{cases}$$

The strong consistency of $\hat{\theta}_{nj},\ 1\le j\le p$ and of $\widehat{\boldsymbol{\theta}}_n$ follows directly from (3.48). Further, whenever the $f_j(\theta_j)$ are positive, $\hat{\theta}_{nj}$ is $\sqrt{n}$-consistent for θ_j by (4.12), and hence $\widehat{\boldsymbol{\theta}}_n$ is $\sqrt{n}$-consistent for $\boldsymbol{\theta}$, without invoking any moment condition on F. Moreover, it follows from Theorem 4.1 (the weak Bahadur-type representation) that for each j $(= 1, \ldots, k)$

$$f_j(\theta_j)n^{1/2}(\hat{\theta}_{nj} - \theta_j) = n^{-1/2}\sum_{i=1}^{n}\{I[X_{ij} > \theta_j] - \tfrac{1}{2}\} + o_p(1),$$

which further implies that, as $n \to \infty$,

$$n^{1/2}\boldsymbol{\Delta}(\widehat{\boldsymbol{\theta}}_n - \boldsymbol{\theta}) = n^{-1/2}\sum_{i=1}^{n}\mathbf{V}_i + o_p(1)$$

$$\text{where} \quad \mathbf{V}_i = (V_{i1}, \ldots, V_{ip})^\top, \ V_{ij} = I[X_{ij} > \theta_j] - \tfrac{1}{2}, \ 1 \leq j \leq p,$$
$$\text{and} \quad \boldsymbol{\Delta} = \operatorname{diag}(\mathrm{f}_1(\theta_1), \ldots, \mathrm{f}_\mathrm{p}(\theta_\mathrm{p})).$$

The $\mathbf{V}_i$ are *i.i.d.* random vectors with each coordinate being binary either $(-\frac{1}{2}$ or $\frac{1}{2})$, $I\!E\mathbf{V}_i = \mathbf{0}$ and $I\!E(\mathbf{V}\mathbf{V}^\top) = \left[F_{jj'}(\theta_j, \theta_{j'}) - \frac{1}{4}\right]_{j,j'=1}^{p} = \boldsymbol{\Gamma}^*$, say. Hence, by the multivariate central limit theorem,

$$n^{1/2}(\widehat{\boldsymbol{\theta}}_n - \boldsymbol{\theta}) \xrightarrow{\mathcal{D}} \mathcal{N}_p\left(\mathbf{0}, \boldsymbol{\Delta}^{-1}\boldsymbol{\Gamma}^*\boldsymbol{\Delta}^{-1}\right)$$

as $n \to \infty$. Because each $\hat{\theta}_{nj}$ is translation and scale equivariant, $\widehat{\boldsymbol{\theta}}_n$ is also translation equivariant but not necessarily affine-equivariant, unless $\boldsymbol{\Lambda}$ in (8.1) is restricted to the class of diagonal positive definite matrices. In spite of its lack of affine-equivariance, the role of $\widehat{\boldsymbol{\theta}}_n$ in the affine-equivariant estimation will be illustrated in Section 8.5.

The coordinatewise median $\widehat{\boldsymbol{\theta}}_n$ is a special case of coordinatewise M, L and R-estimators, studied in Chapters 3–7 for the univariate model. Considering the multivariate analogues of M-, L-, and R-estimators, we usually assume that the marginal distribution functions $F_1, \ldots, F_p$ are symmetric, as well as the marginal score functions. We illustrate the methodology for M-estimators and then append briefly the parallel results for L- and R-estimators.

The univariate M-estimators of location were introduced in Section 3.2. Recalling (3.10)–(3.11), we conceive of a set of score functions $\psi_j(x)$, $x \in \mathbb{R}$, $j = 1, \ldots, p$. As in Section 5.3 (see (5.22)), ψ_j can be written as $\psi_{j1}(x) + \psi_{j2}(x)$, where ψ_{j1} is absolutely continuous. The regularity conditions **A1**–**A4** of Section 5.3 are assumed for each coordinate ψ_j, $1 \leq j \leq p$. Motivated by the assumed symmetry of $F_j, 1 \leq j \leq p$ and by the basic property that $I\!E_{\theta_j}\left\{\frac{\partial}{\partial\theta_j}\ln f_j(x_{ij} - \theta_j)\right\} = 0$, $i = 1, \ldots, n$, it is assumed that for $j = 1, \ldots, p$

(i) ψ_j is nondecreasing

(ii)

$$I\!E_{\theta_j}\psi_j(X_{ij} - \theta_j) = \int_{\mathbb{R}} \psi_j(x - \theta_j)dF_j(x - \theta_j) = 0.$$

(iii) $\psi_j(\cdot)$ is skew-symmetric, i.e. $\psi_j(-x) = -\psi_j(x)$, $x \in \mathbb{R}$.

Let

$$M_{nj}(t) = \frac{1}{n}\sum_{i=1}^{n} \psi_j(X_{ij} - t),\ t \in \mathbb{R},\ j = 1, \ldots, p.$$

Then $M_{nj}(t)$ is nonincreasing in t and $\mathbb{E}_{\theta_j} M_{nj}(\theta_j) = 0,\ 1 \leq j \leq p$. Put for $j = 1, \ldots, p$

$$\begin{aligned} \hat{\theta}_{nj} &= \tfrac{1}{2}(\hat{\theta}_{nj,1} + \hat{\theta}_{nj,2}), \quad \widehat{\boldsymbol{\theta}}_n = (\hat{\theta}_{n1}, \ldots, \hat{\theta}_{np})^\top, \\ \hat{\theta}_{nj,1} &= \sup\{t : M_{nj}(t) > 0\}, \quad \hat{\theta}_{nj,2} = \inf\{t : M_{nj}(t) < 0\}. \end{aligned}$$

Then $\hat{\theta}_{nj}$ is a translation equivariant estimator of θ_j, but generally not scale equivariant; thus, $\widehat{\boldsymbol{\theta}}_n$ is not affine-equivariant. Though $\widehat{\boldsymbol{\theta}}_n$ shares some robustness properties from its coordinates, the interdependence of the coordinates means that the breakdown points and other conventional measures may not transfer to the multivariate case. This drawback is shared by other affine-equivariant estimators, considered in Section 8.5.

Put $\mathbf{M}_n(\boldsymbol{\theta}) = (M_{n1}(\theta_1), \ldots, M_{np}(\theta_p))^\top$. Then by the classical multivariate central limit theorem

$$n^{1/2}\mathbf{M}_n(\boldsymbol{\theta}) \xrightarrow{\mathcal{D}} \mathcal{N}_p\left(\mathbf{0}, \boldsymbol{\Sigma}_\psi\right) \quad \text{as } n \to \infty, \tag{8.9}$$

where $\boldsymbol{\Sigma}_\psi = [\sigma_{j\ell}(\psi)]_{j,\ell=1}^{p}$ is a $p \times p$ matrix with

$$\sigma_{j\ell}(\psi) = \mathbb{E}_{\boldsymbol{\theta}}\left\{\psi_j(X_{ij} - \theta_j)\psi_\ell(X_{i\ell} - \theta_\ell)\right\}, \quad j, \ell = 1, \ldots, p,\ i = 1, \ldots, n. \tag{8.10}$$

The asymptotic representations in (5.28) and (5.45), applied to the components of $\widehat{\boldsymbol{\theta}}_n$, take the form

$$\hat{\theta}_{nj} - \theta_j = \frac{1}{n\gamma_j}\sum_{i=1}^{n} \psi_j(X_{ij} - \theta_j) + \mathrm{O}_p(n^{-\lambda}) \text{ as } n \to \infty \tag{8.11}$$

where $\gamma_j = \int f_j(x) d\psi_j(x),\ j = 1, \ldots, p$, and $\lambda = 1$ if ψ_j is absolutely continuous while $\lambda = \frac{3}{4}$ if ψ_j has jump discontinuities. Let $\boldsymbol{\nu}(\boldsymbol{\psi}) = [\nu_{j\ell}(\boldsymbol{\psi})]_{j,\ell=1}^{p}$ where

$$\nu_{j\ell}(\boldsymbol{\psi}) = \frac{\sigma_{j\ell}}{\gamma_j\gamma_\ell}, \quad j, \ell = 1, \ldots, p. \tag{8.12}$$

Then, by (8.9) and (8.11),

$$n^{1/2}(\widehat{\boldsymbol{\theta}}_n - \boldsymbol{\theta}) \xrightarrow{\mathcal{D}} \mathcal{N}_p(\mathbf{0}, \boldsymbol{\nu}(\boldsymbol{\psi})) \quad \text{as } n \to \infty.$$

Let $\mathbf{I}^* = \left[I^*_{j\ell}\right]_{j,\ell=1}^{p}$ denote the Fisher information matrix, where

$$I^*_{jj} = \int_{\mathbb{R}} \left\{-\frac{f'_j(x - \theta_j)}{f_j(x - \theta_j)}\right\}^2 dF_j(x - \theta_j),\ j = 1, \ldots, p \tag{8.13}$$

$$I^*_{j\ell} = \int\int_{\mathbb{R}_2} \left\{-\frac{f'_j(x - \theta_j)}{f_j(x - \theta_j)}\right\}\left\{-\frac{f'_\ell(y - \theta_\ell)}{f_\ell(y - \theta_\ell)}\right\} dF_{j\ell}(x - \theta_j, y - \theta_\ell),$$
$$j, \ell = 1, \ldots, p.$$

It follows from results in Section 5.3 that

$$\nu_{jj}(\boldsymbol{\psi}) \geq \left(I_{jj}^*\right)^{-1} \quad \forall j = 1, \ldots, p$$

where the equality sign holds only when $\psi_j(x) = -\frac{f_j'(x)}{f_j(x)} = \psi_j^0(x)$, up to a scalar constant. Moreover, if $\psi_j(\cdot) = \gamma_j \psi_j^0(\cdot) + \xi_j(\cdot)$, where $\boldsymbol{\xi}(\cdot) = (\xi_1(\cdot), \ldots, \xi_p(\cdot))^\top$ is orthogonal to $\boldsymbol{\psi}^0(\cdot) = (\psi_1^0(\cdot), \ldots, \psi_p^0(\cdot))^\top$ then

$$\boldsymbol{\nu}(\boldsymbol{\psi}) - (\mathbf{I}^*)^{-1} = \text{positive-semidefinite.}$$

The present discussion has ignored the unknown scale and other nuisance parameters: thus, it has been in a correspondence with the "fixed scale" case of Chapter 5. In fact, the excess of $\nu_{jj}(\boldsymbol{\psi})$ over $\left(I_{jj}^*\right)^{-1}$ for a chosen $\boldsymbol{\psi}$, may be also affected by the unknown scale parameter for $f_j(\cdot)$. Moreover, the $I_{j\ell}^*$ generally depends on the nuisance association matrix, and the same is true for the matrix $\boldsymbol{\Sigma}_\psi$ defined in (8.10).

In the presence of nuisance $\boldsymbol{\Gamma}$, we should consider the information matrix $\mathbf{I}^0$ for (8.6) and partition it as

$$\mathbf{I}^0 = \begin{bmatrix} \mathbf{I}_{\theta\theta} & \mathbf{I}_{\theta\Gamma} \\ \mathbf{I}_{\Gamma\theta} & \mathbf{I}_{\Gamma\Gamma} \end{bmatrix}.$$

Then the MLE $\widehat{\boldsymbol{\theta}}_n^*$ of $\boldsymbol{\theta}$ in model (8.6) has the asymptotic normal distribution

$$\sqrt{n}(\widehat{\boldsymbol{\theta}}_n^* - \boldsymbol{\theta}) \xrightarrow{\mathcal{D}} \mathcal{N}_p\left(\mathbf{0}, \mathbf{I}_{\theta:\Gamma}^{-1}\right), \tag{8.14}$$

where $\mathbf{I}_{\theta:\Gamma} = \mathbf{I}_{\theta\theta} - \mathbf{I}_{\theta\Gamma}\mathbf{I}_{\Gamma\Gamma}^{-1}\mathbf{I}_{\Gamma\theta}$ and the eigenvalues of $\mathbf{I}_{\theta:\Gamma}$ are smaller than or equal to those of $\mathbf{I}_{\theta\theta}$ (the equality holds when $\mathbf{I}_{\theta\Gamma}$ is a null matrix). This together with the multivariate Cramér–Rao inequality leads to a sharper matrix-inequality

$$\boldsymbol{\nu}(\boldsymbol{\psi}) - \mathbf{I}_{\theta:\Gamma}^{-1} = \text{positive semi-definite.} \tag{8.15}$$

A challenging problem is how to use this ,,matrix excess," depending on a fixed choice of $\boldsymbol{\psi}$, for a construction of a scalar measure of the relative efficiency of the estimate. We shall return to this problem in Section 8.6.

The higher robustness of the vector median in (8.8) comparing with the M-estimator is due to its scale-equivariance. On the other hand, the median is less efficient because its $\psi_j(x) = \text{sign(x)}$, j = 1, ..., p, insensitive to the value of $|x|$.

The coordinatewise L-estimators derive their properties from the univariate counterparts studied in Section 3.3 and in detail in Chapter 4. hence, they are coordinatewise translation and scale equivariant. Their first-order representation follows from Theorems 4.3 and 4.5. Working with the representations (4.1), (4.19) and (4.28), we can follow the steps in (8.11) through (8.15) similarly as for the M-estimators.

The *coordinatwise R-estimators* are based on signed-rank statistics described in Section 3.4 and in more detail in Section 6.2. We again assume the symmetry of marginal distribution functions $F_1, \ldots, F_p$, and thus we choose the skew-symmetric score-generating functions $\varphi_j^*(u)$ and the corresponding $\varphi_j(u),\ 0 < u < 1$ (compare (6.2)). The R-estimators are translation-scale equivariant, but generally not affine equivariant. They perform well for the error-contamination models, because the ranks of $|x_i|$ are less sensitive to outliers than $|x_i|$. The asymptotic representation follows from Theorem 6.3. Hence, we can follow the steps in (8.11)–(8.15) and obtain parallel results for R-estimators.

Summarizing, all the L-, M-, and R-estimators have good robustness properties, L- and R-estimators being scale equivariant, but not the M-estimators. Neither of them adopt to the affine-equivariance. Their dispersion matrices simplify, if we can assume the diagonal symmetry of F, besides the marginal symmetry. In the case of elliptically symmetric distributions, the affine equivariance is facilitated by using the studentized Mahalanobis distance; on the other hand, the transformed observations can lose their independence, though they can be still exchangeable.

The Pitman-type estimators, introduced in Sections 3.5 and 3.6, easily extend to the multivariate case. We can also extend the differentiable statistical functionals and their properties, (see Theorem 3.4 and its consequence in (3.120)). The other estimators considered in Section 3.5 are computationally difficult for finite n, though their asymptotic properties can be derived. Generally, the coordinatewise estimators can be robust for a larger class of distributions than the affine equivariant ones.

8.4 Multivariate Regression Estimation

Consider a $p \times n$ random matrix $\mathbb{Y} = [\mathbf{Y}_1, \ldots, \mathbf{Y}_n]$ where the $\mathbf{Y}_i$ are independent p-vectors. Assume that the $\mathbf{Y}_i = (Y_{i1}, \ldots, Y_{ip})^\top$ depends on a nonstochastic regressor $\mathbf{x}_i \in \mathbb{R}_q$ with known components, not all equal, in the form $\mathbf{Y}_i = \boldsymbol{\beta}\mathbf{x}_i + \mathbf{e}_i,\ 1 \leq i \leq n$. Here $\boldsymbol{\beta}$ is a $p \times q$ matrix of unknown regression parameters, and $\mathbf{e}_1, \ldots, \mathbf{e}_n$ are *i.i.d.* random vectors. Denote $\mathbb{X} = [\mathbf{x}_1, \ldots, \mathbf{x}_n]$ (the matrix of order $q \times n$), and write the regression model as

$$\mathbb{Y} = \boldsymbol{\beta}\mathbb{X} + \mathbb{E}; \quad \mathbb{E} = [\mathbf{e}_1, \ldots, \mathbf{e}_n] \tag{8.16}$$

The parameter $\boldsymbol{\beta}$ is often partitioned as $(\boldsymbol{\theta}, \boldsymbol{\beta}^*)$ with $\boldsymbol{\beta}^*$ of order $p \times (q-1)$; then the top row of $\mathbb{X}$ is $\mathbf{1}_n^\top = (1, \ldots, 1)$, and $\boldsymbol{\theta}$ is regarded as the vector of *intercept* parameters, and $\boldsymbol{\beta}^*$ as the matrix of regression parameters. We shall adapt this convention, whenever it does not mean a confusion. If $q = 1$, then (8.16) relates to the multivariate location model treated in Section 8.3. If $q = 2$, then $\mathbb{X}$ is of order $2 \times n$; if its second row is composed of n_1 zeros and of $n_2 = n - n_1$ 1's, then (8.16) reduces to the multivariate two-sample location model. Similarly, the k-sample multivariate location model is a particular case of (8.16), when $q = k$ and $\mathbb{X}$ can be written as

$$\mathbb{X} = \begin{bmatrix} 1, \ldots, 1, \cdots\cdots, 1, \ldots, 1 \\ 0, \ldots, 0, 1, \ldots, 1, 0, \ldots, 0 \\ \cdots\cdots\cdots\cdots\cdots\cdots\cdots \\ 0, \ldots, 0, 0, \ldots, 0, 1, \ldots, 1 \end{bmatrix} \tag{8.17}$$

where row 2 contains the block n_2 of 1's and other components 0's, etc., row k contains the block n_k of 1's and other components 0's; $n = n_1 + \ldots + n_k$.

8.4.1 Normal Multivariate Linear Model

In the conventional multivariate linear model $\mathbf{e}_1, \ldots, \mathbf{e}_n$ are assumed to have a multi-normal distribution with null mean vector and a positive definite dispersion matrix $\mathbf{\Sigma}$. If we denote

$$\|\mathbf{Y}_i - \boldsymbol{\beta}\mathbf{x}_i\|_{\Sigma}^2 = (\mathbf{Y}_i - \boldsymbol{\beta}\mathbf{x}_i)^\top \mathbf{\Sigma}^{-1}(\mathbf{Y}_i - \boldsymbol{\beta}\mathbf{x}_i), \quad i = 1, \ldots, n,$$

then the MLE of $\boldsymbol{\beta}$ for the normal model is the minimizer of $\sum_{i=1}^n \|\mathbf{Y}_i - \boldsymbol{\beta}\mathbf{x}_i\|_{\Sigma}^2$ and is given by

$$\widehat{\boldsymbol{\beta}}_n = (\mathbb{Y}\mathbb{X}^\top)(\mathbb{X}\mathbb{X}^\top)^{-1}. \tag{8.18}$$

$\widehat{\boldsymbol{\beta}}_n$ is a linear estimate of $\boldsymbol{\beta}$ and is unbiased, because $\widehat{\boldsymbol{\beta}}_n = \boldsymbol{\beta} + \mathbb{E}\mathbb{X}^\top(\mathbb{X}\mathbb{X}^\top)^{-1}$ by (8.16) and (8.17). Also, $\widehat{\boldsymbol{\beta}}_n$ is affine-equivariant both in $\mathbb{Y}$ and $\mathbb{X}$. Indeed, consider the dual transformation $\mathbb{Y} = \mathbf{A}\mathbb{Z}$, $\mathbb{X} = \mathbf{B}^{-1}\mathbb{W}$ and the model $\mathbb{Z} = \boldsymbol{\beta}^*\mathbb{Z} + \mathbb{E}^*$, with $\boldsymbol{\beta}^* = \mathbf{A}^{-1}\boldsymbol{\beta}\mathbf{B}^{-1}$ where A, B are symmetric positive definite matrices of orders $p \times p$ and $q \times q$, respectively. Then

$$\begin{aligned} \widehat{\boldsymbol{\beta}}_n &= \mathbf{A}\mathbb{Z}\mathbb{W}^\top(\mathbf{B}^{-1})^\top \left(\mathbf{B}^{-1}\mathbb{W}\mathbb{W}^\top(\mathbf{B}^\top)^{-1}\right)^{-1} \\ &= \mathbf{A}\mathbb{Z}\mathbb{W}^\top(\mathbb{W}\mathbb{W}^\top)^{-1}\mathbf{B} = \mathbf{A}\widehat{\boldsymbol{\beta}}_n^*\mathbf{B}, \end{aligned}$$

where $\widehat{\boldsymbol{\beta}}_n^*$ is the LSE (and thus the MLE) of $\boldsymbol{\beta}^*$. If $\mathbf{e}_1, \ldots, \mathbf{e}_n$ are *i.i.d.* $\mathcal{N}_p(\mathbf{0}, \mathbf{\Sigma})$, then, by (8.18), $\widehat{\boldsymbol{\beta}}_n - \boldsymbol{\beta} \stackrel{\mathcal{D}}{=} \mathcal{N}_p\left(\mathbf{0}, \mathbf{\Sigma} \otimes (\mathbb{X}\mathbb{X}^\top)^{-1}\right)$, where $\otimes$ stands for the Kronecker product of matrices.

8.4.2 General Multivariate Linear Model

Even if $\mathbf{e}_1, \ldots, \mathbf{e}_n$ are not multinormal but *i.i.d.* and $\mathbb{E}(\mathbf{e}_i) = \mathbf{0}$, $\mathbb{E}(\mathbf{e}_i\mathbf{e}_i^\top) = \mathbf{\Sigma}$ (positive definite), and the generalized Noether condition

$$\max_{1 \le i \le n} \mathbf{x}_i^\top(\mathbb{X}\mathbb{X}^\top)^{-1}\mathbf{x}_i \to 0 \quad \text{as} \quad n \to \infty \tag{8.19}$$

is satisfied, the asymptotic normality of $\widehat{\boldsymbol{\beta}}_n$ follows from a multivariate version of the Hájek–Šidák central limit theorem (see (2.85)), due to Eicker (1963). Thus, more precisely, as $n \to \infty$,

$$(\mathbf{\Sigma}^{-1/2} \otimes (\mathbb{X}\mathbb{X}^\top)^{1/2})(\widehat{\boldsymbol{\beta}}_n - \boldsymbol{\beta}) \xrightarrow{\mathcal{D}} \mathcal{MN}_{pq}(\mathbf{0}, \mathbf{\Sigma} \otimes \mathbf{I}_p)$$

where $\mathcal{MN}_{pq}$ stands for a $p \times q$ matrix-valued normal distribution. In this setup, $\widehat{\boldsymbol{\beta}}_n$ is a linear estimator of $\boldsymbol{\beta}$, (double)-affine equivariant and asymptotically normal, but it is highly nonrobust to outliers as well as to structural constraints. In order to diminish nonrobustness while preserving the affine equivariance, we may replace the L_2-norm in (8.18) by the L_1-norm: $\sum_{i=1}^{n} \|Y_i - \boldsymbol{\beta}\mathbf{x}_i\|_\Sigma$ and consider its minimization with respect to $\boldsymbol{\beta}$. This approach will be explored in Section 8.5.

If $\mathbf{e}_i$ in (8.16) has a density $f(\mathbf{e};\boldsymbol{\Gamma})$ with a nuisance (matrix) parameter $\boldsymbol{\Gamma}$, then the log-likelihood function is $\sum_{i=1}^{n} \ln f(\mathbf{Y}_i - \boldsymbol{\beta}\mathbf{x}_i;\boldsymbol{\Gamma})$. The corresponding likelihood score (matrix) function, being the derivative with respect to $\boldsymbol{\beta}$ and $\boldsymbol{\Gamma}$ cannot be generally expressed as a linear combination of score functions derived from the p marginal log-densities $\ln f_j(Y_{ij} - \boldsymbol{\beta}_j^\top \mathbf{x}_i;\boldsymbol{\Gamma}_j)$, $1 \le j \le p$, where $\boldsymbol{\Gamma}_j$ is a subset of $\boldsymbol{\Gamma}$, $1 \le j \le p$, unless $f(\cdot)$ belongs to the elliptically symmetric family. thus, the estimating equations are generally more complex than in the location model. This situation is further explored in Problems 8.4.1 – 8.4.4. For nonelliptically symmetric distributions, the estimating equations can be highly nonlinear, even for marginal score functions, and can yield nonrobust estimators of $\boldsymbol{\beta}$. For this reason, some other coordinatewise robust estimators of $\boldsymbol{\beta}$ are to be considered side by side.

Let us first consider the coordinatewise R-, L- and M-estimators studied in Chapters 3–6 (viz Sections 4.7, 5.5, and 6.3). The finite-sample properties of regression L-, M-, and R-estimators transpire componentwise to the multivariate case. The $\sqrt{n}$-consistency and the asymptotic multi-normality of L-, M-, and R-estimators follow analogously, using the „uniform asymptotic linearity" results. If we assume $\frac{1}{n}\mathbb{X}\mathbb{X}^\top \to \mathbb{Q}$ (positive definite) in addition to (8.19), then for $\widehat{\boldsymbol{\beta}}_n$ being the M-estimator with the score function ψ we get, as $n \to \infty$,

$$\sqrt{n}(\widehat{\boldsymbol{\beta}}_n - \boldsymbol{\beta}) \xrightarrow{\mathcal{D}} \mathcal{MN}(\mathbf{0}, \boldsymbol{\nu}(\psi) \otimes \mathbb{Q}^{-1}) \tag{8.20}$$

with $\boldsymbol{\nu}(\psi)$ defined in (8.12); $\mathcal{MN}(\cdot,\cdot)$ stands for the matrix-valued normal distribution. The parallel results hold for regression R- and L-estimators. Alternatively, we can derive (8.20) using vector notation, converting $\widehat{\boldsymbol{\beta}}$ and $\boldsymbol{\beta}$ into pq-vectors, and applying the vector asymptotic representations and the Cramér–Wold device characterization of the normal laws.

If we use a vector notation for the MLE of $\boldsymbol{\beta}$, then the standardized asymptotic covariance matrix of the MLE will be $(\mathbf{I}_{\theta:\Gamma})^{-1} \otimes \mathbb{Q}^{-1}$ of order $pq \times pq$, where $\mathbf{I}_{\theta:\Gamma}$ is the $p \times p$ defined similarly as in (8.14). The Kronecker product by $\mathbb{Q}$ appears even with different $\boldsymbol{\nu}$ from $\mathbf{I}_{\theta:\Gamma}$, as in the case with R-, L-, and M-estimators of $\boldsymbol{\beta}$. As such, the information matrix inequality in (8.15) applies also to the regression model, and the same is true for the relative efficiencies of the coordinatewise L-, M- and R-estimators.

8.5 Affine-Equivariant Robust Estimation

This area is a evolutionary field and there is an ample room for further developments. We refer to a recent book by Oja (2010) for a nice treatise on this subject area and for an extended bibliography.

The Mahalanobis distance Δ, defined in (8.2), plays a central role in the development of the affine equivariance. It is a natural follower of the Euclidean norm $\|\mathbf{X} - \boldsymbol{\theta}\|$, which is rotation invariant but not affine-invariant. There has been an evolutionary development on affine-equivariant statistical functionals, specially relating to L- and R-functionals; the M-estimators are less likely to be affine-equivariant, because they are not scale equivariant, unless studentized. The robust estimators are nonlinear; hence, they need a special approach to become affine-equivariant. The situation is more complex with the covariance functionals. A novel methodology was developed for the spatial median, spatial rank, and spatial quantile functions; for a nice account of this development, we refer to Oja (2010) and Serfling (2010), and to other references cited therein.

Let $\mathbf{X} \in \mathbb{R}_p$ be a random vector with a distribution function F; an affine transformation $\mathbf{X} \mapsto \mathbf{Y} = \mathbf{A}\mathbf{X} + \mathbf{b}$ with $\mathbf{A}$ nonsingular of order $p \times p$, $\mathbf{b} \in \mathbb{R}_p$ generates a distribution function G also defined on $\mathbb{R}_p$. Denote $G = F_{A,b}$ for a simpler notation. Consider a vector-valued functional $\boldsymbol{\theta}(F)$, designated as a suitable measure of location of F. It is said to be an *affine-equivariant location functional*, provided

$$\boldsymbol{\theta}(F_{A,b}) = \mathbf{A}\boldsymbol{\theta}(F) + \mathbf{b} \quad \forall \mathbf{b} \in \mathbb{R}_p, \ \mathbf{A} \ \text{positive definite.}$$

Let $\boldsymbol{\Gamma}(F)$ be a matrix-valued functional of F, designated as a measure of the *scatter* of F around its location $\boldsymbol{\theta}$ and capturing its *shape* in terms of variation and covariation of the coordinate variables. $\boldsymbol{\Gamma}(F)$ is often termed a *covariance functional*, and a natural requirement is that it be independent of $\boldsymbol{\theta}(F)$. It is termed an *affine-equivariant covariance functional*, provided

$$\boldsymbol{\Gamma}(F_{A,b}) = \mathbf{A}\boldsymbol{\Gamma}(\mathbf{F})\mathbf{A}^\top \quad \forall \mathbf{b} \in \mathbb{R}_p, \ \mathbf{A} \ \text{positive definite.}$$

Note that while location functionals are relatively well defined, the formulation of covariance functionals is less general. Whenever the dispersion matrix $\boldsymbol{\Sigma}$ exists, $\boldsymbol{\Gamma}(F)$ should be proportional to $\boldsymbol{\Sigma}$, though the existence of the second moment may not be generally supposed.

Let us return to the multivariate symmetry considered in Section 8.2. Let $\mathbf{B}$ be a matrix such that $\mathbf{B}\mathbf{B}^\top = \boldsymbol{\Sigma}$; its choice is not unique, as $\mathbf{BP}$ with a matrix $\mathbf{P}$ also satisfies $(\mathbf{BP})(\mathbf{BP})^\top = \mathbf{BB}^\top = \boldsymbol{\Sigma}$. Hence, we can write $\mathbf{X} - \boldsymbol{\theta} = \mathbf{B}\boldsymbol{\varepsilon}$ where $I\!E\boldsymbol{\varepsilon} = \mathbf{0}$ and the covariance matrix of $\boldsymbol{\varepsilon}$ is $\mathbf{I}_p$. However, the coordinates of $\boldsymbol{\varepsilon}$ may be dependent, unless $\boldsymbol{\varepsilon}$ has a multi-normal distribution. If $\mathbf{X}$ has a spherically symmetric distribution, for which $\mathbf{B}$ itself is an orthogonal matrix, then $\frac{\boldsymbol{\varepsilon}}{\|\boldsymbol{\varepsilon}\|}$ has a uniform distribution on $\mathcal{S}_{p-1}$ and $\|\boldsymbol{\varepsilon}\|$ and $\frac{\boldsymbol{\varepsilon}}{\|\boldsymbol{\varepsilon}\|}$ are independent. Hyvärinen at al. (2001), confronting the dilemma between *uncorrelation* and *independence*, considered the *independent component model* for which

$$\mathbf{X} - \boldsymbol{\theta} = \mathbf{B}\boldsymbol{\varepsilon},\ \mathbf{B}\mathbf{B}^\top = \boldsymbol{\Sigma},\ \boldsymbol{\varepsilon} \text{ has } i.i.d. \text{ components.}$$

This is a rather restricted type of the elliptically symmetric distributional model; however, its clause can be justified, because it adds a mathematical convenience, due to which one can obtain results parallel to multinormal distributions. Incidentally, if $\mathbf{B}\mathbf{B}^\top$ is an affine-equivariant functional and one chooses $\mathbf{Y} = \mathbf{B}^{-1}(\mathbf{X} - \boldsymbol{\theta})$, then $\mathbf{Y}$ is affine equivariant. This is also termed an *invariant coordinate* transformation.

To make an affine equivariant transformation, one needs to know the location and covariance functionals. Because they are unknown, a *transformation–retransformation* method has been advocated in the literature (Chakraborty (2001), where other work has been cited). As Serfling (2010) pointed out, due to a data-driven transformation–retransformation scheme, this procedure suffers from some arbitrariness and a possible loss of efficiency. Thus, let us rather introduce some notion of spatial measures which lead to affine-equivariant functionals.

Let $\mathcal{B}_{p-1}(\mathbf{0})$ be the open unit ball and let $\Phi(\mathbf{u}, \mathbf{y}) = \|\mathbf{u}\| + \langle \mathbf{u}, \mathbf{y} \rangle,\ \ \mathbf{u} \in \mathcal{B}_{p-1}(\mathbf{0}),\ \mathbf{y} \in \mathbb{R}_p$. Consider the objective function $\bar{\Psi}(\mathbf{u}) = \mathbb{E}\{\Phi(\mathbf{u}, \mathbf{X} - \boldsymbol{\theta}) - \Phi(\mathbf{u}, \mathbf{X})\},\ \mathbf{X} \in \mathbb{R}_p,\ \mathbf{u} \in \mathcal{B}_{p-1}$. Then the $\mathbf{u}$-th *spatial quantile* $Q_F(\mathbf{u})$ is defined as the minimizer of $\bar{\Psi}(\mathbf{u})$ in $\boldsymbol{\theta}$. More precisely, it is a solution $Q_F(\mathbf{u}) = \boldsymbol{\xi}_u$ of the equation

$$\mathbf{u} = \mathbb{E}\left\{\frac{\boldsymbol{\xi}_u - \mathbf{X}}{\|\boldsymbol{\xi}_u - \mathbf{X}\|}\right\}.$$

thus, $Q_F(\mathbf{0})$ is the spatial median. Further, to any $\mathbf{x} \in \mathbb{R}_p$ there exists a unit vector $\mathbf{u}_\mathbf{X}$ such that $\mathbf{x} = Q_F(\mathbf{u}_\mathbf{X})$. Hence,

$$\mathbf{u}_\mathbf{X} = Q_F^{-1}(\mathbf{x}) \text{ is the } \textit{inverse spatial quantile functional} \text{ at } \mathbf{x}.$$

Note that $\mathbf{u}_\theta = \mathbf{0}$ at $\boldsymbol{\xi} = \boldsymbol{\theta}$, so that $Q_F(\mathbf{0}) = \boldsymbol{\theta}$. The quantile $Q_F(\mathbf{u})$ is central for $\mathbf{u} \sim \mathbf{0}$, and it is extreme for $\mathbf{u} \sim \mathbf{1}$. The values $Q_F^{-1}(\mathbf{x})$ can coincide for multiple $\mathbf{x}$; moreover, Möttönen and Oja (1995) showed that $Q_F(\cdot)$ and $Q_F^{-1}(\cdot)$ are inverse to each other.

The sample counterpart of the population spatial quantile function is defined as a solution $\widehat{\mathbf{x}}_{n,u}$ of the equation

$$\frac{1}{n}\sum_{i=1}^{n}\frac{\mathbf{x} - \mathbf{X}_i}{\|\mathbf{x} - \mathbf{X}_i\|} = \mathbf{u}, \qquad \text{i.e. } \widehat{\mathbf{x}}_{n,u} = \hat{Q}_n^{-1}(\mathbf{u})$$

where $\frac{\mathbf{a}}{\|\mathbf{a}\|}$ is defined as $\mathbf{0}$ if $\mathbf{a} = \mathbf{0}$, similar as in the sign function. The *sample central rank function* [Oja 2010] is defined as

$$R_n^*(\mathbf{x}) = \frac{1}{2n}\sum_{i=1}^{n}\left[\frac{\mathbf{x} - \mathbf{X}_i}{\|\mathbf{x} - \mathbf{X}_i\|} + \frac{\mathbf{x} + \mathbf{X}_i}{\|\mathbf{x} + \mathbf{X}_i\|}\right],\ \mathbf{x} \in \mathbb{R}_p. \tag{8.21}$$

The spatial median is equivariant with respect to the shift and to the orthogonal and homogeneous scale transformations. Indeed, let $\mathbf{y} = \mathbf{A}\mathbf{x} + \mathbf{b}, \ \mathbf{b} \in \mathbb{R}_p$, $\mathbf{A}$ positive definite orthogonal. Then

$$\mathbf{u} \mapsto \mathbf{u}' = \frac{\|\mathbf{u}\|}{\|\mathbf{A}\mathbf{u}\|} \cdot \mathbf{A}\mathbf{u}, \ \ \mathbf{u} \in \mathcal{B}_{p-1}(\mathbf{0})$$

and if $F_{A,b}$ stands for the distribution function of $\mathbf{Y}$, then

$$Q_{F_{A,b}}(\mathbf{u}') = \mathbf{A}Q_F(\mathbf{u}) + \mathbf{b};$$

also $\mathbf{u}' = \mathbf{0}$ if $\mathbf{u} = \mathbf{0}$. This also shows that the spatial quantile function may not be affine-equivariant for all $\mathbf{u}$.

In order to induce the affine equivariance, start with an affine-equivariant covariance functional; let us take the dispersion matrix $\mathbf{\Sigma}_F$ for simplicity. Then the affine-equivariant quantile function is defined as the solution of the equation

$$I\!E\left\{\frac{\mathbf{\Sigma}_F^{-1/2}(\mathbf{x}-\mathbf{X})}{\|\mathbf{\Sigma}_F^{-1/2}(\mathbf{x}-\mathbf{X})\|}\right\} = \mathbf{u}, \quad \mathbf{u} \in \mathcal{B}_{p-1}(\mathbf{0}) \tag{8.22}$$

where $\|\mathbf{\Sigma}_F^{-1/2}(\mathbf{x}-\mathbf{X})\| = \|\mathbf{x}-\mathbf{X}\|_{\Sigma}$ and $\mathbf{\Sigma}_F^{-1/2} = \mathbf{C}$ satisfies $\mathbf{C}\mathbf{C}^\top = \mathbf{\Sigma}_F^{-1}$. To define the sample version, we need affine-equivariant estimators of $\mathbf{\Sigma}_F^{1/2}$ and of $\boldsymbol{\theta}$. This will be done iteratively; in the next subsection, we shall describe a possible iterative construction of an affine-equivariant of a more general sample spatial L-estimator.

8.5.1 Smooth Affine-Equivariant L-Estimation of θ

Note that a *half-space* of $\mathbb{R}_p$ is defined as

$$H(\mathbf{s}, t) = \{\mathbf{x} \in \mathbb{R}_p : \ \langle \mathbf{s}, \mathbf{x} \rangle \le t\}, \quad t \in \mathbb{R}$$

where $\|\mathbf{s}\| = 1$, i.e. $\mathbf{s} \in \mathcal{S}_{p-1} = \{\mathbf{x} : \ \ \|\mathbf{x}\|^2 = 1\}$. If $H(\mathbf{s}, 0)$ under F has the probability $\ge \frac{1}{2}$ for every closed half-space with $\boldsymbol{\theta}$ on the boundary, then F is *half-space symmetric.* When transmitted to the distribution function of $\mathbf{Y} = \mathbf{\Sigma}_F^{-1/2}(\mathbf{X}-\boldsymbol{\theta})$, the half-space symmetry provides a tool for a construction of a sample version through the following iteration procedure:

Start with the sample mean vector $\bar{\mathbf{X}}_n = L_n^{(0)} = \frac{1}{n}\sum_{i=1}^n \mathbf{X}_i$ and sample sum of product matrix $n\widehat{\mathbf{\Sigma}}_n = \mathbf{A}_n^{(0)} = \sum_{i=1}^n (\mathbf{X}_i - \bar{\mathbf{X}}_n)(\mathbf{X}_i - \bar{\mathbf{X}}_n)^\top$, $n > p$. Define

$$d_{ni}^{(0)} = \|\mathbf{X}_i - \bar{\mathbf{X}}_n\|^2_{A_n^{(0)}} = (\mathbf{X}_i - \bar{\mathbf{X}}_n)^\top (\mathbf{A}_n^{(0)})^{-1}(\mathbf{X}_i - \bar{\mathbf{X}}_n), \ \ i = 1, \ldots, n. \tag{8.23}$$

The $d_{ni}^{(0)}$ are nonnegative random variables, not independent but exchangeable, and they are uniformly bounded, because

$$\sum_{i=1}^n d_{ni}^{(0)} = \sum_{i=1}^n \|\mathbf{X}_i - \bar{\mathbf{X}}_n\|^2_{A_n^{(0)}} = \text{ Trace of projection matrix } = p.$$

Let $R_{ni}^{(0)} = \sum_{j=1}^{n} I[d_{nj}^{(0)} \le d_{ni}^{(0)}]$ be the rank of $d_{ni}^{(0)}$ among $d_{n1}^{(0)}, \ldots, d_{nn}^{(0)}$, $i = 1, \ldots, n$, and denote $\mathbf{R}_n^{(0)} = (R_{n1}^{(0)}, \ldots, R_{nn}^{(0)})^\top$ the vector of ranks. Because F is continuous, the probability of ties is 0. Note that the $\mathbf{X}_i$ are affine equivariant and the $R_{ni}^{(0)}$ are affine-invariant. Moreover, the $R_{ni}^{(0)}$ are invariant under any strictly monotone transformation of $d_{ni}^{(0)}$, $i = 1, \ldots, n$. hence, the $R_{ni}^{(0)}$ do not change if the $d_{ni}^{(0)}$ are replaced with $\|\mathbf{X}_i - \bar{\mathbf{X}}_n\|_{A_n^{(0)}}^{\alpha}$ for any $\alpha > 0$; here, we take $\alpha = 2$ for simplicity.

Take a vector of rank scores $\mathbf{a}_n = (a_n(1), \ldots, a_n(n))^\top$, $\sum_{i=1}^{n} a_n(i) = 1$. The first step estimator of $\boldsymbol{\theta}$ is as follows:

$$\mathbf{L}_n^{(1)} = \sum_{i=1}^{n} a_n(R_{ni}^{(0)})\mathbf{X}_i. \tag{8.24}$$

To see that it is an L-estimator, define the following ordering:

$$\mathbf{X}_i \prec \mathbf{X}_j \Leftrightarrow d_{ni}^{(0)} < d_{nj}^{(0)}, \; i \ne j = 1, \ldots, n. \tag{8.25}$$

Then $\mathbf{L}_n^{(1)} = \sum_{i=1}^{n} a_n(i)\mathbf{X}_{n:i}$ where $\mathbf{X}_{n:1} \prec \ldots \prec \mathbf{X}_{n:n}$ are the order statistics corresponding to ordering (8.25). Because the idea of the estimator is to be central around an initial estimator, the typical choice of $a_n(k)$ is to be nonincreasing in k for a fixed n; it is usually assumed that there exists a sequence $\{k_n\}$ such that $\frac{k_n}{n} \to 0$ and $\sum_{k>k_n} |a_n(k)| = o(n^{-2})$ as $n \to \infty$.

In the next step, we put

$$\begin{aligned}
\mathbf{A}_n^{(1)} &= \sum_{i=1}^{n} (\mathbf{X}_i - \mathbf{L}_n^{(1)})(\mathbf{X}_i - \mathbf{L}_n^{(1)})^\top \\
d_{ni}^{(1)} &= \|\mathbf{X}_i - \mathbf{L}_n^{(1)}\|_{\mathbf{A}_n^{(1)}}^2 \\
R_{ni}^{(1)} &= \sum_{j=1}^{n} I[d_{nj}^{(1)} \le d_{ni}^{(1)}], \; i = 1, \ldots, n, \quad \mathbf{R}_n^{(1)} = (R_{n1}^{(1)}, \ldots, R_{nn}^{(1)})^\top.
\end{aligned}$$

The second-step estimator is $\mathbf{L}_n^{(2)} = \sum_{i=1}^{n} a_n(R_{ni}^{(1)})\mathbf{X}_i$. In this way we proceed, so at the r-th step we define $\mathbf{A}_n^{(r)}$, $d_{ni}^{(r)}$, $1 \le i \le n$ and the ranks $\mathbf{R}_n^{(r)}$ analogously, and get the r-step estimator

$$\mathbf{L}_n^{(r)} = \sum_{i=1}^{n} a_n(R_{ni}^{(r-1)})\mathbf{X}_i, \; r \ge 1. \tag{8.26}$$

Note that the d_{ni}^{r} are affine-invariant for every $1 \le i \le n$ and for every $r \ge 0$. Hence, applying an affine transformation $\mathbf{Y}_i = \mathbf{B}\mathbf{X}_i + \mathbf{b}$, $\mathbf{b} \in \mathbb{R}_p$, $\mathbf{B}$ positive definite, we see that

$$\mathbf{L}_n^{(r)}(\mathbf{Y}_1, \ldots, \mathbf{Y}_n) = \mathbf{B}\mathbf{L}_n^{(r)}(\mathbf{X}_1, \ldots, \mathbf{X}_n) + \mathbf{b} \tag{8.27}$$

$\forall r \ge 0$, $\mathbf{b} \in \mathbb{R}_p$, $\mathbf{B}$ positive definite; i.e. the estimating procedure preserves the affine equivariance at each step. thus, $\mathbf{L}_n^{(r)}$ is an affine-equivariant L-estimator of $\boldsymbol{\theta}$ for every r.

The estimator is being smoothed with increasing n and the $a_n(k)$ are tuned according the smoothing method in mind, which is usually either the *kernel method* or the k_n-NN (*nearest neighborhood*) method. In a k_n-NN approach, one can let $a_n(k) = \frac{1}{k_n} I[k \leq k_n]$, $1 \leq k \leq n$. In a kernel smoothing approach, in view of importance of the lower ranks (see Chaubey and Sen (1996)), an exponential or some increasing failure rate smoothing density can be conceived. The $a_n(k)$ rapidly decreases to 0 with increasing k and $a_n(1) \geq a_n(2) \geq \ldots \geq a_n(n) \geq 0$ with a very small contribution from the tail $\sum_{k>k_n} a_n(k)$. For example, one can take $a_n(k) \propto k^{-4}$, $1 \leq k \leq n$ and $k_n = \mathrm{O}(n^{\frac{1}{2}+\varepsilon})$, $\varepsilon > 0$, so that $\sum_{k>k_n} a_n(k) = \mathrm{O}(k_n^{-3}) = \mathrm{O}(n^{-\frac{3}{2}-3\varepsilon}) = o(n^{-\frac{3}{2}})$.

The approximate knowledge of F provides a guide to how to choose the scores $a_n(k)$ in outline of the rank tests theory.

Example 8.1 If $\mathbf{X} \sim \mathcal{N}_p(\boldsymbol{\theta}, \boldsymbol{\Sigma})$, then $D = (\mathbf{X}-\boldsymbol{\theta})^\top \boldsymbol{\Sigma}^{-1}(\mathbf{X}-\boldsymbol{\theta})$ has χ^2 distribution with p degrees of freedom. Moreover, by the equi-probability contours of the multinormal distribution, $\mathbf{X}$ has a uniform distribution on the surface $D = d$. It implies that

$$\mathbb{E}\left(\mathbf{X} - \boldsymbol{\theta} \middle| (\mathbf{X} - \boldsymbol{\theta})^\top \boldsymbol{\Sigma}^{-1}(\mathbf{X} - \boldsymbol{\theta}) = d^2\right) = \mathbf{0},$$

$$\mathbb{E}\Big((\mathbf{X} - \boldsymbol{\theta})(\mathbf{X} - \boldsymbol{\theta})^\top \Big| (\mathbf{X} - \boldsymbol{\theta})^\top \boldsymbol{\Sigma}^{-1}(\mathbf{X} - \boldsymbol{\theta}) = d^2\Big)$$
$$= \mathbb{E}\left(\boldsymbol{\Sigma}^{-1/2}\mathbf{Y}\mathbf{Y}^\top\boldsymbol{\Sigma}^{-1/2} \middle| \mathbf{Y}^\top\mathbf{Y} = d^2\right) = c \cdot d\boldsymbol{\Sigma},$$
$$c = \text{const}, \ \mathbf{Y} = \boldsymbol{\Sigma}^{-1/2}\mathbf{X}.$$

Further, $D_i = (\mathbf{X}_i - \boldsymbol{\theta})^\top \boldsymbol{\Sigma}^{-1}(\mathbf{X}_i - \boldsymbol{\theta})$, $1 \leq i \leq n$ are i.i.d. with central χ_p^2 distribution, and given $(D_1, \ldots, D_1)$, the $\mathbf{X}_i$ are conditionally independent with covariance matrices $\propto D_i\boldsymbol{\Sigma}$, $i = 1, \ldots, n$. This leads to the weighted least squares estimator

$$\mathbf{T} = \arg\min \left\{ \sum_{i=1}^n D_i^{-1} \|\mathbf{X}_i - \boldsymbol{\theta}\|_\Sigma^2 \right\} = \frac{\sum_{i=1}^n D_i^{-1}\mathbf{X}_i}{\sum_{i=1}^n D_i^{-1}}.$$

Hence, under n fixed and $\boldsymbol{\theta}, \boldsymbol{\Sigma}$ known, the resulting rank scores would be

$$a_n(i) = \frac{\xi_{ni}^{-2}}{\sum_{i=1}^n \xi_{ni}^{-2}}, \quad \xi_{ni} = \mathbb{E}(D_{n:i}), \ i = 1, \ldots, n$$

where $D_{n:1} < \ldots, D_{n:n}$ are order statistics corresponding to the sample $D_1, \ldots, D_n$ from the χ_p^2 distribution. However, the unknown $\boldsymbol{\theta}, \boldsymbol{\Sigma}$ are estimated by $\bar{\mathbf{X}}$ and $\widehat{\boldsymbol{\Sigma}}_n = (n-1)^{-1}\mathbf{A}_n^{(0)}$, where $\mathbf{A}_n^{(0)} = \sum_{i=1}^n (\mathbf{X}_i - \bar{\mathbf{X}}_n)(\mathbf{X}_i - \bar{\mathbf{X}}_n)^\top$, and D_i is approximated by $d_{ni}^{(0)}$ in (8.23), $i = 1, \ldots, n$. Hence, we can determine the scores only asymptotically; in this situation, we can use the *approximate scores*

$$a_n^*(i) = \frac{1}{K_n}\left[\Psi^*\left(\frac{i}{n+1}\right)\right]^{-1} \tag{8.28}$$

where

$$K_n = \sum_{k=1}^{n}\left[\Psi^*\left(\frac{k}{n+1}\right)\right]^{-1}$$

and where Ψ^* is the χ_p^2 distribution function. hence, the algorithm proceeds as follows:

(1) Calculate $\bar{\mathbf{X}}_n$ and $\mathbf{S}_n = \frac{1}{n-1}\sum_{i=1}^{n}(\mathbf{X}_i - \bar{\mathbf{X}}_n)(\mathbf{X}_i - \bar{\mathbf{X}}_n)^\top$.

(2) Calculate $d_{ni}^{(0)} = (\mathbf{X}_i - \bar{\mathbf{X}}_n)^\top \mathbf{S}_n^{-1}(\mathbf{X}_i - \bar{\mathbf{X}}_n)$, $1 \le i \le n$.

(3) Determine the rank $R_{ni}^{(0)}$ of $d_{ni}^{(0)}$ among $d_{n1}^{(0)}, \ldots, d_{nn}^{(0)}$, $i = 1, \ldots, n$.

(4) Calculate the scores $a_n^*(i)$, $i = 1, \ldots, n$ according to (8.28), where Ψ^* is the χ_p^2 distribution function.

(5) Calculate the first-step estimator $\mathbf{L}_n^{(1)} = \sum_{i=1}^{n} a_n^*(R_{ni}^{(0)})\mathbf{X}_i$.

(6) $\mathbf{S}_n^{(1)} = \frac{1}{n-1}\sum_{i=1}^{n}(\mathbf{X}_i - \mathbf{L}_n^{(1)})(\mathbf{X}_i - \mathbf{L}_n^{(1)})^\top$.

(7) $d_{ni}^{(1)} = (\mathbf{X}_i - \mathbf{L}_n^{(1)})^\top(\mathbf{S}_n^{(1)})^{-1}(\mathbf{X}_i - \mathbf{L}_n^{(1)})$, $1 \le i \le n$.

(8) $R_{ni}^{(1)}$ = the rank of $d_{ni}^{(1)}$ among $d_{n1}^{(1)}, \ldots, d_{nn}^{(1)}$, $i = 1, \ldots, n$.

(9) $\mathbf{L}_n^{(2)} = \sum_{i=1}^{n} a_n^*(R_{ni}^{(1)})\mathbf{X}_i$.

(10) Repeat the steps (6)–(9).

The procedure is illustrated on three samples of sizes $n = 20$, $n = 50$, and $n = 500$ simulated from the normal distribution $\mathcal{N}_3(\boldsymbol{\theta}, \boldsymbol{\Sigma})$ with

$$\boldsymbol{\theta} = (1,\ 2,\ -1)^\top$$

and

$$\boldsymbol{\Sigma} = \begin{bmatrix} 1 & 1/2 & 1/2 \\ 1/2 & 1 & 1/2 \\ 1/2 & 1/2 & 1 \end{bmatrix}$$

and each time the affine-equivariant estimator of $\boldsymbol{\theta}$ was calculated in 9 iterations of the initial estimator. The following tables illustrate the results and compare them with mean, median, and the different choice of scores (8.29), and provide standard errors of the estimates for better comparison.

Table 8.1. Mean, median, and affine estimator $\widehat{\boldsymbol{\theta}}_n$ of $\boldsymbol{\theta}$: two choices of a_n; $n = 20$

Mean	0.82944	1.68769	-1.34490			
Median	0.67478	1.67190	-1.38401			
	a_n^* (8.28)			a_n (8.29)		
$\mathbf{L}_n^{(1)}$	0.78264	1.5288	-1.2358	0.83047	1.7444	-1.4492
$\mathbf{L}_n^{(2)}$	0.81089	1.5333	-1.2089	0.80415	1.7431	-1.4760
$\mathbf{L}_n^{(3)}$	0.76843	1.5355	-1.2298	0.78251	1.7532	-1.4785
$\mathbf{L}_n^{(4)}$	0.81288	1.5342	-1.2123	0.77906	1.7552	-1.4835
$\mathbf{L}_n^{(5)}$	0.77043	1.5364	-1.2332	0.77906	1.7552	-1.4835
$\mathbf{L}_n^{(6)}$	0.81288	1.5342	-1.2123	0.77906	1.7552	-1.4835
$\mathbf{L}_n^{(7)}$	0.77043	1.5364	-1.2332	0.77906	1.7552	-1.4835
$\mathbf{L}_n^{(8)}$	0.81288	1.5342	-1.2123	0.77906	1.7552	-1.4835
$\mathbf{L}_n^{(9)}$	0.77043	1.5364	-1.2332	0.77906	1.7552	-1.4835
$\mathbf{L}_n^{(10)}$	0.81288	1.5342	-1.2123	0.77906	1.7552	-1.4835

Table 8.2. Mean, median, and affine estimator $\widehat{\boldsymbol{\theta}}_n$ of $\boldsymbol{\theta}$: two choices of a_n; $n = 50$

Mean	1.12676	1.99514	-0.97513			
Median	1.14845	1.96151	-1.01970			
	a_n^* (8.28)			a_n (8.29)		
$\mathbf{L}_n^{(1)}$	0.93519	1.93761	-0.94960	1.19111	2.00991	-0.97243
$\mathbf{L}_n^{(2)}$	1.07811	2.04993	-0.87029	1.21857	2.02786	-0.95625
$\mathbf{L}_n^{(3)}$	0.90929	1.88736	-0.99444	1.23458	2.03597	-0.94716
$\mathbf{L}_n^{(4)}$	1.14172	2.11267	-0.80644	1.24155	2.04133	-0.94154
$\mathbf{L}_n^{(5)}$	0.86033	1.83317	-1.04098	1.24336	2.04428	-0.94026
$\mathbf{L}_n^{(6)}$	1.21143	2.17534	-0.75423	1.24588	2.04513	-0.93877
$\mathbf{L}_n^{(7)}$	0.81820	1.78509	-1.08601	1.24751	2.04834	-0.93725
$\mathbf{L}_n^{(8)}$	1.25390	2.22826	-0.71815	1.24848	2.04841	-0.93682
$\mathbf{L}_n^{(9)}$	0.79188	1.74877	-1.12027	1.24848	2.04841	-0.93682
$\mathbf{L}_n^{(10)}$	1.27853	2.26028	-0.70143	1.24848	2.04841	-0.93682

Table 8.3. Mean, median and affine estimator $\widehat{\boldsymbol{\theta}}_n$ of $\boldsymbol{\theta}$: two choices of a_n; $n = 500$

Mean	0.99163	1.97330	-0.98307			
Median	1.02051	1.96848	-0.94942			
	a_n^* (8.28)			a_n (8.29)		
$\mathbf{L}_n^{(1)}$	0.95907	1.98698	-0.96214	1.01501	1.96454	-0.99307
$\mathbf{L}_n^{(2)}$	0.96759	1.98233	-0.96706	1.02073	1.96221	-0.99466
$\mathbf{L}_n^{(3)}$	0.96527	1.98358	-0.96586	1.02235	1.96193	-0.99480
$\mathbf{L}_n^{(4)}$	0.96629	1.98346	-0.96592	1.02303	1.96194	-0.99460
$\mathbf{L}_n^{(5)}$	0.96524	1.98309	-0.96641	1.02332	1.96203	-0.99448
$\mathbf{L}_n^{(6)}$	0.96676	1.98409	-0.96523	1.02346	1.96219	-0.99436
$\mathbf{L}_n^{(7)}$	0.96479	1.98262	-0.96695	1.02353	1.96227	-0.99428
$\mathbf{L}_n^{(8)}$	0.96722	1.98447	-0.96468	1.02358	1.96233	-0.99424
$\mathbf{L}_n^{(9)}$	0.96356	1.98175	-0.96773	1.02367	1.96242	-0.99422
$\mathbf{L}_n^{(10)}$	0.96821	1.98533	-0.96389	1.02369	1.96246	-0.99423

Note that in this 3-dimensional problem, the trace of the covariance matrix of the estimator is 3 times the marginal variance. This has resulted in greater variation of the estimates compared with the univariate case. The coordinate-wise estimators are quite comparable to the affine-equivariant L-estimators for n=50 or 500. For small samples, variations are more perceptible.

If $f(\cdot)$ is not elliptically symmetric, the choice of $a_n(i)$ can be different. Then, preferring simplicity, we can choose

$$a_n(i) = \frac{2(n-i+1)}{n(n+1)}, \; i = 1, \ldots, n \tag{8.29}$$

though it does not have a similar interpretation as in Example 8.1.

The following alternative procedure leads to a spatial *trimmed mean*: For *simplicity*, let random vectors $\mathbf{X}_1, \ldots, \mathbf{X}_n$, be *i.i.d.* with mean $\boldsymbol{\theta}$ and covariance matrix $\boldsymbol{\Sigma}$. Put $\Delta_i = (\mathbf{X}_i - \boldsymbol{\theta})^\top \boldsymbol{\Sigma}^{-1}(\mathbf{X}_i - \boldsymbol{\theta})$, $i = 1, \ldots, n$. Then the Δ_i are *i.i.d.* nonnegative random variables. By the multivariate central limit theorem and by the Slutzky–Cochran theorem (see (2.91)),

$$n(\mathbf{X}_n - \boldsymbol{\theta})^\top \widehat{\boldsymbol{\Sigma}}_n^{-1}(\mathbf{X}_n - \boldsymbol{\theta}) \xrightarrow{\mathcal{D}} \chi_p^2 \quad \text{whenever } \widehat{\boldsymbol{\Sigma}}_n \to \boldsymbol{\Sigma}.$$

Introduce the following affine-invariant random set $\mathcal{J}_n$ as

$$\mathcal{J}_n = \{i, \; 1 \le i \le n : \; (\mathbf{X}_i - \boldsymbol{\theta})^\top \boldsymbol{\Sigma}^{-1}(\mathbf{X}_i - \boldsymbol{\theta}) \le k_n\}$$

where k_n is nondecreasing in n but $\frac{k_n}{n} \to 0$. Let

$$\bar{\mathbf{X}}_n^* = \frac{1}{k_n} \sum_{i \in \mathcal{J}_n} \mathbf{X}_i. \tag{8.30}$$

Then $\bar{\mathbf{X}}_n^*$ is a robust and affine-equivariant estimator of $\boldsymbol{\theta}$, because $\mathbf{B}\bar{\mathbf{X}}_n^* + \mathbf{b} = \frac{1}{k_n} \sum_{i \in \mathcal{J}_n} (\mathbf{B}\mathbf{X}_i + \mathbf{b})$ $\forall \mathbf{b} \in \mathbb{R}_p$, $\mathbf{B}$ positive definite. When $k_n \approx n(1-\varepsilon), \varepsilon > 0$, this leads to an affine-equivariant trimmed mean estimator of $\boldsymbol{\theta}$. Generally, this approach has a k_n-NN interpretation when $\frac{k_n}{n} \to 0$ at a suitable rate. $\widehat{\boldsymbol{\Sigma}}_n = \mathbf{A}_n^{(0)}$ can be updated in a similar way as in (8.24)–(8.27).

The use of $\bar{\mathbf{X}}_n$ and $\mathbf{A}_n^{(0)}$ in (8.23) implicitly assumes an existence of the finite second moment of F. The initial estimates can be replaced by suitable $\sqrt{n}$-consistent estimators, with an appropriate arrangement. The bounded nature of the $d_{ni}^{(r)}$ makes the choice rather robust. The asymptotic behavior of the procedure is illustrated on the following asymptotic properties of $d_{n1}^{(r)}, \ldots, d_{nn}^{(r)}$ as $n \to \infty$:

Denote $G_n^{(r)}$ the distribution function of $nd_{ni}^{(r)}$, i.e. $G_n^{(r)}(d) = I\!P(nd_{ni}^{(r)} \le d)$; let $\widehat{G}_n^{(r)}(d) = \frac{1}{n}\sum_{i=1}^n I[nd_{ni}^{(r)} \le d]$, $d \in \mathbb{R}^+$ be the empirical distribution function corresponding to $nd_{n1}^{(r)}, \ldots, nd_{nn}^{(r)}$. Note that $d_{ni}^{(r)}$ are bounded nonnegative random variables and $\sum_{i=1}^n d_{ni}^{(r)} = (n-1)p$; then we can use the Hoeffding (1963) inequality [see 2.64] to verify that $I\!P\{|\widehat{G}_n^{(r)}(d) - G_n^{(r)}(d)| > c_n\} \to 0$ as $n \to \infty$, exponentially at the rate nc_n^2, for every $d \ge 0$. Consider the following lemma.

LEMMA 8.1 *As* $n \to \infty$,

$$\sup_{d \in \mathbb{R}^+} \Big\{ |\widehat{G}_n^{(r)}(d) - G_n^{(r)}(d) - \widehat{G}_n^{(r)}(d') + G_n^{(r)}(d')| : |d - d'|$$
$$\le n^{-1/2}\sqrt{2\mathrm{ln}n} \Big\} \overset{a.s.}{=} O(n^{-\frac{3}{4}}\mathrm{ln}n). \tag{8.31}$$

PROOF (outline). Both $\widehat{G}_n^{(r)}(d)$ and $G_n^{(r)}(d)$ are $\nearrow$ in $d \in \mathbb{R}^+$, and $\widehat{G}_n^{(r)}(0) = G_n^{(r)}(0)$ and $\widehat{G}_n^{(r)}(\infty) = G_n^{(r)}(\infty)$. Hence, the proof of the lemma is complete if we proceed similarly as in (4.16). □

Motivated by the above results, we get the following theorem.

THEOREM 8.1 *As* n *increases,*

$$\left\{ \sqrt{n}[\widehat{G}_n^{(r)}(d) - G_n^{(r)}(d)] : c \in \mathbb{R}^+ \right\}$$

converges weakly to a tied-down Brownian motion process, reducible to the Brownian bridge with the time-transformation $t = (G_n^{(r)})^{-1}(d)$ *of* $d \in \mathbb{R}^+$ *to* $t \in [0, 1]$.

PROOF (outline). By virtue of Lemma 8.1, the tightness part of the proof follows readily. For the convergence of finite-dimensional distributions, we appeal to the central limit theorem for interchangeable random variables by Chernoff and Teicher (1958). □

Summarizing, (8.23)–(8.30) exhibit some L-estimators which are affine-equivariant and which can be studied with the standard asymptotic theory under fairly general regularity conditions. The spatial quantile functions (8.22) are quite intuitive; on the other hand, the affine-equivariant estimators lose the simplicity of the coordinatewise estimators.

8.5.2 Affine-Equivariant Regression Estimation

Let us now deal with the scope and limitation of affine-equivariant estimation in the regression model outlined in (8.16) and further elaborated in (8.17). The least squares estimator (LSE) $\widehat{\boldsymbol{\beta}}_n$ of $\boldsymbol{\beta}$ in model with positively definite dispersion matrix $\boldsymbol{\Sigma}$ (though unknown) is the minimizer of the objective function

$$\sum_{i=1}^{n} \|\mathbf{Y}_i - \mathbf{B}\mathbf{x}_i\|_{\Sigma}^2 \quad \text{with respect to } \mathbf{B} \in \mathbb{R}_{p\times q}.$$

$\widehat{\boldsymbol{\beta}}_n$ can be explicitly obtained from the estimating equation

$$\sum_{i=1}^{n} (\mathbf{Y}_i - \mathbf{B}\mathbf{x}_i)\mathbf{x}_i^\top = \mathbf{0},$$

thus, leading to $\widehat{\boldsymbol{\beta}}_n = \mathbf{Y}\mathbf{X}^\top(\mathbf{X}\mathbf{X}^\top)^{-1}$. The LSE $\widehat{\boldsymbol{\beta}}_n$ has a double equivariance property in the sense that if we transform $\mathbf{Y}$ and $\mathbf{X}$ as

$$\mathbf{Y} \mapsto \mathbf{Y}^* = \mathbf{C}_1\mathbf{Y} \quad \text{and} \quad \mathbf{X} \mapsto \mathbf{X}^* = \mathbf{C}_2\mathbf{X}$$

with positive definite matrices $\mathbf{C}_1$ and $\mathbf{C}_2$ of the respective orders $p \times p$ and $q \times q$, then the LSE of $\boldsymbol{\beta}^* = \mathbf{C}_1\boldsymbol{\beta}\mathbf{C}_2$, based on $(\mathbf{Y}^*, \mathbf{X}^*)$ is given by

$$\begin{aligned} \widehat{\boldsymbol{\beta}}_n^* &= (\mathbf{Y}^*\mathbf{X}^{*\top})(\mathbf{X}^*\mathbf{X}^{*\top})^{-1} \qquad (8.32)\\ &= \mathbf{C}_1\mathbf{Y}\mathbf{X}^\top\mathbf{C}_2^\top\left(\mathbf{C}_2^{-1}\mathbf{X}\mathbf{X}^\top(\mathbf{C}_2^{-1})^\top\right)^{-1} \\ &= \mathbf{C}_1\mathbf{Y}\mathbf{X}^\top(\mathbf{X}\mathbf{X}^\top)^{-1}\mathbf{C}_2 = \mathbf{C}_1\widehat{\boldsymbol{\beta}}_n\mathbf{C}_2. \end{aligned}$$

$\widehat{\boldsymbol{\beta}}_n$ coincides with the maximum likelihood estimator in the model with multinormally distributed errors; but it is highly nonrobust. The property (8.32) of an estimator for arbitrary positive definite $\mathbf{C}_1$ and $\mathbf{C}_2$ can be characterized as its affine equivariance. Consider an estimator as a minimizer of the objective function $\sum_{i=1}^{n} \rho(\mathbf{Y}_i - \mathbf{B}\mathbf{x}_i)$, $\mathbf{B} \in \mathbb{R}_{p\times q}$. The function $\rho = \rho_L$ is the squared Mahalanobis distance $\|\mathbf{Y}_i - \mathbf{B}\mathbf{x}_i\|_{\Sigma}^2$ in the case of the LSE. More robust is the L_1-norm $\rho_1 : \sum_{i=1}^{n} \rho_1(\mathbf{Y}_i, \mathbf{x}_i) = \sum_{i=1}^{n} \|\mathbf{Y}_i - \mathbf{B}\mathbf{x}_i\|_{\Sigma}$, and the L_1-norm

(Mahalanobis distance) estimator is affine equivariant, more robust than the LSE. However, as it is highly nonlinear in $\mathbf{B} \in \mathbb{R}_{p \times q}$, it should be computed iteratively.

We can also consider an affine-equivariant version of the Huber (1964) score function, adapted to the multivariate regression. Define $\rho_L(\cdot)$ and $\rho_1(\cdot)$ as above and let

$$\rho_H(\mathbf{Y}_i - \mathbf{B}\mathbf{x}_i) = \begin{cases} \rho_L(\mathbf{Y}_i - \mathbf{B}\mathbf{x}_i) & \text{if} \quad \|\mathbf{Y}_i - \mathbf{B}\mathbf{x}_i\|_\Sigma \leq K \\ K\rho_1(\mathbf{Y}_i - \mathbf{B}\mathbf{x}_i) & \text{if} \quad \|\mathbf{Y}_i - \mathbf{B}\mathbf{x}_i\|_\Sigma > K \end{cases}$$

for some positive constant K. The corresponding estimating equation is $\sum_{i=1}^n \mathbf{T}_H(\mathbf{Y}_i - \mathbf{B}\mathbf{x}_i) = \mathbf{0}$, where

$$\mathbf{T}_H(\mathbf{Y}_i - \mathbf{B}\mathbf{x}_i) = \begin{cases} (\mathbf{Y}_i - \mathbf{B}\mathbf{x}_i)\mathbf{x}_i^\top & \text{if} \quad \|\mathbf{Y}_i - \mathbf{B}\mathbf{x}_i\|_\Sigma \leq K \\ \|\mathbf{Y}_i - \mathbf{B}\mathbf{x}_i\|_\Sigma^{-1}(\mathbf{Y}_i - \mathbf{B}\mathbf{x}_i)\mathbf{x}_i^\top & \text{if} \quad \|\mathbf{Y}_i - \mathbf{B}\mathbf{x}_i\|_\Sigma > K. \end{cases}$$

$\mathbf{T}_H$ is affine equivariant, but the estimating equation is difficult to solve. However, using the coordinatewise Huber's function would not lead to an affine-equivariant estimator.

While the usual R-estimators can be incorporated to multivariate R-estimators of $\boldsymbol{\beta}$, they will not be affine-equivariant. Hence, we are led to define the spatial rank function (Oja 2010):

$$\mathbb{R}(\mathbf{u}) = \sum_{i=1}^n \boldsymbol{\Sigma}^{-1/2} \frac{\mathbf{u}_i}{\|\mathbf{u}_i\|_\Sigma}, \quad \mathbf{u} \in \mathbb{R}_p$$

where $\mathbf{u}_i = \mathbf{Y}_i - \mathbf{B}\mathbf{x}_i - \mathbf{u}$, $1 \leq i \leq n$. The inference will be based on the spatial ranks $\mathbb{R}(\mathbf{Y}_k - \mathbf{B}\mathbf{x}_k)$, $1 \leq k \leq n$.

We need to start with a preliminary estimate of $\boldsymbol{\Sigma}$, since it is unknown. We start with some affine-equivariant estimators $\widehat{\boldsymbol{\beta}}^{(0)}$ and $\widehat{\boldsymbol{\Sigma}}^{(0)}$. For an affine-equivariant R-estimator of $\boldsymbol{\beta}$, we use the spatial ranks based on residuals $(\widehat{\boldsymbol{\Sigma}}^{(0)})^{-1/2}(\mathbf{Y}_i - \widehat{\boldsymbol{\beta}}^{(0)}\mathbf{x}_i)$. The convergence of the iteration procedure for R- and M-estimation is justified by verifying the „uniform asymptotic linearity, in probability" as in Chapters 4–7.

Namely, let $\mathcal{C}$ be a compact subset of $\mathbb{R}_{pq}$, containing $\mathbf{0}$ as an inner point. Let $\mathbf{B}_n(\mathbf{v}) = \boldsymbol{\beta} + n^{-1/2}\mathbf{v}$, $\mathbf{v} \in \mathcal{C}$. Assume that $n^{-1}\sum_{i=1}^n \mathbf{x}_i\mathbf{x}_i^\top = \mathbf{Q}_n \to \mathbf{Q}$ (positive definite) and $\max_{1 \leq i \leq n} \mathbf{x}^\top(\mathbf{X}^\top\mathbf{X})^{-1}\mathbf{x}_i \to 0$ as $n \to \infty$ (Noether condition) and let

$$\widehat{\boldsymbol{\Sigma}}(\mathbf{v}) = (n-q)^{-1}\sum_{i=1}^n (\mathbf{Y}_i - \mathbf{B}(\mathbf{v})\mathbf{x}_i)(\mathbf{Y}_i - \mathbf{B}(\mathbf{v})\mathbf{x}_i)^\top.$$

Then

$$\sup_{\mathbf{v} \in \mathcal{C}} \|\widehat{\boldsymbol{\Sigma}}(\mathbf{v}) - \widehat{\boldsymbol{\Sigma}}(\mathbf{0})\| = O_p(n^{-1});$$

$$\max_{1 \le i \le n} \|\mathbf{Y}_i - \boldsymbol{\beta}\mathbf{x}_i\| = o(\sqrt{n}) \text{ a.s.,} \quad \text{as } n \to \infty.$$

As a result, we have: As n increases,

$$\sup_{\mathbf{v} \in \mathcal{C}} \max_{1 \le i \le n} \left\| (\widehat{\boldsymbol{\Sigma}}(\mathbf{v}))^{-1/2}(\mathbf{Y}_i - \mathbf{B}_n(\mathbf{v})\mathbf{x}_i) - \widehat{\boldsymbol{\Sigma}}(\mathbf{0}))^{-1/2}(\mathbf{Y}_i - \mathbf{B}_n(\mathbf{v})\mathbf{x}_i) \right\|$$
$$= o_p(n^{-1/2}).$$

Using the Taylor expansion, we obtain a first-order representation:

$$n^{-1/2} \sum_{i=1}^{n} \mathbf{t}(\mathbf{Y}_i - \boldsymbol{\beta}\mathbf{x}_i)\mathbf{x}_i^{\top} = \boldsymbol{\Gamma} n^{1/2}(\widehat{\boldsymbol{\beta}}_n - \boldsymbol{\beta})\mathbf{Q} + o_p(1)$$

where $\boldsymbol{\Gamma}$ is a $p \times p$ matrix of constants which depends on the unknown distribution function and on the chosen objective function, and $\mathbf{t}(\mathbf{e}_i)$ is a p-vector with $\mathbf{0}$ mean and covariance matrix $\boldsymbol{\Delta}$. Hence, we obtain by the central limit theorem

$$n^{1/2}(\widehat{\boldsymbol{\beta}}_n - \boldsymbol{\beta}) \xrightarrow{\mathcal{D}} \mathcal{N}_{pq}\Big(\mathbf{0},\ \boldsymbol{\Gamma}^{-1}\boldsymbol{\Delta}\boldsymbol{\Gamma}^{-1} \otimes \mathbf{Q}^{-1}\Big), \quad \text{as } n \to \infty.$$

the estimating equation of the form

$$\mathbb{T}_n(\mathbf{Y} - \mathbf{BX}) = \sum_{i=1}^{n} \mathbb{T}_i(\mathbf{Y}_i - \mathbf{BX}_i) = \mathbf{0}, \quad \mathbf{B} \in \mathbb{R}_{p \times q}$$

where $\mathbb{T}_i,\ i = 1, \ldots, n$ are suitable $p \times q$ matrices, not necessary linear in $\mathbf{Y}$ and $\mathbf{X}$.

8.5.3 Additional Remarks and Comments

We conclude this section with some comments comparing the performance of the coordinatewise approach outlined in Sections 8.2 and 8.3 and of the affine-equivariant approach outlined in the present section. Both approaches were mostly considered for the location model. The affine-equivariant estimates were constructed under stronger regularity conditions, assuming some types of symmetry of F and mainly assuming the affine equivariant structure or shape of the model. They can break down if these conditions are not satisfied. For example, if F has the half-space symmetry property, then the affine-equivariant spatial median coincides with the vector of coordinatewise medians. However, if the affine equivariance is imposed but it is not supported by the shape or structure of F, then the sample affine-equivariant spatial median can estimate some entity without a proper statistical interpretation, while the coordinatewise median does not rest on these conditions and hence, it is more robust.

The high-dimensional multivariate data models are of increasing importance. The real life problems usually put some constraints on the model, under which the parameter space may not remain invariant with respect to the group of affine transformations. For example, this applies to the situation when the location parameter $\boldsymbol{\theta}$ belongs to an orthant or a cone (i.e., $\boldsymbol{\theta} \geq \mathbf{0}$ or $\theta_1 \leq \theta_2 \leq \ldots \leq \theta_p$) which does not remain invariant under affine transformations. Then the affine-equivariant estimator $\bar{\mathbf{X}}_n$ is not optimal; the restricted MLE performs better. Likewise, coordinatewise robust estimates coupled with Roy's (1953) *union-intersection principle* perform better than their affine equivariant counterparts (Sen (2008).

Another aspect relates to the moment conditions. The Mahalanobis distance is the *maximal invariant* for the affine-equivariant model, and thus every affine-equivariant procedure should be its function. However, it demands the finite second-order moments, while the coordinatewise approach may not need stringent moment conditions. The same comment also applies to the general constrained statistical inference (Silvapulle and Sen (2005)).

The affine-equivariant robust estimators should perform better if F supports the affine-equivariant structure. However, we should note that the multivariate estimators are typically not finite-sample admissible, and hence they can be always dominated [Stein (1956), Jurečková and Milhaud (1993), Jurečková and Klebanov (1997, 1998), Jurečková and Sen (2006)]. The efficiency and minimum risk aspects will be considered in the next section.

8.6 Efficiency and Minimum Risk Estimation

If the parameter $\boldsymbol{\theta}$ of interest can be related to a parametric model and its estimator $\mathbf{T}_n$ is asymptotically normal, then the asymptotic relative efficiency of $\mathbf{T}_n$ with respect to the MLE is measured by means of the asymptotic Rao–Cramér inequality. The situation is different in the multivariate case treated in Sections 8.3 and 8.5. The parameter $\boldsymbol{\theta}$ is either a vector or a matrix, which can be reduced to a vector form using a vec-operation. Let $\mathbf{T}_n$ be an estimator of $\boldsymbol{\theta}$ and let $\boldsymbol{\nu}_T$ be the dispersion matrix of $\sqrt{n}(\mathbf{T}_n - \boldsymbol{\theta})$, a $p \times p$ positive semidefinite matrix. However, $\boldsymbol{\nu}_T$'s corresponding to different $\mathbf{T}_n$ may not be mutually proportional. Under Cramér–Rao regularity assumptions,

$$\boldsymbol{\nu}_T - (\mathbb{I}(\boldsymbol{\theta}))^{-1} \;=\; \text{positive semidefinite } \forall\, \mathbf{T}_n \tag{8.33}$$

where $\mathbb{I}(\boldsymbol{\theta})$ is the Fisher information matrix. The marginal asymptotic relative efficiencies ignore the interdependence of the coordinates and hence do not serve the purpose. Generally there is no optimality criterion, and we have a challenge to choose a real-valued function of $\boldsymbol{\nu}_T \cdot \mathbf{I}(\boldsymbol{\theta})$ as a measure of the asymptotic relative efficiency (ARE) and provide it with a rational statistical interpretation.

Note that (8.33) implies that $\mathbf{I}_p - [\boldsymbol{\nu}_T \mathbb{I}(\boldsymbol{\theta})]^{-1}$ is positive semidefinite and the right-hand side is a null matrix when $\boldsymbol{\nu}_T = (\mathbb{I}(\boldsymbol{\theta}))^{-1}$. There exists an orthogonal matrix $\mathbf{P}$ such that $\mathbf{P}\mathbf{P}^\top = \mathbf{I}_p$, which can diagonalize both $\boldsymbol{\nu}_T$ and

$\mathbb{I}(\boldsymbol{\theta})$; we choose $\mathbf{P}$ so that $\mathbf{P}^\top \boldsymbol{\nu}_T \mathbf{P} = \mathbf{D} = \text{diag}(d_1, \ldots, d_p)$ and $\mathbf{P}^\top \mathbb{I}(\boldsymbol{\theta})\mathbf{P} = \mathbf{D}^* = \text{diag}(d_1^*, \ldots, d_p^*)$ where all d_j, d_j^* are nonnegative, and actually positive if both $\boldsymbol{\nu}_T$ and $\mathbb{I}(\boldsymbol{\theta})$ are positive definite. Then

$$\mathbf{P}\mathbf{I}_p\mathbf{P}^\top - [\mathbf{P}^\top \boldsymbol{\nu}_T \mathbf{P}\mathbf{P}^\top \mathbb{I}(\boldsymbol{\theta})\mathbf{P}]^{-1} \text{ is positive semidefinite}$$

thus, $\mathbf{I}_p - (\mathbf{D}\mathbf{D}^*)^{-1}$ is positive semidefinite and $\mathbf{D}\mathbf{D}^* = \text{diag}(d_1 d_1^*, \ldots, d_p d_p^*) = \mathbf{D}^0$ and $d_j d_j^* \geq 1 \; \forall j = 1, \ldots, p$. In other words, $\mathbf{D}^0$ is the diagonal matrix with the eigenvalues of $\boldsymbol{\nu}_T \mathbb{I}(\boldsymbol{\theta})$. These eigenvalues are invariant under affine transformations on $\mathbf{T}_n$ and $\boldsymbol{\theta}$. Driven by this motivation, there are three commonly used efficiency criteria:

(1) *D-efficiency* is defined as the p-th root of the determinant of $(\mathbf{D}^0)^{-1}$, or equivalently $|\boldsymbol{\nu}_T \mathbb{I}(\boldsymbol{\theta})|^{-1/p}$. Because the determinant is the product of the eigenvalues, the D-efficiency equals the reciprocal value of the geometric mean of the eigenvalues of $\mathbf{D}^0$.

(2) *A-efficiency* is the arithmetic mean of the eigenvalues of $(\mathbf{D}^0)^{-1}$, or equivalently it equals to $\frac{1}{p}\sum_{j=1}^{p}(d_j d_j^*)^{-1} = \frac{1}{p}\text{Trace}(\boldsymbol{\nu}_{\mathrm{T}}\mathbb{I}(\boldsymbol{\theta}))^{-1})$.

(3) *E-efficiency* is the largest eigenvalue of $(\mathbf{D}^0)^{-1}$; thus, $\{\min_{1 \leq j \leq p} d_j d_j^*\}^{-1}$.

For $p = 1$ all three efficiency criteria coincide. For $p \geq 2$ we can use the well-known inequality between the arithmetic and geometric means of nonnegative numbers a_j,

$$\max_{1 \leq j \leq p} a_j \geq \frac{1}{p}\sum_{j=1}^{p} a_j \geq \Big(\prod_{j=1}^{p} a_j\Big)^{1/p}$$

and claim that

$$\text{E-efficiency} \geq \text{A-efficiency} \geq \text{D-efficiency}$$

where the equality signs hold when $\mathbf{D}^0 = \mathbf{I}_p$, i.e. $\boldsymbol{\nu}_T = (\mathbb{I}(\boldsymbol{\theta}))^{-1}$. The eigenvalues of $\mathbf{D}^0$ for an affine-equivariant $\mathbf{T}_n$ are all the same in the case of elliptically symmetric family described in Sections 8.2 and 8.5.

The *marginal efficiencies* of estimator $\mathbf{T}_n$ with dispersion matrix $\boldsymbol{\nu}_T$ of $\sqrt{n}(\mathbf{T}_n - \boldsymbol{\theta})$ are defined as

$$e_j(\mathbf{T}_n) = \frac{1}{I_{jj}^* \nu_{jj,T}}, \quad 1 \leq j \leq p$$

where I_{jj}^* are the marginal Fisher informations defined in (8.13) and $\nu_{jj,T}$ are the diagonal elements of matrix $\boldsymbol{\nu}_T$, $1 \leq j \leq p$. Unfortunately, the marginal efficiencies may not have a natural interpretation, because they do not take into account the possible dependence of the coordinates.

Because there is no universal efficiency measure, we can alternatively take into account the loss and risk functions of an estimator and consider the *risk-efficiency*. Let $\mathbf{T}_n$ be an estimator of $\boldsymbol{\theta} \in \Omega \subseteq \mathbb{R}_p$ and $L(\mathbf{T}, \boldsymbol{\theta}) : \mathbb{R}_p \times$

$\mathbb{R}_p \mapsto \mathbf{R}$ be suitable loss function. We may take the Euclidean quadratic loss $L(\mathbf{a}, \mathbf{b}) = \|\mathbf{a} - \mathbf{b}\|^2$, or $L(\mathbf{a}, \mathbf{b}) = [(\mathbf{a} - \mathbf{b})^\top(\mathbf{a} - \mathbf{b})]^{1/2}$ but they again do not reflect the dependence of components of $\mathbf{T}_n$. hence, we should rather take the Mahalanobis norm $L(\mathbf{a}, \mathbf{b}) = \|\mathbf{a} - \mathbf{b}\|_Q^2 = (\mathbf{a} - \mathbf{b})^\top\mathbf{Q}^{-1}(\mathbf{a} - \mathbf{b})$ where for a location parameter $\boldsymbol{\theta}$ we take $\mathbf{Q} = [\mathbf{I}_\theta]^{-1}$, the inverse of the Fisher information matrix on $\boldsymbol{\theta}$. Another possible choice is $L(\mathbf{a}, \mathbf{b}) = I[\|\mathbf{a} - \mathbf{b}\|_Q > c]$ for some $c > 0$, which corresponds to the "large deviation" loss function.

All previous considerations were made under a prefixed (nonrandom) sample size n. If the cost of sampling is also of importance, it should be incorporated into the loss function. The simplest case is that there is a fixed cost $c > 0$ per sample unit, and the whole cost of sampling is $c(n) = c_0 + cn, \ n \geq 1$, where c_0 is some initial administrative cost. More generally, $c(n)$ can be nonlinear, but nondecreasing in n. Taking the cost of sampling into account, we consider a more general loss function of an estimator $\mathbf{T}_n$ of $\boldsymbol{\theta}$ in the form $L^*(\mathbf{T}_n, \boldsymbol{\theta}) = L(\mathbf{T}_n, \boldsymbol{\theta}) + c(n)$ where $L(\mathbf{a}, \mathbf{b})$ is the loss considered before. The associated risk function is $R_n^*(\mathbf{T}_n, \boldsymbol{\theta}) = I\!E\{L^*(\mathbf{T}_n, \boldsymbol{\theta})\} = I\!E_F\{L(\mathbf{T}_n, \boldsymbol{\theta}) + c(n)\}$ and allowing n to be stochastic, we can write it as $R_n^*(\mathbf{T}_n, \boldsymbol{\theta}) = R_F(\mathbf{T}_n, \boldsymbol{\theta}) + I\!E_F c(n)$. Typically $R_F(\mathbf{T}_n, \boldsymbol{\theta}) = I\!E_F L(\mathbf{T}_n, \boldsymbol{\theta})$ is non-increasing in n while $c(n)$ is typically non-decreasing in n. Then we have two allied minimum risk estimation problems:

(i) Let $\mathcal{T} = \{\mathbf{T}_n\}$ be a class of estimators of $\boldsymbol{\theta}$. Then we want to find an estimator $\mathbf{T}_n^0 \in \mathcal{T}$ such that

$$R_F(\mathbf{T}_n^0, \boldsymbol{\theta}) \leq R_F(\mathbf{T}_n, \boldsymbol{\theta}) \quad \forall\, \mathbf{T} \in \mathcal{T}, \quad \text{uniformly in } F \in \mathcal{F}. \tag{8.34}$$

If such an optimal estimator exists, we call it a *uniformly minimum risk estimator* of $\boldsymbol{\theta}$ under the chosen loss $L(\cdot, \cdot)$.

(ii) For a chosen estimator $\mathbf{T}_n \in \mathcal{T}$, we want to find the sample size n_0 such that

$$R_F(\mathbf{T}_{n_0}, \boldsymbol{\theta}) + c(n_0) = \inf_n\{R_F(\mathbf{T}_n, \boldsymbol{\theta}) + c(n)\}. \tag{8.35}$$

In the parametric setup with a fixed $F \in \mathcal{F}$, (8.34) can be attained uniformly over $\boldsymbol{\theta} \in \mathbb{R}_p$. The existence and construction of the minimum risk equivariant estimator for the family of elliptically symmetric distributions has been studied by Eaton (1983). However, in the robust estimation setup F is not specified, thus, (8.34) would lead to the requirement of double uniformity over $\boldsymbol{\theta} \in \mathbb{R}_p$ and $F \in \mathcal{F}$. This can be achieved only asymptotically by an adaptive estimator, which is asymptotically uniformly most powerful under general conditions.

The minimum risk criterion in (8.35) leads to a sequential scheme, and here too this goal can be achieved asymptotically. We shall mainly deal with (8.34), with a brief discussion of (8.35).

If we take $L(\mathbf{T}_n, \boldsymbol{\theta}) = \|\mathbf{T}_n - \boldsymbol{\theta}\|_\nu^2$, then we want to minimize

$$R_F(\mathbf{T}_n, \boldsymbol{\theta}) = \mathbb{E}_F\|\mathbf{T}_n - \boldsymbol{\theta}\|_\nu^2 = \text{Trace}(\mathbb{I}(\boldsymbol{\theta})\ \boldsymbol{\nu}_{F,n}),$$

where $\boldsymbol{\nu}_{F,n} = \mathbb{E}_F(\mathbf{T}_n - \boldsymbol{\theta})(\mathbf{T}_n - \boldsymbol{\theta})^\top$ is the dispersion matrix of $\mathbf{T}_n$. However, for robust estimators with unknown $\boldsymbol{\nu}_{F,n}$ and $\mathbb{I}(\boldsymbol{\theta})$, this task is impossible. thus, we should take recourse to the asymptotics, where under appropriate regularity conditions we have the asymptotic normality

$$\sqrt{n}(\mathbf{T} - \boldsymbol{\theta}) \xrightarrow{\mathcal{D}} \mathcal{N}_p(\mathbf{0}, \boldsymbol{\nu}_F) \qquad \text{as } n \to \infty$$

and can use the *asymptotic dispersion matrix* $\boldsymbol{\nu}_F$. Then we try to minimize $\text{Trace}\big(\mathbb{I}(\boldsymbol{\theta})\boldsymbol{\nu}_F\big)$. This is the *asymptotic risk efficiency criterion*, parallel with the D-, A-, and E-efficiencies, discussed earlier. Then the asymptotic minimum risk efficiency of $\mathbf{T}_n$ is equal to

$$e_F^*(\mathbf{T}_n, \boldsymbol{\theta}) = \frac{p}{\text{Trace}(\mathbb{I}(\boldsymbol{\theta})\boldsymbol{\nu}_F)}. \tag{8.36}$$

Another concept of the *asymptotic distributional efficiency*, simplifying (8.36), will be considered in the next section.

Stein (1945) was the first person who recognized that an exact solution to (8.35) did not exist even in the simplest case of a normal mean with unknown variance. A two-stage procedure or sequential procedures have been advocated in the literature for this situation, and their efficacy properties were extensively studied. We refer to Ghosh et al. (1997) with a rich bibliography on the subject. A linear cost function $c(n) = c_0 + cn,\ c > 0$ is usually adopted with c close to 0. The dispersion matrix of an estimator can be often approximated as $\boldsymbol{\nu}_{F,n} = n^{-1}\boldsymbol{\nu}_F + o_p(n^{-1})$ for moderately large n; hence, the objective function in (8.35) can be written as $c_0 + cn + n^{-1}\text{Trace}(\mathbb{I}(\boldsymbol{\theta})\boldsymbol{\nu}_F) + o(n^{-1})$. For small $c > 0$ it leads to the solution

$$n_0(c, F) = \Big(\frac{\text{Trace}(\mathbb{I}(\boldsymbol{\theta})\boldsymbol{\nu}_F)}{c}\Big)^{1/2} + o(c^{-1/2}) = \mathrm{O}(c^{-1/2})$$

as $c \downarrow 0$ and for large n. The unknown $\boldsymbol{\nu}_F$ is replaced with a consistent estimator, say $\mathbf{V}_n$; such estimators were considered in earlier chapters and earlier in the present chapter. Then a natural estimator of $n_0(c, F)$ for small c and large n is

$$n_c^0 = c^{-1/2}\{\text{Trace}(\hat{\mathbf{I}}_{\theta,n}\mathbf{V}_n)\}^{1/2}$$

where $\hat{\mathbf{I}}_{\theta,n}$ is a consistent estimator of $\mathbf{I}_\theta$ based on the sample of size n.

To solve the contradiction between small c and large n, Stein (1945) proposed a two-stage procedure. Because it may not yet lead to the minimum risk estimator, the Stein procedure was later on extended to *multi-stage* or sequential procedures and studied by many authors. We refer to Ghosh et al. (1997) for detailed references. This procedure starts with an initial sample size n_{00} and continues by determining the stopping time $N_c \geq n_{00}$ in the form

$$N_c = \min_{n \geq n_{00}} \left\{ n : \frac{1}{c}\text{Trace}(\hat{\mathbf{I}}_{\theta,n}\mathbf{V}_n) < n^2 \right\}. \tag{8.37}$$

If $\mathbf{V}_n(\boldsymbol{\nu}_F)^{-1} \to \mathbf{I}_p$ a.s. as $n \to \infty$, then

$$\frac{N_c}{n_0(c,F)} \to 1 \; a.s. \quad \text{as} \;\; c \downarrow 0. \tag{8.38}$$

The a.s. convergence of $\mathbf{V}_n$ to $\boldsymbol{\nu}_F$ can be guaranteed if some lower-order moments of $\sqrt{n}(\mathbf{T}_n - \boldsymbol{\theta})$ converge to the corresponding moments of the asymptotic distribution. In the univariate case, we refer to Chapters 3–6 for some of these results; some further details of the moment convergence can be found in Jurečková and Sen (1982).

Let $\mathbf{T}_n$ be a (robust) estimator of $\boldsymbol{\theta}$ such that $\sqrt{n}(\mathbf{T}_n - \boldsymbol{\theta})$ has the asymptotically multinormal distribution $\mathcal{N}_p(\mathbf{0}, \boldsymbol{\nu}_F)$ and let $\mathbf{V}_n \longrightarrow \boldsymbol{\nu}_F$ a.s. as $n \to \infty$. Then

$$n^{1/2}\mathbf{V}^{-1/2}(\mathbf{T}_n - \boldsymbol{\theta}) \xrightarrow{\mathcal{D}} \mathcal{N}_p(\mathbf{0}, \mathbf{I}_p) \;\; \text{as} \;\; n \to \infty.$$

The crux of the problem is to verify that

$$N_c^{1/2}\mathbf{V}_{N_c}^{-1/2}(\mathbf{T}_{N_c} - \boldsymbol{\theta}) \xrightarrow{\mathcal{D}} \mathcal{N}_p(\mathbf{0}, \mathbf{I}_p) \;\; \text{as} \;\; c \downarrow 0. \tag{8.39}$$

To prove (8.39), we should take into account (8.37), (8.38), and that $\mathbf{V}_n\boldsymbol{\nu}_F^{-1} \to \mathbf{I}_p$. Then it suffices to show that for

$$\sup_{m: |\frac{m}{n} - 1| \leq \eta} \sqrt{n}\|\mathbf{T}_m - \mathbf{T}_n\| \xrightarrow{p} 0$$

as $n \to \infty$ and for η close to 0. This condition is known as the Anscombe (1952) condition in the literature. It can be proved under mild regularity conditions if $\mathbf{T}_n$ admits a first-order asymptotic representation

$$\mathbf{T}_n - \boldsymbol{\theta} = \frac{1}{n}\sum_{i=1}^{n} \boldsymbol{\phi}(\mathbf{X}_i) + \mathbf{R}_n \tag{8.40}$$

where $\boldsymbol{\phi}(\mathbf{X}_i)$ is centered at 0 and $\|\mathbf{R}_n\| = o_p(n^{-1/2})$. The first term on the right-hand side of (8.40) satisfies the Anscombe condition whenever $\phi^2(\mathbf{x})$ is integrable. We can verify $\sup_{m:|m-n|<\eta n} \|\mathbf{R}_n\| = o_p(n^{-1/2})$ with the aid of representations studied in Chapters 4–6.

Let us discuss the regression model $\mathbb{Y} = \boldsymbol{\beta}\mathbb{X} + \mathbb{E}$ studied in Subsection 8.4.1, where $\boldsymbol{\beta}$ is the $p \times q$ matrix of unknown regression parameters and $\mathbb{X} = [\mathbf{x}_1, \ldots, \mathbf{x}_n]$ is the $q \times n$ matrix of known regression constants such that such that $\mathbb{X}\mathbb{X}^\top$ is of full rank q ($< n$). The LSE, the L_1-estimator of $\boldsymbol{\beta}$, is given by (8.18), and its covariance matrix is $\boldsymbol{\Sigma} \otimes (\mathbb{X}\mathbb{X}^\top)^{-1}$. The coordinatewise R-, L- and M-estimators for this multivariate regression model were briefly introduced in Section 8.4, and (8.20) captures their basic asymptotic behavior. As such, (8.36) readily extends to the regression model [see Problem 8.6.4]. Compared to the location model, $(\mathbb{X}\mathbb{X}^\top)$ plays the role of n. Thus, we have for $n \to \infty$,

$$\sqrt{n}(\widetilde{\boldsymbol{\beta}} - \boldsymbol{\beta}) \xrightarrow{\mathcal{D}} \mathcal{MN}_{pq}(\mathbf{0}, \boldsymbol{\nu}_F \otimes (\mathbf{Q})^{-1}),$$

whereas for the MLE $\widehat{\boldsymbol{\beta}}$ of $\boldsymbol{\beta}$ we have

$$\sqrt{n}(\widehat{\boldsymbol{\beta}} - \boldsymbol{\beta}) \xrightarrow{\mathcal{D}} \mathcal{MN}_{pq}(\mathbf{0}, (I\theta)^{-1} \otimes (\mathbf{Q})^{-1}).$$

Noting that

$$[(I\theta)^{-1} \otimes (\mathbf{Q})^{-1}][\boldsymbol{\nu}_F \otimes (\mathbf{Q})^{-1}]^{-1} = (I\theta)^{-1}(\boldsymbol{\nu}_F)^{-1} \otimes \mathbf{I}_q,$$

we obtain that the asymptotic minimum risk agrees with (8.36).

As for the affine-equivariant robust estimators of $\boldsymbol{\beta}$, the spatial ranks are based on the transformed vectors $(\widehat{\boldsymbol{\Sigma}}_n^{(0)})^{-1/2}(\mathbf{Y}_i - \mathbf{B}\mathbf{x}_i)$ where $\mathbf{B}$ lies in a $\sqrt{n}$-neighborhood of $\boldsymbol{\beta}$. This permits an asymptotic representation as in (8.40) and which in turn leads to the asymptotic minimum risk wherein $\mathbf{I}_\theta^{-1}$ is to be replaced by $\boldsymbol{\Sigma}$, the covariance matrix of $\mathbf{Y}_i$. Since $\boldsymbol{\Sigma} - \mathbf{I}_\theta^{-1}$ is positive semidefinite, the minimum risk is greater than that of the MLE whenever $\boldsymbol{\Sigma} - \mathbf{I}_\theta^{-1}$ is positive definite. With the asymptotic representation (8.40) extended to the regression model, the Anscombe (1952) condition can be also verified.

8.7 Stein-Rule Estimators and Asymptotic Minimum Risk Efficiency

Stein (1956) showed that under $p \geq 3$ the sample mean $\bar{\mathbf{X}}_n$ is not an admissible estimator of the center $\boldsymbol{\theta}$ of the normal distribution with respect to the quadratic loss function, though it is the maximum likelihood estimator. Later on, James and Stein (1961) constructed an estimator dominating the MLE in the quadratic risk. There are in fact infinitely many such estimators, called the shrinkage or Stein-rule estimators, and they have been extensively studied during the past five decades. A predecessor of the shrinkage estimators was the preliminary test estimators constructed first in the normal model (Bancroft (1944)). We shall mainly consider the Stein-rule estimators, which have distinct advantages over the preliminary tests estimators. For the latter estimation in the multivariate normal setup, we refer to Judge and Bock (1978) and Saleh (2006), and to Sen (1986) and Saleh (1985, 1987) for the robust preliminary test estimators. The original Stein finite-sample risk dominance was later extended to the asymptotic risk dominance and also to the asymptotic distributional risk dominance (Sen 1986).

The basic difference between the asymptotic risk and the asymptotic distributional risk is that in the former the asymptotes of the exact risk are to be determined, while the latter is based on the asymptotic distributions, requiring less stringent regularity assumptions. Counterexamples show that the two measures do not coincide for robust estimators, which are typically nonlinear and only asymptotically normal. Computation of the asymptotic risk may pose challenging problems whereas the asymptotic distributional risks are much easily handled. We shall therefore confine ourselves to the asymptotic distributional risks with the understanding that it may not represent the asymptotic risk.

8.7.1 Location Model

Consider a vector valued parameter $\boldsymbol{\theta} = (\theta_1, \ldots, \theta_p)^\top$ with the θ_j being considered as functionals of a distribution function F. Let $\mathbf{T}_n = (T_{n1}, \ldots, T_{np})^\top$ be a (robust) estimator, asymptotically multinormal, so that $n^{1/2}(\mathbf{T}_n - \boldsymbol{\theta}) \xrightarrow{\mathcal{D}} \mathcal{N}_p(\mathbf{0}, \mathbf{V})$ with a positive semidefinite matrix $\mathbf{V}$. Assume that $\mathbf{V}$ is estimable; i.e. there exists a sequence $\{\mathbf{V}_n\}$ of positive semidefinite matrices such that $\mathbf{V}_n \to \mathbf{V}$ *a.s.* as $n \to \infty$. The construction of both the Stein-rule and the preliminary test estimator rests on the choice of a special value $\boldsymbol{\theta}_0$ of $\boldsymbol{\theta}$, called *pivot*, whose plausibility plays a basic role in the motivation. Choosing the pivot, we consider the hypothesis $\mathbf{H}_0 : \boldsymbol{\theta} = \boldsymbol{\theta}_0$ vs. $\mathbf{H}_1 : \boldsymbol{\theta} \neq \boldsymbol{\theta}_0$ and construct an asymptotic test statistic for $\mathbf{H}_0$ of the form

$$\mathcal{L}_n = n(\mathbf{T}_n - \boldsymbol{\theta}_0)^\top \mathbf{V}_n^{-1}(\mathbf{T}_n - \boldsymbol{\theta}_0) = n\|\mathbf{T}_n - \boldsymbol{\theta}_0\|^2_{V_n}. \tag{8.41}$$

If $\mathbf{V}$ is positive definite, then the asymptotic distribution of $\mathcal{L}_n$ under $\mathbf{H}_0$ is the central χ^2_p, while under $\mathbf{H}_1$ it is $\mathcal{L}_n$ stochastically larger; under the local (Pitman-type) alternatives, it has an asymptotically noncentral χ^2_p distribution with an appropriate noncentrality parameter. The preliminary test estimator takes on the form

$$\mathbf{T}_n^{PT} = \boldsymbol{\theta}_0 I[\mathcal{L}_n \leq \ell_{n\alpha}] + \mathbf{T}_n I[\mathcal{L}_n > \ell_{n\alpha}],$$

where α is the significance level, and the critical value $\ell_{n\alpha}$ is asymptotically equal to the $(1-\alpha)$-quantile of the χ^2_p distribution. $\mathbf{T}_n^{PT}$ is a convex combination of the pivot $\boldsymbol{\theta}_0$ and of the unrestricted estimator $\mathbf{T}_n$, where the zero-one coefficients depend on the result of the preliminary test. $\mathbf{T}_n^{PT}$ performs much better than $\mathbf{T}_n$ when $\boldsymbol{\theta}$ is actually close to the assumed pivot; for $\boldsymbol{\theta} \neq \boldsymbol{\theta}_0$, $\mathbf{T}_n^{PT}$ becomes equivalent to $\mathbf{T}_n$ in probability as $n \to \infty$. However, $\mathbf{T}_n^{PT}$ may not dominate $\mathbf{T}_n$ for all $\boldsymbol{\theta}$ with respect to the asymptotic risk. In fact the asymptotic distributional risk of $\mathbf{T}_n^{PT}$ may be even greater than that of $\mathbf{T}_n$ outside a small neighborhood of the pivot, although this excess tends to zero as $\|\boldsymbol{\theta} - \boldsymbol{\theta}_0\|$ increases. In conclusion, a preliminary test estimator is not admissible, even asymptotically.

In contrast, the Stein-rule estimator $\mathbf{T}_n^S$ generally has the desired dominance property, exact for normal or for some exponential models and asymptotical for a larger class of models. We measure the performance of $\mathbf{T}_n$ by the quadratic loss $\|\mathbf{T}_n - \boldsymbol{\theta}\|_{\mathbf{Q}} = \{(\mathbf{T}_n - \boldsymbol{\theta})^\top \mathbf{Q}^{-1}(\mathbf{T}_n - \boldsymbol{\theta})\}^{1/2}$ with a positive semidefinite $\mathbf{Q}$, and select a shrinkage factor $k : 0 < k \leq 2(p-2)$, $p \geq 3$.

Then the typical shrinkage version of $\mathbf{T}_n$ is

$$\mathbf{T}_n^S = (\mathbf{I} - kd_n \mathcal{L}_n^{-1} \mathbf{Q}^{-1} \mathbf{V}_n^{-1}) \mathbf{T}_n \tag{8.42}$$

where d_n is the smallest characteristic root of $\mathbf{Q}\mathbf{V}_n$. The role of the pivot $\boldsymbol{\theta}_0$ is hidden in the test statistic $\mathcal{L}_n$; it replaces the dichotomy of the test in (8.41) by a smoother mixture of $\boldsymbol{\theta}_0$ and $\mathbf{T}_n$. Because the matrix $\mathbf{I}_n - kd_n \mathcal{L}_n^{-1} \mathbf{Q}^{-1} \mathbf{V}_n^{-1}$ may be sometimes not positive semidefinite, alternative versions of $\mathbf{T}_n^S$ such as the *positive-rule part* can be considered.

If we look into $\mathbf{T}_n^S$ carefully, we can observe that the shrinkage factor for $\|\mathbf{T}_n^S - \boldsymbol{\theta}\|_{\mathbf{Q}}^2$ involves $\mathcal{L}_n^{-1}$ and $\mathbf{V}_n^{-1}$. Whereas in the normal case, $\mathcal{L}_n$ has noncentral χ^2 distribution and $\mathbf{V}_n$ has rescaled Wishart distribution, none of them may hold for non-normal distributions. Hence, a challenging task is an exact study of the (asymptotic) risk; we bypass this problem by using the asymptotic distributional risk (Sen 1986). The asymptotic distributional risk of $\mathbf{T}_n$ with respect to the quadratic loss $\|\cdot\|_Q^2$ is

$$ADR(\mathbf{T}_n, \boldsymbol{\theta}) = \text{Trace}(\mathbf{Q}\mathbf{V}),$$

and it remains stationary for all $\boldsymbol{\theta} \in \boldsymbol{\Theta}$ provided $\mathbf{V}$ does not depend on $\boldsymbol{\theta}$. In the multinormal case, $\boldsymbol{\Sigma}$ and $\boldsymbol{\theta}$ are orthogonal and $\boldsymbol{\Sigma}$ is not dependent on $\boldsymbol{\theta}$; but it may not hold generally. For example, if $\boldsymbol{\theta}$ stands for the cell probabilities of a multinomial distribution, $\mathbf{V} = \text{Diag}(\boldsymbol{\theta}) - \boldsymbol{\theta}\boldsymbol{\theta}^\top$ depends on $\boldsymbol{\theta}$, the treatment of asymptotic risk does not go to a conclusion, while the asymptotic distributional risk can still have a sense.

However, the behavior of the risk of $\mathbf{T}_n^S$ is quite different; let us illustrate it on the asymptotic behavior of $\mathcal{L}_n$ for $\boldsymbol{\theta}$ far away from the pivot $\boldsymbol{\theta}_0$. Then under $\boldsymbol{\theta}$ being the right parameter value, $n^{-1}\mathcal{L}_n \xrightarrow{P} \|\boldsymbol{\theta} - \boldsymbol{\theta}_0\|_{V^{-1}}^2 > 0$ as $n \to \infty$; hence, $\mathcal{L}_n^{-1} = \mathrm{O}_p(n^{-1})$ and $d_n \to \delta$ a.s., $\mathbf{V}_n^{-1} \to \mathbf{V}^{-1}$ a.s., where δ is the smallest characteristic root of $\mathbf{Q}\mathbf{V}$. thus, $kd_n \ell_n^{-1} \mathbf{Q}^{-1} \mathbf{V}_n^{-1} = \mathrm{O}_p(n^{-1})$, and $\sqrt{n}\|\mathbf{T}_n - \mathbf{T}_n^S\|_Q \xrightarrow{P} 0$ as $n \to \infty$. thus, the Stein-rule estimator $\mathbf{T}_n^S$ is asymptotically risk-equivalent to $\mathbf{T}_n$ whenever $\boldsymbol{\theta} \neq \boldsymbol{\theta}_0$. However, when we consider the local alternatives of the Pitman type (we put $\boldsymbol{\theta}_0 = \mathbf{0}$ for simplicity) $\mathbf{H}_n : \boldsymbol{\theta} = \boldsymbol{\theta}_{(n)} = n^{-1/2}\boldsymbol{\xi},\ \boldsymbol{\xi} \in \boldsymbol{R}_p$, then $\mathcal{L}_n$ under $\mathbf{H}_n$ has an asymptotically noncentral χ_p^2 distribution with the noncentrality parameter $\Delta = \boldsymbol{\xi}^\top \mathbf{V}^{-1} \boldsymbol{\xi}$. Then $n^{1/2}\mathbf{T}_n = \mathbf{Z}_n \xrightarrow{\mathcal{D}} \mathbf{Z} \simeq \mathcal{N}_p(\boldsymbol{\xi}, \mathbf{V})$. As a result,

$$\begin{aligned} \mathcal{L}_n &= \mathbf{Z}_n \mathbf{V}_n^{-1} \mathbf{Z}_n = \mathbf{Z}_n \mathbf{V}^{-1} \mathbf{Z}_n \cdot \frac{\mathbf{Z}_n \mathbf{V}_n^{-1} \mathbf{Z}_n}{\mathbf{Z}_n \mathbf{V}^{-1} \mathbf{Z}_n} \\ &= (\mathbf{Z}_n \mathbf{V}^{-1} \mathbf{Z}_n)\{1 + \mathrm{o}_p(1)\} = (\mathbf{Z}_n \mathbf{V}^{-1} \mathbf{Z}_n) + \mathrm{o}_p(1). \end{aligned} \tag{8.43}$$

hence, by (8.42), and (8.43), we have under $\mathbf{H}_n$

$$n^{1/2}(\mathbf{T}_n^S - \boldsymbol{\theta}_{(n)}) = n^{1/2}(\mathbf{T}_n - \boldsymbol{\theta}_{(n)}) - \frac{k\delta \mathbf{Q}^{-1} \mathbf{V}^{-1} (n^{1/2} \mathbf{T}_n)}{\mathbf{Z}_n \mathbf{V}^{-1} \mathbf{Z}_n} + \mathrm{o}_p(1)$$

$$\xrightarrow{\mathcal{D}} \mathbf{Z} - \boldsymbol{\xi} - (\mathbf{Z}^\top \mathbf{V}^{-1}\mathbf{Z})^{-1}\mathbf{Q}^{-1}\mathbf{V}^{-1}\mathbf{Z}. \tag{8.44}$$

thus, the asymptotic risk efficiency of $\mathbf{T}_n^S$ under H_n is

$$\begin{aligned} ADR(\mathbf{T}_n^S, \boldsymbol{\theta}_{(n)}) &= \mathbb{E}\{(\mathbf{Z} - \boldsymbol{\xi})^\top \mathbf{Q}(\mathbf{Z} - \boldsymbol{\xi})\} \\ &- 2k\delta \mathbb{E}\{(\mathbf{Z} - \boldsymbol{\xi})^\top \mathbf{Q}\mathbf{Q}^{-1}\mathbf{V}^{-1}\mathbf{Z}(\mathbf{Z}\mathbf{V}^{-1}\mathbf{Z})^{-1}\} \\ &+ k^2\delta^2 \mathbb{E}\{(\mathbf{Z}^\top \mathbf{V}^{-1}\mathbf{Q}^{-1}\mathbf{Q}\mathbf{Q}^{-1}\mathbf{V}^{-1}\mathbf{Z}(\mathbf{Z}\mathbf{V}^{-1}\mathbf{Z})^{-2}\}. \end{aligned} \tag{8.45}$$

The first term on the right-hand side of (8.45) is equal to Trace($\mathbf{QV}$); the other terms may be treated with the aid of the Stein identity.

LEMMA 8.2 *(Stein identity). Let* $\mathbf{W} \sim \mathcal{N}_p(\boldsymbol{\theta}, \mathbf{I})$ *and* $\phi : \boldsymbol{R}^+ \mapsto \boldsymbol{R}^+$ *be a real function. Then*

$$\mathbb{E}\{\phi(\|\mathbf{W}\|^2)\mathbf{W}\} = \boldsymbol{\theta}\mathbb{E}\{\phi(\chi^2_{p+2}(\Delta))\}, \quad \Delta = \boldsymbol{\theta}^\top\boldsymbol{\theta} = \|\boldsymbol{\theta}\|^2 \tag{8.46}$$

and

$$\begin{aligned} &\mathbb{E}\{\phi(\|\mathbf{W}\|^2)\mathbf{W}^\top \mathbf{A}\mathbf{W}\} \\ &= \mathrm{Trace}(\mathbf{A})\mathrm{E}\{\phi(\chi^2_{\mathrm{p}+2}(\Delta))\} + \boldsymbol{\theta}^\top \mathbf{A}\boldsymbol{\theta}\mathrm{E}\{\phi(\chi^2_{\mathrm{p}+4}(\Delta))\} \end{aligned} \tag{8.47}$$

for any positive definite symmetric matrix $\mathbf{A}$, *where* $\chi^2_q(\delta)$ *stands for a random variable having the noncentral chi-square distribution with q degrees of freedom and with noncentrality parameter* δ.

PROOF Note that

$$\mathbb{E}\{\phi(\mathbf{W}^\top\mathbf{W})\} = \int_{\boldsymbol{R}_p}\int \phi(\|\mathbf{w}\|^2)(2\pi)^{-p/2}\exp\{-\tfrac{1}{2}\|\mathbf{w} - \boldsymbol{\theta}\|^2\}d\mathbf{w}$$

$$= e^{-\Delta/2}\sum_{r\geq 0}\frac{1}{r!}\Big(\frac{\Delta}{2}\Big)^r 2^{-(\frac{p}{2}+r)}\Big(\Gamma(\tfrac{p}{2}+r)\Big)^{-1}\int_0^\infty \phi(y)e^{-y/2}y^{p/(2+r-1)}dy.$$

Differentiating with respect to $\boldsymbol{\theta}$, we have

$$\int_{\boldsymbol{R}_p}\int \phi(\|\mathbf{w}\|^2)(\mathbf{w} - \boldsymbol{\theta})(2\pi)^{-p/2}\exp\Big\{-\tfrac{1}{2}\|\mathbf{w} - \boldsymbol{\theta}\|^2\Big\}d\mathbf{w}$$

$$= -\boldsymbol{\theta}\mathbb{E}\{\phi(\mathbf{W}^\top\mathbf{W})\} + \boldsymbol{\theta}e^{-\Delta/2}\sum_{r\geq 0}\frac{1}{r!}\Big(\frac{\Delta}{2}\Big)^r 2^{-(\frac{p}{2}+r+1)}\,.$$

$$\cdot\Big(\Gamma(\tfrac{p}{2}+r+1)\Big)^{-1}\int_0^\infty \phi(y)e^{-y/2}y^{\frac{p}{2}+r}dy, \tag{8.48}$$

which yields

$$\begin{aligned} &\mathbb{E}\{\phi(\|\mathbf{W}\|^2)\mathbf{W}\} - \boldsymbol{\theta}\mathbb{E}\{\phi(\|\mathbf{W}\|^2)\} \\ &= -\boldsymbol{\theta}\mathbb{E}\{\phi(\|\mathbf{W}\|^2)\} + \boldsymbol{\theta}\mathbb{E}\{\phi(\chi^2_{p+2}(\Delta))\} \end{aligned}$$

and this directly leads to (8.46). Moreover, using the identity

$$\mathbb{E}\{\phi(\|\mathbf{W}\|^2)\mathbf{W}^\top\mathbf{A}\mathbf{W}\} = \mathrm{Trace}(\mathbf{A}\mathrm{E}\{\phi(\|\mathbf{W}\|^2)\mathbf{W}\mathbf{W}^\top\})$$

and differentiating (8.48) with respect to $\boldsymbol{\theta}$, we obtain (8.47) along parallel lines. □

The following theorem compares the asymptotic distributional risk of the Stein rule estimator with that of the ordinary estimator, whenever $\mathbf{T}_n$ and $\mathcal{L}_n$ have asymptotic distributions for which the Stein identity holds.

THEOREM 8.2 *Under the hypothesis* $\mathbf{H}_n : \boldsymbol{\theta} = \boldsymbol{\theta}_{(n)}$,

$$ADR(\mathbf{T}_n^S, \boldsymbol{\theta}_{(n)}) \leq ADR(\mathbf{T}_n, \boldsymbol{\theta}_{(n)})$$

for any value of the shrinkage factor k *satisfying* $0 \leq k \leq 2(p-2)$. *The equality holds when either* $n\|\boldsymbol{\theta}_{(n)}\|^2 \to \infty$ *as* $n \to \infty$ *or* $\|\boldsymbol{\xi}\| \to \infty$. *hence,* $\mathbf{T}_n$ *is asymptotically dominated by the Stein-rule estimator* $\mathbf{T}_n^S$.

PROOF We make use of (8.44) and of (8.45) and use Lemma 8.1 on $\mathbf{Z}$. Recall that

$$\mathbb{E}\{\chi_q^{-2k}(\Delta)\} = e^{-\Delta/2} \sum_{r \geq 0} \frac{1}{r!}\left(\frac{\delta}{2}\right)^r \mathbb{E}\{\chi_{q+2r}^{-2k}(0)\} = \mathbb{E}_1\Big\{\mathbb{E}\Big[\chi_{q+2\mathcal{K}}^{-2k}(0)\Big|\mathcal{K}\Big]\Big\},$$

for every $k > 0$, $q \geq 2k+1$, $\Delta = \boldsymbol{\xi}^\top \mathbf{V}^{-1}\boldsymbol{\xi}$, where $\mathcal{K}$ is a Poisson random variable with mean $\Delta/2$ and $\mathbb{E}_1$ is the expectation with respect to $\mathcal{K}$. This further implies that $\mathbb{E}\{\chi_q^{-2r}(\Delta) = \mathbb{E}\{[(q-2+2\mathcal{K})\ldots(q-2r+2\mathcal{K})]^{-1}\}$, for every $q > 2r$ and $r \geq 1$, if we recall that $\mathbb{E}\{\chi_q^{-2r}(0)\} = \{(q-2)\ldots(q-2r)\}^{-1}$. Moreover, $\forall\ r \geq 1$

$$\begin{aligned}
&\mathbb{E}\{\chi_q^{-2r}(\Delta) - \chi_{q+2}^{-2r}(\Delta)\} \\
&= \mathbb{E}\Big\{\frac{1}{(q-2+2\mathcal{K})\ldots(q-2r+2\mathcal{K})} - \frac{1}{(q+2\mathcal{K})\ldots(q-2r+2+2\mathcal{K})}\Big\} \\
&= 2r\mathbb{E}\{[(q+2\mathcal{K})\ldots(q-2r+2\mathcal{K})]^{-1}\} = 2r\mathbb{E}\Big(\chi_{q+2}^{-2r-2}(\Delta)\Big)
\end{aligned}$$

Similarly, for $p > 2$ we have an identity

$$\begin{aligned}
\Delta\mathbb{E}\Big(\chi_{p+2}^{-2}(\Delta)\Big) &= e^{-\Delta/2}\sum_{r\geq 0}\frac{1}{r!}\left(\frac{\Delta}{2}\right)^r \frac{\Delta}{p+2r} \\
&= 2e^{-\Delta/2}\sum_{r\geq 0}\frac{1}{(r+1)!}\left(\frac{\Delta}{2}\right)^{r+1}\frac{\Delta}{p+2r} \\
&= (p-2)\Big\{e^{-\Delta/2}\sum_{r\geq 0}\frac{1}{r!}\left(\frac{\Delta}{2}\right)^r\Big[\frac{1}{p-2} - \frac{1}{p+2r-2}\Big]\Big\} \\
&= 1-(p-2)\mathbb{E}(\chi_p^{-2}(\Delta)).
\end{aligned}$$

Moreover, letting $\Delta^\star = \boldsymbol{\xi}^\top\mathbf{V}^{-1}\mathbf{Q}^{-1}\mathbf{V}^{-1}\boldsymbol{\xi}$, we have for $\boldsymbol{\xi} \in \boldsymbol{R}_p$

$$\frac{\Delta^\star}{\Delta} = \frac{(\boldsymbol{\xi}^\top\mathbf{V}^{-1}\mathbf{Q}^{-1}\mathbf{V}^{-1}\boldsymbol{\xi})}{(\boldsymbol{\xi}^\top\mathbf{V}^{-1}\boldsymbol{\xi})} \leq \lambda_1(\mathbf{Q}^{-1}\mathbf{V}^{-1}) = \frac{1}{\lambda_p(\mathbf{Q},\mathbf{V})} = \frac{1}{\delta}$$

where $\lambda_1(\mathbf{A})$ and $\lambda_p(\mathbf{A})$ are the largest and the smallest characteristic roots of matrix $\mathbf{A}$. Using Lemma 8.2, we can modify (8.45) to

$$\begin{aligned}
ADR(\mathbf{T}_n^S, \mathbf{Q}_n) &= \text{Trace}(\mathbf{QV}) - 2\text{k}\delta + 2\text{k}\delta(\boldsymbol{\xi}^\top \mathbf{V}^{-1}\boldsymbol{\xi})\text{E}(\chi_{\text{p}+2}^{-2}(\Delta)) \\
&+k^2\delta^2\Delta(\mathbf{Q}^{-1}\mathbf{V}^{-1})\mathbb{E}(\chi_{p+2}^{-4}(\Delta)) + (\boldsymbol{\xi}^\top \mathbf{V}^{-1}\mathbf{Q}^{-1}\mathbf{V}^{-1}\boldsymbol{\xi})\mathbb{E}(\chi_{p+2}^{-4}(\Delta)) \\
&= \text{Trace}(\mathbf{QV}) - 2k\delta(p-2)\mathbb{E}(\chi_p^{-2}(\Delta)) + k^2\delta^2\Delta(\mathbf{Q}^{-1}\mathbf{V}^{-1})\mathbb{E}(\chi_{p+2}^{-4}(\Delta)) \\
&+k^2\delta^2\Delta^\star \mathbb{E}(\chi_{p+4}^{-4}(\Delta)),
\end{aligned} \tag{8.49}$$

where the last term on the right-hand side of (8.49) is bounded from above by

$$k^2\delta\Delta\mathbb{E}(\chi_{p+4}^{-4}(\Delta)) = k^2\delta\{\mathbb{E}(\chi_{p+2}^{-2}(\Delta)) - (p-2)\mathbb{E}(\chi_{p+2}^{-4}(\Delta))\}.$$

Hence, $ADR(\mathbf{T}_n^S, \boldsymbol{\theta}_{(n)})$ is bounded from above by

$$\begin{aligned}
&\text{Trace}(\mathbf{QV}) - k\delta[2(p-2) - k]\mathbb{E}(\chi_p^{-2}(\Delta))] \\
&-\mathbb{E}(\chi_{p+2}^{-4}(\Delta))\{k^2\delta[p - \delta\text{Trace}(\mathbf{Q}^{-1}\mathbf{V}^{-1})]\},
\end{aligned} \tag{8.50}$$

where

$$\begin{aligned}
\delta\text{Trace}(\mathbf{Q}^{-1}\mathbf{V}^{-1}) &= \delta\sum_{j=1}^{p}\lambda_j(\mathbf{Q}^{-1}\mathbf{V}^{-1}) \\
&= \frac{\sum_{j=1}^{p}\lambda_j(\mathbf{Q}^{-1}\mathbf{V}^{-1})}{\lambda_{\max}(\mathbf{Q}^{-1}\mathbf{V}^{-1})} \leq p.
\end{aligned}$$

Consequently, $ADR(\mathbf{T}_n^S, \boldsymbol{\theta}_{(n)})$ is dominated by $\text{Trace}(\mathbf{QV})$ whenever $0 \leq k \leq 2(p-2)$. □

If $\mathbf{V} = \mathbf{Q}^{-1}$, i.e.when $\text{Trace}(\mathbf{Q}^{-1}\mathbf{V}^{-1}) = p$ and $\delta = 1$, the reduction in (8.50) is maximal for the choice of $k = p - 2$. If generally $\mathbf{V} \neq \mathbf{Q}^{-1}$, the reduction depends on δ, on $\text{Trace}(\mathbf{Q}^{-1}\mathbf{V}^{-1})$, and on Δ. Theorem 8.2 relates to the simple case of a multivariate location model or of a general parameter model based on *i.i.d.* random variables. In the next section, we shall parallelly consider a linear regression model, originally studied by Sen and Saleh (1987).

8.7.2 Extension to the Linear Model

Consider the usual linear model

$$\begin{aligned}
\mathbf{Y}_n &= (Y_1, \ldots, Y_n)^\top = \mathbf{X}_n\boldsymbol{\beta} + \mathbf{e}_n, \\
\mathbf{e}_n &= (e_1, \ldots, e_n)^\top \quad \boldsymbol{\beta} = (\beta_1, \ldots, \beta_p)^\top,
\end{aligned} \tag{8.51}$$

$\mathbf{X}_n$ is an $n \times p$ design matrix with known elements, $p \geq 1$, $n > p$, and $\boldsymbol{\beta} \in \mathbb{R}_p$ is an unknown parameter. The errors e_i are *i.i.d.* random variables with a continuous distribution function F defined on $\boldsymbol{R}_1$. Assume that $\text{Rank}(\mathbf{X}_n) =$

p, without loss of generality. Partition the model in the following way:

$$\boldsymbol{\beta} = \begin{pmatrix} \boldsymbol{\beta}_1 \\ \boldsymbol{\beta}_2 \end{pmatrix} \quad \text{and} \quad \mathbf{X}_n = (\mathbf{X}_{n1}\ ,\ \mathbf{X}_{n2}), \tag{8.52}$$

with $\boldsymbol{\beta}_1, \boldsymbol{\beta}_2, \mathbf{X}_{n1}$, and $\mathbf{X}_{n2}$ of order $p_1 \times 1,\ p_2 \times 1,\ n \times p_1$, and $n \times p_2$, respectively, $p = p_1 + p_2,\ p_1 \geq 0,\ p_2 \geq 0$. Rewrite (8.51) as

$$\mathbf{Y}_n = \mathbf{X}_{n1}\boldsymbol{\beta}_1 + \mathbf{X}_{n2}\boldsymbol{\beta}_2 + \mathbf{e}_n.$$

We are interested in estimating $\boldsymbol{\beta}_1$ while we expect that $\boldsymbol{\beta}_2$ is close to some specified pivot $\boldsymbol{\beta}_2^0$ (which we set $\boldsymbol{\beta}_2^0 = \mathbf{0}$, without loss of generality). For example, in the case of a multifactor design, we may be interested in estimating the vector of main effects $\boldsymbol{\beta}_1$, while there is a question whether the vector of interaction effects $\boldsymbol{\beta}_2$ may be ignored.

Assume that as $n \to \infty$,

$$\begin{aligned} &\mathbf{Q}_n = n^{-1}\mathbf{X}_n^\top \mathbf{X}_n, \to \mathbf{Q} \ \text{ (positive definite) and} \\ &n^{-1/2}\mathbf{D}_n \to \mathbf{D} \ \text{ (positive definite) as } \ n \to \infty \end{aligned} \tag{8.53}$$

where matrix $\mathbf{D}_n$ is defined by

$$\mathbf{D}_n(\mathbf{X}_n^\top \mathbf{X}_n)^{-1}\mathbf{D}_n^\top = \mathbf{I}_p.$$

Let $\mathbf{T}_n$ be an estimator of the whole $\boldsymbol{\beta}$ satisfying

$$\mathbf{D}_n(\mathbf{T}_n - \boldsymbol{\beta}) \xrightarrow{\mathcal{D}} \mathcal{N}_p(\mathbf{0}, \mathbf{V}) \text{ as } \ n \to \infty, \tag{8.54}$$

and suppose that there exists a sequence $\{\mathbf{V}_n\}$ of positive semidefinite matrices such that $\mathbf{V}_n \to \mathbf{V}$ a.s. as $n \to \infty$. The partition (8.52) induces the following decomposition of the matrix $\mathbf{Q}_n$:

$$\mathbf{Q}_n = \begin{pmatrix} \mathbf{Q}_{n11} & \mathbf{Q}_{n12} \\ \mathbf{Q}_{n21} & \mathbf{Q}_{n22} \end{pmatrix}, \ \mathbf{Q}_{nij} \to \mathbf{Q}_{ij} \ \text{ as } \ n \to \infty,\ i, j = 1, 2.$$

Finally, we assume that

$$n^{-1}\sum_{i=1}^{n} \mathbf{x}_i^\top \mathbf{x} = n^{-1}\sum_{i=1}^{n} \|\mathbf{x}_i\|^2 = \mathrm{O}(1) \ \text{ as } n \to \infty, \tag{8.55}$$

where $\mathbf{x}_i$ is the ith row of $\mathbf{X}_n$, $i = 1, \ldots n$. Note that by (8.53) and (8.55),

$$\max_{1 \leq i \leq n} \{\mathbf{x}_i^\top (\mathbf{X}_n^\top \mathbf{X}_n)^{-1}\mathbf{x}_i\}^{1/2} = \mathrm{O}(n^{-1/2}) \text{ as } n \to \infty.$$

Again, consider the hypothesis

$$\mathbf{H}_0 : \boldsymbol{\beta}_2 = \mathbf{0} \text{ vs. } \mathbf{H}_1 : \boldsymbol{\beta}_2 \neq \mathbf{0},\ (\boldsymbol{\beta}_1 \text{ nuisance}).$$

The test statistic $\mathcal{L}_n$ for $\mathbf{H}_0$ can be constructed from the aligned test statistics or from the derived estimators. For example, we may partition $\mathbf{T}_n$ satisfying (8.54) as

$$\mathbf{T}_n^\top = (\mathbf{T}_{n1}^\top\ ,\ \mathbf{T}_{n2}^\top).$$

Then, if we put $\mathbf{D}_n^{-1}\mathbf{V}_n(\mathbf{D}_n^\top)^{-1} = \mathbf{J}_n = \begin{pmatrix} \mathbf{J}_{n11} & \mathbf{J}_{n12} \\ \mathbf{J}_{n21} & \mathbf{J}_{n22} \end{pmatrix}$, we may consider the test statistic

$$\mathcal{L}_n = \mathbf{T}_{n2}^\top \mathbf{J}_{n22}^{-1}\mathbf{T}_{n2}.$$

Alternatively, we can use Wald-type test statistic based on M- or R-estimators of $\boldsymbol{\beta}_1$. Let us illustrate this approach on the M-tests; the case of R-tests is quite analogous. For every $\mathbf{b} \in \boldsymbol{R}_p$, define $\mathbf{M}_n(\mathbf{b}) = \sum_{i=1}^n \mathbf{x}_i\psi(Y_i - \mathbf{x}_i^\top\mathbf{b})$. Then the unrestricted M-estimator $\widetilde{\boldsymbol{\beta}}_n = (\widetilde{\boldsymbol{\beta}}_{n1}^\top, \widetilde{\boldsymbol{\beta}}_{n2}^\top)^\top$ of $\boldsymbol{\beta}$ is a solution of the system of p equations $\mathbf{M}_n(\mathbf{b}) := \mathbf{0}$ with respect to $\mathbf{b} \in \boldsymbol{R}_p$. Partitioning $\mathbf{b}^\top$ further as $(\mathbf{b}_1^\top, \mathbf{b}_2^\top)$ and $\mathbf{M}_n^\top(.)$ as $(\mathbf{M}_{n1}^\top(.)\ \mathbf{M}_{n2}^\top(.))$, we get the restricted M-estimator $\widehat{\boldsymbol{\beta}}_{n1}$ of $\boldsymbol{\beta}_1$ as a solution of the system of p_1 equations

$$\mathbf{M}_{n1}((\mathbf{b}_1^\top, \mathbf{0}^\top)^\top) = \mathbf{0} \text{ with respect to } \mathbf{b} \in \boldsymbol{R}_{p_1}.$$

Then put

$$S_n^2 = (n-p_1)^{-1}\sum_{i=1}^n \psi^2(T_i - \mathbf{x}_{i1}^\top\widehat{\boldsymbol{\beta}}_{n1}),$$

where $\mathbf{x}_i^\top = (\mathbf{x}_{i1}^\top\ \mathbf{x}_{i2}^\top)$, $i = 1, \ldots, n$. Moreover, let

$$\mathbf{Q}_{nkk:\ell} = \mathbf{Q}_{nkk} - \mathbf{Q}_{nk\ell}\mathbf{Q}_{n\ell\ell}^{-1}\mathbf{Q}_{n\ell k},\ k \neq \ell = 1, 2.$$

Now an appropriate aligned M-test statistic is

$$\mathcal{L}_n = \frac{n}{S_n^2}\{\widehat{\mathbf{M}}_{n2}^\top \mathbf{Q}_{n22.1}\widehat{\mathbf{M}}_{n2}\}, \tag{8.56}$$

where

$$\widehat{\mathbf{M}}_{n2} = \mathbf{M}_{n2}(\widehat{\boldsymbol{\beta}}_{n1}, \mathbf{0}).$$

Based on the asymptotic representations in Chapter 5 (Theorem 5.10 and Lemma 5.3), we conclude that under $\mathbf{H}_0 :\ \boldsymbol{\beta}_2 = \mathbf{0}$, $\mathcal{L}_n$ has asymptotically the central $\chi_{p_2}^2$ distribution. hence, we can formulate the preliminary test estimator of $\boldsymbol{\beta}_1$ as follows:

$$\begin{aligned} \widehat{\boldsymbol{\beta}}_{n1}^{PT} &= \begin{cases} \widehat{\boldsymbol{\beta}}_{n1} & \text{if } \mathcal{L}_n < \chi_{p_2,\alpha}^2, \\ \widetilde{\boldsymbol{\beta}}_{n1} & \text{if } \mathcal{L}_n \geq \chi_{p_2,\alpha}^2, \end{cases} \\ &= \widehat{\boldsymbol{\beta}}_{n1} I[\mathcal{L}_n < \chi_{p_2,\alpha}^2] + \widetilde{\boldsymbol{\beta}}_{n1} I[\mathcal{L}_n \geq \chi_{p_2,\alpha}^2]. \end{aligned} \tag{8.57}$$

On the other hand, the Stein-rule estimator $\widehat{\boldsymbol{\beta}}_{n1}^S$ of $\boldsymbol{\beta}_1$ is defined as

$$\widehat{\boldsymbol{\beta}}_{n1}^S = \widehat{\boldsymbol{\beta}}_{n1} + (\mathbf{I}_{p_1} - kd_n\mathcal{L}_n^{-1}\mathbf{W}^{-1}\mathbf{Q}_{n11.2})(\widetilde{\boldsymbol{\beta}}_{n1} - \widehat{\boldsymbol{\beta}}_{n1}),$$

where $d_n =$ smallest characteristic root of $\mathbf{W}\mathbf{Q}_{n11.2}^{-1}$ and $k > 0$ is a shrinkage factor; the other notations have been introduced before.

The natural choice is $\mathbf{W} = \mathbf{Q}_{n11.2}$ (or $\mathbf{Q}_{11.2}$). In this case, the Stein-rule estimator (8.57) reduces to

$$\widehat{\boldsymbol{\beta}}_{n1}^S = \widehat{\boldsymbol{\beta}}n1 + (1 - k\mathcal{L}_n^{-1})(\widetilde{\boldsymbol{\beta}}_{n1} - \widehat{\boldsymbol{\beta}}_{n1}),$$

and this in turn suggests the *positive-rule version*:

$$\widehat{\boldsymbol{\beta}}_{n1}^{S+} = \widehat{\boldsymbol{\beta}}_{n1} + (1 - k\mathcal{L}_n^{-1})^+(\widetilde{\boldsymbol{\beta}}_{n1} - \widehat{\boldsymbol{\beta}}_{n1}),$$

where $a^+ = \max\{0, a\}$, $a \in \boldsymbol{R}$.

We have seen in the multivariate location case that the preliminary test estimator and the Stein-rule estimator become asymptotically isomorphic to the classical version $\mathbf{T}_n$ under a fixed alternative to $\mathbf{H}_0$, as $n \to \infty$. This is also the case in the linear regression model: If $\boldsymbol{\beta}_2 \neq \mathbf{0}$, then $\widehat{\boldsymbol{\beta}}_{n1}^{PT}$ and $\widehat{\boldsymbol{\beta}}_{n1}^{S}$ (or $\widehat{\boldsymbol{\beta}}_{n1}^{S+}$) become asymptotically equivalent to the unrestricted estimator $\widehat{\boldsymbol{\beta}}_{n1}$. The differences appear rather in a small neighborhood of the pivot $\boldsymbol{\beta}_2 = \mathbf{0}$.

Hence, let us consider a sequence $\{\mathbf{H}_n^\star\}$ of local alternatives to the pivot,

$$\mathbf{H}_n^\star : \ \boldsymbol{\beta}_2 = \boldsymbol{\beta}_{2(n)} = n^{-1/2}\boldsymbol{\xi}, \quad \boldsymbol{\xi} \in \boldsymbol{R}_{p_2}$$

and the behavior of the estimators under $\mathbf{H}_n^*$. The null hypothesis means $\mathbf{H}_0 : \ \boldsymbol{\xi} = \mathbf{0}$. For an estimator $\widehat{\boldsymbol{\beta}}_{n1}^\star$ of $\boldsymbol{\beta}_1$, denote

$$G^\star(x) = \lim_{n\to\infty} \mathbb{P}\{n^{1/2}(\widehat{\boldsymbol{\beta}}_{n1}^\star - \boldsymbol{\beta}_1) \le \mathbf{x}|\mathbf{H}_n^\star\}, \ \mathbf{x} \in \boldsymbol{R}_{p_1},$$

its limiting distribution function and assume that it is nondegenerate. Then the asymptotic distributional risk of $\widehat{\boldsymbol{\beta}}_{n1}^\star$ corresponding to the quadratic loss function $n\|\widehat{\boldsymbol{\beta}}_{n1}^\star - \boldsymbol{\beta}_1\|_W^2$ (with a given positive definite matrix $\mathbf{W}$) is defined as

$$R(\widehat{\boldsymbol{\beta}}_1^\star; \mathbf{W}) = \text{Trace}(\mathbf{W})\int \dots \int_{\boldsymbol{R}_{p_1}} \mathbf{x}\mathbf{x}^\top dG^\star(\mathbf{x}) = \text{Trace}(\mathbf{W}\mathbf{V}^\star), \tag{8.58}$$

where $\mathbf{V}^\star$ is the dispersion matrix of the distribution function $G^\star$. Denote

$$\begin{aligned}
\sigma_\psi^2 &= \int_{\boldsymbol{R}} \psi^2(y)dF(y) = \mathbb{E}_F(\psi^2(e_i)), \\
\gamma = \gamma(\psi, F) &= \int_{\boldsymbol{R}} \psi(x)\{-f'(x)/f(x)\}dF(x), \\
\nu^2 &= \sigma_\psi^2/\gamma^2.
\end{aligned}$$

As in Chapter 5, we impose some conditions on the density f and on $\psi(\cdot)$. We assume that $\gamma(\psi, F) > 0$. The asymptotic distribution of the test criterion under $\mathbf{H}_n^*$ and of estimators $\widehat{\boldsymbol{\beta}}_{n1}$, $\widehat{\boldsymbol{\beta}}_{n1}^{PT}$, and $\widehat{\boldsymbol{\beta}}_{n1}^{S}$ of $\boldsymbol{\beta}_1$ is described in the following theorem. It can be proved along the lines of Chapter 5 (Theorem 5.10 and Lemma 5.3). Here $\Phi(\mathbf{x}, \boldsymbol{\mu}, \boldsymbol{\Sigma})$ denotes the distribution function of the multivariate normal distribution $\mathcal{N}_p(\boldsymbol{\mu}, \boldsymbol{\Sigma})$, and $H_p(x, \Delta)$ denotes the distribution function of the noncentral chi-square distribution with p degrees of freedom and with the noncentrality parameter Δ, $x \in \boldsymbol{R}^+$.

THEOREM 8.3 *(i) Let $\widetilde{\boldsymbol{\beta}}_{n1}$, $\widehat{\boldsymbol{\beta}}_{n1}$ and $\widehat{\boldsymbol{\beta}}_{n1}^{PT}$ be the versions of the M-estimators of $\boldsymbol{\beta}_1$ as described above, and let $\mathcal{L}_n$ be the test statistic defined in*

(8.56). Then, under $\mathbf{H}_n^\star$,

$$\lim_{n\to\infty} \mathbb{P}\{\mathcal{L}_n \le x|\mathbf{H}_n^\star\} = H_{p_2}(x,\Delta), \quad x \ge 0, \quad \Delta = \nu^{-2}(\boldsymbol{\xi}^\top \mathbf{Q}_{22.1}\boldsymbol{\xi}),$$

$$\lim_{n\to\infty} \mathbb{P}\{n^{1/2}(\widehat{\boldsymbol{\beta}}_{n1} - \boldsymbol{\beta}_1) \le x|\mathbf{H}_n^\star\} = \Phi_{p_1}(\mathbf{x} + \mathbf{Q}_{11}^{-1}\mathbf{Q}_{12}\boldsymbol{\xi}; \mathbf{0}, \nu^2\mathbf{Q}_{11}^{-1}),$$

$$\begin{aligned}
&\lim_{n\to\infty} \mathbb{P}\{n^{1/2}(\widehat{\boldsymbol{\beta}}_{n1}^{PT} - \boldsymbol{\beta}_1) \le x|\mathbf{H}_n^\star\} \\
&= H_{p_2}(\chi^2_{p_2}, \alpha; \Delta)\Phi_{p_2}(\mathbf{x} + \mathbf{Q}_{11}^{-1}\mathbf{Q}_{12}\boldsymbol{\xi}; \mathbf{0}, \nu^2\mathbf{Q}_{11}^{-1}) \\
&+ \int_{\boldsymbol{E}_\xi} \Phi_{p_1}(\mathbf{x} - \mathbf{B}_{12}\mathbf{B}_{22}^{-2}\mathbf{Z}; \mathbf{0}, \nu^2\mathbf{B}_{11.2}) d\Phi_{p_2}(\mathbf{z}; \mathbf{0}, \nu^2\mathbf{B}_{22}),
\end{aligned}$$

where $\mathbf{B} = \mathbf{Q}^{-1}$ *(so that* $\mathbf{DBD}^\top = \mathbf{I}$*), the* $\mathbf{B}_{ij}$ *and* $\mathbf{B}_{ii.j}$ *correspond to the partition, and*

$$\boldsymbol{E}_\xi = \{\mathbf{z} \in \boldsymbol{R}_{p_2} : (\mathbf{z} + \boldsymbol{\xi})^\top \mathbf{Q}_{22.1}(\mathbf{z} + \boldsymbol{\xi}) \ge \nu^2 \chi^2_{p,\alpha}\}. \tag{8.59}$$

(ii) Let $\widehat{\boldsymbol{\beta}}_{n1}^S$ *be the Stein-rule M-estimator of* $\boldsymbol{\beta}_1$*. Then, under* $\mathbf{H}_n^\star$,

$$n^{1/2}(\widehat{\boldsymbol{\beta}}_{n1}^S - \boldsymbol{\beta}_1) \xrightarrow{\mathcal{D}} \mathbf{B}_1\mathbf{U} + \frac{\nu^2 k \mathbf{Q}_{11}^{-1}\mathbf{Q}_{12}(\mathbf{B}_2\mathbf{U} + \boldsymbol{\xi})}{(\mathbf{B}_2\mathbf{U} + \boldsymbol{\xi})^\top \mathbf{Q}_{22.1}(\mathbf{B}_2\mathbf{U} + \boldsymbol{\xi})},$$

where k *is the shrinkage factor ,* $\mathbf{U}$ *is a random vector with the* $\mathcal{N}_p(\mathbf{0}, \nu^2\mathbf{Q})$ *distribution, and* $\mathbf{B} = (\mathbf{B}_1^\top, \mathbf{B}_2^\top)^\top$ *with* $\mathbf{B}_1$ *and* $\mathbf{B}_2$ *of order* $p_1 \times p$ *and* $p_2 \times p$*, respectively.*

The asymptotic distributional risks of the above estimators are given in the following theorem. It follows from (8.58), from the Stein identity (Lemma 8.2), and from Theorem 8.3.

THEOREM 8.4 *Under* $\{\mathbf{H}_n^\star\}$ *and under the assumed regularity conditions,*

$$R(\widetilde{\boldsymbol{\beta}}_1; \mathbf{W}) = \nu^2 \text{ Trace}(\mathbf{W}\mathbf{Q}_{11.2}^{-1}), \tag{8.60}$$

$$R(\widehat{\boldsymbol{\beta}}_1; \mathbf{W}) = \nu^2 \text{ Trace}(\mathbf{W}\mathbf{Q}_{11}^{-1}) + \boldsymbol{\xi}^\top \mathbf{M}\boldsymbol{\xi},$$

where $\mathbf{M} = \mathbf{Q}_{21}\mathbf{Q}_{11}^{-1}\mathbf{W}\mathbf{Q}_{11}^{-1}\mathbf{Q}_{12}$;

$$\begin{aligned}
R(\widehat{\boldsymbol{\beta}}_1^{PT}; \mathbf{W}) &= \nu^2 \{\text{Trace}(\mathbf{W}\mathbf{Q}_{11.2}^{-1})[1 - H_{p_2+2}(\chi^2_{p_2,\alpha}; \Delta)] \\
&+ \boldsymbol{\xi}^\top \mathbf{M}\boldsymbol{\xi}[2H_{p_2+2}(\chi^2_{p_2,\alpha}; \Delta) - H_{p_2+4}(\chi^2_{p_2,\alpha}; \Delta)]\}
\end{aligned} \tag{8.61}$$

and

$$\begin{aligned}
R(\widehat{\boldsymbol{\beta}}_1^S; \mathbf{W}) &= \nu^2 \{\text{Trace}(\mathbf{W}\mathbf{Q}_{11.2}^{-1} - k\text{Trace}(\mathbf{M}\mathbf{Q}_{22.1}^{-1})[2\mathbb{E}(\chi^{-2}_{p_2+2}(\Delta)) \\
&- k\mathbb{E}(\chi^{-4}_{p_2+2}(\Delta))] + k(k+4)(\boldsymbol{\xi}^\top \mathbf{M}\boldsymbol{\xi})\mathbb{E}(\chi^{-4}_{p_2+4}(\Delta))\},
\end{aligned} \tag{8.62}$$

where Δ*,* $H_q(.)$*, and the other notations were introduced in Theorem 8.3.*

Theorems 8.3 and 8.4 enable comparison of the estimators regarding their asymptotic distributional risks. In the special case $\mathbf{Q}_{12} = \mathbf{0}$, we have $\mathbf{Q}_{11.2} =$

$\mathbf{Q}_{11}$ and $\mathbf{M} = \mathbf{0}$. Then all risks (8.60)–(8.62) are reduced to $\nu^2\text{Trace}(\mathbf{W}\mathbf{Q}_{11}^{-1})$, and hence all estimators are equivalent to each other. If $\mathbf{Q}_{12} \neq \mathbf{0}$ and when $\boldsymbol{\xi} \neq \mathbf{0}$, no one estimator clearly dominates the others under all circumstances, not even under special choice $\mathbf{W} = \mathbf{Q}_{11.2}\nu^{-2}$, corresponding to the Mahalanobis distance.

8.8 Robust Estimation of Multivariate Scatter

For the multi-normal distribution and more generally for the elliptically symmetric distributions, the dispersion matrix $\boldsymbol{\Sigma}$ characterizes the whole complexity of the scatter of the distribution and the interdependence of the components. It is also a matrix-valued parameter of the distribution. Even if the dispersion matrix exists in a more general multivariate distribution, it loses the linear structure and cannot characterize the interdependence of the components. Various alternative approaches to multivariate scatter estimation have been advocated in the literature (see Tyler (1987), Tyler et al. (2009), Oja, (2010), and Serfling (2010)), where other pertinent references have been cited). A useful account of these developments also appears in Huber and Ronchetti (2009). In this section, we provide an outline of these procedures. One possibility is to incorporate the univariate scatter estimators for the marginals and then use the robust correlation matrix for the estimation of the association functionals. In other words, to separate the estimation part for the marginal scale parameters and robust estimation part of the association parameters. The marginal scatters can be estimated by the methodology developed in Chapters 4–6; for instance, one can use the location-invariant L-estimators of the type of interquartile range. In addition, one can consider the Spearman rank correlation matrix for the estimation of association parameters.

Let $\mathbf{X}_1, \ldots, \mathbf{X}_n$ be the observed p-vectors, and let $R_{k1}, \ldots, R_{kn}$ be the ranks of $X_{k1}, \ldots, X_{kn}$, $1 \leq k \leq p$. The rank correlation matrix $\left[r_{gk\ell}\right]_{k,\ell=1}^{p}$ has elements

$$
\begin{aligned}
r_{gk\ell} &= \tfrac{12}{n(n^2-1)} \textstyle\sum_{i=1}^{n} \left(R_{ki} - \tfrac{n+1}{2}\right)\left(R_{\ell i} - \tfrac{n+1}{2}\right) \quad \text{for} \quad k \neq \ell (= 1, \ldots, p) \\
r_{gkk} &= 1 \quad \text{for} \quad k = 1, \ldots, p
\end{aligned}
$$

The Spearman rank correlation can be replaced by a more general score rank correlation: Let $a_{nk}(1), \ldots, a_{nk}(n)$ stand for the scores for the kth coordinate, $k = 1, \ldots, p$. The rank correlation matrix $\left[r_{k\ell}^{*}\right]_{k,\ell=1}^{p}$ is then

$$
\begin{aligned}
r_{k\ell}^{*} &= \tfrac{1}{A_{nk}A_{n\ell}} \textstyle\sum_{i=1}^{n} \left[a_{nk}(R_{ki}) - \bar{a}_{nk}\right]\left[a_{n\ell}(R_{\ell i}) - \bar{a}_{n\ell}\right] \quad \text{for} \quad k \neq \ell (= 1, \ldots, p) \\
r_{kk}^{*} &= 1 \quad \text{for} \quad k = 1, \ldots, p
\end{aligned}
\tag{8.63}
$$

where $\bar{a}_{nk} = \frac{1}{n}\sum_{i=1}^{n} a_{nk}(i)$ and $A_{nk}^2 = \frac{1}{n-1}\sum_{i=1}^{n}[a_{nk}(i) - \bar{a}_{nk}]^2$, $k = 1, \ldots, p$.

However, the L-, R-, and M- statistics are not generally affine-equivariant in the multivariate case. Second, the corresponding association functionals in general do not agree with the corresponding $\boldsymbol{\Sigma}$-matrix. For example, the Spearman rank correlation matrix for the multinormal distribution corresponds to $\left[\frac{6}{\pi}\arcsin(\frac{1}{2}\rho_{k\ell})\right]_{k,\ell=1}^{p}$, where $\rho_{k\ell}$ is the product moment correlation coefficients, but this not true for another distribution. In any case, this matrix differs from $\left[\rho_{k\ell}\right]_{k,\ell=1}^{p}$, and thus, the reconstructed covariance functional may not be proportional to the original $\boldsymbol{\Sigma}$ even if the distribution belongs to the elliptically symmetric family. Even if it is not possible to capture the entire interdependence picture by any correlation matrix (unless the distribution is normal), we at least want to have it affine-equivariant/invariant, to have it comparable to $\boldsymbol{\Sigma}$.

A second approach, motivated by recent advances on spatial ranks and quantile functions, retains the affine equivariance to a greater extent but the question of robustness remains open (see Tyler et al. (2009), Huber and Ronchetti (2009), and others). A broad outline of this approach is presented in Section 8.5. At the base, we start with the classical affine equivariant estimators $\bar{\mathbf{X}}_n = \frac{1}{n}\sum_{i=1}^{n}\mathbf{X}_i$ and $\mathbb{S}_n = \frac{1}{n-1}\sum_{i=1}^{n}(\mathbf{X}_i - \bar{\mathbf{X}}_n)(\mathbf{X}_i - \bar{\mathbf{X}}_n)^\top$ and proceed as in (8.26) and consider affine-equivariant estimators of $\boldsymbol{\theta}$ and $\boldsymbol{\Sigma}$ based on appropriate L-statistics.

Let us define

$$\mathbf{L}_n^{(r)} = \sum_{i=1}^{n} a_n(R_{ni}^{(r-1)})\mathbf{X}_i, \quad \mathbb{S}_n^{(r)} = \frac{1}{n-1}\sum_{i=1}^{n}(\mathbf{X}_i - \mathbf{L}_n^{(r-1)})(\mathbf{X}_i - \mathbf{L}_n^{(r-1)})^\top$$

as in (8.26), $r \geq 1$. Note that $\mathbb{S}_n^{(r)} = \mathbb{S}_n^{(0)} + \frac{n}{n-1}(\bar{\mathbf{X}}_n - \mathbf{L}_n^{(r-1)})(\bar{\mathbf{X}}_n - \mathbf{L}_n^{(r-1)})^\top$ by definition, where the second term is positive semidefinite with rank 1. Then $(\mathbb{S}_n^{(r)})^{-1} = \{(\mathbb{S}_n^{(0)})^{-1} - \text{(positive semidefinite matrix)}\}$, slightly reducing the effect of outliers. Further define $d_{ni}^{(r)}$ and the scores $a_n(i)$, $i = 1, \ldots, n$ as in (8.26); then $a_n(i)$ is $\searrow$ in $i (= 1, \ldots, n)$ and $\sum_{i=1}^{n} a_n(i) = 1$. Then we proceed as follows: Fix an initial $r \geq 1$ and put

$$\widehat{\boldsymbol{\Sigma}}_n^{(r)} = \sum_{i=1}^{n} a_n(R_{ni}^{(r)})(\mathbf{X}_i - \mathbf{L}_n^{(r)})(\mathbf{X}_i - \mathbf{L}_n^{(r)})^\top \tag{8.64}$$

and define

$$\widehat{d}_{ni}^{(r)} = (\mathbf{X}_i - \mathbf{L}_n^{(r)})^\top(\widehat{\boldsymbol{\Sigma}}_n^{(r)})^{-1}(\mathbf{X}_i - \mathbf{L}_n^{(r)}), \quad i = 1, \ldots, n. \tag{8.65}$$

Let $\widehat{R}_{ni}^{(r)}$ be the rank of $\widehat{d}_{ni}^{(r)}$ among $\widehat{d}_{n1}^{(r)}, \ldots, \widehat{d}_{nn}^{(r)}$. Put

$$\mathbf{L}_n^{(r+1)} = \sum_{i=1}^{n} a_n(\widehat{R}_{ni}^{(r)})\mathbf{X}_i, \tag{8.66}$$

$$\widehat{\mathbf{\Sigma}}_n^{(r+1)} = \frac{1}{n-1}\sum_{i=1}^{n}(\mathbf{X}_i - \mathbf{L}_n^{(r+1)})(\mathbf{X}_i - \mathbf{L}_n^{(r+1)})^\top.$$

Repeat this process, if needed. We can use the same scores as in Section 8.5; particularly, for the multinormal distribution we can use the scores described in Example 8.1. The matrix $\widehat{\mathbf{\Sigma}}_n^{(r+1)}$, $r \geq 1$ is affine-equivariant and should be more robust than $\mathbb{S}_n$.

8.9 Some Complementary and Supplementary Notes

The main emphasis in this chapter has been put on the conventional situation where the dimension p of the observed vectors is fixed and usually small, while the sample size n is larger than p, and in the asymptotic setup it is usually assumed that $p/n \to 0$ as $n \to \infty$. A common belief, shared by the researchers in robust statistics, is that one needs a comparatively larger sample size. A rule of thumb is that n should be so large that p^3/n is small (for example, see Portnoy (1984, 1985)). The numerical illustration in Example 8.1 reveals this picture clearly. For $p = 1$ and $n = 20$, the marginal variance of the sample mean is $\sigma^2/n = 1/20 = 0.05$, yielding the standard error as 0.224. On the other hand, for $p = 3$ and $n = 20$, $n^{-1}\text{Trace}(\mathbf{\Sigma})$ is three times more, giving rise to a comparable standard error for a contrast estimator as 0.387. Hence, to have the asymptotic theory appropriate, one should require a much larger sample size when p is not small.

However, during the past 20 years more applied sciences have invaded the realm of statistics, with the result that p/n may not be so small. In bioinformatics and genomic studies, it is often the case that $p \gg n$ with n not so large. This is the so-called *high-dimensional low sample model.* The curse of dimensionality has a significant impact on the applicable statistical methodology. In the high-dimensional low sample models, the robustness should be focused and studied differently than in the one- or low-dimensional setups (for example, see Kang and Sen (2008), with other citations therein). A possible transformation to a symmetric model is not often apparent, and hence we should go beyond the area of symmetric multivariate distributions, with symmetry of some sense. In such a situation, the componentwise robust estimation treated in Section 8.4 can have some advantages over the affine-equivariant estimation treated in Section 8.6.

In the high-dimensional situation, one often conceives an idea that only few of the parameters are relevant while the others are negligible. The *least absolute shrinkage and selection operator* (LASSO) theory has evolved in the past two decades, founded by Tibshirani (1996). Let us describe a modified formulation of LASSO to suit the multivariate location model. Suppose that $\mathbf{X}_1, \ldots, \mathbf{X}_n$ are *i.i.d.* random vectors with a distribution function F defined on $\mathbb{R}_p$, with location parameter $\boldsymbol{\theta} = (\theta_1, \ldots, \theta_p)^\top$, $p \geq 1$, and dispersion matrix

$\mathbf{\Sigma}$. It is anticipated that only a few components of $\boldsymbol{\theta}$ are different from the corresponding components of the selected pivot $\boldsymbol{\theta}^0$ while the rest are close to θ_j^0, $j = 1, \ldots .p$. Then it is intuitive to use the ordinary least squares estimator of $\boldsymbol{\theta}$ subject to some tuning of $\boldsymbol{\theta}$. Consider the minimization of

$$\sum_{i=1}^{n} \|\mathbf{X}_i - \boldsymbol{\theta}\|_{\Sigma} = \sum_{i=1}^{n} (\mathbf{X}_i - \boldsymbol{\theta})^\top \mathbf{\Sigma}^{-1} (\mathbf{X}_i - \boldsymbol{\theta})$$

$$\text{subject to } \sum_{j=1}^{p} |\theta_j - \theta_j^0| \leq t, \tag{8.67}$$

where $t > 0$ is a tuning parameter. Because

$$\sum_{i=1}^{n} (\mathbf{X}_i - \boldsymbol{\theta})^\top \mathbf{\Sigma}^{-1} (\mathbf{X}_i - \boldsymbol{\theta})$$
$$= \sum_{i=1}^{n} (\mathbf{X}_i - \bar{\mathbf{X}}_n)^\top \mathbf{\Sigma}^{-1} (\mathbf{X}_i - \bar{\mathbf{X}}_n) + n(\bar{\mathbf{X}}_n - \boldsymbol{\theta})^\top \mathbf{\Sigma}^{-1} (\bar{\mathbf{X}}_n - \boldsymbol{\theta}),$$

the minimization (8.67) reduces to minimizing

$$n(\bar{\mathbf{X}}_n - \boldsymbol{\theta})^\top \mathbf{\Sigma}^{-1} (\bar{\mathbf{X}}_n - \boldsymbol{\theta}) \text{ subject to } \sum_{j=1}^{p} |\theta_j - \boldsymbol{\theta}_j^0| \leq t,$$

or otherwise minimizing

$$n(\bar{\mathbf{X}}_n - \boldsymbol{\theta})^\top \mathbf{\Sigma}^{-1} (\bar{\mathbf{X}}_n - \boldsymbol{\theta}) + \lambda \sum_{j=1}^{p} |\theta_j - \boldsymbol{\theta}_j^0|, \ \boldsymbol{\theta} \in \mathbb{R}_p$$

where λ is a Lagrangian multiplier. From the computational point of view, (8.67) is a quadratic programming with linear inequality constraints. The parameter $t > 0$ causes a shrinkage of the solutions of the minimization toward $\boldsymbol{\theta}^0$; some components may be exactly equal to θ_j^0.

In line with the development of Sections 8.4 and 8.5, we can consider a parallel minimization of

$$\sum_{i=1}^{n} \rho(\mathbf{X}_i - \boldsymbol{\theta}) = \min \text{ subject to } \sum_{j=1}^{p} |\theta_j - \theta_j^0| \leq t$$

where $\rho(\cdot)$ is a suitable function with derivative $\psi(\cdot)$. Then the desired estimator is a solution of the penalized minimization

$$\sum_{i=1}^{n} \rho(\mathbf{X} - \boldsymbol{\theta}) + \lambda \sum_{j=1}^{p} |\theta_j - \theta_j^0| \leq t, \quad \boldsymbol{\theta} \in \mathbb{R}_p \tag{8.68}$$

with a Lagrangian multiplier λ. The choice of λ can naturally depend on p. In high-dimensional and low sample size models, it is often suspected that a majority of the θ_j are very close to pivot θ_j^0 and only a few are different. In that setup, the tuning factor is often related by

$$\lambda \sum_{j=1}^{p} I(|\theta_j - \theta_j^0| > \varepsilon) \tag{8.69}$$

for some $\varepsilon > 0$ (but small). In that case, the choice of λ may depend on both ε and p. In a multivariate normal distribution setup, (8.67) relates to the so-called penalized likelihood approach. For other than multi-normal F, with appropriate choice of metric $\rho(\cdot)$, the robustness aspect dominates the pictures and LASSO is very much attuned to this scheme.

The LASSO penalty in the context of regression quantiles [with $\rho(\cdot) = \rho_\alpha(\cdot)$, $0 < \alpha < 1$, given in (4.88)] is characterized in Koenker (2005), where also more recent references and computation aspects are also given. For a in-depth account of the LASSO methods, we refer to a recent monograph by Bühlmann and van de Geer (2011).

8.10 Problems

8.2.1 Let F be a multivariate normal distribution with mean vector $\boldsymbol{\theta}$ and dispersion matrix $\boldsymbol{\Sigma}$. Show that (8.4) holds when $\boldsymbol{\Sigma}$ is positive definite. What is the functional form of $h_0(y)$?

8.2.2 In previous problem, suppose that $\boldsymbol{\Sigma}$ is positive semidefinite of rank $q < p$. Show that the density $f(\mathbf{x}; \boldsymbol{\theta}, \boldsymbol{\Sigma})$ is not properly defined though F is uniquely defined (use the one-to-one correspondence between the distribution function and characteristic fucntion). Partition $\mathbf{X}$ into $(\mathbf{X}_1^\top, \mathbf{X}_2^\top,$ where $\mathbf{X}_1$ is a q-vector for which $\boldsymbol{\Sigma}_{11}$ is of full rank q. Then show that $\mathbf{X}_2 - \boldsymbol{\Sigma}_{22}\boldsymbol{\Sigma}_{11}^{-1}\mathbf{X}$ has a degenerated distribution at the point $\boldsymbol{\theta}_2 - \boldsymbol{\Sigma}_{21}\boldsymbol{\Sigma}_{11}^{-1}\boldsymbol{\theta}_1$. Hence, or otherwise, show that (8.4) holds with the norm $\|\mathbf{x} - \boldsymbol{\theta}\|_\Sigma$ replaced by $\|\mathbf{x} - \boldsymbol{\theta}\|_{\Sigma_{11}}$. Work out a characterization of elliptically symmetric distributions bypassing the existence of the density function.

8.2.3 Let F be a multivariate Laplace distribution for which $f(\mathbf{x}; \boldsymbol{\theta}, \boldsymbol{\Sigma}) =$ const. $\exp(\|\mathbf{x} - \boldsymbol{\theta}\|_\Sigma)$. Show that F is elliptically symmetric. Here also, work out the case where $\boldsymbol{\Sigma}$ is of rank $q < p$.

8.2.4 Consider a multivariate t-distribution for which

$$f(\mathbf{x}; \boldsymbol{\theta}) = \text{const. } \left\{1 + (\mathbf{x} - \boldsymbol{\theta})^\top \mathbf{A}(\mathbf{x} - \boldsymbol{\theta})\right\}^{-m/2}, \quad m > 0$$

where $\mathbf{A}$ is an unknown positive definite matrix. Does this distribution function belong to the elliptically symmetric family ? Use the Courant Theorem (on the ratio of two quadratic forms) to show that

$$ch_{\min}\left(\mathbf{A}^{-1}\right) \leq \frac{(\mathbf{x} - \boldsymbol{\theta})^\top (\mathbf{x} - \boldsymbol{\theta})}{(\mathbf{x} - \boldsymbol{\theta})^\top \mathbf{A}(\mathbf{x} - \boldsymbol{\theta})} \leq ch_{\max}\left(\mathbf{A}^{-1}\right),$$

for all $\mathbf{x} \in \mathbb{R}_p$. Hence, or otherwise, show that for $0 < m \leq 3$, the distribution function does not have a finite second-order moment. Hence, comment on the robustness of affine-equivariant estimator of $\boldsymbol{\theta}$ when m is small.

8.2.5 Consider a mixed multi-normal distribution for which the density $f(\mathbf{x};\boldsymbol{\theta})$ can be expressed as

$$f(\mathbf{x};\boldsymbol{\theta}) = \sum_{k\geq 1} \pi_k f_k(\mathbf{x};\boldsymbol{\theta}), \quad \mathbf{x} \in \mathbb{R}_p,$$

where the π_k are nonnegative quantities such that $\sum_{k\geq 1} \pi_k = 1$, and $f_k(\mathbf{x};\boldsymbol{\theta})$ is a multi-normal probability density function with mean vector $\boldsymbol{\theta}$ and dispersion matrix $\boldsymbol{\Sigma}_k$, $k \geq 1$. Does this probability density function belong to the elliptically symmetric family?

8.2.6 Show that for arbitrary $\boldsymbol{\Sigma}$, the distributions in all the preceding problems may not belong to the spherically symmetric family.

8.2.7 Show that the distributions in Problems 8.2.1–8.2.5 are all diagonally symmetric.

8.2.8 Consider a multivariate contamination model, along the lines of Huber (1964), where

$$f(\mathbf{x};\boldsymbol{\theta}) = (1-\varepsilon) f_1(\mathbf{x};\boldsymbol{\theta},\boldsymbol{\Sigma}) + \varepsilon f_2(\mathbf{x};\boldsymbol{\theta}), \quad 0 < \varepsilon < 1,$$

with f_1 being multi-normal while f_2 is heavy-tailed multivariate probability density function, symmetric about $\boldsymbol{\theta}$. Does f belong to the elliptically symmetric family? Does it belong to the diagonally symmetric family?

8.2.9 In the preceding problem, suppose that both f_1 and f_2 are elliptically symmetric but their corresponding functions $h(\cdot)$ are not the same. Is f then elliptically symmetric or diagonally symmetric ?

8.2.10 Let F_0 be a multi-normal distribution function, and let $F(\mathbf{x}-\boldsymbol{\theta}) = [F_0(\mathbf{x}-\boldsymbol{\theta})]^{\lambda}$, for some $\lambda > 0$. For $\lambda \neq 1$, will F belong to the marginal symmetric family? Will it belong to the elliptically symmetric, diagonally symmetric, spherically symmetric family?

8.3.1 F is multi-normal with mean $\boldsymbol{\theta}$ and dispersion matrix $\boldsymbol{\Sigma}$ (positive definite) show that (8.6) leads to the closed expression for the MLE of $\boldsymbol{\theta}$ and $\boldsymbol{\Sigma}$ as $\bar{\mathbf{X}}_n$ and $\mathbf{S}_n$, respectively. Can you obtain the MLE of the $\boldsymbol{\theta}$ in closed form when F has contamination model in Problem 8.2.8? Can you obtain the MLE of $\boldsymbol{\theta}$ in closed expression when F has a multivariate t-distribution (see Problem 8.2.4)?

8.3.2 If F is multi-normal with mean $\boldsymbol{\theta}$ and dispersion matrix $\boldsymbol{\Sigma}$ where $\boldsymbol{\Sigma}$ is q rank $q < p$; can you use (8.6) to obtain the MLE of $\boldsymbol{\theta}$ and $\boldsymbol{\Sigma}$? Will they have closed forms?

8.3.3 In the preceding problem, obtain the marginal MLE of θ_j and σ_{jj} ($1 \leq j \leq p$). Can you obtain the full MLE of $\boldsymbol{\theta}$ and $\boldsymbol{\Sigma}$ from these marginal estimates?

8.3.4 For the multivariate Laplace distribution (see Problem 8.2.3), obtain the MLE of $\boldsymbol{\theta}$ and $\boldsymbol{\Sigma}$ and verify whether or not these can be expressed in terms of marginal MLEs.

8.3.5 Show that the median estimator of $\boldsymbol{\theta}$ in (8.8) is equivariant under any coordinatewise strictly monotone transformation $x_j \to g_j(x_j)$, where $g_1(\cdot), \ldots, g_p(\cdot)$ need not be the same. But this estimator may not be affine-equivariant.

8.3.6 Define $\boldsymbol{\mathcal{V}}(\boldsymbol{\psi})$ and $\mathbf{I}^*$ as in (8.12) and (8.13), and show that $\boldsymbol{\mathcal{V}}(\boldsymbol{\psi}) - (\mathbf{I}^*)^{-1}$ is positive-semidefinite when $\boldsymbol{\psi}^0$ and $\boldsymbol{\xi}$ are orthogonal.

8.4.1 Use (8.16) and (8.18) to rewrite $\widehat{\boldsymbol{\beta}}_n = \boldsymbol{\beta} + \mathbb{E}\mathbb{X}^\top(\mathbb{X}\mathbb{X}^\top)^{-1}$, and show that render (8.19), whenever $\boldsymbol{\Sigma}$ is positive definite,

$$(\boldsymbol{\Sigma}^{-1/2} \otimes (\mathbb{X}\mathbb{X}^\top)^{1/2})(\widehat{\boldsymbol{\beta}}_n - \boldsymbol{\beta}) \xrightarrow{\mathcal{D}} \mathcal{MN}(\mathbf{0}, \mathbf{I}_p \otimes \mathbf{I}_q).$$

8.4.2 For the model in Problem 8.2.8, verify whether or not the MLE based on the joint probability distribution function $f(\cdot)$ can be expressed as a linear compound of the marginal MLE. Hence or otherwise, comment on the mixture model in Problem 8.2.5.

8.4.3 For the errors e_i, consider the multivariate t-distribution (see Problem 8.2.4), where $\mathbf{x} - \boldsymbol{\theta}$ is replaced by $\mathbf{e}$. Examine whether or not the MLE of $\boldsymbol{\beta}$ can be expressed in terms of marginal MLE.

8.4.4 For the marginal MLE of β_j $(1 \leq j \leq p)$, if $f(\cdot)$ has the form in Problem 8.2.5, appraise the nonlinearity of the estimating equations. Justify the use of method of scoring by using the „uniform asymptotic linearity" result, and hence show that the MLE and marginal MLE of $\boldsymbol{\beta}$ can be at least asymptotically tied down to linear complementarity.

8.4.5 For M-, L-, and R-estimators of $\boldsymbol{\beta}$, show that whenever the probability density function has a finite Fisher information (matrix), (8.20) holds. Is this condition true for the multivariate t-distribution or the mixture model in Problems 8.2.4–8.2.8?

8.5.1 Define the $\mathbf{L}_n^{(r)}, r \geq 1$ as in (8.26). For a multivariate t-distribution with $m = 4$ and $m = 3$ (see Problem 8.2.4) when $\mathbb{A}$ is unknown, examine whether or not convergence of the $\mathbf{L}_n^{(r)}$ holds.

8.5.2 For a multivariate t-distribution in Problem 8.2.4, take $\mathbf{A} = \boldsymbol{\Sigma}^{-1}$ and $m = 3.5$. What will be the expected value of $\mathbf{S}_n^{-1}$? Can you justify the scores in (8.28) in this case ?

8.5.3 For the mixture model in Problem 8.2.5, examine the behavior of $\mathbf{S}_n^{-1}$, and hence, the $d_{ni}^{(0)}$, $1 \leq i \leq n$. Will $\mathbf{S}_n^{-1}$ be unbiased for $\boldsymbol{\Sigma}^{-1}$ and will $\mathbf{S}_n^{-1} \xrightarrow{\mathcal{P}} \boldsymbol{\Sigma}^{-1}$ as $n \to \infty$? What about the rate of convergence of $\mathbf{S}_n^{-1}$ to $\boldsymbol{\Sigma}^{-1}$?

8.5.4 Show that for any $r \geq 1$, $\mathbf{S}_n^{(r)} - \mathbf{S}_n^{(0)}$ is positive semidefinite. What can you say about the rate of convergence (as $n \to \infty$) of $\|\mathbf{S}_n^{(r)} - \mathbf{S}_n^{(0)}\|$?

8.6.1 For the representation in (8.40), specify a (sufficient) condition on $\mathbf{R}_n$ under which the Anscombe condition holds.

8.6.2 Show that if in (8.40), $n\mathbf{R}_n$ is a submartingale, then by Hájek–Renyi–Chow inequality, the Anscombe (1952) condition holds, under appropriate moment condition on Rbf_n.

8.6.3 In Chapters 4–6, second order asymptotic representations were considered for M-, L-, and R-estimators. Can they be used to verify the Anscombe (1952) condition?

8.6.4 For the regression model, work out an analogue of the representation in (8.40) and hence, prove the asymptotic normality results.

8.7.1 Let $\mathbf{X} \stackrel{\mathcal{D}}{=} \mathcal{N}_p(\boldsymbol{\theta}, \boldsymbol{\Sigma})$. Show that there exists a transformation $\mathbf{Y} = \mathbf{BX}$, such that $\mathbf{Y} \stackrel{\mathcal{D}}{=} \mathcal{N}_p(\boldsymbol{\xi}, \mathbf{I}_p)$ where $\boldsymbol{\xi} = (\xi_1, \xi_2, \ldots, \xi_p)$ and $\xi_1 = \left(\boldsymbol{\theta}^\top \boldsymbol{\Sigma}^{-1} \boldsymbol{\theta}\right)^{1/2}$ while $\xi_2 = \ldots = \xi_p = 0$.

8.7.2 Let $\mathbf{X} \stackrel{\mathcal{D}}{=} \mathcal{N}_p(\boldsymbol{\theta}, \boldsymbol{\Sigma})$. and $g(\mathbf{x})$ be a function of $\mathbf{x}$, such that $I\!\!E g(\mathbf{X})$ exists. Then $g(\mathbf{x}) \rightarrow g^*(\mathbf{y})$ where y is defined as in Problem 8.7.1.

8.7.3 Let $g_1^*(y_1) = I\!\!E\left[g(Y_1, \ldots, Y_p) | Y_1 = y_1\right], \; y_1 \in \mathbb{R}$. Then show that

$$I\!\!E\left[g(\mathbf{x})\right] = I\!\!E\left[g^*(\mathbf{Y})\right] = I\!\!E\left[g_1^*(Y_1)\right] = \int g_1^*(y) \frac{1}{\sqrt{2\pi}} \exp\left(-\frac{1}{2}(y - \xi_1)^2\right) dy.$$

Hence, or otherwise, provide an alternative derivation of the Stein identity (Lemma 8.2).

8.7.4 If $\mathbf{X} \stackrel{\mathcal{D}}{=} \mathcal{N}_p(\boldsymbol{\theta}, \boldsymbol{\Sigma})$, and $\mathbf{X}_1, \ldots, \mathbf{X}_n$ are i.i.d copies of $\mathbf{X}$, show that for $\mathbf{T}_n = \bar{\mathbf{X}}_n$, the asymptotic risk of $\mathbf{T}_n^S$ is indeed smaller than or equal to that of $\mathbf{T}_n$.

8.8.1 Define the (Spearman) rank correlation matrix $\mathbf{R}_n^S$ as in (8.63). Show that if F has a multivariate normal distribution with correlation matrix $\mathbf{R} = \left(\rho_{jl}\right)_{j,l=1,\ldots,p}$, then $\mathbf{R}_n^S \xrightarrow{\mathcal{P}} \mathbf{R}^S$, where $\rho_{jl}^S = \frac{6}{\pi} \sin^{-1}\left(\frac{1}{2}\rho_{jl}\right)$.

8.8.2 Draw the curve $\frac{6}{\pi} \sin^{-1}\left(\frac{1}{2}\rho\right), \; -1 < \rho < 1$, and examine the nature of the function $\rho - \frac{6}{\pi} \sin^{-1}\left(\frac{1}{2}\rho\right)$.

CHAPTER 9

Robust Tests and Confidence Sets

9.1 Introduction

If we want to test the simple hypothesis P_0 against the simple alternative P_1, then we use the likelihood ratio test, following the Neyman–Pearson lemma. Following Huber (1965), if the true hypothetical distribution may not be exactly P_0 but rather lies in some neighborhood $\mathcal{P}_0$ of an idealized simple hypothesis, and similarly the true alternative distribution lies in some neighborhood $\mathcal{P}_1$ of a simple alternative P_1, one should test robustly $\mathcal{P}_0$ against $\mathcal{P}_1$, using the maximin likelihood ratio test, based on the least favorable distributions of $\mathcal{P}_0$, and $\mathcal{P}_1$. Huber and Strassen (1973) extended the robust testing theory to a general class of testing problems, using the concept of *Choquet capacities*.

However, their elegant theory does not fully cover the problem of testing a composite null hypothesis against a composite alternative, which may not coincide with $\mathcal{P}_0$, and $\mathcal{P}_1$. Without knowing the distributions, we have two possibilities: Either use the robust estimators which are defined implicitly as a solution of a system of equations, and use this system as a test criterion or use the Wald-type test based on robust estimators. The present chapter gives a motivation, and derivation of some robust tests for simple or composite hypotheses, and provides a characteristic of robustness of some statistical tests, interpreted in a local as well as global sense, as in the case of estimates. By no means this chapter constitutes a complete treatise of the robust statistical tests; a full coverage of goes beyond the scope of this book.

There is an intricate connection between the robust hypothesis testing, and the confidence set estimation. The confidence sets, based on asymptotic normality (or even on the Edgeworth expansion) of classical estimators, are even more vulnerable to error contaminations, to outliers, and possible departures from distributional assumptions than the estimators themselves. thus, the robustness considerations are even more important for confidence sets and intervals. For this reason, a part of the current chapter is devoted to robust confidence sets. Affine-equivariant robust tests and confidence sets in the multivariate location and regression model are briefly treated in the last section.

9.2 M-Tests and R-Tests

We shall first motivate the tests through the location problem, and then present their extensions to some linear models.

Let $X_1, \ldots, X_n$ be n *i.i.d.* random variables with a distribution function $F_\theta(x) = F(x - \theta)$, $x \in \boldsymbol{R}$, $\theta \in \Theta \subset \boldsymbol{R}$. Assume that F is symmetric around 0. We want to test the hypothesis $\mathbf{H}_0 : \theta = \theta_0$ against $\mathbf{H}_1 : \theta = \theta_1$, $\theta_1 >$ (or $<$ or $\neq$) θ_0. If the functional form of F is specified, one can use the classical likelihood ratio test which possesses some optimal properties under the assumed model. However, a parametric test is generally very nonrobust, even to small departures from assumed F. If the least favorable distributions in the neighborhoods of the hypothesis and of the alternative can be identified (e.g., with respect to the Lévy–Kolmogorov metric or with respect to the gross errors model), then one can construct a likelihood ratio type statistic for these least favorable laws and show that such a test has a maximin power property (Huber 1965). One conceives of a class $\mathcal{F}_0$ of distribution functions that define an appropriate neighborhood of F, and let $F_0 \in \mathcal{F}_0$ be the least favorable distribution function with respect to the pair ($\mathbf{H}_0$, $\mathbf{H}_1$). Then Huber's suggestion is to construct the usual likelihood ratio test statistic corresponding to the least favorable F_0. If $F \in \mathcal{F}_0$ admits a density f with respect to some sigma-finite measure μ, then instead of the likelihood ratio statistic one may also consider Rao's score statistic

$$L_n^0 = \sum_{i=1}^{n} \frac{\partial \ln f(X_i, \theta)}{\partial \theta}\Big|_{\theta_0}.$$

A robust version of this test is to use the test statistic

$$L_n^{\star 0} = \sum_{i=1}^{n} \frac{\partial \ln f_0(X_i, \theta)}{\partial \theta}\Big|_{\theta_0}$$

where f_0 is the density corresponding to the least favorable F_0. More generally, one can replace the function $\ln f_0(x, \theta)$ with a function generating a selected robust estimator, namely the M- or the R-estimator.

9.2.1 M-Tests of Location

One can conceive of a score function $\psi : \boldsymbol{R} \mapsto \boldsymbol{R}$, and replace $L_n^{\star 0}$ by an M-statistic

$$M_n^0 = \sum_{i=1}^{n} \psi(X_i - \theta_0). \tag{9.1}$$

If $\theta_0 = 0$, then $\int_{\boldsymbol{R}} \psi(x) dF(x) = 0$ and if $\sigma_\psi^2 = \int_{\boldsymbol{R}} \psi^2(x) dF(x) < \infty$, then M_n^0 has zero mean and variance $n\sigma_\psi^2$ under θ_0. On the other hand, if ψ is $\nearrow$, then

$\mathbb{E}_\theta M_n^0$ is $\begin{smallmatrix}\leq\\=\\>\end{smallmatrix}0$ according as θ is $\begin{smallmatrix}\geq\\=\\<\end{smallmatrix}\theta_0$. Moreover, under $\theta = \theta_0$,

$$\frac{M_n^0}{\sigma_\psi\sqrt{n}} \xrightarrow{\mathcal{D}} \mathcal{N}(0,1) \text{ as } n \to \infty,$$

because M_n^0 involves independent summands. If $\hat{\theta}_n$ is the M-estimator of θ based on $X_1, \ldots, X_n$ and the score function ψ, then a convenient form of the test statistic is

$$T_n = \frac{M_n^0}{\hat{\sigma}_n\sqrt{n}}. \tag{9.2}$$

where

$$\hat{\sigma}_n^2 = \int_{\boldsymbol{R}} \psi^2(x - \hat{\theta}_n)dF_n(x) = n^{-1}\sum_{i=1}^n \psi^2(X_i - \hat{\theta}_n).$$

T_n is asymptotically normal under $H_0 : \theta = \theta_0$; thus, for $H_1 : \theta > \theta_0$, the (asymptotic) critical level of test based on (9.2) is $\tau_\alpha = \Phi^{-1}(1-\alpha)$, $0 < \alpha < 1$. The main justification of the test statistic T_n is the choice of a suitable robust ψ. Toward the proper choice of ψ, let us consider a sequence of local (Pitman-type) alternatives $H_{(n)} : \theta = \theta_{(n)} = \theta_0 + n^{-1/2}\lambda$ for some $\lambda \in \boldsymbol{R}$. Denote

$$\delta_\lambda = \lim_{n\to\infty} \int_{\boldsymbol{R}} \sqrt{n}\{\psi(x - \theta_0) - \psi(x - \theta_0 - n^{-1/2}\lambda)\}dF(x - \theta_0),$$

and assume that this limit exists and is positive for all $\lambda > 0$. The usual conditions on ψ, F imposed in Chapters 3 and 5 are sufficient for this purpose. Then, further assuming that

$$\lim_{\delta\to 0} \int_{\boldsymbol{R}} \psi^2(x - \delta)dF(x) = \int_{\boldsymbol{R}} \psi^2(x)dF(x) = \sigma_\psi^2,$$

we obtain that the asymptotic power of the test based on T_n (for H_0 vs. $H_{(n)}$) is given by

$$\beta(\lambda) = \lim_{n\to\infty} \mathbb{P}\{T_n > \tau_\alpha | H_{(n)}\} = 1 - \Phi\Big(\tau_\alpha - \frac{\delta_\lambda}{\sigma_\psi}\Big),\ \lambda \in \boldsymbol{R}.$$

thus, an optimal ψ relates to the maximization of $\delta_\lambda/\sigma_\psi$ for every fixed $\lambda \in \boldsymbol{R}$. By Theorem 5.4, we get $\delta_\lambda = \lambda\gamma$, where $\gamma = \gamma_c + \gamma_s$ (> 0) is given by (5.39) and (5.42). Therefore the problem reduces to minimizing

$$\Big(\frac{\sigma_\psi}{\gamma}\Big)^2 \quad = \quad \text{asymptotic variance of } \sqrt{n}(\hat{\theta}_n - \theta),$$

where $\sigma_\psi^2 = \sigma_\psi^2(F)$, $\gamma = \gamma(\psi, F)$, and $F \in \mathcal{F}_0$. Following Huber (1965), we can look for a particular score function $\psi_0 : \boldsymbol{R} \to \boldsymbol{R}$ such that

$$\sup\Big\{\Big(\frac{\sigma_\psi}{\gamma}\Big)^2 : F \in \mathcal{F}_0\Big\} \text{ is a minimum at } \psi = \psi_0, \tag{9.3}$$

and obtain an asymptotically maximin power M-test of the score type (e.g., ψ_0 is the Huber function provided that $\mathcal{F}_0$ is the family of contaminated normal distributions). The test for H_0 against two-sided alternatives follows on parallel lines.

An alternative is the M-test of the Wald type. Following Section 5.3, select a score function $\psi : \boldsymbol{R} \to \boldsymbol{R}$ as

$$\begin{aligned} \psi(x) &= \rho'(x) \text{ where } \rho : \boldsymbol{R} \mapsto \boldsymbol{R}^+ \qquad (9.4) \\ Q_n(\theta) &= \sum_{i=1}^{n} \rho(X_i - \theta),\ \theta \in \Theta \end{aligned}$$

and denote

$$\begin{aligned} Q_n^\star &= \inf\{Q_n(\theta) :\ \theta \in \Theta\} \\ \hat{\theta}_n^\star &= \arg\min\{Q_n(\theta) : \theta \in \Theta\} \end{aligned}$$

where $\hat{\theta}_n^\star$ is the M-estimator of θ based on the score function ρ. Then let us consider

$$Z_n = Q_n(\theta_0) - \hat{Q}_n^\star = \sum_{i=1}^{n} [\rho(X_i - \theta_0) - \rho(X_i - \hat{\theta}_n^\star)], \qquad (9.5)$$

where Z_n is analogous to the classical log-likelihood ratio type statistic $[\rho(y) \equiv y^2$ for a normal $F]$, so Z_n is reducible to the classical student t-statistic. By Theorem 5.5 (with $\tau = 1/2$), we obtain for large n,

$$Z_n \overset{P}{\sim} \frac{1}{2} n(\hat{\theta}_n^\star - \theta_0)^2 \gamma = \frac{\sigma_\psi^2}{2\gamma} \Big\{ \frac{n\gamma^2(\hat{\theta}_n^\star - \theta_0)^2}{\sigma_\psi^2} \Big\}.$$

Under $\mathbf{H}_0 : \theta = \theta_0$,

$$\frac{2\gamma}{\sigma_\psi^2}\, Z_n \overset{\mathcal{D}}{\to} \chi_1^2.$$

The unknown γ and σ_ψ^2 can be estimated by $\hat{\gamma}_n$ and $\hat{\sigma}_n^2$ as before; the final test statistic is

$$Z_n^\star = \frac{2\hat{\gamma}_n}{\hat{\sigma}_n^2}\, Z_n \qquad (9.6)$$

Under $\mathbf{H}_{(n)} : \theta = \theta_{(n)} = \theta_0 + n^{-1/2}\lambda,\ \lambda \in \boldsymbol{R},\ Z_n^\star$ has asymptotically a noncentral χ^2 squared distribution with 1 degree of freedom and noncentrality parameter

$$\Delta^\star = \frac{\gamma^2\lambda^2}{\sigma_\psi^2}.$$

The choice of ψ_0 is driven by the basic robustness considerations; an asymptotically optimal ψ following a maximin power is similar as in (9.3). If we compare T_n in (9.2) with $Z_n^\star$ in (9.6), then T_n is preferable to $Z_n^\star$, because it only needs to estimate σ_ψ^2, while $Z_n^\star$ needs to estimate both γ and σ_ψ^2. On the other hand, the likelihood ratio-type tests behave better than score-type test under nonlocal alternatives [see Hoeffding (1965a,b)], when $Z_n^\star$ may be preferred to T_n. These considerations have tacitly assumed that F has a fixed scale. It is possible to extend these M-type results to the case where the scale factor is unknown, using the basic results on studentized M-statistics developed in Section 5.4.

9.2.2 M-Tests in Linear Model

Let us next consider the case of linear hypotheses pertaining to the linear model (5.101), for which M-estimators were studied in Section 5.5. The inference concerns the regression parameter $\boldsymbol{\beta} = (\beta_1, \ldots, \beta_p)^\top \in B \subset \boldsymbol{R}_p$, $p \geq 1$. A linear hypothesis may be framed as

$$\mathbf{H}_0 : \mathbf{D}\boldsymbol{\beta} = \mathbf{0} \text{ vs. } \mathbf{H}_1 : \mathbf{D}\boldsymbol{\beta} \neq \mathbf{0}, \tag{9.7}$$

where $\mathbf{D}$ is a specified $q \times p$ matrix, $q \leq p$. Without loss of generality, we can let

$$\boldsymbol{\theta} = \mathbf{D}^0\boldsymbol{\beta} = \begin{pmatrix} \boldsymbol{\theta}_1^{q\times 1} \\ \boldsymbol{\theta}_2^{(p-q)\times 1} \end{pmatrix} \quad \mathbf{D}^0_{p\times p} = \begin{bmatrix} \mathbf{D}^{q\times p} \\ (\mathbf{D}^\star)^{(p-q)\times p} \end{bmatrix}.$$

We assume that $\mathbf{D}^0$ is a nonsingular matrix with the inverse $(\mathbf{D}^0)^{-1}$. Then we write

$$\mathbf{X}\boldsymbol{\beta} = \mathbf{X}(\mathbf{D}^0)^{-1}\mathbf{D}^0\boldsymbol{\beta} = \mathbf{X}^\star\boldsymbol{\theta}.$$

After this reparametrization, we take

$$\mathbf{Y} = \mathbf{X}^\star\boldsymbol{\theta} + \mathbf{E} \tag{9.8}$$

and rewrite the hypothesis and alternative

$$\mathbf{H}_0 : \boldsymbol{\theta}_1 = \mathbf{0} \text{ vs. } \mathbf{H}_1 : \boldsymbol{\theta}_1 \neq \mathbf{0}, \tag{9.9}$$

treating $\boldsymbol{\theta}_2$ as a nuisance parameter. Then under $\mathbf{H}_0$ we have

$$\mathbf{Y} = \mathbf{X}_2^\star\boldsymbol{\theta}_2 + \mathbf{E}, \quad \text{where } \mathbf{X}^\star = \begin{bmatrix} \mathbf{X}_1^\star & \vdots & \mathbf{X}_2^\star \\ n \times q & \vdots & n \times (p-q) \end{bmatrix}. \tag{9.10}$$

For the Wald-type test of $\mathbf{H}_0$, we let parallel to (9.4)

$$Q_n(\boldsymbol{\theta}) = \sum_{i=1}^n \rho(Y_i - \boldsymbol{\theta}^\top \mathbf{x}_i^\star), \ \boldsymbol{\theta} \in \boldsymbol{\Theta} \subset \boldsymbol{R}_p,$$

where $\mathbf{x}_i^{\star\top}$ is the ith row of $\mathbf{X}^\star$, $i = 1, \ldots, n$. Let then

$$\begin{aligned} \hat{\boldsymbol{\theta}}_n^\star &= \arg\min\{Q_n(\boldsymbol{\theta}) : \boldsymbol{\theta} \in \boldsymbol{\Theta}\}, \\ \hat{\boldsymbol{\theta}}_n^{0*} &= \arg\min\{Q_n(\boldsymbol{\theta}) : \boldsymbol{\theta}_1 = \mathbf{0}, \boldsymbol{\theta}_2 \in \boldsymbol{\Theta}_2\}, \end{aligned} \tag{9.11}$$

where $\boldsymbol{\Theta}_2 = \left\{\boldsymbol{\theta}_2 : \begin{pmatrix} \boldsymbol{\theta}_1 \\ \boldsymbol{\theta}_2 \end{pmatrix} \in \boldsymbol{\Theta}, \ \boldsymbol{\theta}_1 = \mathbf{0}\right\}$. Letting $\psi \equiv \rho'$, we have by (9.11)

$$\sum_{i=1}^n \mathbf{x}_i^\star \psi(Y_i - (\hat{\boldsymbol{\theta}}_n^\star)^\top \mathbf{x}_i^\star) = \mathbf{0} \tag{9.12}$$

$$\sum_{i=1}^{n} \mathbf{x}_{i2}^{\star}\psi(Y_i - (\hat{\boldsymbol{\theta}}_{n2}^{o\star})^{\top}\mathbf{x}_{i2}^{\star}) = \mathbf{0},$$

where $\mathbf{x}_i^{\star} = \begin{pmatrix} \mathbf{x}_{i1}^{\star} \\ \mathbf{x}_{i2}^{\star} \end{pmatrix}$, $i = 1, \ldots, n$ and $\hat{\boldsymbol{\theta}}_n^{o\star} = \begin{pmatrix} \mathbf{0} \\ \hat{\boldsymbol{\theta}}_{n2}^{o\star} \end{pmatrix}$. Then, as in (9.5), we let

$$Z_n = Q_n(\hat{\boldsymbol{\theta}}_n^{o\star}) - Q_n(\hat{\theta}_n^{\star}). \tag{9.13}$$

As a direct corollary to Theorem 5.10 (when confined to the fixed scale case), we have

$$\begin{aligned} \hat{\boldsymbol{\theta}}_n^{\star} - \boldsymbol{\theta} &= \gamma^{-1}(\mathbf{X}^{\star\top}\mathbf{X}^{\star})^{-1}\sum_{i=1}^{n}\mathbf{x}_i^{\star}\psi(Y_i - \boldsymbol{\theta}^{\top}\mathbf{x}_i^{\star}) + \mathbf{R}_n^{\star}, \\ \|\mathbf{R}_n^{\star}\| &= \mathrm{o}_p(n^{-1/2}); \end{aligned} \tag{9.14}$$

the rate of $\|\mathbf{R}_n^{\star}\|$ can be improved to $\mathrm{O}_p(n^{-1})$ under additional regularity conditions on ψ. Similarly, under $\mathbf{H}_0 : \boldsymbol{\theta}_1 = \mathbf{0}$, we have

$$\begin{aligned} \hat{\boldsymbol{\theta}}_{n2}^{o\star} - \boldsymbol{\theta}_2 &= \gamma^{-1}(\mathbf{X}_2^{\star\top}\mathbf{X}_2^{\star})^{-1}\sum_{i=1}^{n}\mathbf{x}_{i2}^{\star}\psi(Y_i - \boldsymbol{\theta}_2^{\top}\mathbf{x}_{i2}^{\star}) + \mathbf{R}_n^{0\star}, \\ \|\mathbf{R}_n^{0\star}\| &= \mathrm{o}_p(n^{-1/2}). \end{aligned} \tag{9.15}$$

Assume that

$$n^{-1}\mathbf{X}^{\star\top}\mathbf{X}^{\star} \to \mathbf{C} \ \text{ as } \ n \to \infty, \ \text{ where } \mathbf{C} \ \text{ is positive definite.}$$

Partition $\mathbf{C}$ as

$$\mathbf{C} = \begin{bmatrix} \mathbf{C}_{11} & \mathbf{C}_{12} \\ \mathbf{C}_{21} & \mathbf{C}_{22} \end{bmatrix}$$

where $\mathbf{C}_{11}$ is $q \times q$, $\mathbf{C}_{12} = \mathbf{C}_{12}$ is $q \times (p-q)$, and $\mathbf{C}_{22}$ is $(p-q) \times (p-q)$, and write

$$\mathbf{C}^{\star} = \mathbf{C}_{11} - \mathbf{C}_{12}\mathbf{C}_{22}^{-1}\mathbf{C}_{21}. \tag{9.16}$$

We have from (9.14)

$$\frac{n\gamma^2}{\sigma_{\psi}^2}(\hat{\boldsymbol{\theta}}_n^{\star} - \boldsymbol{\theta})^{\top}\mathbf{C}(\hat{\boldsymbol{\theta}}_n^{\star} - \boldsymbol{\theta}) \xrightarrow{\mathcal{D}} \chi_p^2, \tag{9.17}$$

and similarly we have from (9.15)

$$\frac{n\gamma^2}{\sigma_{\psi}^2}(\hat{\boldsymbol{\theta}}_{n2}^{o\star} - \boldsymbol{\theta}_2)^{\top}\mathbf{C}_{22}(\hat{\boldsymbol{\theta}}_{n2}^{o\star} - \boldsymbol{\theta}_2) \xrightarrow{\mathcal{D}} \chi_{p-q}^2. \tag{9.18}$$

By virtue of (9.14) and (9.15), we have

$$\begin{aligned} n^{1/2}\gamma(\hat{\boldsymbol{\theta}}_n^{\star} - \boldsymbol{\theta}) &= \mathbf{C}^{-1}\mathbf{M}_n(\boldsymbol{\theta}) + \mathbf{o}_p(1), \\ n^{1/2}\gamma(\hat{\boldsymbol{\theta}}_{n2}^{o\star} - \boldsymbol{\theta}_2) &= \mathbf{C}_{22}^{-1}\mathbf{M}_{n2}(\boldsymbol{\theta}) + \mathbf{o}_p(1) \end{aligned} \tag{9.19}$$

where

$$\mathbf{M}_n(\boldsymbol{\theta}) = n^{-1/2}\sum_{i=1}^{n}\mathbf{x}_i^{\star}\psi(Y_i - \boldsymbol{\theta}^{\top}\mathbf{x}_i^{\star}) = \begin{pmatrix} \mathbf{M}_{n1}(\boldsymbol{\theta}) \\ \mathbf{M}_{n2}(\boldsymbol{\theta}) \end{pmatrix}.$$

hence, by (9.17), (9.18), (9.19), and the classical Cochran theorem, we have that under $\mathbf{H}_0$

$$\frac{n\gamma}{\sigma_\psi^2}\Big\{(\hat{\boldsymbol{\theta}}_n^\star-\boldsymbol{\theta})^\top\mathbf{C}(\hat{\boldsymbol{\theta}}_n^\star-\boldsymbol{\theta})-(\hat{\boldsymbol{\theta}}_{n2}^{o\star}-\boldsymbol{\theta}_2)^\top\mathbf{C}_{22}(\hat{\boldsymbol{\theta}}_{n2}^{o\star}-\boldsymbol{\theta}_2)\Big\}\xrightarrow{\mathcal{D}}\chi_q^2.$$

Finally, as a direct corollary to (5.123), we have

$$\sup_{\theta\in K}\left|\mathbf{Q}_n(\boldsymbol{\theta}+n^{-1/2}\mathbf{t})-\mathbf{Q}_n(\boldsymbol{\theta})+\mathbf{t}^\top\mathbf{M}_n(\boldsymbol{\theta})+\tfrac{1}{2}\gamma\mathbf{t}^\top(n^{-1}\mathbf{X}^{\star\top}\mathbf{X}^\star)\mathbf{t}\right|=o_p(1) \tag{9.20}$$

for any compact $K\subset\boldsymbol{R}_p$. Since $n^{1/2}\|\hat{\boldsymbol{\theta}}_n^\star-\boldsymbol{\theta})\|$ and $n^{1/2}\|\hat{\boldsymbol{\theta}}_{n2}^{o\star}-\boldsymbol{\theta}_2)\|$ are both $O_p(1)$ by (9.19), we get from (9.12) and (9.13)

$$Z_n=\frac{1}{2}\gamma\Big\{n(\hat{\boldsymbol{\theta}}_n^\star-\boldsymbol{\theta})^\top\mathbf{C}(\hat{\boldsymbol{\theta}}_n^\star-\boldsymbol{\theta})-n(\hat{\boldsymbol{\theta}}_{n2}^{o\star}-\boldsymbol{\theta}_2)^\top\mathbf{C}_{22}(\hat{\boldsymbol{\theta}}_{n2}^{o\star}-\boldsymbol{\theta}_2)\Big\}+o_p(1). \tag{9.21}$$

thus, under $\mathbf{H}_0:\boldsymbol{\theta}_1=\mathbf{0}$ we have

$$\frac{2\gamma}{\sigma_\psi^2}Z_n\xrightarrow{\mathcal{D}}\chi_q^2.$$

We arrive at the likelihood ratio type test statistic

$$Z_n^\star=\frac{2\hat{\gamma}_n}{\hat{\sigma}_n^2}Z_n=\frac{2\hat{\gamma}_n}{\hat{\sigma}_n^2}[Q_n(\hat{\boldsymbol{\theta}}_n^{o\star})-Q_n(\hat{\boldsymbol{\theta}}_n^\star)]$$

where $\hat{\gamma}_n$ and $\hat{\sigma}_n^2$ are suitable consistent estimators of γ and σ_ψ^2, respectively. This type of statistics was proposed by Schrader and McKean (1977) and Schrader and Hettmansperger (1980). Under the sequence $\{H_{1(n)}\}$ of local (Pitman-type) alternatives

$$H_{1(n)}:\boldsymbol{\theta}_1=n^{-\frac{1}{2}}\boldsymbol{\lambda},\ \boldsymbol{\lambda}\in\boldsymbol{R}_q\ \text{(fixed)}\ , \tag{9.22}$$

then under $\{H_{1(n)}\}$,

$$Z_n^\star\xrightarrow{\mathcal{D}}\chi_{q,\Delta}^2, \tag{9.23}$$

with the noncentrality parameter

$$\Delta=\left(\frac{\gamma}{\sigma_\psi}\right)^2\boldsymbol{\lambda}^\top(\mathbf{C}^\star)^{-1}\boldsymbol{\lambda}. \tag{9.24}$$

Let us now consider the score-type M-tests; they were proposed by Sen (1982). Let $\hat{\boldsymbol{\theta}}_{n2}^{o\star}$ be the M-estimator of $\boldsymbol{\theta}_2$ defined by (9.12) based on the score function $\psi:\ \boldsymbol{R}\mapsto\boldsymbol{R}$. This does not involve the component $\boldsymbol{\theta}_1$ and is therefore computationally less involved. Let us denote

$$\widetilde{\mathbf{M}}_{n1}=n^{-1/2}\sum_{i=1}^n\mathbf{x}_{i1}\psi(Y_i-\mathbf{x}_{i2}^\top\hat{\boldsymbol{\theta}}_{n2}^{o\star})$$
$$\hat{\sigma}_n^2=n^{-1}\sum_{i=1}^n\psi^2(Y_i-\mathbf{x}_{i2}^\top\hat{\boldsymbol{\theta}}_{n2}^{o\star}) \tag{9.25}$$

$$\mathbf{C}_n^* = n^{-1}\Big\{\mathbf{X}_1^{*\top}\mathbf{X}_1^* - \mathbf{X}_1^{*\top}\mathbf{X}_2^*(\mathbf{X}_2^{*\top}\mathbf{X}_2^*)^{-1}\mathbf{X}_2^{*\top}\mathbf{X}_1^*\Big\}.$$

Note that $\mathbf{C}_n^\star \to \mathbf{C}^\star$ of (9.16), as $n \to \infty$. The score M-test criterion has the form

$$\mathcal{T}_n = \frac{1}{\hat{\sigma}_n^2}\,\widetilde{\mathbf{M}}_{n1}^\top(\mathbf{C}_n^\star)^{-1}\widetilde{\mathbf{M}}_{n1}. \tag{9.26}$$

Using Lemma 5.3, the Cochran theorem on quadratic forms in asymptotically normal random variables and the Slutzky theorem on $\frac{\hat{\sigma}_n^2}{\sigma_\psi^2}$, we get

$$\begin{aligned}
&\mathcal{T}_n \xrightarrow{\mathcal{D}} \chi_q^2 \ \text{ under } \mathbf{H}_0, && (9.27)\\
&\mathcal{T}_n \xrightarrow{\mathcal{D}} \chi_{q,\Delta}^2 \ \text{ under } \{\mathbf{H}_{1(n)}\} \ \text{ in (9.22)}
\end{aligned}$$

where Δ is given by (9.24). Thus, $\mathcal{T}_n$ and $Z_n^\star$ are asymptotically equivalent under $\mathbf{H}_0$ as well as local alternatives, while $\mathcal{T}_n$ does not involve the estimation of γ and the computation of $\hat{\theta}_n^\star$.

9.2.3 R-Tests

An excellent treatment of the rank tests is available in the classical text of Hájek and Šidak (1967: *Theory of Rank Tests*), extended in Hájek et al. (1999). Puri and Sen (1985) extended these developments in a more general setup of possibly multivariate observations. thus, we shall only present a brief overview of some recent developments not reported in these earlier texts.

Consider a linear model reduced by reparametrization as in (9.7)–(9.10). We want to test for $\mathbf{H}_0 : \boldsymbol{\theta}_1 = \mathbf{0}$ vs. $\mathbf{H}_1 : \boldsymbol{\theta}_1 \neq \mathbf{0}$, treating $\boldsymbol{\theta}_2$ as a nuisance parameter. As in (6.42)–(6.43), consider a vector of linear rank statistics

$$\mathbf{L}_n(\mathbf{b}) = \sum_{i=1}^n (\mathbf{x}_i^\star - \bar{\mathbf{x}}_n^\star)a_n(R_{ni}(\mathbf{b})),\ \mathbf{b} \in \boldsymbol{R}_p$$

and the test criterion

$$\mathcal{L}_n = \frac{1}{A_n^2}(\mathbf{L}_n(\mathbf{0})^\top\mathbf{V}_n^{-1}\mathbf{L}_n(\mathbf{0}))$$

where $\mathbf{V}_n$ is defined in (6.55) and $A_n^2 = \frac{1}{n-1}\sum_{i=1}^n\{a_n(i) - \bar{a}_n\}^2$. Under $\mathbf{H}_0 : \boldsymbol{\theta} = \mathbf{0}$, $\mathcal{L}_n \xrightarrow{\mathcal{D}} \chi_p^2$ as $n \to \infty$, while under local alternatives it has an asymptotic noncentral χ_p^2 distribution with an appropriate noncentrality parameter. For testing the subhypothesis (9.9) when $\boldsymbol{\theta}_2$ is unknown, we can proceed analogously to the score M-test based on $\mathcal{T}_n$ in (9.26) and use the *aligned rank test*, considered by Sen and Puri (1977) and Adichie (1978). We can proceed as in (6.44)–(6.46) (with $\mathbf{b}_2 \in \boldsymbol{R}_{p-q}$) and consider an R-estimator

$\hat{\boldsymbol{\theta}}^{o\star}_{n2}$ of $\boldsymbol{\theta}_2$. We use the aligned ranks

$$R_{ni}(\hat{\boldsymbol{\theta}}^{o\star}_n) = \tilde{R}_{ni}, \ \hat{\boldsymbol{\theta}}^{o\star}_n = \begin{pmatrix} \mathbf{0} \\ \hat{\boldsymbol{\theta}}^{o\star}_{n2} \end{pmatrix}, \ 1 \leq i \leq n$$

and the subvector of the aligned rank statistics

$$\widetilde{\mathbf{L}}_{n(1)} = \sum_{i=1}^{n} (\mathbf{x}^\star_{i1} - \bar{\mathbf{x}}^\star_{n1}) a_n(\tilde{R}_{ni}).$$

The aligned test criterion of $\mathbf{H}_0 : \boldsymbol{\theta}_1 = \mathbf{0}$ is given by

$$\mathcal{L}^{(1)}_n = \frac{1}{nA^2_n} \widetilde{\mathbf{L}}^\top_{n(1)} (\mathbf{C}^\star_n)^{-1} \widetilde{\mathbf{L}}_{n(1)} \tag{9.28}$$

where $\mathbf{C}^*_n$ is the same as in (9.25). By Theorem 6.5, $\mathcal{L}^{(1)}_n \xrightarrow{\mathcal{D}} \chi^2_q$ under $\mathbf{H}_0$. Under the local alternative $\{H_{1(n)}\}$, defined in (9.22), $\mathcal{L}^{(1)}_n \xrightarrow{\mathcal{D}} \chi^2_{q,\Delta^\star}$, where $\Delta^\star = \frac{\gamma^2}{A^2_\varphi} \boldsymbol{\lambda}^\top (\mathbf{C}^\star)^{-1} \boldsymbol{\lambda}$, and γ and A^2_φ are defined by (6.65) and (6.54), respectively.

Alternatively, the tests of $\mathbf{H}_0 : \boldsymbol{\theta}_1 = \mathbf{0}$ against $\mathbf{H}_1 : \boldsymbol{\theta}_1 \neq \mathbf{0}$ can be based on regression rank scores, developed by Gutenbrunner et al. (1993) and introduced in Section 6.4. However, their asymptotic properties were proved only for densities with tails as the Student t_4 distribution and lighter tails. Write

$$\begin{aligned} \widehat{\mathbf{H}}_2 &= \mathbf{X}^\star_2 (\mathbf{X}^{\star\top}_2 \mathbf{X}^\star_2)^{-1} \mathbf{X}^{\star\top}_2, \ \widehat{\mathbf{X}}^\star_1 = \widehat{\mathbf{H}}_2 \mathbf{X}^\star_1, \\ \mathbf{Q}_n &= n^{-1} (\mathbf{X}^\star_1 - \widehat{\mathbf{X}}^\star_1)^\top (\mathbf{X}^\star_1 - \widehat{\mathbf{X}}^\star_1). \end{aligned} \tag{9.29}$$

As in (6.86) we define a vector of scores based on a score-function $\varphi : (0,1) \mapsto \boldsymbol{R}$, by $\widehat{\mathbf{b}}_n = (\hat{b}_{n1}, \ldots, \hat{b}_{nn})^\top$ and let

$$\mathbf{S}_n = n^{-\frac{1}{2}} (\mathbf{X}^\star_{n1} - \widehat{\mathbf{X}}^\star_{n1})^\top \widehat{\mathbf{b}}_n = n^{-\frac{1}{2}} \mathbf{X}^{\star\top}_{n1} (\mathbf{I} - \widehat{\mathbf{H}}_2)^\top \widehat{\mathbf{b}}_n$$

Then the regression rank scores test criterion is given by

$$\mathcal{T}^\star_n = \frac{1}{A^2_\varphi} (\mathbf{S}^\top_n \mathbf{Q}^{-1}_n \mathbf{S}_n).$$

with A^2_φ as in (ref6.6.14). An important feature of $\mathcal{T}^\star_n$ is that it eliminates the estimation of the nuisance parameter $\boldsymbol{\theta}_2$ by using the projection in (9.29) and the basic regression-invariance property of the regression rank scores (see Section 6.4). Gutebrunner et al. (1993) proved that $\mathcal{T}^\star_n \xrightarrow{\mathcal{D}} \chi^2_q$ under $H_0 : \boldsymbol{\theta}_1 = \mathbf{0}$ and $\mathcal{T}^\star_n \xrightarrow{\mathcal{D}} \chi^\star_{q,\Delta}$, under $\{\mathbf{H}_{1(n)}\}$ in (9.22) with $\Delta^\star = \frac{\gamma^2}{A^2_\varphi} \boldsymbol{\lambda}^\top (\mathbf{Q}^\star)^{-1} \boldsymbol{\lambda}$.

The tests based on regression rank scores involve the parametric linear programming and a calculation of scores $\widehat{\mathbf{b}}_n$. The relevant computational algorithm was elaborated by Koenker and d'Orey (1994).

9.2.4 Robustness of Tests

Similarly as in the estimation theory, the robustness of a test can be characterized by the influence function, by the power or level breakdown points, or by the breakdown function, among others. Rousseeuw and Ronchetti (1979, 1981) extended the influence function to the testing problem. A review of the robustness based on influence function can be found in Markatou and Ronchetti (1997). The breakdown point criteria, along with their connection to the maximal bias function, were studied by He et al. (1990), among others.

We propose an alternative characteristic of robustness of a test, based on the tail of its power function at distant alternatives. Let us illustrate this characteristic on the tests of hypothesis $\mathbf{H}: \theta = \theta_0$ against one-sided alternative $\mathbf{K}: \theta > \theta_0$. The power function of a consistent test tends to 1 as $\theta - \theta_0 \to \infty$. However, the speed of this convergence depends on the tails of the basic distribution; the more robust test, the less sensitive is the tail of its power function with respect to the tails of distribution function F.

Let $X_1, \ldots, X_n$ be *i.i.d.* observations following the continuous distribution function $F(x - \theta)$. Assume that $0 < F(x) < 1$, $x \in \mathbb{R}_1$ and that $F(x) + F(-x) = 1$, $x \in \mathbb{R}_1$, for simplicity. Let φ_n be the test of $\mathbf{H}$ against $\mathbf{K}$ satisfying

$$\varphi_n(\mathbf{X}) = \begin{cases} 1 & \text{if} \quad T_n(X_1 - \theta_0, \ldots, X_n - \theta_0) > C_n(\alpha) \\ 0 & \text{if} \quad T_n(X_1 - \theta_0, \ldots, X_n - \theta_0) < C_n(\alpha) \\ \text{and} & \quad \mathbb{E}_{\theta_0}\, \varphi_n(\mathbf{X}) = \alpha \end{cases} \tag{9.30}$$

for a prescribed α. Among tests satisfying (9.30), consider those for which $T_n(X_1 - t, \ldots, X_n - t)$ is $\searrow$ in t and

$$[X_{n:n} < \theta_0] \implies [T_n(X_1 - \theta_0, \ldots, X_n - \theta_0) < C_n(\alpha)] \tag{9.31}$$

where $X_{n:n} = \max_{1 \le i \le n} X_i$. Introduce the following measure of tail-behavior of a test φ_n:

$$B(T_n, \theta, \alpha) = \frac{-\ln\, \mathbb{E}_{\theta_0}(1 - \varphi_n(\mathbf{X}))}{-\ln\, (1 - F(\theta - \theta_0))}$$

The larger $B(T_n, \theta, \alpha)$, the faster is convergence of the power function to 1, compared with the tail of F. First, notice that for any test φ_n, under the above assumptions,

$$\limsup_{\theta - \theta_0 \to \infty} B(T_n, \theta, \alpha) \le n. \tag{9.32}$$

Indeed,

$$\begin{aligned} \mathbb{E}_\theta(1 - \varphi_n(\mathbf{X})) &\ge \mathbb{P}_\theta\{T_n(X_1 - \theta_0, \ldots, X_n - \theta_0) < C_n(\alpha)\} \\ &\ge \mathbb{P}_\theta\{X_{n:n} \le \theta_0\} = (1 - F(\theta - \theta_0))^n \end{aligned}$$

what implies (9.32).

To illustrate the sensitivity of a classical test to the tails of F, consider the t-test with

$$T_n(\mathbf{X}_n-\theta_0)=\frac{\bar{X}_n-\theta_0}{S_n}\sqrt{n-1},\quad \bar{X}_n=\tfrac{1}{n}\sum_{i=1}^{n}X_i,\ S_n^2=\tfrac{1}{n}\sum_{i=1}^{n}(X_i-\bar{X}_n)^2,$$

where $C_n(\alpha)=t_{n-1}(\alpha)$ is the upper α-percentile of t_{n-1} distribution. The following lemma shows that the t-test is nonrobust; indeed, its upper bound in (9.32) does not exceed 1 under heavy-tailed F, while it is close to n under F normal.

LEMMA 9.1 *(i) Let F be heavy-tailed in the sense that*

$$\lim_{x\to\infty}\frac{-\ln(1-F(x))}{m\ln x}=1\quad \textit{for some}\ \ m>0. \tag{9.33}$$

Then, under $n\geq 3,\ 0<\alpha<1/2$,

$$\limsup_{\theta-\theta_0\to\infty}B(T_n,\theta,\alpha)\leq 1.$$

(ii) If F is normal $\mathcal{N}(\theta,\sigma^2)$, then under $n\geq 3,\ 0<\alpha<1/2$,

$$\liminf_{\theta-\theta_0\to\infty}B(T_n,\theta,\alpha)\geq n\cdot\left[1+\frac{t_{n-1}(\alpha)}{\sqrt{n-1}}\right]^{-2}.$$

PROOF (i) $I\!P_\theta\left\{\frac{\bar{X}_n-\theta_0}{S_n}\sqrt{n-1}<t_{n-1}(\alpha)\right\}$
$\geq I\!P\{\bar{X}_n>\theta-\theta_0\}\geq I\!P_0\{X_1>n(\theta-\theta_0),X_2>0,\ldots,X_n>0\}$. Hence,

$$\limsup_{\theta-\theta_0\to\infty}B(T_n,\theta,\alpha)\leq\limsup_{\theta-\theta_0\to\infty}\frac{-(n-1)\ln 2+m\ln(n(\theta-\theta_0))}{m\ln(\theta-\theta_0)}\leq 1.$$

(ii) The normal distribution has exponentially decreasing tails; more precisely,

$$\lim_{x\to\infty}\frac{-\ln(1-F(x))}{x^2/(2\sigma^2)}=1.$$

Using the Markov inequality, we can write for any $\varepsilon,\ 0<\varepsilon<1$,

$$\begin{aligned}
&I\!E_\theta(1-\varphi_n(\mathbf{X}))\leq I\!P_0\left\{|\bar{X}_n|+S_n\frac{t_{n-1}(\alpha)}{\sqrt{n-1}}>\theta-\theta_0\right\}\\
&\leq I\!P_0\left\{\Big(\frac{1}{n}\sum_{i=1}^{n}X_i^2\Big)^{1/2}\Big(1+\frac{t_{n-1}(\alpha)}{\sqrt{n-1}}\Big)>\theta-\theta_0\right\}\\
&\leq I\!E_0\Big[\exp\Big\{\frac{n(1-\varepsilon)}{2\sigma^2}\tfrac{1}{n}\sum_{i=1}^{n}X_i^2\Big\}\Big]\\
&\qquad\qquad\cdot\exp\Big\{-\frac{n(1-\varepsilon)(\theta-\theta_0)^2}{2\sigma^2}\Big(1+\frac{t_{n-1}(\alpha)}{\sqrt{n-1}}\Big)^{-2}\Big\}\\
&=2^{n/2}\Gamma\Big(\frac{n}{2}\Big)\int_0^\infty \mathrm{e}^{-\varepsilon y/2}\mathrm{y}^{\frac{n}{2}-1}\mathrm{dy}
\end{aligned}$$

$$\cdot \exp\Big\{-\frac{n(1-\varepsilon)(\theta-\theta_0)^2}{2\sigma^2}\Big(1+\frac{t_{n-1}(\alpha)}{\sqrt{n-1}}\Big)^{-2}\Big\}$$

$$= \varepsilon^{-n/2}\cdot \exp\Big\{-\frac{n(1-\varepsilon)}{2\sigma^2}(\theta-\theta_0)^2\Big(1+\frac{t_{n-1}(\alpha)}{\sqrt{n-1}}\Big)^{-2}\Big\},$$

and this implies (ii). □

Unlike the t-test, the following robust version of the probability ratio test is insensitive to the tails of the parent distribution. Let

$$T_n(\mathbf{X}-\theta_0) = \frac{\sum_{i=1}^n \psi(X_i-\theta_0)}{(\psi^2(X_i-\bar{X}_n))^{1/2}} \tag{9.34}$$

where $\psi : \mathbb{R}_1 \mapsto \mathbb{R}_1$ is a nondecreasing, continuous, and odd function such that

$$\psi(x) = \psi(c) \quad \text{for } |x| > c,\ c > 0.$$

The test rejects $\mathbf{H}$ if $T_n(\mathbf{X}-\theta_0) > \tau_\alpha$, where $\tau_\alpha = \Phi(1-\alpha)$ and α is the asymptotic significance level. The following theorem shows that the limit of $B(T_n,\theta,\alpha)$ is always around $\frac{n}{2}$, both for heavy-tailed and exponentially tailed distributions; $\bar{X}_n$ in (9.34) can be replaced by a suitable estimator $\hat{\theta}_n$.

THEOREM 9.1 *Consider the test rejecting* $\mathbf{H}$ *if* $T_n(\mathbf{X}-\theta_0) > \tau_\alpha$. *Let the continuous and symmetric* F *satisfy either (9.33) or*

$$\lim_{x\to\infty} \frac{-\ln(1-F(x))}{bx^r} = 1 \quad \textit{for some } b, r > 0.$$

Then, for $0 < \alpha < 1/2$, $n > \tau_\alpha^2$,

$$A_n(\alpha) \le \liminf_{\theta-\theta_0\to\infty} B(T_n,\theta,\alpha) \le \limsup_{\theta-\theta_0\to\infty} B(T_n,\theta,\alpha) \le C_n(\alpha) \tag{9.35}$$

where

$$A_n(\alpha) = \frac{1}{2}(n-\tau_\alpha\sqrt{n}), \quad C_n(\alpha) = \frac{n+1}{2}.$$

PROOF Denote $q_n = \#\{i : X_i - \theta_0 > c$. Then

$$\begin{aligned} \mathbb{E}_\theta(1-\varphi_n(\mathbf{X})) &\le \mathbb{P}_\theta\Big\{\sum_{i=1}^n \psi(X_i-\theta_0) \le \tau_\alpha\sqrt{n}\cdot\psi(c)\Big\} \\ &\le \mathbb{P}_\theta\Big\{q_n \le \frac{1}{2}(n+\tau_\alpha\sqrt{n})\Big\}. \end{aligned} \tag{9.36}$$

Indeed, if $q_n > \frac{1}{2}(n+\tau_\alpha\sqrt{n}) = s_n$, then

$$\sum_{i=1}^n \psi(X_i-\theta_0) > \frac{1}{2}(n+\tau_\alpha\sqrt{n})\psi(c) - \frac{1}{2}(n-\tau_\alpha\sqrt{n})\psi(c) = \tau_\alpha\sqrt{n}\psi(c)$$

and it follows from (9.36) that $\mathbb{E}_\theta(1-\varphi_n(\mathbf{X})) \le \mathbb{P}_\theta(X_{n:s_n}-\theta_0 < c)$, where $s_n = [\frac{1}{2}(n-\tau_\alpha\sqrt{n})]$; this gives the lower bound in (9.35). On the other hand,

$\mathbb{E}_\theta(1-\varphi_n(\mathbf{X})) \geq \mathbb{P}_\theta\left\{\sum_{i=1}^n \psi(X_i-\theta_0) \leq 0\right\}$, which implies the upper bound in (9.35). $\square$

9.2.5 Some Remarks on the Wald-Type Tests

Wald (1943) advocated the use of maximum likelihood estimators (or BAN estimators) for the construction of tests. We are naturally tempted to replace the BAN estimators by their robust L-, M-, or R-estimator versions. It leads to a natural question: If such robust estimators are derived from robust statistics (as in the case of M- and R-estimators), is it really necessary to use the estimators instead of their providers? We shall discuss the Wald method applied to the general linear model, more precisely the reparametrized linear model in (9.8), and to the null hypothesis $\mathbf{H}_0 : \boldsymbol{\theta}_1 = \mathbf{0}$ against $\mathbf{H}_1 : \boldsymbol{\theta}_1 \neq \mathbf{0}$ with $\boldsymbol{\theta}_2$ nuisance. We start with the usual M-estimator $\widehat{\boldsymbol{\theta}}_{n(M)}$ of $\boldsymbol{\theta}$ studied in Chapter 5. By an appeal to Theorem 5.12 (or Corollary 5.12), we conclude that under the assumed regularity conditions,

$$n^{1/2}(\widehat{\boldsymbol{\theta}}_{n(M)} - \boldsymbol{\theta}) = \boldsymbol{\gamma}^{-1} n^{1/2}(\mathbf{X}^{\star\top}\mathbf{X}^{\star})^{-1}\sum_{i=1}^n \mathbf{x}_i^\star \psi(Y_i - \boldsymbol{\theta}^\top \mathbf{x}_i^\star) + \mathbf{o}_p(1).$$

Let $\tilde{\boldsymbol{\theta}}_{n(M)1}$ be the q-component subvector of $\widehat{\boldsymbol{\theta}}_{n(M)}$. Then if γ and σ_ψ^2 are consistently estimated by $\hat{\gamma}_n$ and $\hat{\sigma}_n^2$, respectively, one may directly use a quadratic form in $\tilde{\boldsymbol{\theta}}_{n(M)1}$; this is asymptotically χ_q^2 under $\mathbf{H}_0$ by the classical Cochran theorem and noncentral χ_q^2 under local alternatives. This holds similarly for L- and R-estimators as well. The asymptotic dispersion matrix is given by a known matrix with a multiplication factor, which depends on the unknown F. To apply a Wald-type test, this multiplication factor needs to be consistently estimated. If $\widehat{\boldsymbol{\theta}}_n$ can be partitioned as $\begin{pmatrix}\widehat{\boldsymbol{\theta}}_{n1}\\ \widehat{\boldsymbol{\theta}}_{n2}\end{pmatrix}$, we have a first-order asymptotic representation and the asymptotic normality

$$\widehat{\boldsymbol{\theta}}_{n1} - \boldsymbol{\theta}_1 = \nu\mathbf{Z}_{n1} + \mathbf{o}_p(n^{-1/2}),$$
$$\sqrt{n}\mathbf{Z}_{n1} \sim \mathcal{N}_q(\mathbf{0}, \mathbf{D}_n)$$

where $\mathbf{D}_n$ is a $q \times q$ matrix of rank q and ν is an unknown functional or parameter. If there exists a consistent estimator $\hat{\nu}_n$ of ν, then we get the typical Wald-type test criterion

$$W_n = \frac{n}{\hat{\nu}_n^2}\,\widehat{\boldsymbol{\theta}}_{n1}^\top \mathbf{D}_n^{-1}\widehat{\boldsymbol{\theta}}_{n1}.$$

which is asymptoticaly χ_q^2 under $\mathbf{H}_0$ and noncentral χ_q^2 under $\{H_{1(n)}\}$ in (9.22). It does not require any reparametrization and is usable whenever robust estimators of ν are available.

9.3 Minimax Tests

Let $\mathcal{P}$ be the class of all probability measures on the σ-algebra $\mathcal{B}$ of Borel sets of the metric space Ω. Let $P_0,\ P_1 \in \mathcal{P},\ P_0 \neq P_1$ be two specific probability measures and $p_0,\ p_1$ be their densities with respect to a σ-finite measure μ ($\mu = P_0 + P_1$ is one possibility). Our goal is to test P_0 against P_1 in a robust manner, considering that the true hypothesis and alternative rather lie in some neighborhoods of P_0 and P_1, respectively. In particular, these neighborhoods will be the families of probability measures dominated by the set functions called *capacities*, and the final test is obtained by the Neyman–Pearson lemma applied to the least favorable probability measures of the respective families. This construction is due to Huber and Strassen (1973), who used the Choquet (1953/54) capacities. Rieder (1977) and Bednarski (1981) constructed special capacities, making it possible to test a contaminated neighborhood of P_0 against a contaminated neighborhood of P_1.

According to Huber and Strassen (1973), 2-alternating capacity is any set function v satisfying

(i) $v(\emptyset) = 0,\ v(\Omega) = 1$

(ii) $A \subseteq B \Longrightarrow v(A) \leq v(B)$

(iii) $A_n \uparrow A \Longrightarrow v(A_n) \uparrow v(A)$

(iv) $F_n \downarrow F,\ F_n$ closed $\Longrightarrow v(F_n) \downarrow v(F)$ if $\mathcal{P}$ is weakly closed.

(v) $v(A \cup B) + v(A \cap B) \leq v(A) + v(B), \quad A,\ B \in \mathcal{B}.$

To a capacity v assign the family $\mathcal{P}_v = \{P \in \mathcal{P}:\ P \leq v\}$. Let $v_0,\ v_1$ be 2-alternating; we look for a minimax test of $\mathcal{P}_0\ (= \mathcal{P}_{v_0})$ vs. $\mathcal{P}_1\ (= \mathcal{P}_{v_1})$ with the critical region A_α such that $v_0(A_\alpha) = \alpha$ and with the maximal guaranteed power $v_1(A_\alpha^c)$. Under additional conditions, there exists the least favorable probability distributions $Q_0,\ Q_1$ of $\mathcal{P}_0$ and $\mathcal{P}_1$, respectively, such that the minimax test is the Neyman–Pearson test of the simple hypothesis $\{Q_0\}$ against the simple alternative $\{Q_1\}$.

However, the compactness conditions in Huber and Strassen (1973) preclude applying their theorems to the ε-contamination and total variation neighborhoods of fixed $P_0,\ P_1$. Rieder (1977), Bednarski (1981) and Buja (1980, 1986) weakened these conditions and defined special types of capacities. They considered a variation neighborhood of $P_\nu \in \mathcal{P}$ in the form

$$\mathcal{P}_\nu = \{Q \in \mathcal{P}:\ Q(A) \leq (1-\varepsilon_\nu)P_\nu(A) + \varepsilon_\nu + \delta_\nu \ \ \forall A \in \mathcal{B};\ \nu = 0,1\}$$

for chosen $\varepsilon_\nu \geq 0,\ \delta_\nu > 0,\ \varepsilon_\nu + \delta_\nu < 1,\ \nu = 0,1$. Buja (1986) pointed out that the condition (iv) above can be modified as

$$\text{(iv}^*) \qquad v(A_n) \downarrow v(A) \qquad \forall A_n \downarrow A \neq \emptyset, \qquad A_n \in \mathcal{B}.$$

Denote $\mathcal{F}$ the set of concave continuous functions $f:\ [0,1] \mapsto [0,1]$ satisfying

$f(1) = 1$. For $f \in \mathcal{F}$ and $P \in \mathcal{P}$ consider the set function

$$v(A) = f(P(A)), \quad A \in \mathcal{B},\ A \neq \emptyset,\ w(\emptyset) = 0.$$

Then $v(\cdot)$ satisfies (i)–(iii), (iv*), (v) (Bednarski (1981) and Buja (1980)). We obtain contamination and total variation norm neighborhoods by specializing

$$f_\nu(x) = [(1-\varepsilon_\nu)x + \varepsilon_\nu + \delta_\nu] \wedge 1, \quad \varepsilon_\nu \geq 0,\ \delta_\nu > 0,\ \varepsilon_\nu + \delta_\nu < 1,\ \nu = 0, 1.$$

Now let $P_0,\ P_1$ be two probability distributions, mutually absolutely continuous, $v_\nu(\cdot) = f_\nu(P_\nu(\cdot))$ and $\mathcal{P}_\nu = \{P \in \mathcal{P}:\ P \leq v_\nu\}$, $\nu = 0, 1$. Assume that $\mathcal{P}_0$ and $\mathcal{P}_1$ are disjoint.
(i) Then there are uniquely determined constants $\Delta_0 \leq \Delta_1$ such that

$$\frac{dv_1}{dv_0} = \begin{cases} \Delta_0 & \text{if } \frac{1-\varepsilon_1}{1-\varepsilon_0}\frac{dP_1}{dP_0} < \Delta_0 \\ \frac{1-\varepsilon_1}{1-\varepsilon_0}\frac{dP_1}{dP_0} & \text{if } \Delta_0 \leq \frac{1-\varepsilon_1}{1-\varepsilon_0}\frac{dP_1}{dP_0} \leq \Delta_1 \\ \Delta_1 & \text{if } \Delta_0 < \frac{1-\varepsilon_1}{1-\varepsilon_0}\frac{dP_1}{dP_0} \end{cases}$$

(ii) There exists a pair $(Q_0,\ Q_1)$, $Q_\nu \in \mathcal{P}_\nu$, $\nu = 0, 1$ such that $\frac{dv_1}{dv_0}$ is a version of $\frac{dQ_1}{dQ_0}$ and

$$Q_0\Big(\frac{dv_1}{dv_0} \leq t\Big) = v_0\Big(\frac{dv_1}{dv_0} \leq t\Big),$$
$$Q_1\Big(\frac{dv_1}{dv_0} \leq t\Big) = v_1\Big(\frac{dv_1}{dv_0} \leq t\Big),\ t \geq 0.$$

(Rieder 1977; Bednarski 1981). The minimax α-test rejects $\mathcal{P}_0$ in favor of $\mathcal{P}_1$ if $\frac{dQ_1}{dQ_0} > k$ and rejects it with probability $\gamma \in (0,1)$ if $\frac{dQ_1}{dQ_0} = k$ where k, γ are determined so that $Q_1\Big(\frac{dQ_1}{dQ_0} > k\Big) + \gamma\, Q_1\Big(\frac{dQ_1}{dQ_0} = k\Big) = \alpha$.

9.4 Robust Confidence Sets

The confidence intervals or generally the confidence sets are often preferred by practitioners. However, the confidence intervals based on asymptotic normality (or even on Edgeworth expansion) are even more vulnerable to departures from the normality, and hence the robustness aspects should be taken into account.

The confidence interval for the parameter θ based on a sample $X_1, \ldots, X_n$ is interval $I_n = [L_n, U_n]$, where $L_n = L(X_1, \ldots, X_n)$ and $U_n = U(X_1, \ldots, X_n)$ are the *lower and upper confidence limits*, such that

$$I\!P_\theta\{L_n \leq \theta \leq U_n\} \geq 1 - \alpha,\ \forall \theta \in \Theta, \tag{9.37}$$

i.e. the probability of coverage of the right value of θ by the interval I_n is at least equal to the *confidence coefficient* $1 - \alpha$, where $\alpha \in (0,1)$ is generally small. In a nonparametric setup where $\theta = \theta(F),\ F \in \mathcal{F}$ is a functional of the distribution function F of the form (9.37), rewritten as

$$I\!P_F\{L_n \leq \theta(F) \leq U_n\} \geq 1 - \alpha,\ \forall F \in \mathcal{F}.$$

If $\boldsymbol{\theta}(F)$ is vector valued, we consider a *confidence region*, a subspace I_n of $\boldsymbol{R}_p$, usually closed, such that

$$I\!P_F\{\boldsymbol{\theta}(F) \in I_n\} \geq 1 - \alpha,\ \ \forall F \in \mathcal{F}. \tag{9.38}$$

A further modification is also possible to cover the case where $\theta(F)$ is possibly infinite-dimensional. Confidence sets are therefore random subsets of the parameter space, for which the coverage probability is preassigned, and which further have some desirable or optimal properties. The crux of the problem is therefore to choose the two confidence limits L_n and U_n in such a way that the coverage probability condition holds, while the length of the interval $\ell_n = U_n - L_n$ is as small as possible. The shortest confidence interval based on a finite sample is generally possible to get in a parametric model, while in the nonparametric setup we should use an asymptotic approach.

In some special cases, like the location parameter of a symmetric distribution or regression parameter in a simple regression model, exact distribution free confidence intervals can be obtained by nonparametric methods (Sen 1968): see some problems at the end of the chapter.

We distinguish two types of confidence intervals (or generally the confidence sets): Those based on the point estimation of parameter θ of interest (Type I) and the intervals obtained by an inversion of tests on θ (Type II). We shall describe both for the one parameter case and later consider an extension to the multiparameter case.

9.4.1 Type I Confidence Intervals

Let $T_n = T(X_1, \ldots, X_n)$ be an estimator of a parameter $\theta = \theta(F)$, based on a sample of n *i.i.d.* observations $X_1, \ldots, X_n$ from the distribution function F. Let

$$G_n(t; F) = I\!P_F\{T_n - \theta \leq t\},\ t \in \boldsymbol{R}, \tag{9.39}$$

be the distribution function of $T_n - \theta$. Given a coverage probability $1 - \alpha$ $(0 < \alpha < 1)$, take a partition $\alpha = \alpha_1 + \alpha_2,\ 0 \leq \alpha_1, \alpha_2 \leq \alpha$, and let

$$\begin{aligned} t^-_{n\alpha_1} &= \sup\{t : G_n(t; F) \leq \alpha_1\} \\ t^+_{n\alpha_2} &= \inf\{t : G_n(t; F) \geq 1 - \alpha_2\}. \end{aligned}$$

Then

$$I\!P_F\{t^-_{n\alpha_1} \leq T_n - \theta \leq t^+_{n\alpha_2}\} \geq 1 - \alpha. \tag{9.40}$$

If $G_n(t; F)$ is symmetric about 0, then one may set $\alpha_1 = \alpha_2 = \alpha/2$ and $t^-_{n\alpha_1} = -t^+_{n\alpha_2} = -t^+_{n,\alpha/2}$. If $t^-_{n\alpha_1}$ and $t^+_{n\alpha_2}$ do not depend on θ (or on F when

F is not specified), then we obtain from (9.40) by inversion that

$$\mathbb{P}_F\{T_n - t^+_{n\alpha_2} \leq \theta \leq T_n + t^-_{n\alpha_1}\} \geq 1 - \alpha \ \forall F \in \mathcal{F}, \tag{9.41}$$

and this provides the desired confidence interval of θ. In practice, this simple prescription may not work out even for the simple parametric models (e.g., θ binomial parameter, Poisson parameter, or the normal mean when the variance σ^2 is not known). In some cases, this drawback may be overcome by using a studentized form $\frac{T_n-\theta}{\sqrt{Z_{nT}}}$ with a suitable Z_{nT} (e.g., the normal mean problem with unknown σ^2). Asymptotically, we have under general regularity assumptions

$$n^{1/2}(T_n - \theta) \xrightarrow{\mathcal{D}} \mathcal{N}(0, \sigma_T^2), \qquad \frac{Z_{nT}}{\sigma_T^2} \xrightarrow{P} 1, \tag{9.42}$$

hence, by the Slutsky theorem,

$$\frac{n^{1/2}(T_n - \theta)}{\sqrt{Z_{nT}}} \xrightarrow{\mathcal{D}} \mathcal{N}(0, 1) \ \text{ as } \ n \to \infty.$$

In this case we have

$$\lim_{n\to\infty} P_F\Big\{\frac{|n^{1/2}(T_n - \theta)|}{\sqrt{Z_{nT}}} \leq \tau_{\alpha/2}\Big\} = 1 - \alpha, \tag{9.43}$$

so

$$I_n = \Big\{T_n - n^{-1/2}\tau_{\alpha/2}\sqrt{Z_{nT}},\ T_n + n^{-1/2}\tau_{\alpha/2}\sqrt{Z_{nT}}\Big\} \tag{9.44}$$

provides an asymptotic confidence interval for θ with coverage probability $1 - \alpha$. This we term a *type I confidence interval* for θ. The length ℓ_{nT} of interval (9.44) satisfies

$$n^{1/2}\ell_{nT} = 2\tau_{\alpha/2}\sqrt{Z_{nT}} \xrightarrow{P} 2\tau_{\alpha/2}\sigma_T, \text{ as } n \to \infty.$$

hence, an optimal choice is T_n with a minimum σ_T^2, that is, T_n is BAN (*best asymptotically normal*) estimator.

A robust confidence interval would be based on robust T_n accompanied by a robust choice of Z_{nT}, leading to a minimum σ_T^2 within a class. The asymptotic optimality of such a confidence interval can be studied through the optimality properties of Z_{nT}. If $\theta = \theta(F)$ is treated as a functional of the underlying distribution function F, this prescription works out well for the general class of estimators discussed in earlier chapters. From the point of view of practical applications, one must examine the adequacy of an asymptotic normal approximation (9.43) for moderate sample sizes. In the parametric case, we can often derive an exact distribution of $n^{1/2}(T_n - \theta)/\sqrt{Z_{nT}}$, and for moderate sample sizes we would find an adequate solution through the Edgeworth-type expansions. Such simple second-order approximations may not be readily adaptable in a robustness setup, although we can try to use some second order asymptotic distribution representations discussed in earlier chapters.

If (9.43) holds, then the basic problem is to choose an appropriate $\{Z_{nT}\}$

that consistently and robustly estimates σ_T^2 and at the same time makes the asymptotic result tenable for finite to moderate sample sizes. In a parametric setup we often use *variance stabilizing transformations of statistics*; the classical examples are the arc sine transformation on the binomial proportion, the log-transformation for the variance parameter, the square-root transformation for Poisson and χ^2 variates, and the tanh inverse transformation for the simple correlation coefficient. However, such a transformation depends explicitly on the structure of the variance σ_T^2 (as a function of the unknown parameters), and in a nonparametric setup such an exact formulation may not be possible. Even when it is feasible, it may not have the simple form for which the Bartlett transformation methodology can be readily adopted. For this reason, a robust choice of Z_{nT} or even estimating the distribution function of $n^{1/2}(T_n - \theta)$ is often made with the aid of *resampling plans*. Among these resampling plans, *jackknifing* and *bootstrapping* deserve special mention. Let us describe their roles in robust interval estimation of a real parameter (Sen (1977a).

Jackknife Method of Estimating σ_T^2

Let T_n be a suitable estimator of θ based on the sample $\{X_1, \ldots, X_n\}$ of size n. Assume weak convergence (9.42), although σ_T^2 may be generally unknown and T_n may not be unbiased for θ. To motivate jackknifing, suppose that

$$\mathbb{E}_F(T_n) = \theta + n^{-1}a + n^{-2}b + \ldots, \tag{9.45}$$

where $a, b, \ldots$ are unknown quantities, eventually depending on the underlying F. Let T_{n-1} be the same estimator of θ but based on a subsample of size $n-1$. Then by (9.45),

$$\mathbb{E}_F(T_{n-1}) = \theta + (n-1)^{-1}a + (n-1)^{-2}b + \ldots,$$

so

$$\mathbb{E}_F\{nT_n - (n-1)T_{n-1}\} = \theta - \frac{b}{n(n-1)} + \mathrm{O}(n^{-3}). \tag{9.46}$$

hence, the effective bias is $\mathrm{O}(n^{-2})$ instead of $\mathrm{O}(n^{-1})$. Motivated by this, we drop the ith observation (X_i) from the base sample $(X_1, \ldots, X_n)$ and denote the estimator based on this subsample of size $n-1$ by $T_{n-1}^{(i)}$, for $i = 1, \ldots, n$. Following (9.46), we define the *pseudovariables* as

$$T_{n,i} = nT_n - (n-1)T_{n-1}^{(i)}, \; i = 1, \ldots, n. \tag{9.47}$$

Then the jackknifed version T_n^J of T_n is defined by

$$T_n^J = n^{-1}\sum_{i=1}^{n} T_{n,i}. \tag{9.48}$$

Note that by (9.46), (9.47) and (9.48) we have

$$\mathbb{E}_F(T_n^J) = \theta - \frac{b}{n(n-1)} + \mathrm{O}(n^{-3}).$$

So the bias of T_n^J is $\mathrm{O}(n^{-2})$ instead of $\mathrm{O}(n^{-1})$ for T_n. Let us further define

$$V_n^J = (n-1)^{-1}\sum_{i=1}^{n}(T_{n,i} - T_n^J)^2 = (n-1)\sum_{i=1}^{n}(T_{n-1}^{(i)} - T_n^{\star})^2, \qquad (9.49)$$

where

$$T_n^{\star} = n^{-1}\sum_{i=1}^{n} T_{n-1}^{(i)}. \qquad (9.50)$$

V_n^J is termed the *jackknifed (or Tukey) variance estimator.* If, in particular $T_n = \bar{X}_n$, then $T_{n,i} = X_i$, $i = 1, \ldots, n$, so $T_n^{\star} = \bar{X}_n$, and hence $V_n^J = (n-1)^{-1}\sum_{i=1}^{n}(X_i - \bar{X}_n)^2 = S_n^2$. Let $\mathcal{C}_n$ be the sigma-field generated by the collection of real $\{X_1, \ldots, X_n\}$ and by X_{n+j}, $j \geq 1$, for $n \geq 1$. Then $\mathcal{C}_n = \mathcal{C}(X_{n:1}, \ldots X_{n:n}; X_{n+j}, j \geq 1)$, where $X_{n:1} \leq \ldots \leq X_{n:n}$ are the order statistics corresponding to $X_1, \ldots, X_n$, and $\mathcal{C}_n$ is nonincreasing in n (≥ 1). further, note that by (9.50),

$$T_n^{\star} = \mathbb{E}\{T_{n-1}|\mathcal{C}_n\}, \; n \geq n_0, \qquad (9.51)$$

where $n_0 = \inf\{m : \; T_{m-1} \text{ is well defined}\}$. As a result

$$\begin{aligned} T_n^J &= T_n + (n-1)\mathbb{E}\{(T_n - T_{n-1})|\mathcal{C}_n\} \\ &= T_n - (n-1)\mathbb{E}\{(T_{n-1} - T_n)|\mathcal{C}_n\}, \; n \geq n_0. \end{aligned}$$

If $\{T_n, \mathcal{C}_n; \; n \geq n_0\}$ is a reverse martingale, then $\mathbb{E}(T_{n-1}|\mathcal{C}_n) = T_n$, so $T_n^J = T_n$, $\forall n \geq n_0$. By (9.49) and (9.51) we have

$$\begin{aligned} V_n^J &= n(n-1)\mathbb{E}\{(T_{n-1} - T_n^{\star})^2|\mathcal{C}_n\} \\ &= n(n-1)\mathbb{E}\{[T_{n-1} - \mathbb{E}(T_{n-1}|\mathcal{C}_n)]^2|\mathcal{C}_n\} \\ &= n(n-1)\mathrm{Var}(T_{n-1}|\mathcal{C}_n), \; n \geq n_0. \end{aligned}$$

If T_n is a reversed martingale, then a sufficient condition for the asymptotic normality (9.42) of T_n is that

$$n(n-1)\mathrm{Var}((T_{n-1} - T_n)|\mathcal{C}_n) \to \sigma_T^2 \text{ a.s.,} \quad \text{as } n \to \infty$$

[see (2.88)]; then V_n^J converges a.s. to σ_T^2 as $n \to \infty$. If $\{T_n\}$ is not a reversed martingale, the convergence $V_n^J \to \sigma_T^2$ a.s. can be proved with the aid of asymptotic representations of robust estimators, derived in earlier chapters. Assume that T_n admits the representation

$$T_n - \theta = n^{-1}\sum_{i=1}^{n}\psi(X_i; \theta) + R_n = n^{-1/2}Z_n^{(1)} + n^{-1}Z_n^{(2)} \qquad (9.52)$$

where $\mathbb{E}_F\psi(X;\theta)=0,\ \mathbb{E}_F[\psi(X_i,\theta)]^2=\sigma_T^2:0<\sigma_T^2<\infty$ and where

$$\begin{aligned} Z_n^{(1)} &= n^{-1/2}\sum_{i=1}^{n}\psi(X_i;\theta)\sim\mathcal{N}(0,\sigma_T^2);\\ Z_n^{(2)} &= nR_n. \end{aligned}$$

Define $R_{n,i}$ and $R_n^\star$ as in (9.47) and (9.50) (with T_n replaced by R_n). Then by (9.52) we have

$$T_{n,i}-\theta=\psi(X_i,\theta)+R_{n,i},\ i=1,\ldots,n. \tag{9.53}$$

Let us denote

$$\begin{aligned} &V_n^\star=(n-1)^{-1}\sum_{i=1}^{n}\{\psi(X_i;\theta)-\frac{1}{n}\sum_{j=1}^{n}\psi(X_j,\theta)\}^2,\\ &V_{n,2}=(n-1)^{-1}\sum_{i=1}^{n}(R_{n,i}-R_n^\star)^2, \qquad (9.54)\\ &V_{n,12}=(n-1)^{-1}\sum_{i=1}^{n}\Big[\psi(X_i;\theta)-\frac{1}{n}\sum_{j=1}^{n}\psi(X_j;\theta)\Big](R_{n,i}-R_n^\star). \end{aligned}$$

By (9.49), (9.53), and (9.54), we have

$$V_n^J=V_n^\star+V_{n,2}+2V_{n,12}.$$

Now $V_n^\star$ is a U-statistic corresponding to a kernel of degree 2, and hence is a reversed martingale, thus,

$$V_n^\star\to\mathbb{E}V_n^\star=\sigma_T^2\ \text{ as } n\to\infty.$$

The crux of the problem is to verify that

$$\begin{aligned} V_{n,2} &= n(n-1)\mathbb{E}\{(R_{n-1}-\mathbb{E}(R_{n-1}|\mathcal{C}_n))^2|\mathcal{C}_n\}\\ &\to 0,\ \text{in probability/a.s. as } n\to\infty \end{aligned} \tag{9.55}$$

which further implies that $V_{n,12}\to 0$ in probability/a.s., too. Recall that

$$\begin{aligned} &\mathbb{E}\{(R_{n-1}-\mathbb{E}(R_{n-1}|\mathcal{C}_n))^2|\mathcal{C}_n\}=\mathrm{Var}(R_{n-1}-R_n)|\mathcal{C}_n)\\ &\le\mathbb{E}\{(R_{n-1}-R_n)^2|\mathcal{C}_n\},\ \forall n\ge n_0, \end{aligned} \tag{9.56}$$

so

$$\begin{aligned} \mathbb{E}V_{n,2} &\le \mathbb{E}\{\mathbb{E}\{(R_{n-1}-R_n)^2|\mathcal{C}_n\}\}\cdot n(n-1)\\ &= n(n-1)\mathbb{E}(R_{n-1}-R_n)^2. \end{aligned}$$

Then we have by (9.52) and (9.54)

$$\begin{aligned} V_{n,2} &= (n-1)^{-1}\sum_{i=1}^{n}(Z_{n-1}^{i(2)}-Z_n^{(2)})^2\\ &= n(n-1)^{-1}\mathbb{E}\{[Z_{n-1}^{(2)}-\mathbb{E}(Z_{n-1}^{(2)}|\mathcal{C}_n)]^2|\mathcal{C}_n\} \end{aligned}$$

$$\leq \; n(n-1)^{-1}\mathbb{E}\{(Z_{n-1}^{(2)} - Z_n^{(2)})^2|\mathcal{C}_n\}.$$

Therefore

$$\begin{aligned} \text{If } \mathbb{E}\{(Z_{n-1}^{(2)} - Z_n^{(2)})^2|\mathcal{C}_n\} &\rightarrow 0 \text{ a.s.,} \\ \text{then } V_{n,2} &\rightarrow 0 \text{ a.s. as } n \to \infty. \end{aligned} \tag{9.57}$$

If T_n is a U-statistics or the von Mises functional, then the classical Hoeffding decompositionyields (9.57). In general, (9.55) holds (a.s.) under (9.56) provided that

$$R_n = R_{n0} + R_{n1}, \; n \geq n_0, \tag{9.58}$$

where $\{R_{n0}, \mathcal{C}_n\}$ is a reversed martingale with

$$n^2\mathbb{E}(R_{n0}) \to c \; (0 \leq c < \infty), \text{ as } n \to \infty,$$

while

$$nR_{n1} \to 0 \text{ a.s., as } n \to \infty, \tag{9.59}$$

which is true for second-order Hadamard differentiable statistical functionals as well as for L-estimators with smooth weight functions. This leads to the following conclusion:

THEOREM 9.2 *Let $\{T_n\}$ satisfy (9.52). Then either of the following conditions is sufficient for $V_n^J \to \sigma_T^2$ in prob/a.s. as $n \to \infty$:*

(i) $n(n-1)(R_n - R_{n-1})^2 \to 0$ *in prob/a.s., as* $n \to \infty$;

(ii) $\{T_n\}$ *is a reversed martingale and*

$$n(n-1)\text{Var}(T_{n-1}|\mathcal{C}_n) \to \sigma_T^2 \text{ a.s. } as\ n \to \infty, \tag{9.60}$$

(iii) $\{T_n\}$ *satisfies (9.58)–(9.59).*

A simple (but illustrative) example where (9.58)–(9.59) and hence (9.60) may not hold, is the simple location model where θ is the population median, the density $f(x,\theta) = f(x-\theta)$ is symmetric, and $f(0) > 0$. Sen (1989a) proposed a modified procedure (called the *delete-k jackknife method*) working for such situations. The nonregular cases should be treated individually, and for some of them there exist alternative methods of solutions. Bootstrapping is one such possibility.

Bootstrap Method of Estimating σ_T^2

Let $(X_1,\ldots,X_n)$ be the original sample with the empirical distribution function F_n and let $T_n = T(X_1,\ldots,X_n) = T(F_n)$ be an estimator. Draw a sample of n observations from the sample distribution function F_n with replacement (under a simple random sampling scheme), and denote them by $X_1^\star,\ldots,X_n^\star$, respectively. Let $F_n^\star$ be the empirical distribution function of these $X_i^\star$, $1 \leq i \leq n$. Then let

$$T_n^\star = T(X_1^\star,\ldots,X_n^\star) \text{ or } T(F_n^\star).$$

The $X_j^\star$ are conditionally independent, given F_n, and

$$\mathbb{P}\{X_j^\star = X_i|F_n\} = 1/n \text{ for every } i = 1,\ldots,n;\ j \geq 1. \tag{9.61}$$

thus, $(X_1^\star,\ldots,X_n^\star)$ can have n^n possible realizations $\{(X_{i_1},\ldots,X_{i_n}):\ i_j = 1,\ldots,n, \text{ for } j = 1,\ldots,n\}$, and each of them has the common (conditional) probability n^{-n}. Therefore

$$\mathbb{E}\{T_n^\star|F_n\} = n^{-n}\sum_{i_1=1}^{n}\cdots\sum_{i_n=1}^{n} T(X_{i_1},\ldots,X_{i_n}),$$

and the conditional mean square error of $T_n^\star$ (given F_n) is

$$\mathbb{E}\{(T_n^\star - T_n)^2|F_n\} = n^{-n}\sum_{i_1=1}^{n}\cdots\sum_{i_n=1}^{n}[T(X_{i_1},\ldots,X_{i_n}) - T_n]^2. \tag{9.62}$$

Under general regularity conditions,

$$n\mathbb{E}\{(T_n^\star - T_n)^2|F_n\} \to \sigma_T^2, \text{ in probability},$$

as $n \to \infty$. To avoid the awkward computation of the conditional mean square error in (9.62), Efron (1979, 1982) proposed the following resampling scheme: Draw independently a large number M of such bootstrap samples from F_n. Denote the observations in the kth bootstrap sample as $(X_{k1}^\star,\ldots,X_{kn}^\star)$ for $k = 1,\ldots,M$. These $X_{ki}^\star$ are conditionally *i.i.d.* random variables with the distribution function F_n, and (9.61) applies to each of them. Let

$$T_{nk}^\star = T(X_{k1}^\star,\ldots,X_{kn}^\star) \text{ or } T(F_{kn}^\star),\ k = 1,\ldots,M,$$

and define

$$V_{nM}^\star = \frac{n}{M}\sum_{k=1}^{M}(T_{nk}^\star - T_n)^2.$$

Then $V_{nM}^\star$ is termed the *bootstrap estimator* of σ_T^2. The resampling plan is thus the simple random sampling with replacement. The M replications not only allow a consistent estimator of σ_T^2, but they also generate the bootstrap empirical distribution function,

$$G_n^\star(y) = \frac{1}{M}\sum_{k=1}^{M} I\{n^{1/2}(T_{kn}^\star - T_n) \leq y\},\ y \in \boldsymbol{R}.$$

Let $G(y)$ denote the limit law $\lim_{n\to\infty} \mathbb{P}_F\{n^{1/2}(T_n - \theta) \leq y\}$, whenever it exists. Then G_n uniformly consistently estimates G under certain regularity conditions. Therefore, the quantiles of $G_n^\star(.)$ estimate the percentile points of $G(.)$ and hence, one may derive an asymptotic confidence interval for θ by using $G_n^\star(.)$. The approximation can be still improved by using the Edgeworth expansion for $G_n^\star(y)$; we refer to Hall (1992) for a detailed account. By Sen (1988b) and Hall (1992), whenever T_n admits a first-order representation [see

(9.52)], then $G(y)$ is a normal distribution function with null mean and variance σ_T^2, to which $G_n^\star(.)$ converges. hence, the (asymptotic) confidence interval for θ can be derived using $V_{nM}^\star$ or $G_n^\star(.)$. This works even for some non-normal $G(.)$; here, the bootstrap has some advantage over jackknifing [which rests on (9.42)], but on the other hand, an ideal choice of $M = n^n$ is unpracticable. It seems that jackknifing is more advantageous in most regular cases where $G(.)$ is Gaussian.

In the regular case, $n^{1/2}(F_n^\star - F_n)$ converges weakly, conditionally given F_n, to a Brownian bridge. This implies that $n^{1/2}(T_n^\star - T_n)$ converges weakly, conditionally given F_n, to the same functional of the Brownian bridge to which $n^{1/2}(T_n - \theta)$ converges in law (unconditionally). However, this equivalence may not generally hold in nonregular cases (see van Zwet 1992).

9.4.2 Type II Confidence Intervals

Let $\{T_n = T(X_1, \ldots, X_n)\}$ be a sequence of real valued estimators of parameter θ, for example, M- and R-estimators of location or regression. We assume the following conditions:

(i) There is a suitable hypothesis $\mathbf{H}_0$ on θ such that the distribution of T_n is independent of θ under validity of $\mathbf{H}_0$.

(ii) There exists a transformation $(X_1, \ldots, X_n)^\top \mapsto (X_1^{(t)}, \ldots, X_n^{(t)})^\top$ for every real t (e.g., $X_j^{(t)} = X_j - t,\ j \geq 1$) such that $(X_1^{(\theta)}, \ldots, X_n^{(\theta)})^\top$ has the same distribution when θ holds as $(X_1, \ldots, X_n)^\top$ has when $\mathbf{H}_0 : \theta = 0$ holds.

If T_n is distribution-free under $\mathbf{H}_0$, then we get a distribution-free confidence interval for θ. If T_n is asymptotically distribution-free and asymptotically normal with asymptotic variance σ_T^2 under $\mathbf{H}_0$ and σ_T^2 can be consistently estimated from the sample, then we obtain an asymptotic confidence interval for θ.

Because of condition (i), there exist two numbers $T_{n,\alpha}^{(1)}$ and $T_{n,\alpha}^{(2)}$ to confidence coefficient $1 - \alpha$ $(0 < \alpha < 1)$, such that

$$I\!P\{T_{n,\alpha}^{(1)} \leq T_n \leq T_{n,\alpha}^{(2)} | \mathbf{H}_0\} \geq 1 - \alpha, \tag{9.63}$$

and $T_{n,\alpha}^{(1)}$, $T_{n,\alpha}^{(2)}$ do not depend on θ. If the distribution function of T_n under $\mathbf{H}_0$ is symmetric (about 0, without loss of generality), we can take $T_{n,\alpha}^{(1)} = -T_{n,\alpha}^{(2)}$, and choose $T_{n,\alpha}^{(2)}$ as the smallest value for which the right-hand side of (9.63) is $\geq 1 - \alpha$. If T_n is not symmetric but asymptotically normal under $\mathbf{H}_0$, then for large n we can take $T_{n,\alpha}^{(1)} = -T_{n,\alpha}^{(2)}$.

Recalling condition (ii), denote

$$T_n(t) = T(X_1^{(t)}, \ldots, X_n^{(t)}),\ t \in \boldsymbol{R}_1 \tag{9.64}$$

and define

$$\begin{aligned} \hat{\theta}_{L,n} &= \inf\left\{\theta :\ T_n(\theta) \in [T_{n,\alpha}^{(1)}, T_{n,\alpha}^{(2)}]\right\}, \\ \hat{\theta}_{U,n} &= \sup\left\{\theta :\ T_n(\theta) \in [T_{n,\alpha}^{(1)}, T_{n,\alpha}^{(2)}]\right\}. \end{aligned} \tag{9.65}$$

Then the interval

$$I_n = \{\theta : \hat{\theta}_{L,n} \le \theta \le \hat{\theta}_{U,n}\}. \tag{9.66}$$

satisfies

$$\mathbb{P}_\theta\{\theta : \hat{\theta}_{L,n} \le \theta \le \hat{\theta}_{U,n}\} \ge 1 - \alpha, \tag{9.67}$$

by virtue of (9.63) and conditions (i) and (ii); so I_n is a distribution-free confidence interval for θ. For the one-sample and two-sample location/scale models, such confidence intervals were proposed by Lehmann (1963) and Sen (1963). A host of research workers, such as Sen (1968a), Jurečková (1969,1971), and Koul (1969), extended the methodology to a much wider class of problems (including the regression model). As an illustration, we shall describe confidence intervals based on ranks and their extensions based on M-statistics. The simplest case is the confidence interval for the population median θ, based on the sign statistic, where even the symmetry of F around θ is not needed. Let $T_n(a) = \sum_{i=1}^n \operatorname{sign}(X_i - a)$, $a \in \mathbb{R}$. Then

(i) $T_n(a)$ is $\searrow$ in a and

(ii) $T_n^*(a) = \frac{1}{2}(T_n(a) + n)$ has the binomial distribution $b(n, F(a))$.

(iii) There exists an $r \le \frac{n}{2}$ such that

$$\mathbb{P}\{r \le T_n^*(\theta) \le n - r\} = \sum_{k=r}^{n-r} \binom{n}{k} \left(\frac{1}{2}\right)^n \ge 1 - \alpha.$$

thus, $\{X_{n:r} \le \theta \le X_{n:n-r+1}\}$ is a distribution-free confidence interval for θ.

Consider a signed-rank statistic

$$S_n = S(X_1, \ldots, X_n) = \sum_{i=1}^n \operatorname{sign} X_i\ a_n(R_{ni}^+),$$

where the scores $a_n(k)$ relate to the score function $\varphi^\star(u) = \varphi((1+u)/2)$, $0 \le u \le 1$, and $\varphi(u)$ is skew-symmetric (about 1/2), square integrable, and monotone. R_{ni}^+ is the rank of $|X_i|$ among $|X_1|, \ldots, |X_n|$, for $i = 1, \ldots, n$. Then, if F is symmetric, S_n is distribution-free under $\mathbf{H}_0 : \theta = 0$ and has a symmetric distribution with median 0. Moreover, $S(X_1 - \theta, \ldots, X_n - \theta)$ has the same distribution function under θ as S_n has under $\mathbf{H}_0$, and $S(X_1 - t, \ldots, X_n - t)$ is nonincreasing in $t \in \boldsymbol{R}_1$. Consequently, we can use (9.63)–(9.67). Choose

$C_{n,\alpha}$ so that $I\!P\{|S_n| \le C_{n,\alpha}|\mathbf{H}_0\} \ge 1-\alpha$, and $C_{n,\alpha}$ is the smallest value for which this inequality holds. Put $\hat{\theta}_{L,n} = \inf\{t : S(X_1-t,\ldots,X_n-t) > C_{n,\alpha}\}$ and $\hat{\theta}_{U,n} = \sup\{t : S(X_1-t,\ldots,X_n-t) < -C_{n,\alpha}\}$. The desired confidence interval for θ is $(\hat{\theta}_{U,n}, \hat{\theta}_{L,n})$. By Theorem 6.2,

$$n^{1/2}(\hat{\theta}_{U,n} - \hat{\theta}_{L,n}) \xrightarrow{P} (2/\gamma)\tau_{\alpha/2}A_\varphi \quad \text{as } n \to \infty \tag{9.68}$$

where

$$A_\varphi^2 = \int_0^1 \varphi^2(u)du, \;\; \gamma = -\int_{-\infty}^{\infty} \varphi(F(x))f'(x)dx. \tag{9.69}$$

If we choose $\varphi(u) = \text{sign}(u-1/2)$, we have the sign-test statistic. Similarly

$$n^{1/2}(\hat{\theta}_{U,n} - \hat{\theta}_n) \xrightarrow{P} \gamma^{-1}\tau_{\alpha/2}A_\phi, \quad n^{1/2}(\hat{\theta}_{L,n} - \hat{\theta}_n) \xrightarrow{P} -\gamma^{-1}\tau_{\alpha/2}A_\phi,$$

$n \to \infty$, where $\hat{\theta}_n$ is the point R-estimator of θ, obtained by equating $S_n(X_1 - t,\ldots,X_n-t)$ to 0. As a result, we have for large n,

$$\begin{aligned}\left(\hat{\theta}_{L,n}, \hat{\theta}_{U,n}\right) &= \Big(\hat{\theta}_n - n^{-1/2}\gamma^{-1}\tau_{\alpha/2}A_\varphi) + o(n^{-1/2}), \\ &\qquad \hat{\theta}_n + n^{-1/2}\gamma^{-1}\tau_{\alpha/2}A_\varphi) + o(n^{-1/2})\Big)\end{aligned} \tag{9.70}$$

which is comparable to (9.44). We neither need to estimate the variance function A_φ^2/γ^2 nor need to use the point estimator $\hat{\theta}_n$. However, the solutions for $\hat{\theta}_{L,n}$ and $\hat{\theta}_{U,n}$ should be obtained iteratively.

Next we shall consider the simple regression model $Y_i = \theta + \beta x_i + e_i$, $i = 1,\ldots,n$, where the x_i are known regression constants (not all equal), θ, β are unknown parameters, and the e_i are *i.i.d.* random variables with a continuous distribution function F (not necessarily symmetric). Our problem is to provide a distribution-free confidence interval for the regression parameter β. A very simple situation crops up with $S_n = \sum_{1\le i<j\le n} \text{sign}(x_i - x_j)\text{sign}(Y_i - Y_j)$, known as *Kendall's tau coefficient.* Under $\mathbf{H}_0 : \beta = 0$ is S_n distribution-free with location 0, and $S(Y_1 - bx_1,\ldots,Y_n - bx_n)$, $b \in \boldsymbol{R}_1$ is nonincreasing in b. Moreover, $S(Y_1 - \beta x_1,\ldots,Y_n-\beta x_n)$ under β has the same distribution function as S_n has under $\mathbf{H}_0$. Equations (9.65)–(9.67) already led us to the desired solution. The two-sample location model is a particular case of the linear model (when the x_i can only assume the values 0 and 1), and S_n then reduces to the classical two-sample Wilcoxon–Mann–Whitney statistic. Detailed asymptotic results may be found in Ghosh and Sen (1971), where sequential confidence intervals were studied.

Consider a linear rank statistics

$$L_n(b) = \sum_{i=1}^n (x_i - \bar{x}_n)a_n(R_{ni}(b))$$

and note that $L_n = L_n(0)$ is distribution-free under $\beta = 0$ and that under β, $L_n(\beta)$ has the same distribution function as L_n under $\mathbf{H}_0 : \beta = 0$. Further,

$L_n(\beta)$ is nonincreasing in $b \in \boldsymbol{R}_1$, so we can again proceed as in (9.63)–(9.67) and define $\hat{\beta}_{L,n}$ and $\hat{\beta}_{U,n}$ corresponding to a confidence coefficient $1-\alpha$. By the results in Section 6.3, we can conclude that

$$T_n(\hat{\beta}_{U,n} - \hat{\beta}_{L,n}) \xrightarrow{P} 2\gamma^{-1}\tau_{\alpha/2}A_\varphi, \tag{9.71}$$

as n increases, where

$$A_\varphi^2 = \int_0^1 (\varphi(u) - \bar{\varphi})^2 du, \ \ \bar{\varphi} = \int_0^1 \varphi(u)du,$$

$\varphi(u)$ is the score function generating the $a_n(k)$ and γ is defined as in (9.69). Additionally, we set

$$T_n^2 = \sum_{i=1}^n (x_i - \bar{x}_n)^2, \ \ \bar{x}_n = n^{-1}\sum_{i=1}^n x_i,$$

and assume that $T_n \to \infty$ and $\max_{1\le i\le n}\left\{T_n^{-2}(x_i - \bar{x}_n)^2\right\} \to 0$ as $n \to \infty$. Using the results of Chapter 6, we conclude that

$$T_n(\hat{\beta}_{U,n} - \hat{\beta}_n) \xrightarrow{P} \gamma^{-1}\tau_{\alpha/2}A_\varphi, \ \ T_n(\hat{\beta}_{L,n} - \hat{\beta}_n) \xrightarrow{P} -\gamma^{-1}\tau_{\alpha/2}A_\varphi$$

as $n \to \infty$. This expression gives us an analogue of (9.70) with n replaced by T_n^2. Here, as well as in (9.70), there is no need to incorporate jackknifing/bootstrapping to estimate the variance function A_φ^2/γ^2. Similarly as the rank statistics, the derived confidence intervals are globally robust.

Next let us consider confidence intervals based on M-statistics. The statistic $T_n = T(X_1, \ldots, X_n)$ in this case satisfies both conditions (i) and (ii), but its distribution under $\mathbf{H}_0$ may not be distribution-free. For example, a typical M-statistic is $T_n(t) = \sum_{i=1}^n \psi(X_i - t)$, $t \in \boldsymbol{R}_1$. Then for the location model with $\theta = 0$, the exact distribution of $T_n = T_n(0)$ depends on the underlying F. If $n^{-1/2}T_n(0)$ is asymptotically normal with mean 0 and variance σ_T^2 under $\mathbf{H}_0$, and if V_n is a consistent estimator of σ_T^2, then for large n we can set

$$n^{-1/2}T_{n,\alpha}^{(j)} = (-1)^j \tau_{\alpha/2} V_n^{1/2} + o_p(1), \ j = 1, 2.$$

With this modification, we proceed as in (9.65)–(9.66) and obtain a confidence interval as in (9.67).

More precisely, consider the M-statistic for the one-sample location model $F(x,\theta) = F(x-\theta), F$ symmetric about 0,

$$M_n(t) = \sum_{i=1}^n \psi(X_i - t), \ t \in \boldsymbol{R}_1. \tag{9.72}$$

Then $M_n(t)$ is monotone in t for monotone $\psi(.)$. Let $\hat{\theta}_n$ be the M-estimator of θ based on (9.72), defined by (5.24)–(5.26). Under $\mathbf{H}_0 : \theta = 0$, the asymptotic distribution of $n^{-1/2}M_n(0)$ is normal $\mathcal{N}(0, \sigma_\psi^2)$, where $\sigma_\psi^2 = \int_{-\infty}^{\infty} \psi^2(x)dF(x)$.

A natural estimator of σ_ψ^2 is

$$V_n = n^{-1} \sum_{i=1}^{n} \psi^2(X_i - \hat{\theta}_n).$$

If we define

$$\begin{aligned} \hat{\theta}_{L,n} &= \inf\{t : n^{-1/2} M_n(t) > \tau_{\alpha/2} V_n^{1/2}\}, \\ \hat{\theta}_{U,n} &= \sup\{t : n^{-1/2} M_n(t) < -\tau_{\alpha/2} V_n^{1/2}\}, \end{aligned}$$

then we obtain a robust (asymptotic) confidence interval for θ, where

$$n^{1/2}(\hat{\theta}_{U,n} - \hat{\theta}_{L,n}) \xrightarrow{P} 2\gamma^{-1}\tau_{\alpha/2}\sigma_\phi, \text{ as } n \to \infty, \tag{9.73}$$

and

$$\gamma = \gamma_1 + \gamma_2, \quad \text{is defined by } \mathbf{A4} \text{ in Section 5.3.}$$

We could have also employed Theorem 5.3 [i.e., the asymptotic linearity result in (5.29)] to formulate an estimator $\hat{\gamma}_n$ of γ, so that σ_ψ/γ might be consistently estimated by $V_n^{1/2}/\hat{\gamma}_n$. The modifications for the simple regression model(but confined to a single parameter) are straightforward.

Finally, let us consider the efficiency criteria for comparing confidence intervals. This is mainly adapted from Sen (1966). Suppose that we have two sequences $\{I_n\}$ and $\{I_n^\star\}$ of confidence intervals for θ based on the same sequence of observations, where

$$I_n = (\theta_{L,n}, \theta_{U,n}), \; I_n^\star = (\hat{\theta}_{L,n}^\star, \hat{\theta}_{U,n}^\star),$$

and

$$\lim_{n\to\infty} I\!P_\theta\{\theta \in I_n\} = 1 - \alpha = \lim_{n\to\infty} I\!P_\theta\{\theta \in I_n^\star\}.$$

Note that I_n and $I_n^\star$ are generally stochastically dependent. Suppose that there exist constants δ and $\delta^\star$: $0 < \delta, \delta^\star < \infty$ such that

$$n^{1/2}(\hat{\theta}_{U,n} - \hat{\theta}_{L,n}) \xrightarrow{P} \delta \text{ and } n^{1/2}(\hat{\theta}_{U,n}^\star - \hat{\theta}_{L,n}^\star) \xrightarrow{P} \delta^\star. \tag{9.74}$$

Then the asymptotic relative efficiency (ARE) of $\{I_n\}$ with respect to $\{I_n^\star\}$ is

$$e(I, I^\star) = \rho^2 = (\delta^\star/\delta)^2.$$

It is also equal to

$$\lim_{n\to\infty} (N_n^\star/N_n) \tag{9.75}$$

where $\{N_n\}$ and $\{N_n^\star\}$ are two sequences of sample sizes leading to equal lengths of I_n, I_n^*. For type I confidence intervals, (9.42) and (9.44) imply that the ARE agrees with the conventional Pitman ARE of the corresponding point estimators. The same applies to confidence intervals of type II, as we can see by (9.68), (9.71), or (9.73), which contain σ_ψ^2/γ^2, the asymptotic variance of the corresponding point estimator.

9.5 Multiparameter Confidence Sets

Consider a vector parameter $\boldsymbol{\theta} = \boldsymbol{\theta}(F) = (\theta_1(F), \ldots, \theta_p(F))^\top \in \boldsymbol{\Theta} \subset \boldsymbol{R}_p$, $p \geq 1$. The sample observations $X_1, \ldots, X_n$ may be real- or vector-valued. The subject of confidence set for $\boldsymbol{\theta}$ includes several important problems:

1. *Multivariate location model.* The $\mathbf{X}_i$ are themselves p-variate random variables, with a distribution function $F(\mathbf{x}, \boldsymbol{\theta}) = F(\mathbf{x} - \boldsymbol{\theta})$, $\mathbf{x} \in \boldsymbol{R}_p$, $\boldsymbol{\theta} \in \boldsymbol{\Theta} \subset \boldsymbol{R}_p$, and the form of $F(\mathbf{y})$ is assumed to be independent of $\boldsymbol{\theta}$. The goal is to construct robust and efficient confidence regions for $\boldsymbol{\theta}$.
2. *General univariate linear model.* Consider the model
$$Y_i = \theta_0 + \mathbf{x}_i^\top \boldsymbol{\beta} + e_i, \quad i = 1, \ldots, n, \tag{9.76}$$
where the $\mathbf{x}_i = (x_{i1}, \ldots, x_{iq})^\top$ are known but not all equal regression vectors, $\boldsymbol{\beta} = (\beta_1, \ldots, \beta_q)^\top$ is an unknown regression parameter, θ_0 is an unknown intercept parameter, $q \geq 1$, and the e_i are *i.i.d.* random variables with a continuous distribution function F defined on $\boldsymbol{R}_1$. thus, $p = q + 1$, and our goal is to provide a simultaneous confidence region for $(\theta_0, \boldsymbol{\beta})$ (or for $\boldsymbol{\beta}$ when $q \geq 2$).
3. *Multivariate linear model.* Let the $\mathbf{Y}_i, \boldsymbol{\theta}_0$, and the $\mathbf{e}_i$ in (9.76) be all r-vectors, $r \geq 1$, and let $\boldsymbol{\beta}^\top$ be a $r \times q$ matrix of unknown regression parameters. Here the $\mathbf{e}_i$ are *i.i.d.* random variables with a distribution function F defined on $\boldsymbol{R}_r$, and the form of F is assumed independent of $(\boldsymbol{\theta}_0, \boldsymbol{\beta}^\top)$. We desire to have robust simultaneous confidence regions for $(\boldsymbol{\theta}_0, \boldsymbol{\beta}^\top)$ (or $\boldsymbol{\theta}_0$ or $\boldsymbol{\beta}$).
4. *Location and scale parameters.* Consider the model
$$F(x, \boldsymbol{\theta}) = F\Big(\frac{x - \theta_1}{\theta_2}\Big), \; x \in \boldsymbol{R}_1, \tag{9.77}$$
where $\boldsymbol{\theta} = (\theta_1, \theta_2) \in \boldsymbol{R}_1 \times \boldsymbol{R}_1^+$. Then θ_1 is termed the location parameter and θ_2 the scale parameter, and F belongs to the class of location-scale family of distribution functions. It is also possible to conceive multivariate extensions of the model in (9.77).

Our goal is to construct robust simultaneous confidence regions for $\boldsymbol{\theta} = (\theta_1, \ldots, \theta_p)^\top$ in a multiparameter setup. As in the preceding section, we have type I and type II confidence regions. Similarly as in (9.39)–(9.41), let $\mathbf{T}_n = (T_{n1}, \ldots, T_{np})^\top$ be a suitable estimator of $\boldsymbol{\theta}$ and let
$$G_n(\mathbf{y}, \boldsymbol{\theta}) = I\!P_{\boldsymbol{\theta}}\{\mathbf{T}_n \leq \mathbf{y}\}, \; \mathbf{y} \in \boldsymbol{R}_p, \tag{9.78}$$
be the distribution function of $\mathbf{T}_n$, defined on $\boldsymbol{R}_p$. Since (9.78) is defined on $\boldsymbol{R}_p$, the question is how to choose the shape of region in $\boldsymbol{R}_p$ as the desired confidence region? For example, we might want a multidimensional rectangle $\mathbf{T}_{L,n} \leq \boldsymbol{\theta} \leq \mathbf{T}_{U,n}$ (where $\mathbf{a} \leq \mathbf{b}$ means coordinatewise inequalities) chosen in such a way that (9.38) holds. Alternatively, we might choose an ellipsoidal (or

spherical) region. If $G_n(.)$ is multivariate, we might be naturally attracted by the Mahalanobis distance, and some alternative criteria exist for the matrix-valued normal distribution. Roy's (1953) largest root criterion leads to some optimal properties of the associated confidence sets (e.g., see Wijsman 1981). The confidence sets corresponding to multivariate normal distributions are affine equivariant. The multivariate robust estimators are generally not affine equivariant; the construction of robust confidence sets is mostly based on the asymptotic (multi-)normality of the allied statistics. In the multivariate case, the distribution function $G_n(.)$ in (9.78) may not be distribution-free even under null hypothesis and for a nonparametric statistic $\mathbf{T}_n$.

thus, suppose that there exists a sequence $\{\mathbf{T}_n\}$ of estimators of $\boldsymbol{\theta}$, asymptotically normal,

$$n^{1/2}(\mathbf{T}_n - \boldsymbol{\theta}) \xrightarrow{\mathcal{D}} \mathcal{N}_p(\mathbf{0}, \boldsymbol{\Sigma}) \quad \text{as } n \to \infty \tag{9.79}$$

and that there exists a sequence $\{\mathbf{V}_n\}$ of consistent estimators of $\boldsymbol{\Sigma}$. If $\boldsymbol{\Sigma}$ is positive definite, we get from (9.79), Slutsky's theorem, along with Cochran's theorem that

$$n(\mathbf{T}_n - \boldsymbol{\theta})^\top \mathbf{V}_n^{-1}(\mathbf{T}_n - \boldsymbol{\theta}) \xrightarrow{\mathcal{D}} \chi_p^2 \quad \text{as } n \to \infty \tag{9.80}$$

where χ_p^2 has the central chi square distribution with p degrees of freedom. By the Courant theorem on the ratio of two quadratic forms,

$$\begin{aligned} &\sup_{\boldsymbol{\lambda}\neq 0}\left[\frac{(\boldsymbol{\lambda}^\top(\mathbf{T}_n - \boldsymbol{\theta}))^2}{\boldsymbol{\lambda}^\top \mathbf{V}_n \boldsymbol{\lambda}}\right] \\ &= \mathrm{Ch}_{\max}\left\{\mathbf{V}_n^{-1}(\mathbf{T}_n - \boldsymbol{\theta})^\top(\mathbf{T}_n - \boldsymbol{\theta})\right\} = (\mathbf{T}_n - \boldsymbol{\theta})^\top \mathbf{V}_n^{-1}(\mathbf{T}_n - \boldsymbol{\theta}), \end{aligned} \tag{9.81}$$

Then we obtain from (9.80) and (9.81) that

$$\begin{aligned} I\!P\Big\{&\boldsymbol{\lambda}\mathbf{T}_n - n^{-1/2}\chi_{p,\alpha}(\boldsymbol{\lambda}^\top \mathbf{V}_n \boldsymbol{\lambda})^{1/2} \leq \boldsymbol{\lambda}^\top \boldsymbol{\theta} \\ &\leq \boldsymbol{\lambda}\mathbf{T}_n + n^{-1/2}\chi_{p,\alpha}(\boldsymbol{\lambda}^\top \mathbf{V}_n \boldsymbol{\lambda})^{1/2}, \forall \boldsymbol{\lambda} \neq \mathbf{0}\Big\} \to 1 - \alpha \end{aligned} \tag{9.82}$$

as $n \to \infty$, where $\chi_{p,\alpha}^2$ is the $(1-\alpha)$ quantile of the chi square distribution with p degrees pf freedom; this provides a simultaneous confidence region for $\boldsymbol{\theta}$, and it follows the classical S-method (Scheffé 1951) of multiple comparison for the multinormal location/regression model. The width of the interval for $\boldsymbol{\lambda}^\top \boldsymbol{\theta}$ in (9.82) is equal to

$$2n^{-1/2}\chi_{p,\alpha}(\boldsymbol{\lambda}^\top \mathbf{V}_n \boldsymbol{\lambda})^{1/2}$$

and it depends on both $\mathbf{V}_n$ and $\boldsymbol{\lambda}$. Then $\boldsymbol{\lambda}$ is usually normalized as $\|\boldsymbol{\lambda}\| = (\boldsymbol{\lambda}^\top \boldsymbol{\lambda})^{1/2} = 1$ and

$$\sup\left\{\frac{\boldsymbol{\lambda}^\top \mathbf{V}_n \boldsymbol{\lambda}}{\boldsymbol{\lambda}^\top \boldsymbol{\lambda}} : \|\boldsymbol{\lambda}\| = 1\right\} = \mathrm{Ch}_{\max}(\mathbf{V}_n), \tag{9.83}$$

so we have by (9.82) and (9.83)

$$\lim_{n\to\infty} \mathbb{P}\Big\{\boldsymbol{\lambda}^\top \mathbf{T}_n - n^{-1/2}\chi_{p,\alpha}\mathrm{Ch}_{\max}(\mathbf{V}_n)]^{1/2} \le \boldsymbol{\lambda}^\top\boldsymbol{\theta} \le \boldsymbol{\lambda}^\top \mathbf{T}_n$$
$$+n^{-1/2}\chi_{p,\alpha}\mathrm{Ch}_{\max}(\mathbf{V}_n)]^{1/2},\ \forall\boldsymbol{\lambda} : \|\boldsymbol{\lambda}\| = 1\Big\} \ge 1-\alpha. \tag{9.84}$$

The diameter of the confidence set in (9.84) is equal to

$$2n^{-1/2}\chi_{p,\alpha}\{\mathrm{Ch}_{\max}(\mathbf{V}_n)\}^{1/2} \tag{9.85}$$

which also gives the maximum diameter for the confidence set in (9.82) when $\boldsymbol{\lambda}^\top\boldsymbol{\lambda} = 1$. Since $\mathbf{V}_n$ is a consistent estimator of $\boldsymbol{\Sigma}$ in (9.79), $n^{1/2}$ times the diameter in (9.85) stochastically converges to

$$2\chi_{p,\alpha}\{\mathrm{Ch}_{\max}(\boldsymbol{\Sigma})\}^{1/2}. \tag{9.86}$$

Therefore, from the shortest confidence set point of view, we should choose a sequence $\{\mathbf{T}_n^\star\}$ of estimators for which the asymptotic covariance matrix $\boldsymbol{\Sigma}_{T^\star}$ satisfies the minimax condition:

$$\mathrm{Ch}_{\max}(\boldsymbol{\Sigma}_{T^\star}) = \inf\Big\{\mathrm{Ch}_{\max}(\boldsymbol{\Sigma}_T : \mathbf{T} \subset \mathcal{C}\Big\}, \tag{9.87}$$

where $\mathcal{C}$ is a nonempty class of competing estimators of $\boldsymbol{\theta}$. This explains the relevance of the Roy (1953) largest-root criterion. Thus, to obtain robust multiparameter confidence sets based on point estimators and their estimated covariance matrix, we need a robust and asymptotically multi-normal estimator $\mathbf{T}_n$ and a consistent and robust and robust estimator $\mathbf{V}_n$. However, $\mathbf{T}_n$ in the multivariate location model is the vector of sample means, and $\mathbf{V}_n$ is the corresponding sample covariance matrix; both are known to be highly nonrobust, $\mathbf{V}_n$ being relatively worse.

The multiparameter confidence set estimation problem can be handled more conveniently for the univariate linear model in (9.76). For example, if we desire to have a confidence set for the regression parameter (vector) $\boldsymbol{\beta}$, then we can consider a simplified model that

$$\boldsymbol{\Sigma} = \sigma_0^2\mathbf{V}_0, \quad \text{where } \mathbf{V}_0 \text{ is known positive definite matrix,}$$

and σ_0^2 is a nonnegative parameter, for which a convenient estimator v_n^2 exists. The model is affine equivariant with respect to $\mathbf{x}_i \mapsto \mathbf{x}_i^\star = \mathbf{B}\mathbf{x}_i + \mathbf{b}$, $\mathbf{B}$ nonsingular, and $\mathbf{b} \in \boldsymbol{R}_p$, $(i = 1, \ldots, n)$. A robust estimator of $\boldsymbol{\beta}$ satisfies (9.79) and we can also find a robust estimation of σ_0^2. In this case, we have

$$\mathrm{Ch}_{\max}(\boldsymbol{\Sigma}) = \sigma_0^2\mathrm{Ch}_{\max}(\mathbf{V}_0).$$

and the situation is isomorphic to the uniparameter model treated in Section 9.2.

Let us proceed to consider type II confidence sets in the multiparameter case. Consider the same conditions (i) and (ii) as in Section 9.2 but with both $\mathbf{T}_n$

and $\boldsymbol{\theta}$ being p-vectors. Instead of the interval in (9.63), we will look for a suitable set K_n such that

$$\mathbb{P}_\theta\{\mathbf{T}_n \in K_n\} \geq 1 - \alpha, \tag{9.88}$$

where K_n contains $\mathbf{0}$ as an inner point and does not depend on $\boldsymbol{\theta}$. Then parallel to (9.64)–(9.66), we define

$$I_n = \{\boldsymbol{\theta} : \mathbf{T}_n \in K_n\}. \tag{9.89}$$

Whenever K_n and I_n are well defined, we have

$$\mathbb{P}_\theta\{\boldsymbol{\theta} \in I_n\} \geq 1 - \alpha, \tag{9.90}$$

which can be taken as the desired solution. There are two basic issues involved in the formulation of the confidence set I_n in (9.89).

(a) How to choose K_n in an optimal and robust way?

(b) Is it always possible to get a closed convex subset I_n of $\boldsymbol{\Theta}$?

The answer to the first question is relatively easier than that to the second question. We may consider $\mathbf{T}_n$ as a vector of rank test statistics for testing a null hypothesis $H_0 : \boldsymbol{\theta} = \boldsymbol{\theta}_0$ $(= \mathbf{0})$. The test criterion would be the quadratic form of $\mathbf{T}_n$ with respect to its dispersion matrix $\mathbf{W}_n$:

$$\mathcal{L}_n = \mathbf{T}_n^\top \mathbf{W}_n^{-1} \mathbf{T}_n,$$

and $\mathcal{L}_n \xrightarrow{\mathcal{D}} \chi_p^2$ under $\mathbf{H}_0$. This leads to the following choice of K_n in (9.88):

$$K_n = \{\mathbf{T}_n : \ \mathbf{T}_n \mathbf{W}_n^{-1} \mathbf{T}_n \leq \chi_{p,\alpha}^2\}, \tag{9.91}$$

and this provides an ellipsoidal region with the origin $\mathbf{0}$. Although $\mathcal{L}_n$ is analogous to the Hotelling T^2-statistic in the multinormal mean estimation problem, the ellipsoidal shape of K_n in (9.91) may not be precisely justified in the finite sample case. Often, a rectangular region is also contemplated. For example, we may set

$$K_n = \{\mathbf{T}_n : a_{nj} \leq T_{nj} \leq b_{nj}, \ 1 \leq j \leq p\}, \tag{9.92}$$

where the a_{nj}, b_{nj} are chosen so that (9.88) holds.

The second question depends on the nature of $\mathbf{T}_n(\boldsymbol{\theta})$. Recall that $\mathbf{T}_n(\boldsymbol{\theta})$ is itself a p-vector and that each coordinate of $\mathbf{T}_n(\boldsymbol{\theta})$ may typically involve the vector argument $\boldsymbol{\theta}$ (or suitable subsets of it). Inversions of the region (9.88) with respect to $\boldsymbol{\theta}$ may be laborious, if at all possible. For these reasons, in multiparameter confidence set estimation problems, type II regions, despite having good robustness property, have not gained much popularity in usage.

As an interesting compromise, we can use an alternative method for constructing confidence sets for location and regression parameters based on R- and

M-estimators. We can appeal the asymptotic linearity, which for $\mathbf{T}_n$ in the sense that for any compact $K \subset \boldsymbol{R}_p$,

$$\sup\{n^{-1/2}\|\mathbf{T}_n(\boldsymbol{\theta}+n^{-1/2}\mathbf{b})-\mathbf{T}_n(\boldsymbol{\theta})+\gamma\mathbf{V}_n\mathbf{b}\| : \mathbf{b}\in K\} \xrightarrow{p} 0 \tag{9.93}$$

as $n \to \infty$, where γ is a suitable nonzero constant (may depend on F) and $\mathbf{V}_n$ is a suitable matrix. (For rank statistics and M-statistics, such results have been treated in Chapters 5 and 6.) By (9.89) and (9.91), we may conclude that for large n,

$$I_n \sim \{\boldsymbol{\theta} : n\gamma^2(\boldsymbol{\theta}-\hat{\boldsymbol{\theta}}_n)^\top \mathbf{V}_n^\top \mathbf{W}_n^{-1}\mathbf{V}_n(\boldsymbol{\theta}-\hat{\boldsymbol{\theta}}_n) \le \chi^2_{p,\alpha}\}.$$

If $\mathbf{V}_n, \mathbf{W}_n$ are known, it suffices to estimate γ^2. The estimate of γ can also be based on (9.93) by choosing appropriate $\mathbf{b}$. However, the use of asymptotic linearity and asymptotic $\chi^2_{p,\alpha}$ may induce a slower rate of convergence and thereby also some nonrobustness.

9.6 Affine-Equivariant Tests and Confidence Sets

Affine-equivariant point estimators of multivariate location and regression parameters were discussed in Section 8.5, where the L_2- and L_1-norm estimators were presented side by side with nonparametric procedures. Now we shall treat the dual problems of hypothesis testing and confidence set estimation.

The ordinary multivariate sign and signed-rank tests, investigated in detail in Puri and Sen (1971, Chapter 4), are not affine-equivariant. If $\mathbb{X} = (\mathbf{X}_1, \ldots, \mathbf{X}_n)$ constitute the sample and we interpret $\boldsymbol{\theta}$ as the spatial median of distribution function F_0, we want to test the hypothesis $H_0 : \ \boldsymbol{\theta} = \mathbf{0}$ vs. $H_1 : \ \boldsymbol{\theta} \neq \mathbf{0}$. Define a covariance functional $\mathcal{C}_n = \mathcal{C}(\mathbb{X})$ which is affine equivariant and $\mathcal{C}\Big((-1)^{i_1}\mathbf{X}_1 \ldots, (-1)^{i_n}\mathbf{X}_n\Big) = \mathcal{C}(\mathbf{X}_1, \ldots, \mathbf{X}_n)$ $\forall i_j = 0, 1;\ 1 \le j \le n$. For example, we may set $\mathcal{C} = n^{-1}\sum_{i=1}^n \mathbf{X}_i\mathbf{X}_i^\top$. Denote $\mathcal{C}^{-1/2} = \mathbf{B}$, and set $\mathbf{Y}_i = \mathbf{B}\mathbf{X}_i,\ i = 1, \ldots, n$. Then denote $S_{ij} = \frac{Y_{ij}}{\|\mathbf{Y}_i\|},\ 1 \le j \le p;\ i = 1, \ldots, n$ and $\mathbf{S}_i = (S_{i1}, \ldots, S_{ip})^\top$ and further, $\mathbf{V}_n = \frac{1}{n}\sum_{i=1}^n \mathbf{S}\mathbf{S}^\top$. Such $\mathbf{V}_n$ is symmetric in $\mathbf{Y}_1, \ldots, \mathbf{Y}_n$ and is invariant to sign-inversions of the $\mathbf{Y}_i$. In this setup, a Wald-type test statistic based on these spatial signs (Oja 2010) is based on the statistic

$$Q_n = (\sum_{i=1}^n \mathbf{S}_i)^\top (n\mathbf{V}_n)^{-1} (\sum_{i=1}^n \mathbf{S}_i). \tag{9.94}$$

Let $\mathcal{G}_n$ be the group of 2^n sign-inversions: $\mathbf{i}_n = \Big((-1)^{i_1}\mathbf{Y}_1 \ldots, (-1)^{i_n}\mathbf{Y}_n\Big)$ where each i_j is either 0 or 1 with probability $\frac{1}{2}$, $1 \le j \le n$. Since $\mathbf{V}_n$ is $\mathcal{G}_n$ invariant, in this way we generate a permutation distribution of $Q_n(\mathbf{i}_n)$, $\mathbf{i}_n \in \mathcal{G}_n$; each element has the permutational (conditional) probability 2^{-n}. This parallels the classical sign test, discussed in Puri and Sen (1971, Chapter 4). Tyler (1987) suggested a choice of $\mathbf{B}$ such that $\mathbf{V}_n \propto \mathbf{I}_p$, thus adding more convenience to the computation of Q_n.

To construct the spatial signed-rank test, define the $\mathbf{Y}_i$ as above, and let $\mathbf{R}_n^*(\mathbf{x})$ be defined as in (8.21). Denote $\bar{\mathbf{R}}_n^* = \frac{1}{n}\sum_{i=1}^n \mathbf{R}_n^*(\mathbf{Y}_i)$ and $\mathbf{V}_n^* = \frac{1}{n}\sum_{i=1}^n [\mathbf{R}_n^*(\mathbf{Y}_i)][\mathbf{R}_n^*(\mathbf{Y}_i)]^\top$. Then the spatial signed-rank test statistic is

$$\mathcal{L}_n^* = n(\bar{\mathbf{R}}_n^*)^\top (\mathbf{V}_n^*)^{-1}(\bar{\mathbf{R}}_n^*). \tag{9.95}$$

Since $\mathbf{Y}_1, \ldots, \mathbf{Y}_n$ are interchangeable random vectors, we can consider a group $\mathcal{G}_n^*$ of 2^n sign-inversions coupled with $n!$ permutations of the $\mathbf{Y}_i$, [i.e., $\left((-1)^{j_1}\mathbf{Y}_{i_1}, \ldots, (-1)^{j_n}\mathbf{Y}_{i_n}\right)$, $j_k = 0,1;\ 1 \le k \le n$ and $(i_1, \ldots, i_n)$ being a permutation of $(1, \ldots, n)$] and generate the permutational distribution of $\mathcal{L}_n^*$ under H_0 by reference to these $2^n(n!)$ possible realizations. The χ^2-approximation of this distribution follows by similar arguments as in Puri and Sen (1971, Chapter 4).

It is also possible to formulate the spatial M-tests based on the $\frac{\mathbf{Y}_i}{\|\mathbf{Y}_i\|}$, using appropriate score functions. In view of the assumed elliptical symmetry, we can use a common score function $\psi(\cdot)$ for all p components. Then denote $\boldsymbol{\psi}_i = \left(\psi(\|\mathbf{Y}_i\|^{-1}\mathbf{Y}_{i1}), \ldots, \psi(\|\mathbf{Y}_i\|^{-1}\mathbf{Y}_{ip})\right)^\top$, $i = 1, \ldots, n$. Let $\widetilde{\mathbf{M}}_n = \frac{1}{n}\sum_{i=1}^n \boldsymbol{\psi}_i$ and $\widetilde{\mathbf{V}}_n = \frac{1}{n}\sum_{i=1}^n (\boldsymbol{\psi}_i)(\boldsymbol{\psi}_i)^\top$, and consider the test statistic

$$\widetilde{\mathcal{L}}_n = n[\widetilde{M}_n]^\top \widetilde{\mathbf{V}}_n^{-1}[\widetilde{M}_n]. \tag{9.96}$$

Similarly as above, we can construct a permutationally (conditionally) distribution-free test for $H_0:\ \boldsymbol{\theta} = \mathbf{0}$ vs. $H_1:\ \boldsymbol{\theta} \neq \mathbf{0}$, using the permutation distribution of $\widetilde{\mathcal{L}}_n$ under the group of $2^n(n!)$ permutations.

Let us now turn to affine equivariant confidence sets. Let $\mathbf{Y}_i(\boldsymbol{\theta}) = \mathbf{B}(\mathbf{X}_i - \boldsymbol{\theta})$ and $\mathbf{S}_i(\boldsymbol{\theta}) = \frac{\mathbf{Y}_i(\boldsymbol{\theta})}{\|\mathbf{Y}_i(\boldsymbol{\theta})\|}$, $i = 1, \ldots, n$. As an estimator of $\boldsymbol{\theta}$, take the solution $\widehat{\boldsymbol{\theta}}_n$ of $\left|\sum_{i=1}^n \mathbf{S}_i(\boldsymbol{\theta})\right| = \min$. Take a translation-invariant covariance functional $\mathbf{V}$, and let $\widehat{\mathbf{V}}_n$ be the same based on the $\mathbf{Y}_i(\widehat{\boldsymbol{\theta}}_n)$. Then we can construct the test criterion of the hypothesis $H_{\boldsymbol{\theta}}$ that $\boldsymbol{\theta}$ is the true parameter value:

$$\widehat{\mathbf{Q}}_n(\boldsymbol{\theta}) = \frac{1}{n}\Big(\sum_{i=1}^n \mathbf{S}_i(\boldsymbol{\theta})\Big)^\top \widehat{\mathbf{V}}_n^{-1}\Big(\sum_{i=1}^n \mathbf{S}_i(\boldsymbol{\theta})\Big)$$

and define the set

$$I\!\!P\left\{\widehat{\mathbf{Q}}(\boldsymbol{\theta}) \le Q_{n,\alpha}\Big|\boldsymbol{\theta}\right\} \approx 1 - \alpha \tag{9.97}$$

where $Q_{n,\alpha}$ is either the permutational critical value or its $\chi^2_{p,\alpha}$ approximation. Then we can define the confidence set $\mathcal{I}_n$ by

$$\mathcal{I}_n = \left\{\boldsymbol{\theta}:\ \widehat{\mathbf{Q}}(\boldsymbol{\theta}) \le Q_{n,\alpha}\right\}. \tag{9.98}$$

For $\frac{1}{n}\sum_{i=1}^n \mathbf{S}_i(\boldsymbol{\theta})$ one can use again the „uniform asymptotic linearity" in a $n^{-1/2}$-neighborhood of the true parameter point, and claim that $\mathcal{I}_n$ relates to a closed convex hull, in probability. A similar procedure can be based on the sign-rank and M-statistics.

Because this approach can be computationally intensive, as an alternative we can use the pivot estimator of $\boldsymbol{\theta}$ and its asymptotic multinormality, to se tup a large sample Scheffé (1953) type confidence set on $\boldsymbol{\theta}$.

Finally, consider the multivariate regression model $\mathbb{Y} = \boldsymbol{\beta}\mathbb{X}+\mathbb{E}$ treated in Section 8.5 [see (8.16)]. Consider first the test for the null hypothesis $H_0 : \boldsymbol{\beta} = \mathbf{0}$ vs $H_1 : \boldsymbol{\beta} \neq \mathbf{0}$. Recall that $\mathbb{Y} = (\mathbf{Y}_1, \ldots, \mathbf{Y}_n)$ is of order $p \times n$ where $\mathbf{Y}_i = \boldsymbol{\beta}\mathbf{x}_i + \mathbf{e}_i$, $1 \leq i \leq n$. thus, $\mathbf{Y}_i - \boldsymbol{\beta}\mathbf{x}_i = \mathbf{e}_i$ are *i.i.d.* random vectors with an unspecified elliptically symmetric F, defined on $\mathbb{R}_p$. We define a covariance functional $\mathcal{C}_n$ as in the location model. For the current hypothesis and alternative, we may take $\mathcal{C}_n = \frac{1}{n}\sum_{i=1}^n \mathbf{Y}_i\mathbf{Y}_i^\top$. However, keeping the estimation of $\boldsymbol{\beta}$ or possible subhypothesis testing in mind, we prefer to choose $\mathcal{C}_n$ to be location-regression invariant. If $\widehat{\boldsymbol{\beta}}_n$ is the least-squares estimator of $\boldsymbol{\beta}$, we may take $\mathcal{C}_n = \frac{1}{n-q}\sum_{i=1}^n(\mathbf{Y}_i - \widehat{\boldsymbol{\beta}}\mathbf{x}_i)(\mathbf{Y}_i - \widehat{\boldsymbol{\beta}}\mathbf{x}_i)^\top$, which is also affine-equivariant. As in the location model, we consider the standardized vectors $\mathcal{C}_n^{-1/2}\mathbf{Y}_i$, $i = 1, \ldots, n$, on which we define either a coordinatewise linear rank statistic or an M-statistic based on suitable score functions. In view of the assumed elliptically symmetric family, the score function can be the same for all p coordinates, and the $\mathbf{x}_i$ can be involved in the same manner as in Chapters 4–6. The resulting tests are of similar structure as those in Sections 9.2.2 and 9.2.3 but extended to the multivariate case. As such, a permutation distribution can be still incorporated to obtain some conditionally (permutationally) distribution-free tests, asymptotically χ^2_{pq} distributed for a large number of observations.

To construct a confidence set, we replace $\mathcal{C}_n^{-1/2}\mathbf{Y}_i$ by $\mathcal{C}_n^{-1/2}(\mathbf{Y}_i - \mathbf{B}\mathbf{x}_i)$, $\mathbf{B} \in \mathbb{R}_{pq}$ for $i = 1, \ldots, n$, and use the same rank $[L_n(\mathbf{B})]$ or M-statistics $[M_n(\mathbf{B})]$ on these aligned vectors. The corresponding estimator $\widetilde{\boldsymbol{\beta}}_n$ of $\boldsymbol{\beta}$ is determined by estimating equation, either $L_n(\mathbf{B}) \equiv 0$ or $M_n(\mathbf{B}) \equiv 0$.

If we partition $\boldsymbol{\beta}$ into $[\boldsymbol{\beta}_1, \boldsymbol{\beta}_2]$ where $\boldsymbol{\beta}_1$ is $p \times q_1$ and $\boldsymbol{\beta}_2$ is $p \times q_2$, $q_1 + q_2 = q$, and want to test for $H_0 : \boldsymbol{\beta}_2 = \mathbf{0}$ vs $H_1 : \boldsymbol{\beta}_2 \neq \mathbf{0}$, then $\boldsymbol{\beta}_1$ plays the role of nuisance parameter. We estimate $\boldsymbol{\beta}_1$ under validity of H_0 by the above method and denote it by $\widetilde{\boldsymbol{\beta}}^0_{1n}$. Then consider the aligned vectors $\mathbf{Y}_i - \widetilde{\boldsymbol{\beta}}^0_{1n}\mathbf{x}_i^{(1)}$, where $\mathbf{x}_i = \begin{pmatrix} \mathbf{x}_i^{(1)} \\ \mathbf{x}_i^{(1)} \end{pmatrix}$, $i = 1, \ldots, n$. We perform the R- or M-tests on these aligned vectors as in Section 9.1.

The confidence sets can be obtained from the $L_n(\mathbf{B})$ or $M_n(\mathbf{B})$ by the same procedure as in Section 9.5.

9.7 Problems

9.2.1 Use (9.20) and asymptotic representation for $\hat{\boldsymbol{\theta}}_n^{\star}$ and $\widehat{\boldsymbol{\theta}}_n^{o\star}$ to show that (9.23) holds under $\{\mathbf{H}_{1(n)}\}$.

9.2.2 Try to construct a T_n-test for a general parameter θ under the regularity conditions of Theorem 5.1.

9.2.3 For the simple regression model $Y_i = \theta_1 + \theta_2 x_i + e_i,\ i \geq 1$, consider an aligned signed-rank test for $\mathbf{H}_0 : \theta_1 = 0$ vs. $H_1 : \theta_1 \neq 0$ when θ_2 is treated as a nuisance parameter.

9.2.4 Consider the linear model $Y_i = \theta_0 + \mathbf{x}_{i1}^{\top}\boldsymbol{\theta}_1 + \mathbf{x}_{i2}^{\top}\boldsymbol{\theta}_2 + e_i,\ i \geq 1$. For testing $\mathbf{H}_0 : \boldsymbol{\theta}_1 = \mathbf{0}$ vs. $\mathbf{H}_1 : \boldsymbol{\theta}_1 \neq \mathbf{0}$, the aligned rank statistic can be used.

9.2.5 (Continuation). Consider the hypothesis $\mathbf{H}_0 : (\theta_0, \boldsymbol{\theta}_1^{\top})^{\top} = \mathbf{0}$ vs. $\mathbf{H}_1 : (\theta_0, \boldsymbol{\theta}_1^{\top})^{\top} \neq \mathbf{0}$. For this testing, the signed-rank statistics are usable, but not the linear rank statistics.

9.2.6 For an L-estimator with smooth weights, verify that (9.58)–(9.59) hold under appropriate regularity conditions.

9.2.7 The distribution of the sample mean $\bar{X}_n$ from a Cauchy distribution is also Cauchy and jackknifing is not viable. Why?

9.2.8 Suppose that the density $f(x;\theta)$ has a jump discontinuity at θ (as the population median). Bootstrapping may not work out in that case, although it can be modified to do so.

9.4.1 Obtain M-confidence intervals for the regression parameter β in a simple regression model. Is it robust?

Appendix

Uniform Asymptotic Linearity

We outline the proof of Lemma 5.3 in Section 5.5 which provides the central concept behind the derivation of the basic uniform asymptotic linearity results. Let $\psi = \psi_a + \psi_c + \psi_s : \boldsymbol{R} \mapsto \boldsymbol{R}$, where ψ_a, ψ_c, and ψ_s denote the absolutely continuous, continuous, and step-function components, respectively.

Case I: $\psi \equiv \psi_s$ **(i.e.,** $\psi_a \equiv \psi_c \equiv 0$).
Without loss of generality, we assume that there is a single jump-point. We set $\psi_s(y)$ as 0 or 1 according as y is $\leq$ or $>$ 0. Further, we assume the regularity assumptions in Section 5.5 (a subset of $[M1]--[M5]$ and $[X1]--[X3]$) to be true. Using (5.116) or (5.117), we set the scale factor $S = 1$ and write

$$\mathbf{S}_n(\mathbf{t}, u) \tag{9.99}$$
$$= \sum_{i=1}^{n} \mathbf{x}_i \{\psi(e^{-u/\sqrt{n}}(E_i - n^{-1/2}\mathbf{x}_i^\top \mathbf{t})) - \psi(\mathbf{E}_i)\},$$

for $(\mathbf{t}, u) \in C = [-K, K]^{p+1}$ for some $K :\ 0 < K < \infty$. We let

$$\mathbf{S}_n^0(\mathbf{t}, u) = \mathbf{S}_n(\mathbf{t}, u) - \mathbb{E}\mathbf{S}_n(\mathbf{t}, u),\ (\mathbf{t}, u) \in C.$$

By the vector structure in (5.117) and (9.99), it suffices to show that for each coordinate of $\mathbf{S}_n^0(\mathbf{t}, u)$, the uniform asymptotic linearity result holds for $(\mathbf{t}, u) \in C$. To simplify the proof, we consider only the first coordinate and drop the subscript 1 in $S_{n1}(.)$ or $S_{n1}^0(.)$:

$$S_n^0(\mathbf{t}, u) \tag{9.100}$$
$$= \sum_{i=1}^{n} x_{i1}\{I(E_i \leq 0) - I(E_i \leq n^{-1/2}\mathbf{x}_i^\top t) - F(0) + F(n^{-1/2}\mathbf{x}_i^\top t)\}.$$

Because of the special nature of ψ_s, $S_n^0(\mathbf{t}, u) \equiv S_n^0(\mathbf{t}, 0)\ \forall u$. For a more general ψ_s, we can use a similar expression involving u, and proceed similarly. Because (9.100) involves independent summands, we get by the central limit theorem

$$n^{-1/4}S_n^0(\mathbf{t}, u) \xrightarrow{\mathcal{D}} \mathcal{N}(0, \gamma^2(\mathbf{t}, u)),\ \ (\mathbf{t}, u) \in C, \tag{9.101}$$

where

$$\gamma^2(\mathbf{t}, u) = f(0) \cdot \lim_{n\to\infty} \left\{ \frac{1}{n} \sum_{i=1}^{n} x_{i1}^2 |\mathbf{x}_i^\top \mathbf{t}|) \right\} \tag{9.102}$$

for each $(\mathbf{t}, u) \in C$, and $\gamma^2(\mathbf{t}, u) < \infty \; \forall (\mathbf{t}, u) \in C$ by **[X1]**–**[X3]**. Therefore, if we choose an arbitrary positive integer M and a set $(\mathbf{t}_i, u_i)$, $i = 1, \ldots, M$ of points in C, then by (9.101)–(9.102)

$$\max_{1\leq i\leq M} \left| n^{-1/4} S_n^0(\mathbf{t}_i, u_i) \right| = O_p(1) \text{ as } n \to \infty,$$

what relates to the finite-dimensional distributions. To establish the uniform asymptotic linearity, we still need to establish the compactness or tightness of $n^{-1/4} S_n^0(\mathbf{t}, u)$, $(\mathbf{t}, u) \in C$. This can be done either by an appeal to weak convergence of $n^{-1/4} S_n^0(\mathbf{t}, u)$ on C to a Wiener function (as done by Jurečková and Sen 1989)) or by the following simpler decomposition:

Consider the set of signs: $\text{sign}(x_{i1} x_{ij})$, $j = 0, 2, 3, \ldots, p$, where $x_{i0} = 1$. Then consider a set of 2^p subsets of $\{1, \ldots, n\}$ such that $x_{i1} x_{ij}$ have the same sign within each subset (which may differ from j to j). Thus, we write $S_n^0(.) = \sum_{j\leq 2^p} S_{n,j}^0(.)$. For each $j(= 1, \ldots, 2^p)$, $S_{n,j}^0(\mathbf{t}, u)$ can be expressed as a difference of two functions, each of which is monotone in each argument $\mathbf{t}, u$, although these may not be concordant. By this monotonicity, the Bahadur (1966) representation (studied in Chapter 4) extends directly to $S_{n,j}^0(.)$, and hence it follows that

$$\sup \left\{ n^{-1/4} |S_{n,j}^0(\mathbf{t}, u)| : \; (\mathbf{t}, u) \in C \right\} = O_p(1) \tag{9.103}$$

for $j = 1, \ldots, 2^p$. Hence (9.103) implies that

$$\sup \left\{ n^{-1/4} |S_n^0(\mathbf{t}, u)| : \; (\mathbf{t}, u) \in C \right\} = O_p(1),$$

and this implies (5.117).

Case II: ψ **absolutely continuous, but** ψ' **step function.**

We know that ψ is continuous, piecewise linear, and that it is a constant for $x \leq r_1$ or $x \geq r_k$, where $-\infty < r_1 < r_k < \infty$, and there are k jump points for ψ' (denoted by $r_1, \ldots, r_k$). Again, take $k = 2$ for simplicity (which corresponds to the classical Huber score function). As in Case I, we set $S = 1$, and also consider only the first coordinate of $\mathbf{S}_n(\mathbf{t}, u)$. Since ψ is bounded and satisfies a Lipschitz condition of order 1, it can be shown that

$$\sum_{i=1}^{n} x_{i1} \left\{ \psi[e^{-u/\sqrt{n}}(E_i - n^{-1/2} \mathbf{x}_i^\top \mathbf{t})] - \psi(E_i - n^{-1/2}(uE_i + \mathbf{x}_i^\top \mathbf{t})) \right\} = O_p(1)$$

uniformly in $(\mathbf{t}, u) \in C$. Hence, it is sufficient to prove the proposition for

$$S_n(\mathbf{t}, u) = \sum_{i=1}^{n} x_{i1} \left\{ \psi[E_i - n^{-1/2}(uE_i + \mathbf{x}_i^\top \mathbf{t})] - \psi(E_i) \right\}.$$

Now it can be shown that for any pair $(\mathbf{t}_1, u_1)$ and $(\mathbf{t}_2, u_2)$ of distinct points,

$$\begin{aligned}&\mathrm{Var}\{S_n(\mathbf{t}_1, u_1) - S_n(\mathbf{t}_2, u)\} \qquad (9.104)\\ &\leq \sum_{i=1}^{n} x_{i1}^2 I\!E\Big\{\Big[\psi(E_i - n^{-1/2}(u_1 E_i + \mathbf{x}_i^\top \mathbf{t}_1))\\ &\qquad -\psi(E_i - n^{-1/2}(u_2 E_i + \mathbf{x}_i^\top \mathbf{t}_2))\Big]^2\Big\}\\ &\leq K^\star\{(u_1 - u_2)^2 + \|\mathbf{t}_1 - \mathbf{t}_2\|^2\}, \;\; K^\star < \infty,\end{aligned}$$

uniformly in $(\mathbf{t}_k, u_k) \in C, \; k = 1, 2$. The boundedness of ψ' implies that

$$\begin{aligned}&\Big| I\!E\Big[S_n(\mathbf{t}_1, u_1) - S_n(\mathbf{t}_2, u_2) \qquad (9.105)\\ &\qquad + n^{-1/2} \sum_{i=1}^{n} \Big\{x_{i1}\mathbf{x}_i^\top(\mathbf{t}_1 - \mathbf{t}_2)\gamma_1 + (u_1 - u_2)\gamma_2\Big\}\Big]\Big|\\ &\leq K^{\star\star}\{\|\mathbf{t}_1 - \mathbf{t}_2\| + |u_1 - u_2|\},\end{aligned}$$

hence by (9.104) and (9.105)

$$\begin{aligned}&I\!E\Big\{\Big[S_n(\mathbf{t}_1, u_1) - S_n(\mathbf{t}_2, u_2) \qquad (9.106)\\ &\qquad + n^{-1/2} \sum_{i=1}^{n} x_{i1}(\gamma_1(\mathbf{t}_1 - \mathbf{t}_2)^\top \mathbf{x}_i + \gamma_2(u_1 - u_2))\Big]^2\Big\}\\ &\leq K_0\{\|\mathbf{t}_1 - \mathbf{t}_2\|^2 + (u_1 - u_2)^2\}, \; K_0 < \infty\end{aligned}$$

uniformly in $(\mathbf{t}_k, u_k) \in C, \; k = 1, 2$. Therefore, pointwise $S_n^0(\mathbf{t}, u) = \mathrm{O}_p(1)$. To prove the compactness, we shall consider increments of $S_n(\mathbf{t}, u)$ over small blocks. We present only the case $p = 1$ because it imparts the full generality of considerations. For $t_2 > t_1$, and $u_2 > u_1$, the increment of $S_n(.)$ over the block $B = B(t_1, u_1; t_2, u_2)$ is

$$\begin{aligned}S_n(B) &= S_n(t_2, u_2) - S_n(t_1, u_2) - S_n(t_2, u_1) + S_n(t_1, u_1)\\ &= \sum_{i=1}^{n} x_i \psi_i(E_i; B), \qquad (9.107)\end{aligned}$$

where

$$\begin{aligned}&\psi_i(E_i; B) \qquad (9.108)\\ &= \psi\Big(E_i - n^{-1/2}(u_2 E_i + x_i t_2)\Big) - \psi\Big(E_i - n^{-1/2}(u_2 E_i + x_i t_1)\Big)\\ &\quad -\psi\Big(E_i - n^{-1/2}(u_1 E_i + x_i t_2)\Big) + \psi\Big(E_i - n^{-1/2}(u_1 E_i + x_i t_1)\Big)\end{aligned}$$

for $i = 1, \ldots, n$. By [**X2**] in Section 5.5, $\max\{|x_i| : 1 \leq i \leq n\} = \mathrm{O}(n^{1/4})$, while ψ is piecewise linear and bounded a.e., so $\psi_i(E_i, B)$ is $\mathrm{O}(n^{-1/4})$ a.e. On the other hand, in view of (9.108) and piecewise linearity of ψ, $\psi_i(E_i, B) \equiv 0$ if all arguments in (9.108) lay in the same interval among

$(-\infty, \mu_1)$, (μ_1, μ_2), (μ_2, ∞). Thus, setting $t_1 = -K$, $u_1 = -K$, $t_2 \in [-K, K]$, $u_2 \in [-K, K]$, we obtain from (9.107) and (9.108) that

$$\sup\Big\{|S_n(B(-K,-K;t_2,u_2))| : -K \le t_2 \le K,\ -K \le u_2 \le K\Big\}$$
$$\le n^{-1/2}\textstyle\sum_{i=1}^n |x_{i1}|K^\star(|r_1|+|r_2|+\|\mathbf{x}_i\|)I_i,$$

where the I_i, $i = 1, \ldots, n$, are independent nonnegative indicator variables with

$$I\!\!E I_i \le K^\star(|r_1|+|r_2|+\|\mathbf{x}_i\|)n^{-1/2},\ i = 1,\ldots,n, \tag{9.109}$$

and $0 < K^\star < \infty$. Hence,

$$\mathrm{Var}\Big(\sup\Big\{|S_n(B(-K,-K;t_2,u_2))| : \ -K \le t_2, u_2 \le K\Big\}\Big) = \mathrm{O}(1),$$

and this, combined with (9.105), gives the desired result.

Case III: ψ **and** ψ' **are both absolutely continuous.**

Because ψ' or ψ may be unbounded, [**M3**] is introduced to control their unboundedness. For every $(\mathbf{t}, u) \in C$, by a second-order Taylor's expansion,

$$\begin{aligned}
&\psi(e^{-u/\sqrt{n}}(E_i - n^{-1/2}\mathbf{t}\mathbf{x}_i)) \\
&= \psi(E_i) - n^{-1/2}(uE_i + \mathbf{t}^\top\mathbf{x}_i)\psi'(E_i) \\
&+\frac{1}{2n}u^2\Big(E_i\psi'(E_i) + E_i^2\psi''(E_i)\Big) + \frac{1}{2n}E_i\psi''(E_i)u(\mathbf{x}_i^\top\mathbf{t}) \\
&+\frac{1}{2n}(\mathbf{t}^\top\mathbf{x}_i\mathbf{x}_i^\top\mathbf{t})\psi''(E_i) + \text{ a remainder term.}
\end{aligned} \tag{9.110}$$

The remainder term involves ψ' and ψ'' as well as the E_i, $\mathbf{x}_i$, $\mathbf{t}$, and u. With a block B defined as in Case II and $S_n(B)$ as in (9.105)–(9.106), we have by (9.107) and (9.109),

$$\psi_i(E_i, B) = 0 - 0 + \frac{1}{n}E_i\psi''(E_i)(u_2 - u_1)\mathbf{x}_i^\top(\mathbf{t}_2 - \mathbf{t}_1) \tag{9.111}$$
$$+\frac{1}{2n}(\mathbf{t}_2 - \mathbf{t}_1)^\top\mathbf{x}_i\mathbf{x}_i^\top(\mathbf{t}_2 - \mathbf{t}_1)\psi''(E_i) + \text{ a remainder term.}$$

Therefore, we have

$$S_n(B) = \frac{1}{n}\sum_{i=1}^n x_{i1}[E_i\psi''(E_i)(u_2 - u_1)\mathbf{x}_i^\top(\mathbf{t}_2 - \mathbf{t}_1)$$
$$+ \ \frac{1}{2}(\mathbf{t}_2 - \mathbf{t}_1)^\top\mathbf{x}_i\mathbf{x}_i^\top(\mathbf{t}_2 - \mathbf{t}_1)\psi''(E_i)] + \text{ a remainder term.}$$

Letting $S_n^\star = \sup\{|S_n(B) : B \subset C\}$, we have

$$S_n^\star \le n^{-1}\sum_{i=1}^n |x_{i1}|\{|\psi''(E_i)|\}[\|\mathbf{x}_i\|(2k)^{p+1}|E_i| + \frac{1}{2}(2k)^p\|\mathbf{x}_i\|^2]$$
$$+ \text{ remainder term.}$$

By [**X2**] and [**M3**] of Section 5.5, and the Markov law of large numbers, the first term on the right-hand side of (9.111) converges to a (nonnegative) finite limit. A very similar treatment holds for the remainder term (which $\xrightarrow{P} 0$ as $n \to \infty$). Therefore, we obtain that

$$S_n^\star = O_p(1), \quad \text{as } n \to \infty.$$

On the other hand, $S_n^0(\mathbf{t}, u) = O_p(1)$ for any fixed $(\mathbf{t}, u) \in C$ follows as in Case II, using (9.109), so the desired result follows. This completes the proof of Lemma 5.3.

□

References

Adichie, J. N. (1967). Estimate of regression parameters based on rank tests. *Ann. Math. Statist. 38*, 894–904.

Adichie, J. N. (1978). Rank tests for subhypotheses in the general linear regression. *Ann. Statist. 6*, 1012–1026.

Adichie, J. N. (1984). Rank tests in linear models. *Handbook of Statististics, Vol. 4: Nonparametric Methods* (eds. P. R. Krishnaiah and P. K. Sen), North Holland, Amsterdam, pp. 229–257.

Adrover, J., Maronna, R. A., and Yohai, V. J. (2004). Robust regression quantiles. *J. Statist. Planning Infer. 122*, 187–202.

Aerts, M., Janssen, P., and Veraverbeke, N. (1994). Asymptotic theory for regression quantile estimators in the heteroscedastic regression model. *Asymptotic Statistics; Proc. 5th Prague Cofer.* (eds. M. Hušková and P. Mandl), Physica-Verlag, Heidelberg, pp. 151–161.

Agulló, J., Croux C., and Van Aelst S. (2008). The multivariate least-trimmed squares estimator. *J. Multivariate Analysis 99*, 311–338.

Akahira, M. (1975a). Asymptotic theory for estimation of location in nonregular cases I: Order of convergence of consistent estimators. *Rep. Stat. Appl. Res.JUSE, 22*, 8–26.

Akahira, M. (1975b). Asymptotic theory for estimation of location in nonregular cases II: Bounds of asymptotic distributions of consistent estimators. *Rep. Stat. Appl. Res., JUSE, 22*, 101–117

Akahira, M., and Takeuchi, K. (1981). *Asymptotic Efficiency of Statistical Estimators: Concepts and Higher Order Asymptotic Efficiency. Springer Lecture Notes in Statistics, 7*, Springer-Verlag, New York.

Amari, S. (1985). *Differential Geometric Methods in Statistics.* Springer-Verlag, New York.

Andersen, P. K. and Gill, R. D. (1982). Cox's regression model for counting processes: a sample study. *Ann. Statist. 10*, 1100–1120.

Andersen, P. K., Borgan, O., Gill, R. D., and Keiding, N. (1993). *Statistical Models Based on Counting Processes.* Springer-Verlag, New York.

Anderson, J. R. (1978). Use of M-estimator theory to derive asymptotic results for rank regression estimators. *Biometrics 34*, 151. (Abstract.)

Andrews, D. F., Bickel, P. J., Hampel, F. R., Huber, P. J., Rogers, W. H., and Tukey, J. W. (1972) *Robust Estimates of Location: Survey and Advances.* Princeton Univ. Press, Princeton.

Andrews, D. W. K. (1986). A note on the unbiasedness of feasible GLS, quasimaximum likelihood, robust adaptive, and spectral estimators of the linear model. *Econometrica 54*, 687–698.

Anscombe, F. J. (1952). Large sample theory of sequential estimation. *Proc. Cambridge Phil. Soc. 48*, 600–607.

Antille, A. (1972). Linearité asymptotique d'une statistique de rang. *Zeit. Wahrsch. verw. Geb. 32*, 147–164.
Antille, A. (1976). Asymptotic linearity of Wilcoxon signed-rank statistics. *Ann. Statist. 4*, 175–186.
Antoch, J. (1984). *Collection of Programs for Robust Regression Analysis.* Charles Univ., Prague.
Antoch, J., Collomb, G., and Hassani, S. (1984). Robustness in parametric and nonparametric regression estimation: An investigation by computer simulation. *COMPSTAT 1984.* Physica Verlag, Vienna, pp. 49–54.
Antoch, J. and Jurečková, J. (1985). Trimmed least squares estimator resistant to leverage points. *Comp. Statist. Quarterly 4*, 329–339.
Arcones, M. A. (1996). The Bahadur-Kiefer representation of L_p regression estimators. *Econometric Theory 12*, 257–283.
Atkinson, A. C. (1985). *Plots, Transformations, and Regression.* Clarendon Press, Oxford.
Aubuchon, J. C. and Hettmansperger, T. P. (1984). On the use of rank tests and estimates in the linear model. *Handbook of Statistics. Vol. 4 : Nonparametric Methods* (eds. P. R. Krishnaiah and P. K. Sen), North Holland, Amsterdam, pp. 259–274.
Azencott, F. J., Birgé, L., Costa, V., Dacunha-Castelle, D., Deniau, C., Deshayes, P., Huber-Carol, C., Jolivaldt, P., Oppenheim, G., Picard, D., Trécourt, P., and Viano, C. (1977). Théorie de la robustesse et estimation d'un paramétre. *Astérisque,* 43–44.
Bahadur, R. R. (1960). Stochastic comparison of tests. *Ann. Math. Statist. 31*, 276–295.
Bahadur, R. R. (1966). A note on quantiles in large samples. *Ann. Math. Statist. 37*, 557–580.
Bahadur, R. R. (1971). *Some Limit Theorems in Statistics.* SIAM, Philadelphia.
Bai, Z. D., Rao, C. R., and Wu, Y. (1990). Recent contributions to robust estimation in linear models. *Probability, Statistics and Design of Experiment: R.C. Bose Mem. Confer.*,(ed. R. R. Bahadur). Wiley Eastern, New Delhi, pp. 33–50.
Bai, Z. D., Rao, C. R., and Wu, Y. (1992). M-estimation of multivariate linear regression parameter under a convex discrepancy function. *Statist. Sinica 2*, 237–254.
Bancroft, T. A. (1944)). On biases in estimation due to use of preliminary test of significance. *Ann. Math. Statist. 15*, 190–204.
Barndorff-Nielsen, O. E., and Cox, D. R. (1989). *Asymptotic Techniques for Use in Statistics.* Chapman and Hall, New York.
Barrodale, I., and Roberts, F. D. K. (1975). Algorithm 478: Solution of an overdetermined system of equations in the L_1-norm. *Commun. ACM 17*, 319–320.
Bassett, G., and Koenker, R. (1978). Asymptotic theory of least absolute error regression. *J. Amer. Statist. Assoc. 73*, 618–622.
Bassett, G., and Koenker, R. (1982). An empirical quantile function for linear models with iid errors. *J. Amer. Statist. Assoc. 77*, 407–415.
Bednarski, T. (1981). On the solution of minimax test problem for special capacities. *Z. Wahrscheinlichkeitstheorie verw. Gebiete 58*, 397–405.
Bednarski, T. (1994). Fréchet differentiability and robust estimation. *Asymptotic Statistics; Proc. 5th Prague Symp.* (eds., M. Hušková and P. Mandl), Physica-Verlag, Heidelberg, pp. 49–58.
Bednarski, T., Clarke, B. R., and Kolkiewicz, W. (1991). Statistical expansions and locally uniform Fréchet differentiability. *J. Austral. Math. Soc. Ser. A 50*, 88–97.

Bennett, C. A. (1952). Asymptotic properties of ideal linear estimators. Ph.D. dissertation, Univ. of Michigan, Ann Arbor.

Beran, R. J. (1977). Minimum Hellinger distance estimators for parametric models. *Ann. Statist. 5*, 445–463.

Beran, R. J. (1978). An efficient and robust adaptive estimator of location. *Ann. Statist. 6*, 292–313.

Beran, R. J. (1982). Robust estimation in models for independent non-identically distributed data. *Ann. Statist. 12*, 415–428.

Beran, R. J. (1984). Minimum distance procedures. *Handbook of Statistics, Vol. 4: Nonparametric Methods* (eds. P. R. Krishnaiah and P. K. Sen), North-Holland, Amsterdam, pp. 741–754.

Bergesio, A., and Yohai, V. J. (2011). Projection estimators for generalized linear models. *J. Amer. Statist. Assoc. 106*, 662–671.

Bhandary, M. (1991). Robust M-estimation of Dispersion matrix with a structure. *Ann. Inst. Statist. Math. 43* , 689–705.

Bhattacharya, P. K., and Gangopadhyay, A. K. (1990). Kernel and nearest neighbour estimation of a conditional quantile. *Ann. Statist. 17*, 1400–1415.

Bhattacharya, R. N., and Ghosh, J. K. (1978). On the validity of the formal Edgeworth expansion. *Ann. Statist. 6*, 434–451.

Bhattacharya, R. N., and Ghosh, J. K. (1988). On moment conditions for valid formal Edgeworth expansions. *J. Multivar. Anal. 27*, 68–79.

Bickel, P. J. (1965). On some robust estimates of location. *Ann. Math. Statist. 36*, 847–858.

Bickel, P. J. (1973). On some analogues to linear combinations of order statistics in the linear model. *Ann. Statist. 1*, 597–616.

Bickel, P. J. (1975). One-step Huber estimates in the linear model. *J. Amer. Statist. Assoc. 70*, 428–433.

Bickel, P. J. (1981). Quelques aspects de la statistque robuste. *Ecole dété de St. Flour, Lecture Notes in Mathematics 876*, Springer-Verlag, New York, pp. 1–72.

Bickel, P. J. (1982). On adaptive estimation. *Ann. Statist. 10*, 647–671.

Bickel, P. J., Klaassen, C. A., Ritov, Y., and Wellner, J. A. (1993). *Efficient and Adaptive Estimation for Semiparametric Models*, Johns Hopkins Univ. Press, Baltimore, MD.

Bickel, P. J., and Lehmann, E. L. (1975a). Descriptive statistics for nonparametric model. I Introduction. *Ann. Statist. 3*, 1038–1044.

Bickel, P. J., and Lehmann, E. L. (1975b). Descriptive statistics for nonparametric model. II Location. *Ann. Statist. 3*, 1045–1069.

Bickel, P. J., and Lehmann, E. L. (1976). Descriptive statistics for nonparametric model. III Dispersion. *Ann. Statist. 4*, 1139–1158.

Bickel, P. J., and Lehmann, E. L. (1979). Descriptive statistics for nonparametric model. IV. Spread. *Contributions to Statistics: Jaroslav Hájek Memorial Volume* (ed. J. Jurečková). Acadmia, Prague, pp 33–40.

Bickel, P. J., and Wichura, M. J. (1971). Convergence criteria for multi-parameter stochastic process and some applications. *Ann. Math. Statist. 42*, 1656–1670.

Billingsley, P. (1968). *Convergence of Probability Measures.* Wiley, New York.

Birkes, D., and Dodge, Y. (1993). *Alternative Methods of Regression.* Wiley, New York.

Birnbaum, A., and Laska, E. (1967). Optimal robustness: A general method with applications to linear estimates of location. *J. Amer. Statist. Assoc. 62*, 1230–1240.

Blackman, J. (1955). On the approximation of a distribution function by an empirical distribution. *Ann. Math. Statist. 26*, 256–267.

Blom, G. (1956). On linear estimates with nearly minimum variance. *Ark. Mat., 3 (31)*, 365–369.

Bloomfield, P., and Stieger, W. L. (1983). *Least Absolute Deviations: Theory, Applications and Algorithms.* Birhäuser, Boston.

Boos, D. (1979). A differential for L-statistics. *Ann. Statist. 7*, 955–959.

Boos, D. (1982). A test of symmetry associated with the Hodges-Lehmann estimator. *J. Amer. Statist. Assoc. 77*, 647–651.

Boos, D., and Serfling, R. J. (1979). On Berry-Esseen rates for statistical functions, with application to L-estimates. *Technical report*, Florida State Univ., Tallahassee.

Box, G. E. P. (1953) Non-normality and tests of variance. *Biometrika 40*, 318–335.

Box, G. E. P., and Anderson, S. L. (1955). Permutation theory in the derivation of robust criteria and the study of departures from assumption. *J. Royal Statist. Soc. Ser. B 17*, 1–34.

Breckling, J., Kokic, P., and Lübke, O. (2001). A note on multivariate M-quantiles. *Statist. Probab. Letters 55*, 39-44.

Brillinger, D. R. (1969). An asymptotic representation of the sample distribution. *Bull. Amer. Math. Soc. 75*, 545–547.

Brown, B. M. (1971). Martingale central limit theorems. *Ann. Math. Statist. 42*, 59–66.

Buja, A. (1980). Sufficiency, least favorable experiments and robust tests. Ph.D. thesis, Swiss Federal Institute of Technology, ETH Zurich.

Buja, A. (1986). On the Huber-Strassen Theorem. *Probab. Th. Rel. Fields 73*, 149–152.

Bustos, O. H. (1982). General M-estimates for contaminated p-th order autoregression processes: consistency and asymptotic normality. *Zeit. Wahrsch. verw. Geb. 59*, 491–504.

Butler, R. W. (1982). Nonparametric interval and point prediction using data trimmed by a Grubbs-type outlier rule. *Ann. Statist. 10*, 197–204.

Bühlmann, P. and van de Geer, S. (2011). *Statistics for High-Dimensional Data. Methods, Theory and Applications.* Springer-Verlag, Berlin-Heidelberg.

Cantoni, E. and Roncheetti, E. (2001). Robust Inference for Generalized Linear Models. *J. Amer. Statist. Assoc. 96*, 1022-1030.

Carroll, R. J. (1977a). On the uniformity of sequential procedures. *Ann. Statist. 5*, 1039–1046.

Carroll, R. J. (1977b). On the asymptotic normality of stopping times based on robust estimators. *Sankhyā, Ser.A 39*, 355–377.

Carroll, R. J. (1978). On almost sure expansions for M-estimates. *Ann. Statist. 6*, 314–318.

Carroll, R. J. (1979). On estimating variances of robust estimators when the errors are asymmetric. *J. Amer. Statist. Assoc. 74*, 674–679.

Carroll, R. J. (1982). Robust estimation in certain heteroscedastic linear models when there are many parameters. *J. Statist. Plan. Infer. 7*, 1–12.

Carroll, R. J., and Gallo, P. P. (1982). Some aspects of robustness in the functional errors-in-variables regression models. *Com. Statist. Theor. Meth. A 11*, 2573–2585.

Carroll, R. J., and Ruppert, D. (1980). A comparison between maximum likelihood and generalized least squares in a heteroscedastic model. *J. Amer. Statist. Assoc. 77*, 878–882.

Carroll, R. J., and Ruppert, D. (1982). Robust estimation in heteroscadastic linear models. *Ann. Statist. 10*, 429–441.

Carroll, R. J., and Ruppert, D. (1988). *Transformation and Weighting in Regression.* Chapman and Hall, London.

Carroll, R.J., and Welsh, A. H. (1988). A note on asymmetry and robustness in linear regression. *Amer. Statist. 42*, 285–287.

Chakraborty, B. (2001). On affine equivariant multivariate quantiles. *Ann. Inst. Statist. Math. 53*, 380–403.

Chakraborty, B. (2003). On multivariate quantile regression. *J. Statist. Planning Infer. 110*, 109-132.

Chanda, K. C. (1975). Some comments on the asymptotic probability laws of sample quantiles. *Calcutta Statist. Assoc. Bull. 24*, 123–126.

Chatterjee, S., and Hadi, A.S. (1986). Influential observations, high leverage points and outliers in linear regression (with discussion). *Statist. Sci. 1*, 379–416.

Chatterjee, S. K., and Sen, P. K. (1973). On Kolmogorov-Smirnov type test for symmetry. *Ann. Inst. Statist. Math. 25*, 288–300.

Chaubey, Y. P., and Sen, P. K. (1996). On smooth estimation of survival and density functions. *Statist. Decisions 14*, 1–22.

Chaudhuri, P. (1991). Nonparametric estimators of regression quantiles and their local Bahadur representation. *Ann. Statist. 19*, 760–777.

Chaudhuri, P. (1992). Generalized regression quantiles: Forming a useful toolkit for robust linear regression. *L_1-Statistical Analysis and Related Methods.* (ed. Y. Dodge). North-Holland, Amsterdam, pp. 169–185.

Chaudhuri, P. (1992) Multivariate location estimation using extension of R-estimators through U-statistics-type approach. *Ann. Statist. 20*, 897–916.

Chaudhuri, P. (1996). On a geometric notion of quantiles for multivariate data. *J. Amer. Statist. Assoc. 91,* 862-872.

Chaudhuri, P., and Sengupta, D. (1993). Sign tests in multi-dimension: Inference based on the geometry of the data cloud. *J. Amer. Statist. Assoc. 88*, 1363–1370.

Cheng, K. F. (1984). Nonparametric estimation of regression function using linear combinations of sample quantile regression functions. *Sankhyā, Ser. A 46*, 287–302.

Chernoff, H., Gastwirth, J. L., and Johns, M. V. (1967). Asymptotic distribution of linear combinations of order statistics. *Ann. Math. Statist. 38*, 52–72.

Chernoff, H., and Teicher, H. (1958). A central limit theorem for sums of interchangeable random variables. *Ann. Math. Statist. 29*, 118–130.

Choquet, G. (1953/54). Theory of capacities. *Ann. Inst. Fourier 5*, 131–292.

Čížek, P. (2005). Least trimmed squares in nonlinear regression under dependence. *J. Statist. Planning Infer. 136*, 3967–3988.

Čížek, P. (2010). Reweighted least trimmed squares: an alternative to one-step estimators. CentER discussion paper 2010/91, Tilburg University, The Netherlands.

Čížek, P. (2011). Semiparametrically weighted robust estimation of regression models. *Comp. Statist. Data Analysis* 55/1, 774–788

Clarke, B. R. (1983). Uniqueness and Fréchet differentiability of functional solutions to maximum likelihood type equations. *Ann. Statist. 11*, 1196–1205.

Clarke, B. R. (1986). Nonsmooth analysis and Fréchet differentiability of M-functionals. *Probab. Th. Rel. Fields 73*, 197–209.

Collins, J. R. (1976). Robust estimation of a location parameter in the presence of asymmetry. *Ann. Statist. 4*, 68–85.

Collins, J. R. (1977). Upper bounds on asymtotic variances of M-estimators of location. *Ann. Statist. 5*, 646–657.

Collins, J. R. (1982). Robust M-estimators of location vectors. *J. Mult. Anal. 12*, 480–492.

Collins, J. R., and Portnoy, S. (1981). Maximizing the variance of M-estimators using the generalized method of moment spaces. *Ann. Statist. 9*, 567–577.

Collins, J. R., Sheahan, J. N., and Zheng, Z. (1985). Robust estimation in the linear model with asymmetric error distribution. *J. Mult. Anal. 20*, 220–243.

Cook, R. D., and Weisberg, S. (1982). *Residuals and Influence in Regression.* Chapman and Hall, New York.

Coakley, C. W., and Hettmansperger, T. P. (1988). A Bounded influence, high breakdown, efficient regression estimator. *J. Amer. Statist. Assoc. 88*, 872–880.

Croux, C., and Rousseeuw, P. J. (1992). A class of high-breakdown scale estimators based on subranges. *Commun. Statist. Theor. Meth. A21*, 1935–1951.

Csörgő, M., and Horváth, L. (1993). *Weighted Approximations in Probability and Statistics.* Wiley, New York.

Csörgő, M., and Révész, P. (1981). *Strong Approximations in Probability and Statistics.* Akadémiai Kiadó, Budapest.

Daniell, P. J. (1920). Observations weighted according to order. *Ann. Math. 42*, 222–236.

David, H. A. (1981). *Order Statistics.* Second Edition. Wiley, New York.

Davidian, M., and Carroll, R. J. (1987). Variance function estimation. *J. Amer. Statist. Assoc. 82*, 1079–1091.

Davies, L. (1994). Desirable properties, breakdown and efficiency in the linear regression model. *Statist. Probab. Letters 19*, 361–370.

Davies, P. L. (1994). Aspects of robust linear regression. *Ann. Statist. 21*, 1843–1899.

de Haan, L., and Taconis-Haantjes, E. (1979). On Bahadur's representation of sample quantiles. *Ann. Inst. Statist. Math. A31*, 299–308.

Denby, L., and Martin, D. (1979). Robust estimation of the first order autoregressive parameters. *J. Amer. Statist. Assoc. 74*, 140–146.

Dionne, L. (1981). Efficient nonparametric estimators of parameters in the general linear hypothesis. *Ann. Statist. 9*, 457–460.

Dodge, Y., and Jurečková, J. (1995). Estimation of quantile density function based on regression quantiles. *Statist. Probab. Letters 23*, 73–78.

Dodge, Y., and Jurečková, J. (2000). *Adaptive Regression.* Springer, New York.

Donoho, D. L., and Gasko, M. (1992). Breakdown propertties of location estimates based on halfspace depth and projected outlyingness. *Ann. Statist. 20*, 1803–1827.

Donoho, D. L., and Huber, P. J. (1983). The notion of breakdown point. *Festschrift for E. L. Lehmann* (eds. P. J. Bickel et al.). Wadsworth, Belmont, Calif., 157–184.

Donoho, D. L., and Liu, R. C. (1988). The "automatic" robustness of minimum distance functionals. *Ann. Statist. 16*, 552–586.

Doob, J. L. (1949). Heurustic approach to the Kolmogorov-Smirnov theorems. *Ann. Math. Statist. 20*, 393–403.

Doob, J. L. (1967). *Stochastic Processes.* Wiley, New York.

Dudley, R. M. (1984). A course on empirical Processes. *Lecture Notes in Mathematics 1097*, Springer-Verlag, New York, pp. 1–142.

Dudley, R. M. (1985). An extended Wichura theorem, definitions of Donsker classes, and weighted empirical distributions. *Lecture Notes in Mathematics 1153*, Springer-Verlag, New York, pp. 141–148.

Dupač, V., and Hájek, J. (1969). Asymptotic normality of linear rank statistics. *Ann. Math. Statist. 40*, 1992–2017.

Dutter, R. (1977). Numerical solutions of robust regression problems: Computational aspects, a comparison. *J. Statist. Comp. Simul. 5*, 207–238.

Dvoretzky, A. (1972). Central limit theorem for dependent random variables. *Proc. 6th Berkeley Symp. Math. Statist. Probab.*, vol. 2 (eds. L. LeCam et al.). Univ. Calif. Press, Los Angeles, pp. 513–555.

Dvoretzky, A., Kiefer, J., and Wolfowitz, J. (1956). Asymptotic minimax character of the sample distribution function and the classical multinomial estimator. *Ann. Math. Statist. 27*, 642–669.

Eaton, M. (1983). *Multivariate Statistics.* Wiley, New York.

Eicker, F. (1970). A new proof of the Bahadur-Kiefer representation of sample quantiles. *Nonparametric Techniques in Statistical Inference.* (ed. M. L. Puri), Cambridge Univ. Press, pp. 321–342.

Fabian, M., Habala, P., Hájek, P., Santalucía, V. M., Pelant, J., and Zízler, V. (2001). *Functional Analysis and Infinite-Dimensional Geometry.* Springer, New York.

Falk, M. (1986). On the estimation of the quantile density function. *Statist. Probab. Letters 4*, 69–73.

Fernholtz, L. T. (1983). *von Mises' Calculus for Statistical Functionals. Lecture notes in Statistics, 19*, Springer-Verlag, New York.

Field, C. A., and Wiens, D. P. (1994). One-step M-estimators in the linear model, with dependent errors. *Canad. J. Statist. 22*, 219–231.

Fillippova, A. A. (1961). Mises' theorem on the asymptotic behavior of functionals of empirical distribution functions and its applications. *Teor. Veroyat. Primen. 7*, 24–57.

Fisher, R. A. (1925). Theory of statistical estimation. *Proc. Cambridge Phil. Soc. 22*, 700–725.

Freedman, D. A., and Diaconis, P. (1982). On inconsistent M-estimators. *Ann. Statist. 10*, 454–461.

Fu, J. C. (1975). The rate of convergence of consistent point estimators. *Ann. Statist. 3*, 234–240.

Gallant, A. R. (1989). On Asymptotic Normality when the Number of Regressors Increases and the Minimum Eigenvalue of X'X/n Decreases. Class notes, Department of Economics, University of North Carolina, Chapel Hill NC 27599-3305 USA.

Gangopadhyay, A. K., and Sen, P. K. (1990). Bootstrap confidence intervals for conditional quantile functions. *Sankhyā Ser. A 52*, 346–363.

Gangopadhyay, A. K., and Sen, P. K. (1992). Contiguity in nonparametric estimation of a conditional functional. In *Nonparameteric Statistics and Related Topics* (ed. A. K. M. E. Saleh), North Holand, Amsterdam, pp. 141–162.

Gangopadhyay, A. K., and Sen, P. K. (1993). Contiguity in Bahadur-type representations of a conditional quantile and application in conditional quantile process. In *Statistics and Probability: R. R. Bahadur Festschrift* (eds. J. K. Ghosh et al.), Wiley Eastern, New Delhi, pp. 219–232.

Gardiner, J. C., and Sen, P. K. (1979). Asymptotic normality of a variance estimator of a linear combination of a function of order statistics. *Zeit. Wahrsch. verw. Geb. 50*, 205–221.

Geertsema, J. C. (1970). Sequential confidence intervals based on rank tests. *Ann. Math. Statist. 41*, 1016–1026.

Geertsema, J. C. (1987). The behavior of sequential confidence intervals under contaminations. *Sequent. Anal. 6*, 71–91.

Geertsema, J. C. (1992). A comparison of nonparametric sequential procedures for estimating a quantile. *Order Statistics and Nonparametrics: Theory and Applications* (eds. P. K. Sen and I. A. Salama), North Holland, Amsterdam, pp. 101–113.

Ghosh, J. K. (1971). A new proof of the Bahadur representation of quantiles and an application. *Ann. Math. Statist. 42*, 1957–1961.

Ghosh, J. K. (1994). *Higher Order Asymptotics*, NSF-CBMS Reg. Confer. Ser. Probab. Statist., Vol. 4, Inst. Math. Statist., Hayward, Calif.

Ghosh, J. K., Mukerjee, R., and Sen, P. K. (1995). Second-order Pitman admissibility and Pitman closeness: The multiparameter case and Stein-rule estimators. *J. Multivar. Anal. 57*, 52–68.

Ghosh, J. K., and Sen, P. K. (1985). On the asymptotic performance of the log-likelihood ratio statistic for the mixture model and related results. In *Proceedings of the Berkeley Coference in Honor of J. Neyman and J. Kiefer*, Vol. 2 (eds. L. M. LeCam and R. A. Olshen), Wadsworth, Belmont, Calif., pp. 789–806.

Ghosh, J. K., Sen, P. K., and Mukerjee, R. (1994). Second-order Pitman closeness and Pitman admissibility. *Ann. Statist. 22*, 1133–1141.

Ghosh, M. (1972). On the representation of linear functions of order statistics. *Sankhyā Ser. A 34*, 349–356.

Ghosh, M., Mukhopadhyay, N., and Sen, P. K. (1997). *Sequential Estimation.* Wiley, New York.

Ghosh, M., Nickerson, D. M., and Sen, P. K. (1987). Sequential shrinkage estimation. *Ann. Statist. 15*, 817–829.

Ghosh, M., and Sen, P. K. (1970). On the almost sure convergence of von Mises' differentiable statistical functions. *Calcutta Statist. Assoc. Bull. 19*, 41–44.

Ghosh, M., and Sen, P. K. (1971a). Sequential confidence intervals for the regression coefficients based on Kendall's tau. *Calcutta Statist. Assoc. Bull. 20*, 23–36.

Ghosh, M., and Sen, P. K. (1971b). A class of rank order tests for regression with partially informed stochastic predictors. *Ann. Math. Statist. 42*, 650–661.

Ghosh, M., and Sen, P. K. (1972). On bounded width confidence intervals for the regression coefficient based on a class of rank statistics. *Sankhyā Ser. A 34*, 33–52.

Ghosh, M., and Sen, P. K. (1973). On some sequential simultaneous confidence intervals procedures. *Ann. Inst. Statist. Math. 25*, 123–135.

Ghosh, M., and Sen, P. K. (1977). Sequential rank tests for regression. *Sankhyā Ser. A 39*, 45–62.

Ghosh, M., and Sen, P. K. (1989). Median unbiasedness and Pitman closeness. *J. Amer. Statist. Assoc. 84*, 1089–1091.

Ghosh, M., and Sukhatme, S. (1981). On Bahadur representation of quantiles in nonregular cases. *Comm. Statist. Theor. Meth. A10*, 169–182.

Gill, R. D. (1989). Non- and semi-parametric maximum likelihood estimators and the von Mises method. Part 1. *Scand. J. Statist. 16*, 97–128.

Giloni, A., and Padberg, M. (2002). Least trimmed squares regression, least median squares regression, and mathematical programming. *Math. Computer Modelling* 35, 1043–1060.

Gutenbrunner, C. (1986). Zur Asymptotik von Regression Quantil Prozesen und daraus abgeleiten Statistiken. Ph.D. dissertation, Universitat Freiburg, Freiburg.

Gutenbrunner, C., and Jurečková, J. (1992). Regression rank scores and regression quantiles. *Ann. Statist. 20*, 305–330.

Gutenbrunner, C., and Jurečková, J., Koenker, R., and Portnoy, S. (1993). Tests of linear hypotheses based on regression rank scores. *J. Nonpar. Statist. 2*, 307–331.

Hájek, J. (1961). Some extensions of the Wald-Wolfowitz-Noether theorem. *Ann. Math. Statist. 32*, 506–523.

Hájek, J. (1962). Asymptotically most powerful rank order tests. *Ann. Math. Statist. 33*, 1124–1147.

Hájek, J. (1965). Extensions of the Kolmogorov-Smironov tests to regression alternatives. *Bernoull-Bayes-Laplace Seminar* (ed. L. LeCam), Univ. Calif. Press, Los Angeles, Calif., pp. 45–60.

Hájek, J. (1968). Asymptotic normality of simple linear rank statistics under alternatives. *Ann. Math. Statist. 39*, 325–346.

Hájek, J., and Šidák, Z. (1967). *Theory of Rank Tests.* Academia, Prague.

Hájek, J., and Šidák, Z and Sen (1999). *Theory of Rank Tests.* Academic Press, New York.

Hall, P. (1992). *The Bootstrap and Edgeworth Expansion.* Springer Verlag, New York.

Hallin, M., and Jurečková, J. (1999). Optimal tests for autoregressive models based on autoregression rank scores. *Ann. Statist. 27*, 1385–1414.

Hallin, M., Paindaveine, M., and Šiman, M. (2010). Multivariate quantiles and multiple-output regression quantiles: From L_1 optimization to halfspace depth. *Ann. Statist. 38*, 635–669.

Halmos, P. R. (1946). The theory of unbiased estimation. *Ann. Math. Statist. 17*, 34–43.

Hampel, F. R. (1968). Contributions to the theory of robust estimators. Ph.D. dissertation, Univ. California, Berkeley.

Hampel, F. R. (1971). A general qualitative definition of robustness. *Ann. Math. Statist. 42*, 1887–1896.

Hampel, F. R. (1974). The influence curve and its role in robust estimation. *J. Amer. Statist. Assoc. 62*, 1179–1186.

Hampel, F. R. (1997). Some additional notes on the Princeton Robustness Year. In *The Practice of Data Analysis: Essays in Honor of John W. Tukey* (eds. D. R. Brillinger and L. T. Fernholz), pp. 133-153. Princeton University Press.

Hampel, F. R., Rousseeuw, P. J., Ronchetti, E., and Stahel, W. (1986). *Robust Statistics – The Approach Based on Influence Functions.* Wiley, New York. Second Edition: (2005).

Hampel, F. R., Hennig, C., and Ronchetti, E. (2011). A smoothing principle for the Huber and other location M-estimators. *Comp. Statist. Data Analysis 55* 324–337.

Hanousek, J. (1988). Asymptotic relations of M- and P-estimators of location. *Comp. Statist. Data Anal. 6*, 277–284.

Hanousek, J. (1990). Robust Bayesian type estimators and their asymptotic representation. *Statist. Dec. 8*, 61–69.

Harrell, F., and Davis, C. (1982). A new distribution-free quantile estimator. *Biometrika 69*, 635–640.

He, X., Jurečková, J., Koenker, R., and Portnoy, S. (1990). Tail behavior of regression estimators and their breakdown points. *Econometrica 58*, 1195–1214.

He, X., and Shao, Qi-Man (1996). A general Bahadur representation of M-estimators and its application to linear regression with nonstochastic designs. *Ann. Statist. 24*, 2608–2630.

He, X., Simpson, D. G., and Portnoy (1990). Breakdown point of tests. *J. Amer. Statist. Assoc. 85*, 446–452.

He, X., and Wang, G. (1995). Law of the iterated logarithm and invariance principle for M- estimators. *Proc. Amer. Math. Soc. 123*, 563–573.

Heiler, S. (1992). Bounded influence and high breakdown point regression with linear combinations of order statistics and rank statistics. L_1-*Statistical Analysis and Related Methods* (ed. Y. Dodge), North Holland, Amsterdam, pp. 201–215.

Heiler, S., and Willers, R. (1988). Asymptotic normality of R-estimates in the linear model. *Statistics 19*, 173–184.
Helmers, R. (1977). The order of normal approximation for linear combinations of order statistics with smooth weight function. *Ann. Probab. 5*, 940–953.
Helmers, R. (1980). Edgeworth expansions for linear combinations of order statistics. *Math. Centre Tract 105*, Amsterdam.
Helmers, R. (1981). A Berry-Esseen theorem for linear combination of order statistics. *Ann. Probab. 9*, 342–347.
Hodges, J. L. Jr. (1967). Efficiency in normal samples and tolerance of extreme values for some estimate of location. *Proc. 5th Berkeley Symp. Math. Statist. Probab.*, Univ. Calif. Press, Los Angeles, vol. 1, pp. 163–186.
Hodges, J. L., and Lehmann, E. L. (1963). Estimation of location based on rank tests. *Ann. Math. Statist. 34*, 598–611.
Hoeffding, W. (1948). A class of statistics with asymptotically normal distribution. *Ann. Math. Statist. 19*, 293–325.
Hoeffding, W. (1961). The strong law of large numbers for U-statistics. *Univ. North Carolina, Institute of Statistics Mimeo Series*, No 302, Chapel Hill.
Hoeffding, W. (1963). Probability inequalities for sums of bounded random variables. *J. Amer. Statist. Assoc. 58*, 13–30.
Hoeffding, W. (1965a). Asymptotically optimal tests for multinomial distributions (with discussion). *Ann. Math. Statist. 36*, 369–408.
Hoeffding, W. (1965b). On probabilities of large deviations. *Proc. 5th Berkeley Symp. Math. Stat. Probab.*, Univ. Calif. Press, Los Angeles, vol 1, pp. 203–219.
Hofmann, M., and Kontoghiorghes, E. J. (2010). Matrix strategies for computing the least trimmed squares estimation of the general linear and SUR models. *Comput. Stat. Data Anal. 54*, 3392–3403.
Hössjer, O. (1994). Rank-based estimates in the linear model with high breakdown point. *J. Amer. Statist. Assoc. 89*, 149–158.
Huber, P. J. (1964). Robust estimator of a location parameter. *Ann. Math. Statist. 35*, 73–101.
Huber, P. J. (1965). A robust version of the probability ratio test. *Ann. Math. Statist. 36*, 1753–1758.
Huber, P. J. (1968). Robust confidence limits. *Zeit. Wahrsch. verw. Geb. 10*, 269–278.
Huber, P. J. (1969). Théorie de l'inférence statistique robuste. *Seminar de Math. Superieurs*, Univ. Montréal.
Huber, P. J. (1972). Robust statistics: A review. *Ann. Math. Statist. 43*, 1041-1067.
Huber, P. J. (1973). Robust regression: Asymptotics, conjectures and Monte Carlo. *Ann. Statist. 1*, 799–821.
Huber, P. J. (1981). *Robust Statistics.* Wiley, New York.
Huber, P. J. (1984). Finite sample breakdown of M- and P-estimators. *Ann. Statist. 12*, 119–126.
Huber, P. J., and Dutter, R. (1974). Numerical solutions of robust regression problems. *COMPSTAT 1974* (eds. G. Bruckmann et al.). Physica Verlag, Vienna, pp. 165–172.
Huber, P. J., and Ronchetti, E. (2009). *Robust Statistics.* Second Edition. Wiley, New York.
Huber, P. J., and Strassen, V. (1973). Minimax tests and the Neyman-Pearson lemma for capacities. *Ann. Statist. 1*, 251–263.
Huber-Carol, C. (1970). Etude asymptotique de tests robustes. Ph.D. dissertation, ETH Zűrich.

Humak, K. M. S. (1983). *Statistische Methoden der Modellbildung. Band II, Nichtlineare Regression, Robuste Verfahren in linearen Modellen, Modelle mit Fehlern in den Variablen.* Akademie Verlag, Berlin. [English Translation (1989): *Statistical Methods of Model Building, Vol. II: Nonlinear Regression, Functional Relations and Robust Methods* (eds. H. Bunke and O. Bunke). Wiley, New York.]

Hušková, M. (1979). The Berry-Esseen theorem for rank statistics. *Comment. Math. Univ. Caroliae 20*, 399–415.

Hušková, M. (1981). On bounded length sequential confidence interval for parameter in regression model based on ranks. *Coll. Nonpar. Infer. Janos Bolyai Math. Soc.* (ed. B. V. Gnedenko et al.), North Holland, Amsterdam, pp. 435–463.

Hušková, M. (1991). Sequentially adaptive nonparametric procedures. *Handbook of Sequential Analysis* (eds. B. K. Ghosh and P. K. Sen), Dekker, New York, pp. 459–474.

Hušková, M., and Janssen, P. (1993). Consistency of generalized bootstrap for degenerate U-statistics. *Ann. Statist. 21,* 1811–1823.

Hušková, M., and Jurečková, J. (1981). Second order asymptotic relations of M–estimators and R-estimators in two-sample location model. *J. Statist. Planning Infer. 5*, 309–328.

Hušková, M., and Jurečková, J. (1985). Asymptotic representation of R-estimators of location. *Proc. 4th Pannonian Symp.*, 145–165.

Hušková, M., and Sen, P. K. (1985). On sequentially adaptive asymptotically efficient rank statistics. *Sequen. Anal. 4*, 225–251.

Hušková, M., and Sen, P. K. (1986). On sequentially adaptive signed rank statistics. *Sequen. Anal. 5*, 237–251.

Hyvärinen,A. , Karhunen, J., and Oja, E. (2001). *Independent Component Analysis.* J. Wiley, New York.

Ibragimov, I. A., and Hasminski, R. Z. (1970). On the asymptotic behavior of generalized Bayes estimators. *Dokl. Akad. Nauk SSSR 194*, 257–260.

Ibragimov, I. A., and Hasminski, R. Z. (1971). On the limiting behavior of maximum likelihood and Bayesian estimators. *Dokl. Akad. Nauk SSSR 198*, 520–523.

Ibragimov, I. A., and Hasminski, R. Z. (1972). Asymptotic behavior of statistical estimators in the smooth case I. *Theor. Probab. Appl. 17*, 445–462.

Ibragimov, I. A., and Hasminski, R. Z. (1981). *Statistical Estimation: Asymptotic Theory.* Springer-Verlag, New York.

Inagaki, N. (1970). On the limiting distribution of a sequence of estimators with uniformity property. *Ann. Inst. Statist. Math. 22*, 1–13.

Jaeckel, L. A. (1971). Robust estimation of location: Symmetry and asymmetric contamination. *Ann. Math. Statist. 42*, 1020–1034.

Jaeckel, L. A. (1972). Robust estimates of location. Ph.D. dissertation, University of California, Berkeley.

Jaeckel, L. A. (1972). Estimating regression coefficients by minimizing the dispersion of the residuals. *Ann. Math. Statist. 43*, 1449–1458.

James, W., and Stein, C. (1961). Estimation with quadratic loss. *Proc. 4th Berkeley Symp. Math. Statist. Probab.* (ed. J. Neyman), Univ. Calif. Press, Los Angeles, vol. 1, pp. 361–380.

Janssen, P., Jurečková, J., and Veraverbeke, N. (1985). Rate of convergence of one- and two-step M-estimators with applications to maximum likelihood and Pitman estimators. *Ann. Statist. 13*, 1222–1229.

Janssen, P., and Veraverbeke, N. (1987). On nonparametric regression estimators based on regression quantiles. *Comm. Statist. Theor. Meth. A16*, 383–396.

Johns, M. V. (1979). Robust Pitman-like estimators. *Robustness in Statistics* (eds: R. L. Launer and G. N. Wilkinson). Academic Press, New York.

Jung, J. (1955). On linear estimates defined by a continuous weight function. *Arkiv für Mathematik 3*, 199–209.

Jung, J. (1962). Approximation to the best linear estimates. *Contributions to Order Statistics* (eds. A. E. Sarhan and B. G. Greenberg). Wiley, New York, 28–33.

Jurečková, J. (1969). Asymptotic linearity of a rank statistic in regression parameter. *Ann. Math. Statist. 40*, 1889–1900.

Jurečková, J. (1971a). Nonparametric estimate of regression coefficients. *Ann. Math. Statist. 42*, 1328–1338.

Jurečková, J. (1971b). Asymptotic independence of rank test statistic for testing symmetry on regression. *Sankhyā, Seer. A 33*, 1–18.

Jurečková, J. (1973a). Almost sure uniform asymptotic linearity of rank statistics in regression parameter. *Trans. 6th Prague Conf. on Inform. Theor., Random Proc. Statist. Dec. Funct.*, pp. 305–313.

Jurečková, J. (1973b). Central limit theorem for Wilcoxon rank statistics process. *Ann. Statist. 1*, 1046–1060.

Jurečková, J. (1977a). Asymptotic relations of least squares estimate and of two robust estimates of regression parameter vector. *Trans. 7th Prague Conf. and 1974 European meeting of Statisticians*, Academia, Prague, pp. 231–237.

Jurečková, J. (1977b). Asymptotic relations of M-estimates and R-estimates in linear regression model. *Ann. Statist. 5*, 464–472.

Jurečková, J. (1978). Bounded-length sequential confidence intervals for regression and location parameters. *Proc. 2nd Prague Symp. Asympt. Statist.*, Academia, Prague, pp. 239–250.

Jurečková, J. (1979). Finite-sample comparison of L-estimators of location. *Comment. Math. Univ. Carolinae 20*, 507–518.

Jurečková, J. (1980). Rate of consistency of one sample tests of location. *J. Statist. Planning Infer. 4*, 249–257.

Jurečková, J. (1981a). Tail behavior of location estimators. *Ann. Statist. 9*, 578–585.

Jurečková, J. (1981b). Tail behavior of location estimators in nonregular cases. *Comment. Math. Univ. Carolinae 22*, 365–375.

Jurečková, J. (1982). Tests of location and criterion of tails. *Coll. Math. Soc. J.Bolai 32*, 469–478.

Jurečková, J. (1983a). Robust estimators of location and regression parameters and their second order asymptotic relations. *Trans. 9th Prague Conf. on Inform. Theor., Rand. Proc. and Statist. Dec. Func.* (ed. J. A. Višek), Reidel, Dordrecht, 19–32.

Jurečková, J. (1983b). Winsorized least-squares estimator and its M-estimator counterpart. *Contributions to Statistics: Essays in Honour of Norman L. Johnson* (ed. P. K. Sen), North Holland, Amsterdam, pp. 237–245.

Jurečková, J. (1983c). Trimmed polynomial regression. *Comment. Math. Univ. Carolinae 24*, 597–607.

Jurečková, J. (1983d). Asymptotic behavior of M-estimators of location in nonregular cases. *Statist. Dec. 1*, 323–340.

Jurečková, J. (1984a). Rates of consistency of classical one-sided tests. *Robustness of Statistical Methods and Nonparametric Statistics* (eds. D. Rasch and M. L. Tiku), Deutscher-Verlag, Berlin, pp. 60–62.

Jurečková, J. (1984b). Regression quantiles and trimmed least squares estimator under a general design. *Kybernetika 20*, 345–357.

Jurečková, J. (1985). Robust estimators of location and their second-order asymptotic relations. *A Celebration of Statistics* (eds. A. C. Atkinson and S. E. Fienberg), Springer-Verlag, New York, pp. 377–392.

Jurečková, J. (1986). Asymptotic representations of L-estimators and their relations to M-estimators. *Sequen. Anal. 5*, 317–338.

Jurečková, J. (1989). Robust statistical inference in linear models. *Nonlinear Regression, Functional Relations and Robust Methods: Statistical Methods of Model Building* (eds. Bunke, H. and Bunke, O.), Wiley, New York. vol. 2, pp. 134–208.

Jurečková, J. (1992a). Uniform asymptotic linearity of regression rank scores process. *Nonparametric Statistics and Related Topics* (ed. A. K. Md. E. Saleh), North Holland, Amsterdam, pp. 217–228.

Jurečková, J. (1992b). Estimation in a linear model based on regression rank scores. *J. Nonpar. Statist. 1*, 197–203.

Jurečková, J. (1995a). Jaroslav Hájek and asymptotic theory of rank tests. *Kybernetika 31,(2)*, 239–250.

Jurečková, J. (1995b). Affine- and scale-equivariant M-estimators in linear model. *Probab. Math. Statist. 15*, 397–407.

Jurečková, J. (1995c). Trimmed mean and Huber's estimator: Their difference as a goodness-of-fit criterion. *J. Statistical Science 29*, 31–35.

Jurečková, J. (1995d). Regression rank scores: Asymptotic linearity and RR-estimators. *Proc. MODA'4* (eds. C. P. Kitsos and W. G. Müller), Physica-Verlag, Heidelberg, pp. 193–203.

Jurečková, J. (1999). Equivariant estimators and their asymptotic representations. *Tatra Mountains Math. Publ.* 17, 1–9.

Jurečková, J. (2008). Regression rank scores in nonlinear models. In: Beyond parametrics in interdisciplinary research: Festschrift in honor of Professor Pranab K. Sen (eds. N. Balakrishnan, E. A. Peña and M. J. Silvapulle). *Institute of Mathematical Statistics Collections*, Vol. 1, 173-183.

Jurečková, J. (2010). Finite sample distribution of regression quantiles. *Stat. Probab. Lett. 80*, 1940–1946.

Jurečková, J. and Kalina (2012). Nonparametric multivariate rank tests and their unbiasedness. *Bernoulli 18*, 229-251.

Jurečková, J., and Klebanov, L. B. (1997). Inadmissibility of robust estimators with respect to L1 norm. In: *L_1-Statistical Procedures and Related Topics* (Y. Dodge, ed.). IMS Lecture Notes - Monographs Series 31, 71–78.

Jurečková, J., and Klebanov, L. B. (1998). Trimmed, Bayesian and admissible estimators. *Stat. Probab. Lett. 42*, 47-51.

Jurečková, J., Koenker, R., and Portnoy, S. (2001). Tail behavior of the least squares estimator. *Stat. Probab. Lett. 55*, 377–384.

Jurečková, J., Koenker, R., and Welsh, A. H. (1994). Adaptive choice of trimming. *Ann. Inst. Statist. Math. 46*, 737–755.

Jurečková, J., and Malý, M. (1995). The asymptotics for studentized k-step M-estimators of location. *Sequen. Anal. 14*, 229–245.

Jurečková, J., and Milhaud, X. (1993). Shrinkage of maximum likelihood estimator of multivariate location. *Asymptotic Statistics [Proc. 5th Prague Symp.]* (P. Mandl and M. Hušková, eds.), pp. 303–318. Physica-Verlag, Heidelberg.

Jurečková, J., and Omelka M. (2010). Estimator of the Pareto index based on nonparametric test. *Commun. Stat.-Theory Methods 39*, 1536–1551.

Jurečková, J., and Picek, J. (2005). Two-step regression quantiles. *Sankhya 67/2*, 227-252.

Jurečková, J., and Picek, J. (2006). *Robust Statistical Methods with R.* Chapman & Hall/CRC.

Jurečková, J., and Picek, J. (2009). Minimum risk equivariant estimators in linear regression model. *Stat. Decisions 27*, 1001–1019.

Jurečková, J., and Picek, J. (2011). Finite-sample behavior of robust estimators. In: *Recent Researches in Instrumentation, Measurement, Circuits and Systems* (eds. S. Chen, N. Mastorakis, F. Rivas-Echeverria, V. Mladenov), pp. 15–20.

Jurečková, J., and Picek, J., and Saleh, A. K. Md. E. (2010). Rank tests and regression rank scores tests in measurement error models. *Comput. Stat. Data Anal. 54*, 3108–3120.

Jurečková, J., and Portnoy, S. (1987). Asymptotics for one-step M-estimators in regression with application to combining efficiency and high breakdown point. *Commun. Statist. Theor. Meth. A16*, 2187–2199.

Jurečková, J., and Puri, M. L. (1975). Order of normal approximation of rank statistics distribution. *Ann. Probab. 3*, 526–533.

Jurečková, J., Saleh, A. K. Md. E., and Sen, P. K. (1989). Regression quantiles and improved L-estimation in linear models. *Probability, Statistics and Design of Experiment: R. C. Bose Memorial Conference* (ed. R. R. Bahadur), Wiley Eastern, New Delhi, pp. 405–418.

Jurečková, J., and Sen, P. K. (1981a). Sequential procedures based on M- estimators with discontinuous score functions. *J. Statist. Plan. Infer. 5*, 253–266.

Jurečková, J., and Sen, P. K. (1981b). Invariance principles for some stochastic processes related to M-estimators and their role in sequential statistical inference. *Sankhyā, Ser. A 43*, 190–210.

Jurečková, J., and Sen, P. K. (1982). M-estimators and L-estimators of location: Uniform integrability and asymptotically risk-efficient sequential version. *Commun. Statist. C 1*, 27–56.

Jurečková, J., and Sen, P. K. (1982a). Simultaneous M-estimator of the common location and of the scale-ratio in the two sample problem. *Math. Operations. und Statistik, Ser. Statistics 13*, 163–169.

Jurečková, J., and Sen, P. K. (1982b). M-estimators and L-estimators of location: Uniform integrability and asymptotically risk-efficient sequential versions. *Sequen. Anal. 1*, 27–56.

Jurečková, J., and Sen, P. K. (1984). On adaptive scale-equivariant M-estimators in linear models. *Statist. Dec. Suppl. 1*, 31–46.

Jurečková, J., and Sen, P. K. (1989). Uniform second order asymptotic linearity of M-statistics in linear models. *Statist. Dec. 7*, 263–276.

Jurečková, J., and Sen, P. K. (1990). Effect of the initial estimator on the asymptotic behavior of one-step M-estimator. *Ann. Inst. Statist. Math. 42*, 345–357.

Jurečková, J., and Sen, P. K. (1993). Asymptotic equivalence of regression rank scores estimators and R-estimators in linear models. *Statistics and Probability: A R. R. Bahadur Festschrift* (eds. J. K. Ghosh et al.). Wiley Eastern, New Delhi, pp. 279–292.

Jurečková, J., and Sen, P. K. (1994). Regression rank scores statistics and studentization in the linear model. *Proc. 5th Prague Conf. Asympt. Statist.* (eds. M. Hušková and P. Mandl), Physica-Verlag, Vienna, pp. 111–121.

Jurečková, J. and Sen P. K. (2000). Goodness-of-fit tests and second order asymptotic relations. *J. Statist. Planning Infer.* 91, 377–397.

Jurečková, J. and Sen, P. K. (2006). Robust multivariate location estimation, admissibility and shrinkage phenomenon. *Statist. Decisions 24*, 273-290.

Jurečková, J., and Víšek, J. A. (1984). Sensitivity of Chow-Robbins procedure to the contamination. *Sequen. Anal. 3*, 175–190.

Jurečková, J., and Welsh, A. H. (1990). Asymptotic relations between L- and M-estimators in the linear model. *Ann. Inst. Statist. Math. 42*, 671–698.

Kagan, A. M. (1970). On ϵ-admissibility of the sample mean as an estimator of location parameter. *Sankhyā, Ser. A 32*, 37–40.

Kagan, A. M., Linnik, J. V., and Rao, C. R. (1967). On a characterization of the normal law based on a property of the sample average. *Sankhyā, Ser A 27*, 405–406.

Kagan, A. M, Linnik, J. V., and Rao, C. R. (1973). *Characterization Problems in Mathematical Statistics.* Wiley, New York.

Kaigh, W. D., and Lachebruch, P. A. (1982). A generalized quantile estimator. *Commun. Statist., Theor. Meth. A11*, 2217–2238.

Kallianpur, G., and Rao, C. R. (1955). On Fisher's lower bound to asymptotic variance of a consistent estimate. *Sankhyā, Ser. A 16*, 331–342.

Kang, M. and Sen, P. K. (2008). Kendall's tau-type rank statistics in genome data. *Appl. Math. 53*, 207–221.

Kang-Mo Jung (2005). Multivariate least-trimmed squares regression estimator. *Comp. Statist. Data Anal. 48*, 307–316.

Karmous, A. R., and Sen, P. K. (1988). Isotonic M-estimation of location: Union-intersection principle and preliminary test versions. *J. Multivar. Anal. 27*, 300–318.

Keating, J. P., Mason, R. L., and Sen, P. K. (1993). *Pitman's Measure of Closeness: A Comparison of Statistical Estimators.* SIAM, Philadelphia.

Kendall, M. G. (1938). A new measure of rank correlation. *Biometrika 30*, 81–93.

Kiefer, J. (1961). On large deviations of the empirical D.F. of the vector chance variable and a law of iterated logarithm. *Pacific J. Math. 11*, 649–660.

Kiefer, J. (1967). On Bahadur's representation of sample quantiles. *Ann. Math. Statist. 38*, 1323–1342.

Kiefer, J. (1970). Deviations between the sample quantile process and the sample cdf. *Nonparametric Techniques in Statistical Inference* (ed. M. L. Puri), Cambridge Univ. Press, Cambridge, pp. 299–319.

Kiefer, J. (1972). Skorokhod embedding of multivariate RV's and the sample DF. *Zeit. Wahrsch. verw. Geb. 24*, 1–35.

Koenker, R. (2005). *Quantile Regression.* Cambridge University Press, Cambridge.

Koenker, R., and Bassett, G. (1978). Regression quantiles. *Econometrika 46*, 33–50.

Koenker, R., and d'Orey, V. (1993). A remark on computing regression quantiles. *Appl. Statist. 36*, 383–393.

Koenker, R., and Portnoy, S. (1987). L-estimation for linear models. *J. Amer. Statist. Assoc. 82*, 851–857.

Koenker, R., and Portnoy, S. (1990) M-estimate of multivariate regression. *J. Amer. Statist. Assoc. 85*, 1060–1068.

Komlós, J., Májor, P., and Tusnády, G. (1975). An approximation of partial sums of independent R.V.'s and the DF.I. *Zeit. Wahrsch. verw. Geb. 32*, 111–131.

Koul, H. L. (1969). Asymptotic behavior of Wilcoxon type confidence regions in multiple linear regression. *Ann. Math. Statist. 40*, 1950–1979.

Koul, H. L. (1971). Asymptotic behavior of a class of confidence regions in multiple linear regression. *Ann. Math. Statist. 42*, 42–57.

Koul, H. L. (1992). *Weighted Empiricals and Linear Models.* Inst. Math. Statist. Lect. Notes Monographs, vol 21, Hayward, Calif. Second Edition (2002): *Weighted Empirical Processes in Dynamic Nonlinear Models.* Lecture Notes in Statistics 166, Springer.

Koul, H. L., and Mukherjee, K. (1994). Regression quantiles and related processes under long range dependent errors. *J. Multivar. Anal. 51*, 318–337.

Koul, H. L., and Saleh, A. K. Md. E. (1995). Autoregression quantiles and related rank scores processes. *Ann. Statist. 23*, 670–689.

Koul, H. L., and Sen, P. K. (1985). On a Kolmogorov-Smirnov type aligned test in linear regression. *Statist. Probab. Letters 3*, 111–115.

Koul, H. L., and Sen, P. K. (1991). Weak convergence of a weighted residual empirical process in autoregression. *Statist. Dec. 9*, 235–261.

Kraft, C. and van Eeden, C. (1972). Linearized rank estimates and signed rank estimates for the general linear hypothesis. *Ann. Math. Statist. 43*, 42–57.

Krasker, W. S. (1980). Estimation in linear regression models with disparate data points. *Econometrica 48*, 1333–1346.

Krasker, W. S., and Welsch, R. E. (1982). Efficient bounded-influence regression estimation. *J. Amer. Statist. Assoc. 77*, 595–604.

Kudraszow, N. L. and Maronna, R. A. (2011). Estimates of MM type for the multivariate linear model. *J. Multiv. Analysis 102*, 1280–1292.

LeCam, L. (1953). On some asymptotic properties of maximum likelihood estimates and related Bayes' estimates. *Univ. Calif. Publ. Statist. 1*, 277–330.

LeCam, L. (1960). Locally asymptotically normal families of distributions. *Univ. Calif. Publ. Statist. 3*, 37–98.

LeCam, L. (1986). *Asymptotic Methods in Statistical Decision Theory.* Springer-Verlag, New York.

LeCam, L., and Yang, G. L. (1990). *Asymptotics in Statistics – Some Basic Concepts.* Springer-Verlag, New York.

Lecoutre, J. P., and Tassi, P. (1987). *Statistique non parametrique et robustesse.* Economica, Paris.

Lehmann, E. L. (1963a). Asymptotically nonparametric inference: an alternative approach to linear models. *Ann. Math. Statist. 34*, 1494–1506.

Lehmann, E. L. (1963b). Nonparametric confidence interval for a shift parameter. *Ann. Math. Statist. 34*, 1507–1512.

Lehmann, E. L. (1975). *Nonparametrics.* Holden-Day, San Francisco.

Lehmann, E. L. (1983). *Theory of Point Estimation.* Wiley, New York.

Lehmann, E. L. and Romano, J. P. (2005). *Testing Statistical Hypotheses.* Third Edition. Springer, New York.

Levit, B. Ya. (1975). On the efficiency of a class of nonparametric estimates. *Theor. Prob. Appl. 20*, 723–740.

Liese, F. and Vajda, I. (1994). Consistency of M-estimates in general regression models. *J. Multivariate Anal. 50*, 93–114.

Lindwall, T. (1973). Weak convergence of probability measures and random functions in the function space $D[0,\infty]$. *J. Appl. Probab. 10*, 109–121.

Lloyd, E. H. (1952). Least squares estimation of location and scale parameters using order statistics. *Biometrika 34*, 41–67.

Lopuhaä, H. P. (1992). Highly efficient estimators of multivariate location with high breakdown point. *Ann. Statist.* 20, 398–413.

Malý, M. (1991). Studentized M-estimators and their k-step versions. Ph.D. dissertation (in Czech). Charles Univ., Prague.

Marazzi, A. (1980). ROBETH, a subroutine library for robust statistical procedures. *COMPSTAT 1980*, Physica Verlag, Vienna.

Marazzi, A. (1986). On the numerical solutions of bounded influence regression problems. *COMPSTAT 1986*, Physica Verlag, Heidelberg, pp. 114–119.

Marazzi, A. (1987). Solving bounded influence regression problems with ROBSYS. *Statistical data Analysis Based on L_1-Norm and Related Methods* (ed. Y. Dodge), North Holland, Amsterdam, pp. 145–161.

Markatou, M. and Ronchetti, E. (1997). Robust inference: The approach based on influence function. *Handbook of Statistics 15*, (eds. G. S. Maddala and C. R. Rao), North-Holland, Amsterdam, pp. 49–75.

Maronna, R. A., Bustos, O. H., and Yohai, V. J. (1979). Bias- and efficiency-robustness of general M-estimators for regression with random carriers. *Smoothing Techniques for Curve Estimation, Lecture Notes in Mathematics,* 757, Springer-Verlag, Berlin, pp. 91–116.

Maronna, R. A., and Yohai, V. J. (1981). Asymptotic behavior of general M-estimators for regression and scale with random carriers. *Zeit. Wahrsch. verw. Geb. 58*, 7–20.

Maronna, R., and Yohai, V. J. (1991). The breakdown point of simulataneous general M-estimates of regression and scale. *J. Amer. Statist. Assoc. 86*, 699–716.

Maronna, R. A., Martin, R. D., and Yohai, V. J. (2006). *Robust Statistics. Theory and Methods.* J. Wiley, Chichester.

Martin, R. D., Yohai, V. J., and Zamar, R. H. (1989). Min-max bias robust regression. *Ann. Statist. 17*, 1608–1630.

Martin, R. D., and Zamar, R. H. (1988). High breakdown-point estimates of regression by means of the minimization of the efficient scale. *J. Amer. Statist. Assoc. 83*, 406–413.

Martinsek, A. (1989). Almost sure expansion for M-estimators and S-estimators in regression. Technical Report 25, Dept. Statistics, Univ. Illinois.

McKean, J. W., and Hettmansperger, T. P. (1976). Tests of hypotheses based on ranks in general linear model. *Commun. Statist. A5*, 693–709.

McKean, J. W., and Hettmansperger, T. P. (1978). A robust analysis of the general linear model based on one-step R-estimates. *Biometrika 65*, 571–579.

McKean, J. W., and Schrader, R. M. (1980). The geometry of robust procedures in linear models. *J. Royal Statist. Soc. B 42*, 366–371.

Millar, P. W. (1981). Robust estimation via minimum distance methods. *Zeit. Wahrsch. verw. Geb. 55*, 73–89.

Millar, P. W. (1985). Nonparametric applications of an infinite dimensional convolution theorem. *Zeit. Wahrsch. verw. Geb. 68*, 545–556.

Miller, R. G. Jr., and Sen, P. K. (1972). Weak convergence of U-statistics and von Mises differentiable statistical functionals. *Ann. Math. Statist. 43*, 31–41.

Mizera, I. (1996). Weak continuity of redescending M-estimators of location with an unbounded objective function. In: *PROBASTAT '94*, Proceedings of the Second International Conference on Mathematical Statistics (eds. A. Pázman and V. Witkovský), 343–347.

Mizera, I. and Müller, C. H. (1999). Breakdown points and variation exponents of robust M-estimators in linear models. *Ann. Statist.* 27, 1164–1177.

Mizera, I. and Müller, C. H. (2002). Breakdown points of Cauchy regression-scale estimators. *Statist. Probab. Letters* 57, 79–89.

Moses, L. E. (1965). Confidence limits from rank tests. *Technometrics 3*, 257–260.

Möttönen, J. and Oja, H. (1995). Multivariate spatial sign and rank methods. *J. Nonparametric Statist. 5*, 201-213.

Möttönen, J., Nordhausen, K. and Oja, H. (2010). Asymptotic theory of the spatial median. *IMS Collection 7* (eds. J. Antoch, M. Hušková, P. K. Sen), 182-193.

Müller, Ch. H. (1994). Asymptotic behaviour of one-step M-estimators in contaminated nonlinear models. *Asymptotic Statistics; Proc. 5th Prague Confer.* (eds. M. Hušková and P. Mandl), Physica-Verlag, Heidelberg, pp. 394–404.

Müller, Ch. H. (1995). Breakdown points for designed experiments. *J. Statist. Plann. Inference 45*, 413-427.

Müller, Ch. H. (1997). *Robust Planning and Analysis of Experiments. Lecture Notes in Statist. 124.* Springer, New York.

Niemiro, W. (1992). Asymptotics for M-estimators defined by convex minimization. *Ann. Statist. 20*, 1514–1533.

Niinimaa, A., Oja, H. and Tableman (1990). The finite-sample breakdown point of the Oja bivariate median and the corresponding half-samples version *Statist. Probab. Lett.* 10, 325–328.

Oja, H. (2010). *Multivariate Nonparametric Methods with R. An Approach Based on Spatial Signs and Ranks. Lecture Notes in Statistics 199*, Springer, New York.

Omelka, M. (2010). Second-order asymptotic representation of M-estimators in a linear model. *IMS Collections 7* (eds. J. Antoch, M. Hvsková, P. K. Sen, eds.), 194-203.

Ortega, J. M., and Rheinboldt, W. C. (1970). *Iterative Solution of Nonlinear Equations in Several Variables.* Academic Press, New York.

Parr, W. C. (1981). Minimum distance estimation: A bibliography. *Commun. Statist. Theor. Meth. A10*, 1205–1224.

Parr, W. C., and Schucany, W. R. (1982). Minimum distance estimation and components of goodness-of-fit statistics. *J. Roy. Statist. Soc. B*, 178–189.

Parthasarathy, K. R. (1967). *Probability Measures on Metric Spaces.* Academic Press, New York.

Parzen, E. (1979). Nonparametric statistical data modelling. *J. Amer. Statist. Assoc. 74*, 105–131.

Pearson, E. S. (1931). The analysis of variance in cases of nonnormal variation. *Biometrika 23*, 114–133.

Pfanzagl, J., and Wefelmeyer, W. (1982). *Contributions to a General Asymptotic Statistical Theory, Lecture Notes in Statistics, 13*, Springer-Verlag, New York.

Pitman, E. J. G. (1937). The closest estimates of statistical parameters. *Proc. Cambridge Phil. Soc. 33*, 212–222.

Pitman, E. J. G. (1939). Tests of hypotheses concerning location and scale parameters. *Biometrika 31*, 200–215.

Pollard, D. (1984). *Convergence of Stochastic Processes.* Wiley, New York.

Pollard, D. (1990). *Empirical Processes: Theory and Applications.* NSF-CBMS Reg. Conf. Series in Probab. and Statist., vol. 2, IMS, Hayward, California.

Pollard, D. (1991). Asymptotics for least absolute deviation regression estimators. *Econometric Theory 7*, 186–199.

Portnoy, S. (1984a). Tightness of the sequence of c.d.f. processes defined from regression fractiles. *Robust and Nonlinear Time Series Analysis* (eds. J. Franke et al.), Springer Verlag, New York, pp. 231–246.

Portnoy, S. (1984b). Asymptotic behavior of M-estimators of p regression parameters when p^2/n is large. I. Consistency. *Ann. Statist. 12*, 1298–1309.

Portnoy, S. (1985). Asymptotic behavior of M-estimators of p regression parameters when p^2/n is large; II. Normal approximation. *Ann. Statist. 13*, 1403–1417.

Portnoy, S. and Jurečková, J. (1999). On extreme regression quantiles. *Extremes* 2:3, 227–243.

Portnoy, S., and Koenker, R. (1989). Adaptive L-estimation for linear models. *Ann. Statist. 17*, 362–381.

Powell, J. L. (1984). Least absolute deviations estimation for the censored regression model. *J. Econometr. 25(3)*, 303–325.

Puri, M. L., and Sen, P. K. (1985). *Nonparametric Methods in General Linear Models.* Wiley, New York.

Rao, C. R. (1945). Information and accuracy attainable in the estimation of statistical parameters. *Bull. Calcutta Math. Soc. 37*, 81–91.

Rao, C. R. (1959). Sur une caracterisation de la distribution normal étable d'apres une propriéte optimum des estimations linéaries. *Colloq. Intern. C.N.R.S. France 87*, 165–171.

Rao, P. V., Schuster, E. F., and Littell, R. C. (1975). Estimation of shift and center of symmetry based on Kolmogorov-Smirnov statistics. *Ann. Statist. 3*, 862–873.

Reeds, J. A. (1976). On the definitions of von Mises functions. Ph.D. dissertation, Harvard Univ., Cambridge, Mass.

Ren, J. J. (1994). Hadamard differentiability and its applications to R-estimation in linear models. *Statist. Dec. 12*, 1–22.

Ren, J. J., and Sen, P. K. (1991). Hadamard differentiability of extended statistical functionals. *J. Multivar. Anal. 39*, 30–43.

Ren, J. J., and Sen, P. K. (1992). Consistency of M-estimators: Hadamard differentiability approaches. *Proc. 11th Prague Conf. on Inf. Th. Stat. Dec. Funct. Random Proc.*, 198–211.

Ren, J. J., and Sen, P. K. (1994). Asymptotic normality of regression M-estimators: Hadamard differentiability approaches. *Asymptotic Statistics, Proc. 5th Prague Conf.* (eds. M. Hušková and P. Mandl), Physica Verlag, Heidelberg, pp. 131–147.

Ren, J. J., and Sen, P. K. (1995a). Second order Hadamard differentiability with applications. (to appear).

Ren, J. J., and Sen, P. K. (1995b). Hadamard differentiability of functionals on $D[0,1]^p$. *J. Multivar. Anal. 55*, 14–28.

Rey, W. J. J. (1983). *Introduction to Robust and Quasi-Robust Statistical Methods.* Springer-Verlag, New York.

Rieder, H. (1977). Least favorable pairs for special capacities. *Ann. Statist. 5*, 909–9221.

Rieder, H. (1978). A robust asymptotic testing model. *Ann. Statist. 6*, 1080–1094.

Rieder, H. (1980). Estimation derived from robust tests. *Ann. Statist. 8*, 106–115.

Rieder, H. (1981). On local asymptotic minimaxity and admissibility in robust estimation. *Ann. Statist. 9*, 266–277.

Rieder, H. (1987). Robust regression estimators and their least favorable contamination curves. *Statist. Dec. 5*, 307–336.

Rieder, H. (1994). *Robust Asymptotic Statistics*, Springer-Verlag, New York.

Rivest, L. P. (1982). Some asymptotic distributions in the location-scale model. *Ann. Inst. Statist. Math. A 34*, 225–239.

Rockafellar, R. T. (1970). *Convex Analysis.* Princeton Univ. Press. Princeton.

Rousseeuw, P. J. (1984). Least median of squares regression. *J. Amer. Statist. Assoc. 79*, 871–880.

Rousseeuw, P. J. (1985): Multivariate estimation with high breakdown point. In: *Mathematical Statistics and Applications* (eds. W. Grossman, G. Pflug, I. Vincze, and W. Wertz), vol. B, Reidel, Dordrecht, Netherlands, 283–297.

Rousseeuw, P. J. (1994). Unconventional features of positive-breakdown estimators. *Statist. Probab. Letters* 19, 417–431.

Rousseeuw, P. J., and Bassett, G. W., Jr. (1990). The Remedian: A robust averaging method for large data sets. *J. Amer. Statist. Assoc. 85*, 97–104.

Rousseeuw, P. J., and Croux, C. (1993). Alternatives to the median absolute deviation. *J. Amer. Statist. Assoc. 88*, 1273–1283.

Rousseeuw, P. J., and Croux, C. (1994). The bias of k-step M-estimators. *Statist. Probab. Letters 20*, 411–420.

Rousseeuw, P. J. and Hubert, M. (1999). Regression depth. *J. Amer. Statist. Assoc. 94*, 388-402.

Rousseeuw, P. J., and Leroy, A. M. (1987). *Robust Regression and Outlier Detection.* Wiley, New York. Second Edition (2003), Wiley.

Rousseeuw, P. J., and Ronchetti (1979). The influence curve for tests. Research Report 21, Fachgruppe für Statistik, ETH Zürich.

Rousseeuw, P. J., and Ronchetti (1981). Influence curve for general statistics. *J. Comp. Att. Math. 7*, 161–166.

Rousseeuw, P. J., Van Driessen, K. S., Van Aelst. S., and Agulló, J. (2004). Robust multivariate regression. *Technometrics 46*, 293-305.

Rousseeuw, P. J., and Yohai, V. J. (1984). Robust regression by means of S-estimates. *Robust and Nonlinear Time Series Analysis, Lecture Notes in Statistics, 26* (eds. J. Franke et al.), Springer-Verlag, Berlin, pp. 256–272.

Roy, S. N. (1953). On a heuristic method of test construction and its use in multivariate analysis. *Ann. Math. Statist. 24*, 220–238.

Ruppert, D., and Carroll, R. J. (1980). Trimmed least squares estimation in the linear model. *J. Amer. Statist. Assoc. 75*, 828–838.

Ruschendorf, L. (1988). *Asymptotische Statistik.* B.G. Teubner, Stuttgart.

Sacks, J., and Ylvisacker, D. (1978). Linear estimation for approximately linear models. *Ann. Statist. 6*, 1122–1137.

Saleh, A. K. Md. E. (2006). *Theory of Preliminary Test and Stein-Type Estimation with Applications.* Wiley, New York.

Salibian-Barrera M. (2006). The asymptotics of MM-estimators for linear regression with fixed designs. *Metrika 63*, 283-294.

Sarhan, A. E., and Greenberg, B. G. (eds.) (1962). *Contributions to Order Statistics.* Wiley, New York.

Scheffé, H. (1951). *An Analysis of Variance for Paired Comparisons.* New York: Columbia University.

Schrader, R. M., and Hettmansperger, T. P. (1980). Robust analysis of variance based upon a likelihood ratio criterion. *Biometrika 67*, 93–101.

Schrader, R. M., and McKean, J. W. (1977). Robust analysis of variance. *Comm. Statist. Theor. Meth. 46*, 879–894.

Schuster, E. F., and Narvarte, J. A. (1973). A new nonparametric estimator of the center of a symmetric distribution. *Ann. Statist. 1*, 1096–1104.

Scott, D. J. (1973). Central limit theorems for martingales and processes with stationary increments using a Skorokhod representation approach. *Adv. Appl. Probab. 5*, 119–137.

Sen, P. K. (1959). On the moments of sample quantiles. *Calcutta Statist. Assoc. Bull. 9*, 1–20.

Sen, P. K. (1960). On some convergence properties U-statistics. *Calcutta Statist. Assoc. Bull. 10*, 1–18.

Sen, P. K. (1963). On the estimation of relative potency in dilution (-direct) essays by distribution-free methods. *Biometrics 19*, 532–552.

Sen, P. K. (1964). On some properties of the rank-weighted means. *J. Indian Soc. Agric. Statist. 16*, 51–61.

Sen, P. K. (1966a). On a distribution-free method of estimating asymptotic efficiency of a class of nonparametric tests. *Ann. Math. Statist. 37*, 1759–1770.

Sen, P. K. (1966b). On nonparametric simultaneous confidence regions and tests in the one-criterion analysis of variance problem. *Ann. Inst. Statist. Math. 18*, 319–366.

Sen, P. K. (1967). U-statistics and combination of independent estimates of regular functionals. *Calcutta Statist. Assoc. Bull. 16*, 1–16.

Sen, P. K. (1968a). Estimates of regression coefficients based on Kendall's tau. *J. Amer. Statist. Assoc. 63*, 1379–1389.

Sen, P. K. (1968b). On a further robustness property of the test and estimator based on Wilcoxon's signed rank statistic. *Ann. Math. Statist. 39*, 282–285.

Sen, P. K. (1968c). Robustness of some nonparametric procedures in linear models. *Ann. Math. Statist. 39*, 1913–1922.

Sen, P. K. (1969). On a class of rank order tests for the parallelism of several regression lines. *Ann. Math. Statist. 40*, 1668–1683.

Sen, P. K. (1970a). A note on order statistics from heterogeneous distributions. *Ann. Math. Statist. 41*, 2137–2139.

Sen, P. K. (1970b). On some convergence properties of one-sample rank order statistics. *Ann. Math. Statist. 41*, 2140–2143.

Sen, P. K. (1970c). The Hăjek-Renyi inequality for sampling from a finite population. *Sankhyā, Ser. A 32*, 181–188.

Sen, P. K. (1971a). Robust statistical procedures in problems of linear regression with special reference to quantitative bioassays, I. *Internat. Statist. Rev. 39* , 21–38.

Sen, P. K. (1971b). A note on weak convergence of empirical processes for sequences of ϕ-mixing random variables. . *Ann. Math. Statist. 42*, 2132–2133.

Sen, P. K. (1972a). On the Bahadur representation of sample quantiles for sequences of mixing random vbariables. *J. Multivar. Anal. 2*, 77–95.

Sen, P. K. (1972b). Weak convergence and relative compactness of martingale processes with applications to nonparametric statistics. *J. Multivar. Anal. 2*, 345–361.

Sen, P. K. (1972c). Robust statistical procedures in problems of linear regression with special reference to quantitative bioassays, II. *Internat. Statist. Rev. 40*, 161–172.

Sen, P. K. (1973a). On weak convergence of empirical processes for random number of independent stochastic vectors. *Proc. Cambridge Phil. Soc. 73*, 135–140.

Sen, P. K. (1973b). An almost sure invariance principle for multivariate Kolmogorov-Smirnov statistics. *Ann. Probab. 1*, 488–496.

Sen, P. K. (1974a). Almost sure behavior of U-statistics and von Mises' differentiable statistical functions. *Ann. Statist. 2*, 387–395.

Sen, P. K. (1974b). Weak convergence of generalized U-statistics. *Ann. Probab. 2*, 90–102.

Sen, P. K. (1974c). On L_p-convergence of U-statistics. *Ann. Inst. Statist. Math. 26*, 55–60.

Sen, P. K. (1975). Rank statistics, martingales and limit theorems. In *Statistical Inference and Related Topics* (ed. M. L. Puri), Academic Press, New York, pp. 129–158.

Sen, P. K. (1976a). Weak convergence of a tail sequence of martingales. *Sankhyā, Ser. A 38*, 190–193.

Sen, P. K. (1976b). A two-dimensional functional permutational central limit theorem for linear rank statistics. *Ann. Probab. 4*, 13–26.

Sen, P. K. (1976c). A note on invariance principles for induced order styatistics. *Ann. Probab. 4*, 474–479.

Sen, P. K. (1977a). Some invariance principles relating to jackknifing and their role in sequential analysis. *Ann. Statist. 5*, 315–329.

Sen, P. K. (1977b). On Wiener process embedding for linear combinations of order statistics. *Sankhyā, Ser. A5*, 1107–1123.

Sen, P. K. (1977c). Tied down Wiener process approximations for aligned rank order statistics and some applications. *Ann. Statist. 5*, 1107–1123.

Sen, P. K. (1978a). An invariance principle for linear combinations of order statistics. *Zeit. Wahrsch. verw. Geb. 42*, 327–340.

Sen, P. K. (1978b). Invariance principles for linear rank statistics revisited. *Sankhyā, Ser.A 40*, 215–236.

Sen, P. K. (1979a). Weak convergence of some quantile processes arising in progressively censored tests. *Ann. Statist. 7*, 414–431.

Sen, P. K. (1979b). Asymptotic properties of maximum likelihood estimators based on conditional specifications. *Ann. Statist. 7*, 1019–1033.

Sen, P. K. (1980a). On almost sure linearity theorems for signed rank order statistics. *Ann. Statist. 8*, 313–321.

Sen, P. K. (1980b). On nonparametric sequential point estimation of location based on general rank order statistics. *Sankhyā, Ser. A 42*, 223–240.

Sen, P. K. (1981a). *Sequential Nonparametrics: Invariance Principles and Statistical Inference.* Wiley, New York.

Sen, P. K. (1981b). Some invariasnce principles for mixed-rank statistics and induced order statistics and some applications. *Commun. Statist. Theor. Meth. A 10*, 1691–1718.

Sen, P. K. (1981c). The Cox regression model, invariance principles for some induced quantile processes and some repeated significance tests. *Ann. Statist. 9*, 109–121.

Sen, P. K. (1982a). On M-tests in linear models. *Biometrika 69*, 245–248.

Sen, P. K. (1982b). Invariance principles for recursive residuals. *Ann. Statist. 10*, 307–312.

Sen, P. K. (1983a). On the limiting behavior of the empirical kernel distribution function. *Calcutta Statist. Assoc. Bull. 32*, 1–8.

Sen, P. K. (1983b). On permutational central limit theorems for general multivariate linear rank statistics. *Sankhyā, Ser A 45*, 141–49.

Sen, P. K. (1984a). Jackknifing L-estimators: Affine structure and asymptotics. *Sankhyā, Ser. A 46*, 207–218.

Sen, P. K. (1984b). Nonparametric procedures for some miscellaneous problems. *Handbook of Statistics, vol. 4: Nonparametric Methods* (eds. P. R. Krishnaiah and P. K. Sen), North Holland, Amsterdam, pp. 699–739.

Sen, P. K. (1984c). Invariance principles for U-statistics and von Mises functionals in the non-i.d. case. *Sankhyā, Ser A 46*, 416–425.

Sen, P. K. (1984d). On a Kolmogorov-Smirnov type aligned test. *Statist. Probab. Letters 2* , 193–196.

Sen, P. K. (1984e). A James-Stein detour of U-statistics. *Commun. Statist. Theor. Metd. A 13*, 2725–2747.

Sen, P. K. (1985). *Theory and Applications of Sequential Nonparametrics.* SIAM, Philadelphia.

Sen, P. K. (1986a). Are BAN estimators the Pitman-closest ones too? *Sankhyā, Ser A 48*, 51–58.

Sen, P. K. (1986b). On the asymptotic distributional risks of shrinkage and preliminary test versions of maximum likelihood estimators. *Sankhyā, Ser A 48*, 354–371.

Sen, P. K. (1986c). Whither jackknifing in Stein-rule estimation? *Commun. Statist. Theor. Meth. A 15*, 2245–2266.

Sen, P. K. (1988a). Functional jackknifing: Rationality and general asymptotics. *Ann. Statist. 16*, 450–469.

Sen, P. K. (1988b). Functional approaches in resampling plans: A review of some recent developments. *Sankhyā, Ser. A 40*, 394–435.

Sen, P. K. (1988c). Asymptotics in finite population sampling. In *Handbook of Statistics, vol. 6: Sampling* (eds. P. R. Krishnaiah and C. R. Rao), North Holland, Amsterdam, pp. 291–331.

Sen, P. K. (1989a). Whither delete-k jackknifing for smooth statistical functionals. *Statistical Data Analysis and Inference* (ed. Y. Dodge), North Holland, Amsterdam, pp. 269–279.

Sen, P. K. (1989b). Asymptotic theory of sequential shrunken estimation of statistical functionals. *Proc. 4th Prague Confer. Asympt. Statist.* (eds. M. Hušková and P. Mandl), Academia, Prague, pp. 83–100.

Sen, P. K. (1990a). Optimality of BLUE and ABLUE in the light of the Pitman closeness of statistical estimators. *Coll. Math. Soc. Janos Bolyai, 57: Limit Theorems in Probability and Statistics*, pp. 459–476.

Sen, P. K. (1990b). Statistical functionals, stopping times and asymptotic minimum risk property. *Probability Theory and Mathematical Statistics.* (eds. B. Gregelionis et al.), World Sci. Press, Singapore, vol. 2, pp. 411–423.

Sen, P. K. (1991a). Nonparametrics: Retrospectives and perspectives (with discussion). *J. Nonparamet. Statist. 1*, 1–68.

Sen, P. K. (1991b). Nonparametric methods in sequential analysis. *Handbook of Sequential Analysis* (eds. B. K. Ghosh and P. K. Sen), Dekker, New York, pp. 331–360.

Sen, P. K. (1991c). Asymptotics via sub-sampling Bahadur type representations. *Probability, Statistics and Design of Experiment: R. C. Bose Memorial Conference* (ed. R. R. Bahadur), Wiley Eastern, New Delhi, pp. 653–666.

Sen, P. K. (1992). Some informative aspects of jackknifing and bootstrapping. In *Order statistics and Nonparametrics: Theory and Applications* (eds. P. K. Sen and I. A. Salama). North Holland, Amsterdam, pp. 25–44.

Sen, P. K. (1993a). Multivariate L_1-norm estimation and the vulnerable bootstrap. *Statistical Theory and Data Analysis* (eds. K. Matusita et al.), North Holland, Amsterdam, pp. 441–450.

Sen, P. K. (1993b). Perspectives in multivariate nonparametrics: conditional functionals and Anocova models. *Sankhyā, Ser. A 55*, 516–532.

Sen, P. K. (1994a). Isomorphism of quadratic norm and PC ordering of estimators admitting first order AN representation. *Sankhyā, Ser.B 56*, 465–475

Sen, P. K. (1994b). Regression quantiles in nonparametric regression. *J. Nonparamet. Statist. 3*, 237–253.

Sen, P. K. (1994c). The impact of Wassily Hoeffding's research on nonparametrics. *The Collected Works of Wassily Hoeffding* (eds. N. I. Fisher and P. K. Sen), Springer-Verlag, New York, pp. 29–55.

Sen, P. K. (1995a). Robust and Nonparametric Methods in linear models with mixed-effects. *Tetra Mount. Math. J., Sp. Issue on Probastat'94 Confer.* Bratislava, Slovak., in press.

Sen, P. K. (1995b). Regression rank scores estimation in ANOCOVA. *Ann. Statist.* (to appear).

Sen, P. K. (1995c). The Hájek asymptotics for finite population sampling and their ramifications. *Kybernetrika 31*, 251–268.

Sen, P. K. (1995d). Censoring in theory and practice: statistical perspectives and controversies. *Analysis of Censored Data, IMS Lecture Notes Mon. Ser. 27* (eds. H. Koul and J. Deshpande), Hayward, Calif. pp. 177–192.

Sen, P. K. (1995e). Bose-Einstein statistics, generalized occupancy problems and allied asymptotics. *Rev. Bull. Calcutta Math. Soc. 2*, 1–12.

Sen, P. K. (1996). Statistical functionals, Hadamard differentiability and Martingales. *A Festschrift for Professor J. Medhi* (eds. A.C. Borthakur and H. Chaudhury), New Age Press, Delhi, pp. 29-47.

Sen, P. K. (1998). Multivariate median and rank sum tests. In *Encyclopedia of Biostatistics,* vol. IV, (eds. P. Armitage and T. Colton eds.), J. Wiley, Chichester, pp. 2887-2900.

Sen, P. K., and Bhattacharyya, B. B. (1976). Asymptotic normality of the extrema of certain sample functions. *Zeit. Wahrsch. verw Geb. 34*, 113–118.

Sen, P. K., Bhattacharyya, B. B., and Suh, M. W. (1973). Limiting behavior of the extrema of certain sample functions. *Ann. Statist. 1*, 297–311.

Sen, P. K., and Ghosh, M. (1971). On bounded length sequential confidence intervals based on one-sample rank order statistics. *Ann. Math. Statist. 42*, 189–203.

Sen, P. K., and Ghosh, M. (1972). On strong convergence of regression rank statistics. *Sankhyā, Ser. A 34*, 335–348.

Sen, P. K., and Ghosh, M. (1973a). A Chernoff-Savage representation of rank order statistics for stationary ϕ-mixing processes. *Sankhyā, Ser. A 35*, 153–172.

Sen, P. K., and Ghosh, M. (1973b). A law of iterated logarithm for one sample rank order statistics and some applications. *Ann. Statist. 1*, 568–576.

Sen, P. K., and Ghosh, M. (1974). On sequential rank tests for location. *Ann. Statist. 2*, 540–552.

Sen, P. K., and Ghosh, M. (1981). Sequential point estimation of estimable parameters based on U-statistics. *Sankhyā, Ser. A 43*, 331–344.

Sen, P. K., Jurečková, J., and Picek, J. (2003). Goodness-of-fit test of Shapiro-Wilk type with nuisance regression and scale. *Austrian J. Statist.* 32, No. 1 & 2, 163–177.

Sen, P. K., Kubokawa, T., and Saleh, A. K. Md. E. (1989). The Stein paradox in the sense of the Pitman measure of closeness. *Ann. Statist. 17*, 1375–1386.

Sen, P. K., and Puri, M. L. (1977). Asymptotically distribution-free aligned rank order tests for composite hypotheses for general multivariate linear models. *Zeit. Wahrsch. verw. Geb. 39*, 175–186.

Sen, P. K., and Saleh, A. K. Md. E. (1985). On some shrinkage estimators of multivariate location. *Ann. Statist. 13*, 272–281.

Sen, P. K., and Saleh, A. K. Md. E. (1987). On preliminary test and shrinkage M-estimation in linear models. *Ann. Statist. 15*, 1580–1592.

Sen, P. K., and Saleh, A. K. Md. E. (1992). Pitman closeness of Pitman estimators. *Gujarat Statist. Rev., C. G. Khatri Memorial Volume*, 198–213.

Sen, P. K., and Singer, J. M. (1985). M-methods in multivariate linear models. *J. Multivar. Anal. 17*, 168–184.

Sen, P. K., and Singer, J. M. (1993). *Large sample Methods in Statistics: An Introduction with Applications.* Chapman and Hall, New York.

Seneta, E. (1976). *Regularly Varying Functions, Lectures Notes Math. 508*, Springer-Verlag, New York.

Serfling, R. J. (1980). *Approximation Theorems of Mathematical Statistics.* Wiley, New York. Reprinted 2002.

Serfling, R. J. (1984). Generalized L-, M-, and R-statistics. *Ann. Statist. 12*, 76–86.

Serfling, R. J. (2002). Quantile functions for multivariate analysis: Approaches and applications. *Statist. Neerlandica 56,* 214-232.

Serfling, R. J. (2010). Equivariance and invariance properties of multivariate quantile and related functions, and the role of standardization. *J. Nonparametric Statist. 22*, 915–936.

Shapiro, S. S., and Wilk, M. B. (1965). An analysis of variance for normality (complete samples). *Biometrika 52*, 591-611.

Shevlyakov, G., Morgenthaler, S., and Shurygin, A. M. (2008). Redescending M-estimators. *J. Stat. Plann. Inference 138*, 2906–2917.

Shorack, G. R., and Wellner, J. A. (1986). *Empirical Processes with Applications to Statistics.* Wiley, New York.

Siegel, A. F. (1982). Robust regression using repeated medians. *Biometrika 69*, 242–244.

Sievers, G. L. (1978). Weighted rank statistics for simple linear regression. *J. Amer. Statist. Assoc. 73*, 628–631.

Sievers, G. L. (1983). A weighted dispersion function for estimation in linear models. *Commun. Statist. A 12*, 1161–1179.

Silvapulle, M. J., and Sen, P. K. (2005). *Constrained Statistical Inference: Inequality, Order, and Shape Restrictions.* Wiley, New York.

Silverman, B. W. (1983). Convergence of a class of empirical distribution functions of dependent random variables. *Ann. Probab. 11*, 745–751.

Silverman, B. (1986). *Density Estimation for Statistics and Data Analysis.* Chapman and Hall, New York.

Simpson, D. G., Ruppert, D., and Carroll, R. J. (1992). On one step GM estimates and stability of inferences in linear regression. *J. Amer. Statist. Assoc. 87*, 439–450.

Singer, J. M., and Sen, P. K. (1985). Asymptotic relative efficiency of multivariate M-estimators. *Commun. Statist. Sim. Comp. B 14*, 29–42.

Skorokhod, A. V. (1956). Limit theorems for stochastic processes. *Theo. Probab. Appl. 1*, 261–290.

Staudte, R. J., and Sheather, S. J. (1990). *Robust Estimation and Testing.* Wiley, New York.

Stefanski, L. A., Carroll, R. J., and Ruppert, D. (1986). Otimally bounded score functions for generalized linear models with applications to logistic regression. *Biometrika 73*, 413–424.

Stein, C. (1945). A Two-sample test for a linear hypothesis whose power is independent of the variance *Ann. Math. Statist. 16*, 243–258.

Stein, C. (1956). Inadmissibility of the usual estimator for the mean of a multivariate normal distribution. *Proc. 3rd Berkeley Symp. Math. Statist. Probab.* (ed. J. Neyman), vol. 1, Univ. Calif. Press, Los Angeles, pp. 187–195.

Steyn, H. S., and Geertsema, J. C. (1974). Nonparametric confidence sequence for the centre of a symmetric distribution. *South Afr. J. Statist. 8*, 24–34.

Stigler, S. M. (1969). Linear functions of order statistics. *Ann. Math. Statist. 40*, 770–784.

Stigler, S. M. (1973a). The asymptotic distribution of the trimmed mean. *Ann. Statist. 1*, 472–477.

Stigler, S. M. (1973b). Simon Newcomb, Percy Daniell, and the history of robust estimation 1885–1920. *J. Amer. Statist. Assoc. 68*, 872–879. Reprinted in *Studies in the History of Statistics and Probability*, vol. 2, (eds. M. G. Kendall and R. L. Plackett), London: Griffin, pp. 410-417, 1977.

Stigler, S. M. (1974). Linear functions of order statistics with smooth weight function. *Ann. Statist. 2*, 676–693.

Stigler, S. M. (1977). Do robust estimators work with real data? *Ann. Statist. 5*, 1055-1098.

Stigler, S. M. (1980). Studies in the history of probability and statistics XXXVIII. R. H. Smith, a Victorian interested in robustness. *Biometrika 67*, 217–221.

Stigler, S. M. (2010). The changing history of robustness. *The American Statistician 64/4*, 277-281.

Stone, C. (1974). Asymptotic properties of estimators of a location parameter. *Ann. Statist. 2* 1127–1137.

Stout, W. F. (1974). *Almost Sure Convergence.* Academic Press, New York.

Strassen, V. (1964). An invariance principle for the law of iterated logarithm. *Zeit. Wahrsch. verw. Geb. 3*, 211–226.

Strassen, V. (1967). Almost sure behavior of sums of independent random variables and martingales. *Proc. 5th Berkeley Symp. Math. Statist. Probab.* (eds. L. Lecam et al.), Univ. Calif. Press, Los Angeles, vol. 2, pp. 315–343.

Tableman, M. (1990). Bounded influence rank regression – a one step estimator based on Wilcoxon scores. *J. Amer. Statist. Assoc. 85*, 508–513.

Tableman, M. (1994). The asymptotic of the least trimmed absolute deviations (LTAD) estimator. *Statist. Probab. Letters 14*, 387–398.

Tatsuoka, K. S., and Tyler, D. E. (2000). The uniqueness of S- and M-functionals under non-elliptical distributions. *Ann. Statist. 28,* 1219–1243.

Tibshirani, R. (1996). Regression shrinkage and selection via the Lasso. *J. Royal Statist. Soc. B 58*, 267–288.

Tsiatis, A. A. (1990). Estimating regression parameters using linear rank tests for censored data. *Ann. Statist. 18*, 354–372.

Tukey, J. W. (1958). Bias and confidence in not quite large samples. (Abstract). *Ann. Math. Statist. 29*, 614.

Tukey, J. W. (1960). A survey of sampling from contaminated distributions. In: *Contributions to Probability and Statistics* (eds. I. Olkin et al.) Stanford University Press, Stanford, California, pp. 448–485.

Tukey, J. W. (1977). *Exploratory Data Analysis.* Addison-Wesley, Reading, Mass.

Tyler, D. E. (1987). A distribution-free M-estimator of multivariate scatter. *Ann. Statist. 15*, 234–251.

Tyler, D. E., Critchley, F., Dümbgen, L., and Oja, H. (2009). Invariant co-ordinate selection. *J. Royal Statist. Soc. B 71*, Part 3, 549-592.

Vajda, I. (1984). Minimum weak divergence estimators of structural parameters. *Proc. 3rd Prague Symp. on Asympt. Statist.* (eds. M. Hušková and P. Mandl), Elsevier, Amsterdam, pp. 417–424.

van de Geer. S. (2010). The Lasso with within group structure. *IMS Collections 7* (eds. J. Antoch, M. Hušková, P. K. Sen), 235–244.

van der Vaart, A. W. (1991). Efficiency and Hadamard differentiability. *Scand. J. Statist. 18*, 63–75.

van der Vaart, A. W., and Wellner, J. A. (1996, 2000). *Weak Convergence and Empirical Processes, with Applications to Statistics.* Springer, New York.

van Eeden, C. (1972). Ananalogue for signed rank statistics, of Jurečková's asymptotic linearity theorem for rank statistics. *Ann. Math. Statist. 43*, 791–802.

van Eeden, C. (1983). On the relation between L-estimators and M-estimators and asymptotic efficiency relative to the Cramér-Rao lower bound. *Ann. Statist. 11*, 674–690.

van Zwet, W. R. (1980). A strong law for linear functions of order statistics. *Ann. Probab. 8*, 986–990.

van Zwet, W. R. (1984). A Berry-Esseen bound for symmetric statistics. *Zeit. Wahrsch. verw. Geb. 66*, 425–440.

van Zwet, W. R. (1985). Van de Hulstx and robust statistics: A historical note. *Statist. Neerlandica 32*, 81–95.

van Zwet, W. R. (1992). *Wald Memorial Lecture*, IMS, Boston, Mass.

Víšek, J. Á. (2010). Robust error-term-scale estimate. *IMS Collections 7* (eds. J. Antoch, M. Hušková, P. K. Sen), 254-267.

von Mises, R. (1936). Les lois de probabilité pour des fonctions statistiques. *Ann. Inst. H.Poincaré 6*, 185–212.

von Mises, R. (1947). On the asymptotic distribution of differentiable statistical functions. *Ann. Math. Statist. 18*, 309–348.

Wald, A. (1943). Test of statistical hypothesis concerning several parameters when the number of observations is large. *Trans. Amer. Math. Soc. 54*, 426–482.

Wald, A. (1947). *Sequential Analysis*, Wiley, New York.

Walker, H. M. and Lev, J. (1953). *Statistical Inference.* Holt, New York.

Welsh, A. H. (1986). Bahadur representation for robust scale estimators based on regression residuals. *Ann. Statist. 14*, 1246–1251.

Welsh, A. H. (1987a). One-step L-estimators for the linear model. *Ann. Statist. 15*, 626–641.

Welsh, A. H. (1987b). The trimmed mean in the linear model. *Ann. Statist. 15*, 626–641.

Welsh, A. H. (1989). On M-processes and M-estimation. *Ann. Statist. 17*, 337–361.

Welsh, A. H., and Ronchetti, E. (2002). A journey in single steps: Robust one-step M-estimation in linear regression. *J. Statist. Planning Infer. 103*, 287–310.

Whitt, W. (1970). Weak convergence of probability measures on the function space $C[0,\infty)$. *Ann. Math. Statist. 41*, 939–944.

Wiens, D. P. (1990). Minimax-variance L- and R-estimators of locations. *Canad. J. Statist. 18*, 47–58.

Wiens, D. P., and Zhou, J. (1994). Bounded-influence rank estimator in the linear model. *Canad. J. Statist. 22*, 233–245.

Wijsman, R. A. (1979). Constructing all smallest simultaneous confidence sets in a given class with applications to MANOVA. *Ann. Statist. 7*, 1003–1018.

Wolfowitz, J. (1957). The minimum distance method. *Ann. Math. Statist. 28*, 75–88.

Yanagawa, T. (1969). A small sample robust competitor of Hodges-Lehmann estimate. *Bull. Math. Statist. 13*, 1–14.

Yohai V. J. (1987). High breakdown-point and high efficiency robust estimates for regression. *Ann. Stat. 15*, 642-656.

Yohai, V. J., and Maronna, R. A. (1976). Location estimators based on linear combinations of modified order statistics. *Comm. Statist. Theor. Meth. A5*, 481–486.

Yohai, V. J., and Zamar, R. (1988). High breakdown point of estimates of regression by means of the minimization of an efficient scale. *J. Amer. Statist. Assoc. 83*, 406–413.

Yoshizawa, C. N., Davis, C. E., and Sen, P. K. (1986). Asymptotic equivalence of the Harrell-Davis estimator and the sample median. *Commun. Statist., Theor. Meth. A 14*, 2129–2136.

Zuo, Y. (2003). Finite sample tail behavior of multivariate location estimators. *J. Multiv. Analysis 85*, 91-105.

Zuo, Y., and Serfling, R. J. (2000a). On the performance of some robust nonparametric location measures relative to a general notion of multivariate symmetry. *J. Statist. Planning Infer. 84*, 55-79.

Zuo, Y., and Serfling, R. J. (2000b). Nonparametric notions of multivariate scatter measure and more scattered based on statistical depth functions. *J. Multiv. Analysis 75*, 62-78.

Subject index

Author index